Physics and Technology for Future Presidents

Physics and Technology for Future Presidents

An Introduction to the Essential Physics Every World Leader Needs to Know

Richard A. Muller

PRINCETON UNIVERSITY PRESS   PRINCETON AND OXFORD

Published by Princeton University Press, 41 William Street, Princeton,
New Jersey 08540
In the United Kingdom: Princeton University Press, 6 Oxford Street,
Woodstock, Oxfordshire OX20 1TW
press.princeton.edu

Library of Congress Cataloging-in-Publication Data

Muller, R. (Richard)
    Physics and technology for future presidents : an introduction to the
essential physics every world leader needs to know / Richard A. Muller.
        p.   cm.
    Includes index.
    ISBN 978-0-691-13504-5 (cloth : alk. paper)   1. Physics—Popular
works.   2. Technology—Popular works.   I. Title.
    QC24.5.M85   2010
    530—dc22
                                                    2009028490

British Library Cataloging-in-Publication Data is available

This book has been composed in Sabon text with ITC Stone Sans Family
Display
Printed on acid-free paper. ∞

10 9 8

Contents

Preface

Physics Is the Liberal Arts of High-Tech

*P*hysics and Technology for Future Presidents? Yes, that is a serious title. Energy, global warming, terrorism and counterterrorism, health, Internet, satellites, remote sensing, ICBMs and ABMs, DVDs and HDTVs—economic and political issues increasingly have a strong high-tech content. Misjudge the science; make a wrong decision. Yet many of our leaders never studied physics and do not understand science and technology. Even my school, the University of California at Berkeley, doesn't require physics. This book is designed to address that problem. Physics is the liberal arts of high technology. Understand physics, and you will never again be intimidated by technological advances. *Physics and Technology for Future Presidents* is designed to attract students and to teach them the physics and technology they need to know to be effective world leaders.

Is science too hard for world leaders to learn? No, it is just badly taught. Think of an analogous example: Charlemagne was only half-literate—he could read but not write. Writing was a skill considered too tough even for world leaders, just as physics is today. And yet now most of the world is literate. Many children learn to read before kindergarten. Literacy in China is 84% (according to the OECD). We can, and must, achieve the same level with scientific literacy, especially for our leaders.

This course is based on several decades of experience I've had presenting tough scientific issues to top leaders in government and business. My conclusion is that these people are smarter than most physics professors. They readily understand complex issues, even though they don't relax by doing integrals. (I know a physics professor who does.) *Physics and Technology for Future Presidents* is not Physics for Poets, Physics for Jocks, or Physics for Dummies. It is the physics you need to know be an effective world leader.

Can physics be taught without math? Of course! Math is a tool for computation, but it is not the essence of physics. We often cajole our advanced students,

"Think physics, not math!" You can understand and even compose music without studying music theory, and you can understand light without knowing Maxwell's equations. The goal of this course is not to create mini-physicists. It is to give future world leaders the knowledge and understanding that they need to make decisions. If they need a computation, they can always hire a physics professor. But the knowledge of physics will help them judge, on their own, whether the physicist is right. Let me illustrate what can be taught by telling a short story that I share with my students in the first lecture. It tells them what I want from them.

An Ideal Student

Liz, a former student of my class, came to my office hour, eager to share a wonderful experience she had had a few days earlier. Her family had invited a physicist over for dinner, someone who worked at the Lawrence Livermore National Laboratory. He regaled them through the dinner with his stories of controlled thermonuclear fusion and its great future for the power needs of our country. According to Liz, the family sat in awe of this great man describing his great work. Liz knew more about fusion than did her parents, because we had covered it in our class.

There was a period of quiet admiration at the end. Finally, Liz spoke up. "Solar power has a future too," she said.

"Ha!" the physicist laughed. (He didn't mean to be patronizing, but this is a typical tone physicists affect.) "If you want enough power just for California," he continued, "you'd have to plaster the whole state with solar cells!"

Liz answered right back. "No, you're wrong," she said. "There is a gigawatt in a square kilometer of sunlight, and that's about the same as a nuclear power plant."

Stunned silence from the physicist. Liz said he frowned. Finally, he said, "Hmm. Your numbers don't sound wrong. Of course, present solar cells are only 15% efficient . . . but that's not a huge factor. Hmm. I'll have to check my numbers."

Yes! That's what I want my students to be able to do. Not integrals, not roller-coaster calculations, not pontifications on the scientific method or the deep meaning of conservation of angular momentum. Liz was able to shut up an arrogant physicist who hadn't done his homework! She hadn't just memorized facts. She knew enough about the subject of energy that she could confidently present her case under duress when confronted by a supposed expert. Her performance is even more impressive when you recognize that solar power is only a tiny part of this course. She remembered the important numbers because she had found them fascinating and important. She hadn't just memorized them but had thought about them and discussed them with her classmates. They had become part of her, a part that she could bring out and use when she needed it, even a year later.

Physics for the Future Leader

Physics and Technology for Future Presidents is not watered-down physics. It is advanced physics. It covers the most interesting and most important topics. Students recognize the value of what they are learning and are naturally motivated to do well. In every chapter, they find material they want to share with their friends, roommates, and parents. Rather than keep the students beneath the math glass ceiling, I take them way above it. "You don't have the time or the inclination to learn the math," I tell them. "So we'll skip over that part and get to the important stuff right away." I then teach them things that ordinary physics students don't learn until *after* they earn their PhD.

Typical physics majors, even typical PhDs, do *not* know the material in this book. They know little to nothing about nukes, optics, fluids, batteries, lasers, IR and UV, x-rays and gamma rays, and MRI, CAT, and PET scans. Ask a physics major how a nuclear bomb works, and you'll hear what the student learned in high school. For that reason, at Berkeley we have now opened this course to physics majors. The predecessor course, *Physics for Poets*, was considered too easy for them, but most of the material in the *Future Presidents* version is completely new to them. It is not baby physics. It is advanced physics.

I must confess that I made one major concession to my Berkeley students. They really do want to learn about relativity and cosmology, subjects superfluous for world leadership but fascinating to thinking people. So I added two chapters at the end. They cover subjects that every educated person should know, but they won't help the president make key decisions.

The response to this new approach has been fantastic. Enrollment grew mostly by word-of-mouth from 34 students (spring 2001) to over 500 (fall 2006). The class now fills up the largest physics-ready lecture hall at Berkeley. Many of my students previously hated physics and swore (after their high school class) never to take it again. But they are drawn, like moths to a flame, to a subject that they find not only intriguing but also highly relevant to current world affairs. My job is to make sure that their craving is fulfilled and that they won't be burned again. These students come to college to learn, and they are happiest when they sense their knowledge and abilities growing.

Students don't take the course because it is easy; it isn't. It covers an enormous amount of material. But every chapter is full of information that is evidently important. That's why students sign up. They don't want to be entertained. They want a good course, well taught, that fills them with important information and the ability to use it well. They are proud to take this course, but more importantly, they are very proud that they enjoy it.

Physics and Technology for Future Presidents

Energy and Power

and the Physics of Explosions

> At the end of the Cretaceous period, the golden age of dinosaurs, an asteroid or comet about 10 miles in diameter headed directly towards the Earth with a velocity of about 20 miles per *second*, over ten times faster than our speediest bullets. Many such large objects may have come close to the Earth, but this was the one that finally hit. It hardly noticed the air as it plunged through the atmosphere in a fraction of a second, momentarily leaving a trail of vacuum behind it. It hit the Earth with such force that it and the rock near it were suddenly heated to a temperature of over a million degrees Centigrade, several hundred times hotter than the surface of the sun. Asteroid, rock, and water (if it hit in the ocean) were instantly vaporized. The energy released in the explosion was greater than that of a hundred million megatons of TNT, 100 *teratons*, more than ten thousand times greater than the total U.S. and Soviet nuclear arsenals. . . . Before a minute had passed, the expanding crater was 60 miles across and 20 miles deep. It would soon grow even larger. Hot vaporized material from the impact had already blasted its way out through most of the atmosphere to an altitude of 15 miles. Material that a moment earlier had been glowing plasma was beginning to cool and condense into dust and rock that would be spread worldwide.
>
> —from Richard A. Muller, *Nemesis*

Few people are surprised by the fact that an asteroid the size of Mount Everest could do a lot of damage when it hits the Earth. And it is not really surprising that such bodies are out there (figure 1.1). The danger has been the subject of many movies, including *Deep Impact, Meteor,* and *Armageddon.* Asteroids and comets frequently come close to the Earth. Every few years, we see a newspaper headline about a "near miss" in which an object misses the Earth by "only a few million miles." That is hardly a near miss. The

Figure 1.1 Comet Shoemaker-Levy crashes into Jupiter. This explosion is much smaller than the one that occurred when an asteroid or a comet crashed into the Earth 65 million years ago. (Image taken by Peter McGregor at the ANU 2.3m telescope at Siding Spring, Research School of Astronomy and Astrophysics. Copyright Australian National University. Used with permission.)

radius of the Earth is about 4000 miles. So a miss by, say, four million miles would be a miss by a thousand Earth radii. Hitting the Earth is comparable to hitting an ant on a dartboard.

Although the probability of an asteroid impact during your lifetime is small, the consequences could be huge, with millions or maybe even billions of people killed. For this reason, the U.S. government continues to sponsor both asteroid searches, to identify potential impactors, and research into ways to deflect or destroy such bodies.

But why should an asteroid impact cause an explosion? The asteroid was made of rock, not dynamite. And why would it cause such a big explosion? But then what is an explosion, after all?

Explosions and Energy

An *explosion* occurs when a great deal of stored energy is suddenly converted to heat in a confined space. This is true for a grenade, an atomic bomb, or an asteroid hitting the Earth. The heat is enough to vaporize the matter, turning it into an extremely hot gas. Such a gas has enormous pressure—that is, it puts a great force on everything that surrounds it. Nothing is strong enough to resist this pressure, so the gas expands rapidly and pushes anything near it out of the way. The flying debris is what does the damage in an explosion. It doesn't matter what the original form of the energy is—it could be kinetic energy (the result of motion), like the energy of the asteroid, or chemical energy, like the energy in the explosive trinitrotoluene (TNT). It is the rapid conversion of this energy to *heat* that is at the heart of most explosions.

You may have noticed that I used a lot of common terms in the preceding paragraph that I didn't explain. Words such as *energy* and *heat* have everyday meanings, but they also have precise meanings when used in physics. Physics can be derived in a deductive way, just like geometry, but it is hard to learn in that manner. So our approach will be to start with intuitive definitions and then make them more precise as we delve deeper into the physics. Here are some beginning definitions that you may find helpful. The precise meanings of these definitions will become clearer over the next three chapters.

Definitions (Don't Memorize)

- **Energy** is the ability to do work. (*Work* is defined numerically as the magnitude of a force multiplied by the distance that the force moves in the direction of the force.) Alternative definition for energy: anything that can be turned into heat.[1]
- **Heat** is something that raises the temperature of a material, as measured by a thermometer. (It will turn out that heat is actually the microscopic energy of motion of vibrating molecules.)

These definitions sound great to the professional physicist, but they might be somewhat mysterious to you. They don't really help much since they involve other concepts (work, force, energy of motion) that you may not precisely understand. I'll talk more about all these concepts in the coming pages. In fact, it is very difficult to understand the concept of energy just from the definitions alone. Trying to do so is like trying to learn a foreign language by memorizing a dictionary. So be patient. I'll give lots of examples, and those will help you to feel your way into this subject. Rather than read this chapter slowly, I recommend that you read it quickly, and more than once. You learn physics by iteration—that is, by going over the same material many times. Each time you do that, the material makes a little more sense. That's also the best way to learn a foreign language: total immersion. So don't worry about understanding things just yet. Just keep on reading.

Amount of Energy

Guess: How much energy is there in a pound of an explosive, such as dynamite or TNT, compared to, say, a pound of chocolate chip cookies? Don't read any more until you've made your guess.

Here's the answer: The chocolate chip cookies have the greater energy. Not only that, but the energy is *much* greater—eight times greater in the cookies than in TNT! That fact surprises nearly everybody, including many physics professors. Try it out on some of your friends who are physics majors.

---

[1] It is likely, as the Universe evolves, that virtually *all* energy will be converted to heat. This idea has spawned numerous essays by philosophers and theologians. It is sometimes called the "heat death" of the Universe, since heat energy cannot always be converted back to other forms.

How can it be? Isn't TNT famous for the energy it releases? We'll resolve this paradox in a moment. First, let's list the energies in various different things. There are a lot more surprises coming, and if you are investing in a company, or running the U.S. government, it is important that you know many of these facts.

To make the comparisons, let's consider the amount of energy in 1 gram of various materials. (A *gram* is the weight of a cubic centimeter of water; a penny weighs 3 grams, and there are 454 grams in a pound.) I'll give the energy in several units: the Calorie, the calorie, the watt-hour, and the kilowatt-hour.

CALORIE

The unit you might feel most familiar with is the *Calorie*. That's the famous "food calorie" used in dieting. It is the one that appears on the labels of food packages. A chocolate chip (just the chip, not the whole cookie) contains about 3 Calories. A 12-ounce can of Coca Cola has about 150 Calories.

> **Beware:** If you studied chemistry or physics, you may have learned about the unit called the *calorie*. That is different from the Calorie! A food Calorie (usually capitalized) is 1000 little physics calories. That is a terrible convention, but it is not my fault. Physicists like to refer to food Calories as kilocalories. Food labels in Europe and Asia frequently list kilocalories, but not in the United States. So 1 Cal = 1000 cal = 1 kilocalorie.[2]

KILOWATT-HOUR

Another famous unit of energy is the *kilowatt-hour*, abbreviated kWh. (The W is capitalized, some say, because it stands for the last name of James Watt, but that doesn't explain why we don't capitalize it in the middle of the word kilowatt.) What makes this unit so well known is that we buy electricity from the power company in kWh. That's what the meter outside the house measures. One kWh costs between 5 and 25 cents, depending on where you live. (Electric prices vary much more than gasoline prices.) We'll assume an average price of 10 cents per kWh in this text.

It probably will not surprise you that there is a smaller unit called the *watt-hour*, abbreviated Wh. A kilowatt-hour consists of a thousand watt-hours. This unit isn't used much, since it is so small; however, my computer battery has its capacity marked on the back as *60 Wh*. Its main value is that a Wh is approximately 1 Calorie.[3] So for our purposes, it will be useful to know that:

$$Wh = 1 \text{ Calorie (approximately)}$$
$$1 \text{ kWh} = 1000 \text{ Calories}$$

---

[2] I got into trouble in a cake recipe once because I didn't know the difference between a Tsp and a tsp of baking powder. In fact, 1 Tsp = 3 tsp. Ask a cook! (1 Tsp is the standard abbreviation for a tablespoon; 1 tsp is the abbreviation for a teaspoon.)

[3] To an accuracy of 16%.

JOULE

Physicists like to the use energy unit called the *joule* (named after James Joule) because it makes their equations look simpler. There are about 4200 joules in a Calorie, 3600 in a Wh, 3.6 million in a kWh.

Table 1.1 shows the approximate energies in various substances. I think you'll find that this table is one of the most interesting ones in this entire textbook. It is full of surprises. The most interesting column is the rightmost one.

Table 1.1   Energy per Gram

Object	Calories (or watt-hours)	Joules	Compared to TNT
Bullet (at sound speed, 1000 ft/s)	0.01	40	0.015
Battery (auto)	0.03	125	0.05
Battery (rechargeable computer)	0.1	400	0.15
Flywheel (at 1 km/s)	0.125	500	0.2
Battery (alkaline flashlight)	0.15	600	0.23
TNT (the explosive trinitrotoluene)	0.65	2700	1
Modern high explosive (PETN)	1	4200	1.6
Chocolate chip cookies	5	21,000	8
Coal	6	27,000	10
Butter	7	29,000	11
Alcohol (ethanol)	6	27,000	10
Gasoline	10	42,000	15
Natural gas (methane, CH_4)	13	54,000	20
Hydrogen gas or liquid (H_2)	26	110,000	40
Asteroid or meteor (30 km/s)	100	450,000	165
Uranium-235	20 million	82 billion	30 million

Note: Many numbers in this table have been rounded off.

Stop reading now, and ponder this energy table. Concentrate on the rightmost column. Look for the numbers that are surprising. How many can you find? Circle them. I think all of the following are surprises:

- The very large amount of energy in chocolate chip cookies
- The very small amount of energy in a battery (compared to gasoline!)
- The high energy in a meteor, compared to a bullet or to TNT
- The enormous energy available in uranium (compared to anything else in the table)

Try some of these facts on your friends. Even most physics majors will be surprised. These surprises and some other features of the table are worthy of much further discussion. They will play an important role in our energy future.

Discussion of the Energy Table

Let's pick out some of the more important and surprising facts shown in the energy table and discuss them in more detail.

TNT VERSUS CHOCOLATE CHIP COOKIES

Both TNT and chocolate chip cookies store energy in the forces between their atoms. That's like the energy stored in compressed springs—we'll discuss atoms in more detail soon. Some people like to refer to such energy as *chemical energy*, although this distinction isn't really important. When TNT is exploded, the forces push the atoms apart at very high speeds. That's like releasing the springs so that they can suddenly expand.

One of the biggest surprises in the energy table is that chocolate chip cookies (CCCs) have eight times the energy as the same weight of TNT. How can that be true? Why can't we blow up a building with CCCs instead of TNT? Almost everyone who hasn't studied the subject assumes (incorrectly) that TNT releases a great deal more energy than cookies. That includes most physics majors.

What makes TNT so useful for destructive purposes is that it can release its energy (transfer its energy into heat) very, very quickly. The heat is so great that the TNT becomes a gas that expands so suddenly that it pushes and shatters surrounding objects. (We'll talk more about the important concepts of *force* and *pressure* in the next chapter.) A typical time for 1 gram of TNT to release all of its energy is about one *millionth* of a second. Such a sudden release of energy can break strong material.[4] *Power* is the rate of energy release. CCCs have high energy, but the TNT explosion has high power. We'll discuss power in greater detail later in this chapter.

Even though chocolate chip cookies contain more energy than a similar weight of TNT, the energy is normally released more slowly, through a series of chemical processes that we call *metabolism*. This requires several chemical changes that occur during digestion, such as the mixing of food with acid in the stomach and with enzymes in the intestines. Last, the digested food reacts with oxygen taken in by the lungs and stored in red blood cells. In contrast, TNT contains all the molecules it needs to explode; it needs no mixing, and as soon as part of it starts to explode, that triggers the rest. If you want to destroy a building, you can do it with TNT. Or you could hire a group of teenagers, give them sledgehammers, and feed them cookies. Since the energy in chocolate chip cookies exceeds that in an equal weight of TNT, each gram of chocolate chip cookies will ultimately do more destruction than would each gram of TNT.

Note that we have cheated a little bit. When we say there are 5 Calories per gram in CCCs, we are ignoring the weight of the air that combines with the CCCs. In contrast, TNT contains all the chemicals needed for an explosion, whereas CCCs need to combine with air. Although air is "free" (you don't have

---

[4] As you'll see in chapter 3, to calculate the force, you can take the energy of a substance such as TNT and divide it by the distance over which it is released (from chemical to kinetic energy).

to buy it when you buy the CCCs), part of the reason that CCCs contain so much energy per gram is that the weight of the air was not counted. If we were to include the weight of the air, the energy per gram would be lower, about 2.5 Calories per gram. That's still almost four times as much as for TNT.

THE SURPRISINGLY HIGH ENERGY OF GASOLINE

As table 1.1 shows, gasoline contains significantly more energy per gram than cookies, butter, alcohol, or coal. That's why it is so valuable as fuel. This fact will be important when we consider alternatives to gasoline for automobiles.

Gasoline releases its energy (turns it into heat) by combining with oxygen, so it must be well mixed with air to explode. In an automobile, this is done by a special device known as a *fuel injector*; older cars use something called a *carburetor*. The explosion takes place in a cylindrical cavity known, appropriately, as the *cylinder*. The energy released from the explosion pushes a piston down the axis of the cylinder, and that is what drives the wheels of the car. An internal "combustion" engine can be thought of as an internal "explosion" engine.[5] The *muffler* on a car has the job of making sure that the sound from the explosion is muffled and not too bothersome. Some people like to remove the muffler—especially some motorcyclists—so that the full explosion is heard; this can give the illusion of much greater power. Removing the muffler also lowers the pressure just outside the engine, so the power to the wheels is actually increased, although not by very much. We'll talk more about the gasoline engine in the next chapter.

The high energy per gram in gasoline is the fundamental physics reason why gasoline is so popular. Another reason is that when it burns, all the residues are gas (mostly carbon dioxide and water vapor), so there is no residue to remove. In contrast, for example, most coal leaves a residue of ash.

THE SURPRISINGLY LOW ENERGY IN BATTERIES

A battery also stores its energy in chemical form. It can use its energy to release electrons from atoms (we'll discuss this more in chapters 2 and 6). Electrons can carry their energy along metal wires and deliver their energy at another place; think of wires as *pipes* for electrons. The chief advantage of electric energy is that it can be easily transported along wires and converted to motion with an electric motor.

A car battery contains 340 times *less* energy than an equal weight of gasoline! Even an expensive computer battery is about 100 times worse than gasoline. Those are the physics reasons why most automobiles use gasoline instead of batteries as their source of energy. Batteries are used to start the engine because they are reliable and fast.

---

[5] Engineers like to make a distinction between an *explosion*, in which an abrupt front called a shock wave is generated that passes through the rest of the material and ignites it, and a *deflagration*, in which there is no shock wave. There is no shock wave in the detonation of gasoline in an automobile, so by this definition, there is no explosion in an automobile engine. Newspapers and the general public do not make this fine distinction, and in this book, neither will I.

Battery-powered cars

A typical automobile battery is also called a *lead–acid battery*, because it uses the chemical reaction between lead and sulfuric acid to generate electricity. Table 1.1 shows that such batteries deliver 340 times less energy than gasoline. However, the electric energy from a battery is very convenient. It can be converted to wheel energy with 85% efficiency—put another way, only 15% is lost in running the electric motor. A gasoline engine is much worse: only 20% of the energy of gasoline makes it to the wheels; the remaining 80% is lost as heat. When you put in those factors, the advantage of gasoline is reduced from 340 down to a factor of 80. So for automobiles, batteries are *only* 80 times worse that gasoline. That number is small enough to make battery-driven autos feasible. In fact, every so often you'll read in the newspaper about someone who has actually built one. A typical automobile fuel tank holds about 100 pounds of gasoline. (A gallon of gasoline weighs about 6 pounds.) To have batteries that carry the energy in 100 pounds of gasoline would take 80 times that weight—that is, 8000 pounds of lead–acid batteries. But if you are willing to halve the range of the car, from 300 miles to 150, then the weight is reduced to 4000 pounds. If you need only 75 miles to commute, then the lead–acid battery weight is only 2000 pounds. (We'll discuss lighter lithium–ion batteries in a moment.)

Why would you trade a gasoline car for a car that could go only 75 miles? The usual motivation is to save money. Electricity bought from the power company, used to charge the battery, costs only 10 cents per kWh. Gasoline costs (as of this writing) about $2.50 per gallon. When you translate that into energy delivered to the wheels, that works out to about 40 cents per kWh. So electricity is four times cheaper! Actually, it isn't quite that good. When most people work out those numbers, they ignore the fact that standard lead–acid car batteries have to be replaced after, typically, 700 charges. When you include the battery expense, the cost per kWh is about 20 cents per kWh. It beats the cost of gasoline by a factor of two. But because batteries take so much space, it's not an attractive option for people who value trunk space.

Batteries have additional advantages in some circumstances. In World War II, when submarines had to submerge and could not obtain oxygen, their energy source was a huge number of batteries stored beneath the decks. When on the surface, or at "snorkeling depth," the submarines ran on diesel fuel, a form of gasoline. The diesel fuel also ran generators that recharged the batteries. So during World War II, most submarines spent most of their time on the surface, recharging their batteries. Watch an old World War II movie, and they don't show that; you get the misimpression that the subs were always below water. Modern nuclear submarines don't require oxygen, and they can remain submerged for months. That greatly increases their security against detection.

Electric car hype

Suppose that we use better batteries, ones that hold more energy per gram. Let's look at the Tesla Roadster, a battery car that has received a lot of attention. It is powered by 1000 pounds of rechargeable lithium-ion batteries, similar to those found in laptop computers. The car range is 250 miles. Tesla Motors claims that if you charge the batteries from your home power plug, driving the car costs 1 to 2 cents per mile. Top speed: 130 miles per hour. Wow!

Can't wait to get one? The cars are built in a factory in England and are currently being sold for about $100,000.

The catch is in the cost of the batteries. Lead–acid batteries, the ones that we considered for the electric car calculation earlier, have a retail cost of about a dollar per pound of battery: $50 for 50 pounds. A good computer battery has a retail price of about $100 per pound—$100,000 for the 1000 pounds in the Tesla Roadster. (When you buy batteries in bulk, the price is about half, so the Roadster batteries cost only $50,000.) When we included replacement costs, the lead–acid batteries cost 10 cents per kWh; a similar calculation shows that computer batteries cost about $4 per kWh. That's 10 times as much as the cost of gasoline! So, when you consider the cost of replacing the batteries, electric cars are far more expensive to operate than our standard gasoline cars. A great deal of research is going into battery improvement, so it is likely that the cost of batteries will come down and that in the future batteries will be made that last longer before they have to be replaced.

There is a lot of hype about "who killed the electric car." Some people say it was the oil industry, because they didn't want a cheaper alternative. But the electric car is not cheaper, unless you are willing to live with the very short range (and heavy) version that uses lead–acid batteries.

HYBRID AUTOS

Despite the limitations of batteries, there is a fascinating technology called *hybrid automobiles*. In a hybrid, a small gasoline engine provides energy to charge a battery; the car then gets its energy from the battery. This has more value than you might guess: the gasoline engine can be run at a constant rate, under ideal conditions, and as a result, it is two to three times as efficient as the engine in ordinary cars. In addition, hybrid engines can convert some of the mechanical motion of the automobile (e.g., its extra speed picked up when descending a hill) back to stored chemical energy in the rechargeable battery. It does this instead of using brakes—which only turn the energy of motion into heat. Hybrid engines are becoming very popular, and in a few years, they may be the most common type of automobile, particularly if gasoline costs go back up to the very high prices of 2008. Hybrid autos can get about 50 miles per gallon (that's what I get with my Toyota Prius, if I drive with low accelerations), considerably better than the 30 miles per gallon that similar nonhybrid autos get.

Many people complain that their hybrids do not have the facility to charge up from the wall plug. The first American version of the Prius got its energy only from its own gasoline engine. In Japan, people can charge the battery from the electric grid, and on the Internet you can find clubs that show you how to change your older Prius to accomplish just that too. The people who do this mistakenly think that they are saving money. They aren't, for the same reason I articulated when discussing all electric autos. It is likely that the batteries in the hybrid can be charged only about 500 to 1000 times. After that, they will have to be replaced, and that will make the average cost per mile much higher. In the current Prius, the batteries are used only during moments when the gasoline engines would be inefficient, such as during rapid acceleration. With this limited use, the batteries will last much longer. It is not clear how much longer that will be, but it could conceivably affect older cars.

HYDROGEN VERSUS GASOLINE—AND THE FUEL CELL

Notice in table 1.1 that hydrogen gas has 2.6 times more chemical energy per gram than gasoline. Popular articles about the future "hydrogen economy" are partially based on this fact. In 2003, President George W. Bush announced a major program with the goal of making hydrogen into a more widely used fuel. But within two years, most of the hydrogen economy programs were cancelled, for the physics reasons we will discuss in a moment.

Another attractive feature of hydrogen is that the only waste product it produces is water, created when the hydrogen is chemically combined with oxygen from the air to make H_2O (water). Moreover, the conversion can be done with high efficiency by using an advanced technology called a *fuel cell* to convert the chemical energy directly to electricity.

A fuel cell looks very much like a battery, but it has a distinct advantage. In a battery, once the chemical is used up, you have to recharge it with electricity produced elsewhere or throw it away. In a fuel cell, all you have to do is provide more fuel (e.g., hydrogen and oxygen). Figure 1.2 shows a setup to demonstrate *electrolysis*, in which electricity is passed between two terminals through water, and hydrogen and oxygen gas are produced at the terminals.

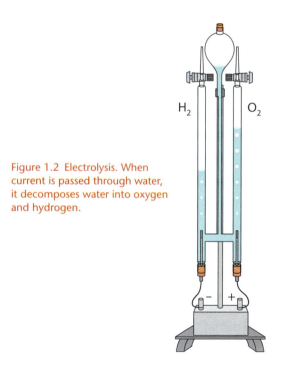

Figure 1.2 Electrolysis. When current is passed through water, it decomposes water into oxygen and hydrogen.

A fuel cell is very similar to an electrolysis apparatus, but it is run backward. Hydrogen and oxygen gas are compressed at the electrodes, they combine to form water, and that makes electricity flow through the wires that connect one terminal to the other. So figure 1.2 can also represent a fuel cell.

The main technical difficulty of the hydrogen economy is that hydrogen is not very dense. Even when liquefied, it has a density of only 0.071 grams per cubic centimeter (cc), a factor of 10 times less than gasoline. As you saw in table 1.1, per gram, hydrogen has 2.6 times more energy than gasoline. Put these together, and we find that liquid hydrogen stores only $0.071 \times 2.6 = 0.18$ times as much energy per cubic centimeter (or per gallon) as gasoline. That is a factor of 5 times worse. However, many experts say that the factor is only 3 times worse, since hydrogen can be used more efficiently than gasoline. It is useful to remember the following approximate numbers; you will find them valuable when discussing the hydrogen economy with other people.

Remember: Compared to gasoline, *liquid hydrogen* has about

$3 \times$ *more* energy per gram (or per pound)
$3 \times$ *less* energy per gallon (or per liter)

Here's another approximate rule that is easy to remember. In terms of energy that can be delivered to a car:

1 kilogram of hydrogen ≈ 1 gallon of gasoline

Hydrogen liquid is dangerous to store since it expands by a factor of a thousand if warmed. If you protect against that with a thick-walled tank, you might as well store the hydrogen as a high-pressure gas. At a pressure of 10,000 pounds per square inch (66 times atmospheric pressure), the gas is almost half as dense as hydrogen liquid. But that factor of half makes it even harder to fit hydrogen into a reasonable space.

Compared to gasoline, compressed gas hydrogen has
$6 \times$ *less* energy per gallon (or per liter).

And the tank to contain the hydrogen typically weighs 10 to 20 times as much as the hydrogen itself. That takes away the weight advantage too.

Because hydrogen takes up so much space (even though it doesn't weigh much), it may be used for buses and trucks before it is used for automobiles. It is also possible that hydrogen will be more valuable as a fuel for airplanes, since for large airplanes the low weight of the hydrogen may be more important than the fact that it takes more volume than gasoline. Fuel cells first achieved prominence in the space program as the energy storage method used by the astronauts (figure 1.3). For the mission to the Moon, low weight was more important than the space that could be saved in the capsule. Moreover, the water that was produced could be used by the astronauts, and there was no waste carbon dioxide to eject.

A technical difficulty with liquid hydrogen is that it boils at a temperature of *minus* 423 degrees Fahrenheit. This means that it must be transported in special thermos bottles (technically known as *dewars*). Either that, or it can be transported in a form in which it is chemically or physically combined with other materials at room temperature, although that greatly increases the weight per Calorie. A more practical alternative may be to transport it as compressed gas, but then the weight of the pressure tank actually exceeds the weight of the hydrogen carried.

Figure 1.3 Fuel cell developed by NASA. Hydrogen gas enters through the inlet on the top. Air enters through some of the circular openings, and carbon dioxide leaves through the others. The electric power comes from the wires in the back. (Photo courtesy of NASA.)

You can't mine hydrogen! There is virtually no hydrogen gas (or liquid) in the environment. There's lots of hydrogen in water and in fossil fuels (hydrocarbons)—but not "free" hydrogen, the molecule H_2. That's what we want for the hydrogen economy. Where can the hydrogen we need come from? The answer is that we have to "make" it—that is, release it from the compounds of water or hydrocarbons. Hydrogen gas must be obtained by electrolysis of water, by reacting fossil fuels (methane or coal) with water to produce hydrogen gas and carbon monoxide. Doing any of these takes energy.

A typical hydrogen production plant of the future would start with a power plant fueled by coal, gasoline, nuclear fuel, or solar energy. That power plant might use this energy to convert ordinary water to hydrogen and oxygen (through electrolysis, or through a series of chemical reactions known as "steam reforming"). Then, for example, the hydrogen could be cooled until it is turned into a liquid and then transported to the consumer. When hydrogen is obtained in this manner, you get back out of the hydrogen only some of the energy that you put in to make it. A reasonable estimate is that the fraction of the original energy (used to create the hydrogen) that gets to the wheels of the car is about 20%. Thus:

> Hydrogen is not a *source* of energy.
> It is only a means for *transporting* energy.

Many people who favor the hydrogen economy believe that the source of hydrogen will be methane gas. Methane molecules consist of one carbon atom and four hydrogens. That's why the chemical formula is CH_4. When methane is heated with water to high temperatures, the hydrogen in the methane is released, along with carbon dioxide. Since carbon dioxide is considered an air pollutant (see chapter 11), this method of production may not be optimum, but it is probably the cheapest way to make hydrogen.

Although the fuel cell produces no pollution (only water), it is not quite right to say that a hydrogen-based economy is pollution-free unless the plant that used energy to produce the hydrogen is also pollution-free. Nevertheless, the use of hydrogen as a fuel is expected to be environmentally less harmful than gasoline for two reasons: a power plant can, in principle, be made more efficient than an automobile (so less carbon dioxide is released); and the power plant can have more elaborate pollution-control devices than an automobile. If we use solar or nuclear power to produce the hydrogen, then no carbon dioxide, the most problematic global warming gas, is released.

Hydrogen can also be produced as a by-product of "clean coal" conversion. In some modern coal plants, coal is reacted with water to make carbon monoxide and hydrogen. These are then burned. In such a plant, the hydrogen could be transported to serve as fuel elsewhere, but the energy stored in it originated from the coal.

Other people like the idea of hydrogen as fuel because it moves the sources of pollution away from the cities, where a high concentration of pollutants can be more dangerous to human health. Of course, it is hard to predict all environmental effects. Some environmentalists argue that significant hydrogen gas could leak into the atmosphere and drift to high altitudes. There it could combine with oxygen to make water vapor, and that could affect both the Earth's temperature and delicate atmospheric structure such as the ozone layer (see chapter 9).

The United States has enormous coal reserves. About 2 trillion tons of coal are "known" reserves, but with more extensive searching, geologists expect that about twice as much is likely to be present. Coal could be used to produce all the energy that we would need (at current consumption rates) for hundreds of years. Of course, the environmental consequences from strip mining and carbon dioxide production could be very large. Coal can be converted to liquid fuel for easy pumping and use in automobiles by a technology known as the *Fisher-Tropsch process*; we'll discuss that more in chapter 11.

GASOLINE VERSUS TNT

In most movies, when a car crashes, it explodes. Does this happen in real life? Have you ever witnessed the scene of a car crash? Did an explosion taken place? The answer is: usually not. Cars explode in movies only because they have been loaded with TNT or other explosives for dramatic visual effects. Unless mixed with air in just the right ratio (done in the automobile by the fuel injector or carburetor), gasoline burns but doesn't explode.

In the Spanish revolution, the rebels invented a device that later became known as a "Molotov cocktail." It was a bottle filled with gasoline, with a rag stuck in the neck. The rag was soaked with gasoline and ignited, and then the bottle was thrown at the enemy. It broke upon impact. It usually didn't explode, but it spread burning gasoline, and that was pretty awful to the people who were the targets. This weapon quickly achieved a strong reputation as an ideal weapon for revolutionaries.

I hesitate to give examples from the unpleasant subject of war, but it is important to future presidents and citizens to know of these. On 6 November 2002, the United States started dropping "fuel–air explosives" on Taliban soldiers in Afghanistan. You can probably guess that this was a liquid fuel similar to gasoline. Fifteen thousand pounds of fuel is dropped from an airplane in a large container (like a bomb) that descends slowly on a parachute. As it nears the ground, a small charge of high explosive (probably only a few pounds worth) explodes in the center, destroying the container and dispersing the fuel and mixing it with air—but not igniting it. Once the fuel is spread out and well-mixed with air, it is ignited by a second explosion. The explosion is spread out over a large area, so it doesn't exert the same kind of intense force that it takes to break through a concrete wall, but it has enough energy released to kill people and other "soft" targets. What makes it so devastating is the fact that 15,000 pounds of fuel, like gasoline, contains the energy equivalent of 225,000 pounds of TNT. So although 15,000 pounds sounds bad, in fact it is much worse than it sounds. Once the soldiers had seen the fuel–air explosive from a distance, the mere approach of a parachute induced panic.

URANIUM VERSUS TNT

The most dramatic entry in table 1.1 is the enormous energy in the form of uranium known as *U-235*. The amount of energy in U-235 is 30 million times that of the energy found in TNT. We will discuss this in detail in chapters 4 and 5. For now, there are only a few important facts to know. The enormous forces inside the uranium atom's nucleus provide the energy. For most atoms, this energy cannot be easily released, but for U-235 (a special kind of uranium that makes up only 0.7% of natural uranium), the energy can be released through a process called a *chain reaction* (discussed in detail in chapter 5). This enormous energy release is the principle behind nuclear power plants and atomic bombs. Plutonium (the kind known as *Pu-239*) is another atom capable of releasing such huge energy.

Compared to gasoline, U-235 can release 2 million times as much energy per gram. Compared to chocolate chip cookies, it releases about 3 million times as much. The following approximation is so useful that it is worth memorizing:

> For the same weight of fuel, nuclear reactions release about
> a *million times* more energy than do chemical or food reactions.

More Surprises: Coal Is Dirt Cheap

There are also some amazing surprises in the cost of fuel. Suppose you want to buy a Calorie of energy, to heat your house. What is the cheapest source? Let's forget all other considerations, such as convenience, and just concentrate on the cost of the fuel. It is not easy for the consumer to compare. Prices are constantly changing, so let's just use some average prices from the last few years: Coal costs about $40 per ton, gasoline costs about $2.50 per gallon, natural gas (methane) costs about $3 per thousand cubic feet, and electricity costs about 10¢ per kilowatt-hour. So which gives the most Calories per dollar? It isn't obvious, since the different fuels are measured in different units, and they provide different amounts of energy. But if you put all the numbers together, you get table 1.2. This table also shows the cost of the energy if it is converted to electricity. For fossil fuels, that increases the cost by a factor of 3, since motors convert only about 1/3 of the heat energy to electricity.

The wide disparity of these prices is quite remarkable. Concentrate on the third column, the cost per kWh. Note that it is 25 times more expensive to heat

Table 1.2   Cost of Energy

Fuel	Market cost	Cost per kWh (1000 Cal)	Cost if converted to electricity
Coal	$40 per ton	0.4¢	1.2¢
Natural gas	$3 per thousand cubic feet	0.9¢	2.7¢
Gasoline	$2.50 per gallon	7¢	21¢
Electricity	$0.10 per kWh	10¢	10¢
Car battery	$50 to buy battery	21¢	21¢
Computer battery	$100 to buy battery	$4.00	$4.00
AAA battery	$1.50 per battery	$1000.00	$1000.00

your home with electricity than with coal! Gasoline costs over 2 times as much as natural gas. That has led some mechanics to modify their autos to enable them to use compressed natural gas instead of gasoline.

Note that for heating your home, natural gas *not converted to electricity* is 3 times cheaper than electricity. Back in the 1950s, many people thought that the "all-electric home" was the ideal—since electricity is convenient, clean, and safe. But most such homes have now been converted to use coal or natural gas, just because the energy is considerably cheaper.

Most dramatic on this list is the low price of coal. If energy per dollar were the only criterion, we would use coal for all our energy needs. Moreover, in many countries that have huge energy requirements, including the United States, China, Russia, and India, the reserves of coal are huge—enough to last for hundreds of years. We may run out of oil in the next few decades, but that does not mean that we are running out of cheap fossil fuel.

So why do we use oil instead of coal in our automobiles? The answer isn't physics, so I am only guessing. But part of the reason is that gasoline is very convenient. It is a liquid, and that makes it easy to pump into your tank, and from the tank to the engine. It was once much cheaper than it is now, and so in the past the cost was not as important an issue as convenience, and once we have optimized our auto designs and fuel delivery systems for gasoline, it's not easy to switch. It does contain more energy per gram than coal, so you don't have to carry as many pounds—although it is less dense, so it takes more space in the tank. Coal also leaves behind a residue of ash that has to be removed.

The low price of coal presents a very serious problem for people who believe that we need to reduce the burning of fossil fuels. Countries with substantial numbers of poor people may feel that they cannot switch to more expensive fuels. So the incredibly low price of coal is the real challenge to alternative fuels, including solar, biofuels, and wind. Unless the cost of these fuels can match the low cost of coal, it may be very difficult to convince developing countries that they can afford to switch.

It is odd that energy cost depends so much on the source. If the marketplace were "efficient," as economists sometimes like to postulate, then all these different fuels would reach a price at which the cost would be the same. This hasn't happened, because the marketplace is *not* efficient. There are large investments in energy infrastructure, and the mode of delivery of the energy is important. We are willing to spend a lot more for energy from a flashlight battery than from a wall plug because the flashlight is portable and convenient. Locomotives once ran on coal, but gasoline delivers more energy per pound, and it does so without leaving behind a residue of ash, so we switched from steam to diesel locomotives. Our automobiles were designed during a period of cheap oil, and we became accustomed to using them as if the price of fuel would never go up. Regions of the world with high gas prices (such as the countries of Europe) typically have more public transportation. The United States has suburbs—a luxury that is affordable when gas is cheap. Much of our way of living has been designed around cheap gasoline. The price we are willing to pay for fuel depends not only on the energy that it delivers, but also on its convenience.

The real challenge for alternative energy sources is to be more economically viable than coal. When we talk about global warming (in chapter 11), we'll discuss how coal is one of the worst carbon dioxide polluters that we use. To reduce our use of coal, we could, of course, tax it. But doing that solely in the developed nations would not accomplish much, since the ultimate problem will be energy use by nations such as China and India. Leaders of such countries might choose to get their energy in the cheapest possible way so that they can devote their resources to improving the nutrition, health, education, and overall economic well-being of their people.

Forms of Energy

We have talked about food energy and chemical energy. The energy in a moving bullet or an asteroid is called energy of motion, or *kinetic energy*. The energy stored in a compressed spring is called stored energy or *potential energy*. (Despite its name, potential energy does *not* mean that it is something that can "potentially" be converted to energy; potential energy *is* energy that is stored, just as food that is stored is still food.) *Nuclear energy* is the energy stored in the forces between parts of the atomic nucleus, released when the nucleus is broken. *Gravitational energy* is the energy that an object has at high altitude; when it falls, this energy is converted to kinetic energy. As we will discuss in chapter 2, the heat in an object is a form of energy. All these energies can all be measured in Calories or joules.

Many physics texts like to refer to chemical, nuclear, and gravitational energy as different forms of potential energy. This definition lumps together in one category all the kinds of energy that depend on shape and position—e.g., whether the spring is compressed, or how the atoms in a chemical are arranged. This lumping is done in order to simplify equations; there is no real value in doing it in this text, as long as you realize that all energy is energy, regardless of its name.

In popular usage, the term *energy* is used in many other ways. Tired people talk about having "no energy." Inspirational speakers talk about the "energy of the spirit." Be clear: they have the right to use *energy* in these nontechnical ways. Physicists stole the word *energy* from the English language and then redefined it in a more precise way. Nobody gave physicists the right to do this. But it is useful to learn the precise usage and to be able to use the term in the way physicists do. Think of this as "physics as a second language." The more precise definition is useful when discussing physics.

In the same precise physics language, *power* is defined as the energy used per second. It is the *rate* of energy release, as I mentioned early in this chapter. In equation form:

$$power = energy/time$$

Note that in popular usage, the terms of power and energy are often used interchangeably. You can find examples of this if you pay attention when reading newspaper articles. In our precise use of these terms, however, we can say that the value of TNT is that even though it has less energy per gram than chocolate chip cookies, it has greater power (since it can convert its limited energy to heat in a few millionths of a second). Of course, it can't deliver this power for very long because it runs out of energy.

As I mentioned earlier, the most common unit for power is the watt, named after James Watt, who truly developed the science of the steam engine. It was the most powerful motor of its time, and the "high-tech" of the late 1700s and early 1800s. The watt is defined as one joule per second:

$$1\ W = 1\ watt = 1\ joule\ per\ second$$
$$1\ kW = 1\ kilowatt = 1000\ joules\ per\ second$$

As you saw earlier, the term *kilowatt* is usually abbreviated as kW since Watt is a person's name, even though *watt* is usually not capitalized. The same logic (or lack of logic) applies to the kilojoule, abbreviated kJ.

There is a physics joke about the watt, inspired by an Abbott and Costello routine called "Who's on First" about baseball names. I relegate it to a footnote.[6] The original "Who's on First" routine is available on the Internet.

Energy Is "Conserved"

When the chemical energy in TNT or gunpowder is suddenly turned into heat energy, the gases that come out are so hot that they expand rapidly and push the bullet out of the gun. In doing this pushing, they lose some of their energy (they cool off); this energy goes into the kinetic energy of the bullet. Remarkably, if you add up all this energy, the total is the same. Chemical energy is converted to heat energy and kinetic energy, but the number of Calories (or joules) after the gun is fired is exactly the same as was stored in the gunpowder. This is the meaning behind the physics statement that "energy is conserved."

The conservation of energy is one of the most useful discoveries ever made in science. It is so important that it has earned a fancy name: the *first law of*

---

[6] Two people are talking. Costello: "What is the unit of power?" Abbott: "Watt." Costello: "I said, 'What is the unit of power?'" Abbott: "I said, 'Watt.'" Costello: "I'll speak louder. WHAT is the unit of power?" Abbott: "That's right." Costello: "What do you mean, 'that's right?' I asked you a question." Abbott: "Watt is the unit of power." Costello: "That's what I asked." Abbott: "That's the answer."—You can extend this dialog as long as you want.

thermodynamics. Thermodynamics is the study of heat, and we'll talk a lot about that in the next chapter. The first law points out the fact that any energy that appears to be lost isn't really lost; it is usually just turned into heat.

When a bullet hits a target and stops, some of the kinetic energy is transferred to the object (ripping it apart), and the rest is converted to heat energy. (The target and the bullet each get a little bit warmer when they collide.) This fact, that the total energy is always the same, is another example of the *conservation of energy*. It is one of the most useful laws of physics.[7] It is particularly valuable to people doing calculations in physics and engineering. Use of this principle allows physicists to calculate how rapidly the bullet will move as it emerges from the gun; it allows us to calculate how fast objects will move as they fall.

But if energy conservation is a law of physics, why are we constantly admonished by our teachers, by our political leaders, and by our children that we should conserve energy? Isn't energy automatically conserved?

Yes it is, but not all forms of energy have equal economic value. It is easy to convert chemical energy to heat, and very difficult to convert it back. When you are told to conserve energy, what is really meant is "conserve useful energy." The most useful kinds are chemical (e.g., in gasoline) and potential energy (e.g., the energy stored in water that has not yet run through a dam to produce electric power). The least useful form is heat, athough some (but not all) of heat energy can be converted to more useful forms.

Measuring Energy

The easiest way to measure energy is to convert it to heat and then see how much it raises the temperature of water. The original definition of the Calorie was actually based on this kind of effect: one Calorie is the energy it takes to raise one kilogram of water by one degree Celsius (1.8 degrees Fahrenheit). One "little" calorie is the energy to raise one *gram* of water by one degree Celsius. There are about 4200 joules in a Calorie. Another unit of energy that is widely used is the kilowatt-hour (kWh). This is the unit that you pay for when you buy electric energy from a utility company. A kWh is the energy delivered when you get a thousand watts for an hour. That's 1000 joules per second for 3600 seconds (an hour), i.e., 3.6 million joules = 860 Calories. You can remember this as 1 watt-hour (Wh) is approximately 1 Calorie. It is tedious and unnecessary to memorize all these conversions, and you probably shouldn't bother (except for the cases that I specifically recommend). Table 1.3 shows the conversions.

Although you shouldn't bother memorizing this table, it is useful to refer to it often so that you can get a feel for the amount of energy in various issues. For example, if you become interested in the energy usage of countries, then you will read a lot about *quads* and will find them a useful unit. U.S. energy

---

[7] When Einstein's theory of relativity (see chapter 12) predicted that mass can be converted to energy, the law was modified to say that the total of mass and energy is conserved.

Table 1.3   Common Energy Units

Energy unit	Definition and equivalent
calorie (lowercase)	Heats 1 gram of water by 1°C
Calorie (capitalized), the food calorie, also called kilocalorie	Heats 1 kg of water by 1°C 1 Calorie = 4182 joules ≈ 4 kJ
Joule	1/4182 Calories
	≈ Energy to lift 1 kg by 10 cm
	≈ Energy to lift 1 lb by 9 in
Kilojoule	1000 joules = 1/4 Calorie
Megajoule	1000 kilojoules = 10^6 joules Costs about 5 cents from electric utility
Kilowatt-hour (kWh)	861 Calories ≈ 1000 Calories = 3.6 megajoules
	Costs 10 cents from electric utility
British Thermal Unit (BTU)	1 BTU = 1055 joules ≈ 1 kJ = 1/4 Calorie
Quad	A quadrillion BTUs = 10^{15} BTU ≈ 10^{18} J
	Total U.S. energy use ≈ 100 quads per year;
	total world use ≈ 400 quads per year

Note: The symbol ≈ means "approximately equal to."

use is about 100 quads per year. (Notice that quads per year is actually a measure of power.)

Power

As we discussed earlier, *power* is the rate of energy transfer. The rate at which something happens is the "something" divided by the time—for example, miles/hour = miles per hour, or births/year = births per year. Thus, when 1 gram of TNT releases 0.651 Calories in 0.000001 second (one millionth of a second), the power is 651,000 Calories per second.

Although power can be measured in Calories per second, the two other units that are far more commonly used are the watt (one joule per second) and the horsepower. The horsepower was originally defined as the power that a typical horse could deliver, i.e., how much work the horse could do every second. These days, the most common use of the term is to describe the power of an automobile engine—a typical auto delivers 50 to 400 horsepower. James Watt, in the 1700s, was the first to actually determine how big one horsepower is. One horsepower turned out to be 0.18 Calorie per second. (Does that sound small to you? Or does it illustrate that a Calorie is a big unit?) Watts are the most commonly used unit to measure electric power.

James Watt found that a horse could lift a 330-pound weight vertically for a distance of 100 feet in one minute. He defined this rate of work to be one horsepower (hp). It turns out that 1 hp is about 746 W, which you can think of as approximately 1000 W, or 1 kilowatt (kW). (By now, I hope you are getting used to my approximations, such as 746 is approximately 1000.) Common units are as follows:

> kilowatt (1 kW = 1000 watts),
> megawatt (1 MW = 1 million watts)
> gigawatt (1 GW = 1 billion watts = 10^9 watts = 1000 MW)

The abbreviation for million (*mega-*) is capital M, and for billion (*giga-*) is capital G. So, for example, 1000 kW = 1 MW = 0.001 GW. One Calorie per second is about 4 kilowatts.

Only if you need to do engineering calculations do you need to know that 1 hp is 746 W. I do not recommend that you try to remember this; you can always look it up if you really need it. Instead, remember the approximate equation:

$$1 \text{ horsepower} \approx 1 \text{ kilowatt}$$

It is far more useful to remember this approximate value than it is to try unsuccessfully to remember the exact value.

Power usage is so important (for future presidents and knowledgeable citizens) that it is worthwhile learning some key numbers. These are given in table 1.4. Learn the approximate values by visualizing the examples.

Table 1.4 Power Examples

Value	Equivalent	Examples of that much power use
1 watt (1 W)	1 joule per second	Flashlight
100 watts		Bright lightbulb; heat from a sitting human
1 horsepower (1 hp)	≈ 1 kilowatt[a]	Typical horse (for extended time); human running fast up flight of stairs
1 kilowatt (1 kW)	≈ 1 hp[b]	Small house (not including heat); power in 1 square meter of sunlight
20 horsepower	≈ 20 kW[c]	Small automobile
1 megawatt (1 MW)	1 million (10^6) watts	Electric power for a small town
45 megawatts		747 airplane; small power plant
1 gigawatt (1 GW)	1 billion (10^9) watts	Large coal, gas, or nuclear power plant
400 gigawatt (0.4 terawatts)		Average electric power use for United States
2 terawatts	$= 2 \times 10^{12}$ watts	Average electric power for world

[a] More precise value: 1 hp = 746 watts
[b] More precise value: 1 kW = 1.3 hp
[c] More precise value: 20 hp = 14.9 kW

Power Examples

Since energy is conserved, the entire energy industry never actually produces or generates energy, it only converts it from one form to another and transports it from one location to another. Nevertheless, the popular term for this is "generating power." (It is an interesting exercise to read the words used in newspaper articles and then translate them into a more precise physics version.)

To give a sense of how much power is involved in important uses, we'll now describe some examples in more detail. Many of these numbers are worth knowing, because they affect important issues, such as the future of solar power.

Here is a brief description of what happens between the power plant and the lighting of a lightbulb in your home. The original source of the energy may be chemical (oil, gas, or coal), or nuclear (uranium). In a power plant, energy is converted to heat, which boils water, creating hot compressed steam. The expanding steam blows past a series of fans called a *turbine*. These fans rotate the crank of a device called an *electric generator*. We'll discuss how electric generators work in more detail in chapter 6, but they turn the mechanical rotation into electric current—that is, into electrons that move through metal. The main advantage of electric energy is that it is easily transported over thousands of miles, just using metal wires, to your home.

A typical large power generating station produces electric power at the rate of about one gigawatt = one billion watts = 10^9 watts = 1 GW (see table 1.4). This is a useful fact to remember. It is true for both nuclear and oil/coal burning plants. If each house or apartment required one kilowatt (that would light ten 100-watt bulbs), then one such power plant could provide the power for one million houses. Smaller power plants typically produce 40 to 100 MW (megawatts). These are often built by small towns to supply their own local needs. One hundred MW will provide power for about 100,000 homes (fewer if we include heating or air conditioning). The state of California is large, and on a hot day it uses 50 GW, so it needs the equivalent of about 50 large electric power plants.

In an electric power plant, not all the fuel energy goes into electricity; in fact, about two-thirds of the energy is lost when it turns into heat. That's because the steam does not cool completely, and because much of the heat escapes into the surroundings. Sometimes this heat is used to warm surrounding buildings. When this is done, the plant is said to be "co-generating" both electricity and useful heat.

Table 1.4 gives the typical power of important devices, ranging from a flashlight (1 watt) up to the total world power (2 terawatts, equal to 2 million million watts).

LIGHTBULBS

Ordinary household lightbulbs, sometimes called *incandescent* or *tungsten bulbs*, work by using electricity to heat a thin wire inside the bulb. This wire, called the *filament*, is heated until it glows white-hot. (We'll discuss the glow of such filaments in more detail in chapters 2 and 8.) All of the visible light comes from the hot filament, although the bulb itself can be made frosted so that it spreads the light out, making it less harsh to look at. The glass bulb (which gives the lightbulb its name) protects the filament from touch (its temperature is over $1000°C \approx 1800°F$) and keeps away oxygen, which would react with the hot tungsten and weaken it.

The brightness of the bulb depends on how much power it uses—that is, on how much electricity is converted to heat each second. A tungsten light that uses 100 watts is brighter than one that uses 60 watts. Because of this, many people mistakenly believe that a watt is a unit of brightness, but it isn't. A 13-watt *fluorescent* lightbulb (we'll discuss these in chapters 9 and 10) is as bright as a 60-watt conventional (incandescent) bulb. Does that mean that a conventional bulb wastes more electricity than does a fluorescent bulb? Yes. The extra electric power used just heats the bulb. That's why tungsten bulbs are much hotter to the touch than equally bright fluorescent bulbs. One kilowatt, the amount of power used by ten 100-watt bulbs, will illuminate your home brightly, assuming that you have an average-size house and are using conventional bulbs.

> **Memory trick:** Imagine that it takes one horse to light your home (one horsepower ≈ 1 kilowatt).

A new kind of light source called a *light-emitting diode*, or *LED*, is now coming on the market. It is almost as efficient as a fluorescent bulb, but it is not yet as cheap. That could change in the near future. LEDs are already being used for traffic lights and flashlights.

SUNLIGHT AND SOLAR POWER

How much power is in a square meter of sunlight? The energy of sunlight is about 1 kilowatt per square meter. So the sunlight hitting the roof of a car (about 1 square meter) is about 1 kilowatt ≈ 1 horsepower. And all of that energy is in the form of light. When the light hits the surface, some bounces off (that's why you can see it), and some is converted to heat (making the surface warm).

Suppose that you placed a kilowatt tungsten bulb in every square meter of your home. Would the home then be as brightly lit as it would be by sunlight? Hint: Recall that a watt is not a unit of brightness, but of energy delivered per second. In sunlight, all of that energy is in the form of light. In an electric bulb, most of the energy goes into heat. Does your answer match what you think would happen with this much light?

Many environmentalists believe that the best source of energy for the long-term future is sunlight. It is "sustainable" in the sense that sunlight keeps coming as long as the Sun shines, and the Sun is expected to have many billions of years left. Solar energy can be converted to electricity by using silicon solar cells, which are crystals that convert sunlight directly to electricity. (We'll discuss these in more detail in chapters 10 and 11.) The power available in sunlight is about one kilowatt per square meter. So if we could harness all of the solar energy falling on a square meter for power production, that energy would generate one kilowatt. But a cheap solar cell can only convert about 15% of the power, or about 150 watts per square meter. The rest is converted to heat, or reflected. A more expensive solar cell (such as those used on satellites) is about 40% efficient, i.e., it can produce about 400 watts per square meter. A square kilometer contains a million square meters, so a square kilometer of sunlight has a gigawatt of power. If 15% is converted to solar cells, then that is 150 megawatts per square kilometer, or about 1 gigawatt for 7 square kilometers. That is about the same as the energy produced by a large modern nuclear power plant.

Here is a summary of the important numbers for solar power:

1 square meter	1 kilowatt of sunlight
	150–400 watts electric using solar cells
1 square kilometer	1 gigawatt of sunlight
	150–400 megawatts electric

Some people say that solar power is not practical. Even educated people sometimes say that to get enough solar even for a state such as California, you would have to cover the entire country with solar cells.

Is that true? Look at table 1.4. A gigawatt, the output of a typical nuclear power plant, would take 7 square kilometers. This may sound big, but it really isn't. California has a typical peak power use (during the day, largely to run air conditioners) of about 50 gigawatts of electrical power; to produce this would take 350 square kilometers of solar cells. This would take less than one thousandth—that's one-tenth of one percent—of the 400,000-square-kilometer area of California. Besides, the solar plants could be placed in a nearby state, such as Nevada, that gets less rain and doesn't need the power itself.

Others complain that solar energy is available only during the day. What do we do at night? Of course, it is during the day that we have the peak power demand, to run our factories and our air conditioners. But if we are to convert completely to solar cells, then we will need an energy storage technology. Many people think that batteries, compressed air, or flywheels might provide that.

Right now, solar power costs more than other forms, largely because the solar cells are expensive and don't last forever. See what you can find about the costs of solar cells and the cost of building such a plant. (I've talked to contractors who have told me that installation of anything costs $10 per square foot.) Would solar power be more feasible in underdeveloped regions of the world, where construction costs are usually lower?

SOLAR-POWERED AUTOMOBILES AND AIRPLANES

There is an annual race across Australia for solar-powered automobiles. The fundamental problem with such a vehicle can be seen from the fact that one square meter of sunlight has about one kilowatt of power, which is equal to about one horsepower. Since expensive solar cells are only about 40% efficient, that means that you need 2.5 square meters of solar cells just to get one horsepower, whereas typical automobiles use 50 to 400 hp.[8] The race is obviously among very low-powered vehicles!

[8] To read more about the annual race, go to their Web page, at http://www.wsc.org.au/index.html.

Given that low power, it is surprising to discover that a solar-powered airplane has successfully flown. Actually, the vehicle isn't truly an airplane—it doesn't have a pilot or passengers, so it is called an *aircraft*, a *drone*, or an *UAV* (for "unmanned aerial vehicle"). The aircraft was named the *Centurion* (figure 1.4). The solar cells are on the upper and lower surfaces of the wings; the cells on the undersides use light reflected off the Earth. The solar cells have to be big to gather solar power, and yet they also have to be light in weight. The *Centurion* has a wingspan of 206 feet, greater than for a Boeing 747. The total power from the solar cells is only 28 horsepower. The entire weight of the *Centurion* is 1100 pounds. It has already set an altitude record for airplanes of 96,500 feet. (Commercial airplanes fly at about 40,000 ft.) The *Centurion* was built by AeroVironment, a company started by engineer Paul McCready, who designed the *Gossamer Condor* and the *Gossamer Albatross*. We'll talk more about the *Gossamer Albatross* in a moment.[9]

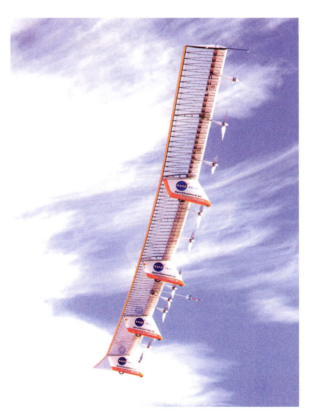

Figure 1.4 *Centurion*, a solar-powered aircraft. (Photo courtesy of NASA.)

HUMAN POWER

If you weigh 140 pounds and you run up a 12-foot flight of stairs in 3 seconds, your muscles are generating about 1 horsepower. (Remember: *Generating* means

[9] For more information, see the AeroVironment Web page at http://www.avinc.com.

converting from one form to another. The muscles store energy in chemical form and convert it to energy of motion.) If you can do this, does that make you as powerful as a horse? No. One horsepower is about as much power as most people can produce briefly, but a horse can produce one horsepower for a sustained period, and several horsepower for short bursts.

Over a sustained period of time, a typical person riding a bicycle can generate power at the rate of about 1/7 = 0.14 = 14% of a horsepower. (Does that seem reasonable? How much does a horse weigh compared to a person?) A world-class cyclist (Tour de France competitor) can do better: about 0.67 horsepower for more than an hour, or 1.5 horsepower for a 20-second sprint.[10] In 1979, cyclist Bryan Allen used his own power to fly a superlight airplane, the *Gossamer Albatross*, across the 23-mile-wide English Channel (figure 1.5).

Figure 1.5 Bryan Allen, about to pedal the *Gossamer Albatross*, with its 96-foot wingspan. It weighed only 66 pounds. Allen was both the pilot and the engine. (Photo courtesy of NASA.)

The *Gossamer Albatross* had to be made extremely light and yet stable enough to control. A key aspect of the design was that it had to be made easy to *repair*. Paul McCready, the engineer who designed it, knew that such a lightweight airplane would crash frequently—for example, whenever there was a large gust of wind. It flew only a few feet above the surface.

DIET VERSUS EXERCISE

How much work does it take to lose weight? We have most of the numbers that we need to calculate this. We saw in the last section that a human can generate a sustained effort of 1/7 horsepower. According to measurements made on such people, the human body is about 25% efficient—i.e., to generate work of 1/7 horsepower uses fuel at the rate of 4/7 horsepower. Put another way, if you can do useful work at 1/7 horsepower, the total power you use including heat generated is four times larger.

[10] I thank bicyclist Alex Weissman for these numbers. Here is a reference: http://jap.physiology.org/cgi/content/full/89/4/1522.

That's good if you want to lose weight. Suppose that you do continuous strenuous exercise and burn fat at the rate of 4/7 horsepower. Since one horsepower is 746 watts (I'm using the more exact value here), that means that in strenuous exercise you use $(4/7) \times 746 = 426$ watts = 426 joules per second. In an hour (3600 seconds), you will use 426×3600 joules = 1,530,000 joules = 367 Calories.

Coca-Cola, for example, contains about 40 grams of sugar in one 12-ounce can. That endows it with about 155 Calories of "food energy." That can be "burned off" with about a half hour of continuous, strenuous exercise. That does not mean jogging. It means running, or swimming, or cleaning stables.[11]

Exercise vigorously for a half hour, or jog for an hour, and drink a can of Coke. You've replaced all the calories "burned" in the exercise. You will neither gain nor lose weight (not counting short-term loss of water). Milk and many fruit juices contain even more Calories per glass. So don't think you can lose weight by drinking "healthy" instead of Coke. They may contain more vitamins, but they are high in Calories.

A typical human needs about 2000 Calories per day to sustain constant weight. Fat (e.g., butter) contains about 7 Calories per gram. So if you cut back by 500 Calories per day—that is, you reduce your consumption by a quarter of the 2000 you otherwise would have eaten—you will consume about 70 grams of your own fat per day, 500 grams per week, equal to a little more than a pound per week. That seems slow, for such a severe diet, and it is, and that's why so many people give up on their diets.

Alternatively, you can lose that pound per week by working out at 1/7 horsepower for one hour every day, seven days per week. Activities that do this include racquetball, skiing, jogging, or very fast walking. Swimming, dancing, or mowing grass uses about half as many Calories per hour. So, to lose a pound per week, exercise vigorously for an hour every day, or moderately for two hours, or cut your food consumption by 500 Calories. Or find some combination.

But don't exercise for an hour, and then reward yourself by drinking a bottle of Coke. If you do, you'll gain back every Calorie you worked off.

WIND POWER

Wind is generated from solar energy, when different parts of the surface of the Earth are heated unevenly. Uneven heating could be caused by many things, such as differences in absorption, differences in evaporation, or differences in cloud cover. Windy places have been used as sources of power for nearly a thousand years. The windmill was originally a mill (a factory for grinding flour) driven by wind power, although early windmills were also used by the Dutch for pumping water out from behind their dikes. Many people are interested in wind power again these days as an alternative source of electricity. Pilot wind generation plants were installed at the Altamont Pass in California in the 1970s. These

---

[11] Books on exercise physiology tabulate these numbers; the readers tend to be athletes and farm operators—hence the interest in cleaning stables.

Figure 1.6 Wind turbines. (Photo courtesy of New Mexico Wind Energy Center.)

are more commonly called *wind turbines* (figure 1.6), since they no longer mill flour.

Modern wind turbines are much more efficient at removing energy from wind when they are large. This is, in part, because then they can get energy from winds blowing at higher elevations. Some wind turbines are taller than the Statue of Liberty.

Wind power ultimately derives from solar, since it is differences in temperature that drive the winds. We'll discuss this further in the next chapter, in the section "Heat Flow." The wind turbines cannot be spaced too closely, because when they take energy from the wind, the wind velocity is decreased, and the wind is made turbulent—i.e., it is no longer flowing in a smooth pattern. A "forest" of wind turbines has been proposed for construction on the ocean, off the coast of Massachusetts, to supply commercial power (figure 1.7). In case you are interested, here are some of the details: There would be 170 large windmills in a 5-mile-by-5-mile square, connected to land via an undersea cable. Each windmill would rise 426 feet, from water level to the tip of the highest blade (the height of a 40-story building). They would be spaced 1/2 mile from each other. The maximum power this forest can deliver will be 0.42 gigawatts. The major opposition to the idea appears to be coming from environmentalists who argue that the array would destroy a wilderness area, kill birds, and create noise that could disturb marine animals.

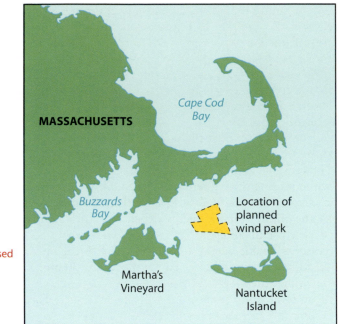

Figure 1.7 A map of the proposed offshore wind turbine park in Massachusetts.

Kinetic Energy

Let's go back to table 1.1 again and discuss another surprising fact from that table: The energy of motion of a typical meteor is 150 times greater than the chemical energy of an equal mass of TNT.

Unlike chemical energy, which usually has to be measured (not calculated), there is a simple equation for kinetic energy:

The kinetic energy equation:
$$E = \tfrac{1}{2} mv^2$$

To use this equation, v must be in meters per second, and m in kilograms, and the energy will be in joules. To convert energy to Calories, divide by 4200. Here are useful (approximate) conversions[12]:

1 meter per second (mps) = 2 miles per hour (mph)
1 kilogram (kg) = 2 pounds (lb)

The equation for kinetic energy is optional. It is useful to see it, but because the units may be unfamiliar, the equation is tricky to use.

---

[12] In many textbooks, kilograms are used solely as a measure of mass. I may be accused of being "sloppy" in not following that physics convention. In fact, scales in both Europe and the United States "weigh" in kilograms. *Kilogram* has become, in common use, a term that denotes the weight of one kilogram of mass.

But notice how similar the kinetic equation is to Einstein's famous equation of special relativity, $E = mc^2$. In the Einstein equation, c is the speed of light in a vacuum: 3×10^8 meters per second. The similarity is not a coincidence, as you will see when we discuss relativity in chapter 12. Einstein's equation states that the energy hidden in the mass of an object is approximately equal to the classical kinetic energy that object would have if it moved at the speed of light. For now, Einstein's famous equation might help you to remember the less famous kinetic energy equation.

Let's take a closer look at what the kinetic energy equation tells us about the relation of kinetic energy to mass and speed. First, the kinetic energy is proportional to the object's mass. This is very useful to remember, and can give you insights even without using the equation. For example, a 2-ton SUV has twice as much kinetic energy as a 1-ton Volkswagen Beetle traveling at the same speed.

In addition, the object's kinetic energy depends on the square of its velocity. This is also a very useful thing to remember. If you double your car's speed, you will increase its kinetic energy by a factor of 4. A car moving at 60 mph has 4 times the kinetic energy as a similar car moving at 30 mph. At 3 times the speed, there is 9 times the kinetic energy.

Now let's plug some numbers into the kinetic equation and see what we get for a very fast object, a meteor. We will express mass in kilograms and velocity in meters/second. We'll do the calculation for a 1-gram meteor traveling at 30 kilometers per second. First, we must convert these numbers: the mass $m = 0.001$ kg; the velocity $v = 30$ km/sec = 30,000 meters/sec. If we plug these numbers into the equations, we get

$$E = \tfrac{1}{2}\, mv^2$$
$$= \tfrac{1}{2}\, (0.001)(30000)^2$$
$$= 450000 \text{ joules} = 450 \text{ kJ} \approx 100 \text{ Calories}$$

SMART ROCKS AND BRILLIANT PEBBLES

For over two decades, the U.S. military has seriously considered a method of destroying nuclear missiles (an "anti-ballistic-missile," or ABM, system) that would not use explosives. Instead, a rock or other chunk of heavy material is simply placed in the missile's path. In some formulations, the rock is made "smart" by putting a computer on it, so that if the missile tries to avoid it, the rock will maneuver to stay in the path

How could a simple rock destroy a nuclear warhead? The warhead is moving at a velocity of about 7 kilometers per second—i.e., $v = 7000$ meters per second. From the point of view of the missile, the rock is approaching it at 7000 meters per second. (Switching point of view like this is called *classical relativity*.) The kinetic energy of each gram (0.001 kg) of the rock, relative to the missile, is

$$E = \tfrac{1}{2}\, (0.001)(7000)^2 = 25000 \text{ J} = 6 \text{ Cal}$$

Thus, the kinetic energy of the rock (seen from the missile) is 6 Calories. That is 9 times the energy it would have if it were made from TNT. It is hardly necessary to make it from explosives; the kinetic energy by itself will destroy the missile. In fact, making the rock out of TNT would provide only a little additional energy, and it would have very little additional effect.

The military likes to refer to this method of destroying an object as "kinetic energy kill" (as contrasted with "chemical energy kill"). A later invention that used even smaller rocks and smarter computers was called "brilliant pebbles." (I'm not kidding. Try looking it up on the Internet.)

Here is an interesting question: How fast must a rock travel so that its kinetic energy is the same as the chemical energy in an equal mass of TNT? According to table 1.1, the energy in 1 gram of TNT is 0.651 Calories = 2,723 joules. We set $1/2\, mv^2 = 2723$ J. Use 1-gram rock for m, so $m = 0.001$ kilograms (getting the units right is always the hardest part of these calculations!). Then,

$$v^2 = 5446000$$
$$v = \text{sqrt}(5446000)$$
$$= 2300 \text{ m/sec}$$
$$= 2.3 \text{ km/sec}$$

That's 7 times the speed of sound.

THE DEMISE OF THE DINOSAURS

Now let's think about the kinetic energy of the asteroid that hit the Earth and killed the dinosaurs. The velocity of the Earth around the Sun is 30 km/sec,[13] so it is reasonable to assume that the impact velocity was about that much. (It would have been more in a head-on collision, and less if the asteroid approached from behind.)

If the asteroid had a diameter of 10 kilometers, its mass would be about 1.6×10^{12} tons (1.6 teratons).[14] From table 1.1, we see that its energy was 165 times greater than the energy of a similar amount of TNT. So it would have had the energy of $(165) \times (1.6 \times 10^{12}) = 2.6 \times 10^{14}$ tons $= 2.6 \times 10^8$ megatons of TNT. Taking a typical nuclear bomb to be 1 megaton of TNT,[15] this says that the impact released energy equivalent to over 10^8 nuclear bombs. That's 10,000 times the entire Russian–U.S. nuclear arsenal at the height of the Cold War.

The asteroid made a mess, but it stopped. The energy was all turned to heat, and that resulted in an enormous explosion. However, an explosion of that size is still large enough to have very significant effects on the atmosphere. (Half of the air is within three miles of the surface of the Earth.) A layer of dirt

[13] The Earth–Sun distance is $r = 93 \times 10^6$ miles $= 150 \times 10^6$ kilometers. The total distance around the circumference is $C = 2\pi r$. The time it takes to go around is one year $t = 3.16 \times 10^7$ seconds. Putting these together, we get the velocity of the Earth is $v = C/t = 30$ km/sec. (Note that the number of seconds in a year is very close to $t \approx \pi \times 10^7$. That is a favorite approximation used by physicists.)

[14] Taking the radius to be 5 km $= 5 \times 10^5$ cm, we get the volume $V = (4/3)\,\pi r^3 = 5.2 \times 10^{17}$ cubic centimeters. The density of rock is about 3 grams per cubic centimeter, so the mass is about 1.6×10^{18} grams $= 1.6 \times 10^{12}$ metric tons.

[15] The Hiroshima bomb had an energy equivalent of 13 kilotons $= 0.013$ megatons of TNT. The largest nuclear weapon ever tested was a Soviet test in 1961 that released energy equivalent to 58 megatons of TNT.

thrown up into the atmosphere probably blocked sunlight over the entire Earth for many months. The absence of sunlight stopped plant growth, and that meant that many animals starved.

Would that kind of impact knock the Earth out of its orbit? We assumed that the asteroid was about 10 kilometers across—that's about one-thousandth the diameter of the Earth—which makes it one billionth the mass of the Earth. The asteroid hitting the Earth is comparable to a mosquito hitting a truck. The impact of a mosquito doesn't change the velocity of the truck (at least not very much), but it sure makes a mess on the windshield. In this analogy, the windshield represents the Earth's atmosphere. (We'll do a more precise calculation in chapter 3, when we discuss momentum.)

Most of the energy of the asteroid was converted to heat, and that caused the explosion. The impact of a smaller comet (about 1 km in diameter) on the planet Jupiter is shown at the start of this chapter in figure 1.1. Look at it again. It looks pretty dramatic, but the explosion that killed the dinosaurs was a thousand times larger.

But what is heat, really? What is temperature? Why does enormous heat result in an explosion? These are the questions we will address in the next chapter.

Chapter Review

Energy is the ability to do work. It can be measured in food Calories (Cal, also called kilocalories or kcal), kilowatt-hours (kWh), and joules (J). Gasoline has about 10 Cal per gram, cookies have about 5, TNT has about 0.65, and expensive batteries hold about 0.1 Cal. The very high energy in gasoline explains why it is used so widely. The high energy in cookies explains why it is difficult to lose weight. The relatively low energy stored in batteries makes it difficult to use them for electric cars. Hybrid automobiles consist of efficient gasoline engines combined with batteries. The batteries can absorb energy when the car slows down, without forcing it to be wasted as heat. Fuel cells produce electricity like batteries, but they are recharged by adding chemicals (such as hydrogen) rather than by plugging them into the wall. Uranium has 20 million Calories per gram, but requires nuclear reactors or bombs to release it in large amounts.

Coal is the cheapest form of fossil fuel, and it can be converted to gasoline. The major countries that use energy have abundant coal supplies.

Power is the rate of energy delivery and can be measured in Cal/sec or in watts, where 1 watt = 1 J/sec. TNT is valued not for its energy, but for its power—i.e., its ability to deliver energy quickly. A horsepower is about 1 kilowatt (kW). A typical small house uses about 1 kW. Humans can deliver 1 horsepower for a short interval, but only about 1/7 horsepower over an extended period.

Large nuclear power plants can create electricity with a power of about 1 billion watts, also called 1 gigawatt (GW). The power in 1 square kilometer of sunlight is about the same: 1 GW. Solar cells can extract 10% to 40% of that, but the better solar cells are very expensive. A solar car is not practical, but there are uses for solar airplanes, particularly in spying.

Sugar and fat are high in Calories. A half hour of vigorous exercise uses the Calories in one can of soft drink.

Kinetic energy is the energy of motion. To have the same energy as TNT, a rock has to move at about 1.5 miles per second. To destroy an enemy missile, all you have to do is put a rock in its way, since from the point of view of the missile, the rock is moving very fast with lots of energy. If the rock has 10 times the velocity, then it will have $10 \times 10 = 100$ times the energy. The rock that hit the Earth 65 million years ago was moving about 15 miles per second, so it had 100 times the energy of TNT. When it hit, that kinetic energy was converted to heat. The heat caused the object to explode, and we believe that's what resulted in the death of the dinosaurs.

Discussion Topics

These questions involve issues that are not discussed in the text and so they are recommended as discussion questions. You are welcome to express your personal opinions, but try to back your statements with facts and (when appropriate) technical arguments. You might want to discuss these topics with friends before writing your answers.

1. Oil efficiency and national security. Right now, the United States is extremely dependent on oil for its automobile. Our dependence on oil has turned the Middle East into one of the most important areas in the world. If our automobiles were 40% efficient rather than 20% efficient, we would not have to import any oil. The global consequences of our oil inefficiency can reach as far as war in the Middle East. Getting more efficient use of oil is both a technological and a social issue. Who should pay for the research? The U.S. government? Private industry? Is this an economic question or is it a national security one?

2. Automobiles typically carry 100 lb of gasoline. That has the energy content of 1500 lb of TNT. Is gasoline really as dangerous as this makes it sound? If so, why do we accept it in our automobiles? If not, why not? Do we accept gasoline only because it is a "known" evil?

Internet Research Topics

1. Asteroid impacts are rare; a big one hits the Earth only about once every 25 million years. But small ones occur more frequently. In 1908, a small piece of a comet hit the Tunguska region of Siberia and exploded with an energy equivalent to that of about a million tons of TNT. Look on the Internet and find out about the Tunguska impact.

2. What is the current status of hybrid automobiles? How much more efficient are they than gasoline automobiles (in miles per gallon)? What kinds of improvements are expected in the next few years? Are all hybrids fuel efficient? Do any "standard" cars have better mpg?

3. Verify the area that it would take for solar cells to provide sufficient power for the state of California. Look on the Web to see what you can find out about the current cost of solar cells and their expected lifetime. Are there companies working to lower the cost of solar cells? What alternative ways are there to convert solar energy to electricity? Do you think that solar power would be more or less feasible in underdeveloped regions of the world?

4. Look up "smart rocks" and "brilliant pebbles" on the Internet. Are there current programs to develop these for defense purposes? What are the arguments used in favor and against these programs?[16]

5. What is the status of wind power around the world? How large are the current largest wind turbines? How much energy can be obtained from a single wind turbine? Are wind turbines being subsidized by the government, or are they commercially viable?

6. What can you find about electric automobiles? What is their range? Are they less expensive than gasoline autos, when the replacement of batteries is taken into account?

7. Find examples in newspapers and magazines in which the writer uses *power* and *energy* interchangeably, and not in the technical sense we use in this book.

8. Look up the Fisher-Tropsch process for converting coal to diesel fuel. What countries have used it? Are new plants being planned?

Essay Questions

1. Read an article that involves physics or technology that appeared in the last week or two. (You can usually find one in the *New York Times* in the Tuesday Science section.) Describe the article in one to three paragraphs, with emphasis on the technological aspects—not on business or political aspects. If you don't understand the article, then you can get full credit by listing the things that you don't understand. For each of these items, state whether you think the writer understood it.

2. Describe in a page what aspects of this chapter you think are most important. What would you tell your friends, parents, or children are the key points? Which points are important for future presidents or just good citizens?

3. In his 2003 State of the Union address, President George W. Bush announced that the United States will develop a "hydrogen economy." Describe what this means. What mistaken ideas do some people have about such an economy? How will hydrogen be used?

4. When the numbers matter, the confusion between energy and power can be problematical. For example, here is a quote I found on

---

[16] A particularly useful Web site for national defense technology is run by the Federation of American Scientists at www.fas.org.

the Web site for Portland General Electric: "One very large industrial plant can use as much power in one hour as 50 typical residences use in a month."[17] Can you see the reason for confusion? What do you guess the author means by the "amount of power in one hour"? Do you suppose that they really meant the "amount of energy in one hour"? Do your best to describe what the author meant. What impression was the author trying to leave? Was it an accurate impression?

5. A friend tells you that in 30 years we will all be driving automobiles powered by solar energy. You say to him, "It's hard to predict 30 years ahead. But let me give you a more likely scenario." Describe what you would say. Back up your predictions with relevant facts and numbers whenever they would strengthen your analysis.

6. When an automobile crashes, the kinetic energy of the vehicle is converted to heat, crushed metal, injury, and death. From what you have seen (in real life and in movies), consider two crashes, one at 35 mph, and another at 70 mph. Is it plausible that a crash of the faster automobile is 4 times worse? What other factors besides speed could affect the outcome of the crash? Airplane velocities are typically 600 mph except during take-off and landing, when they are closer to 150 mph. Does the kinetic equation explain why there are few survivors in an airplane crash?

7. Energy is conserved—that is a law of physics. Why then do our leaders beseech us to "conserve energy"?

8. Although TNT has very little relative energy per gram, it is a highly effective explosive. Explain why, briefly.

9. Some people say that the United States is "addicted" to gasoline. Compare gasoline to alternative ways of powering an automobile. Describe both the advantages and disadvantages of the alternatives you discuss, compared to a gasoline engine.

Multiple-Choice Questions

1. "Smart rocks" are considered for
   A. geologic dating
   B. ballistic missile defense
   C. nuclear power
   D. solar power

2. One watt is equivalent to:
   A. one joule/second
   B. one coulomb/second

---

[17] Since there are typically 30 days per month, that means that there are $30 \times 24 = 720$ hours per month. So the industrial plant uses 720 times as much energy as 50 houses. As stated in the text, 50 houses typically use 50 kilowatts. So this would imply that the power plant uses 720×50 kilowatts = 36 megawatts. Recall that a typical large power plant produces 1 gigawatt = 1000 megawatts. The usage of the industrial plant seems quite small compared to this. Yet the original statement made the usage appear quite large (at least that was my interpretation).

C. one calorie/second

D. one horsepower

3. The asteroid that killed the dinosaurs exploded because

A. it was made of explosive material

B. it was made out of U-235

C. it got very hot from the impact

D. It didn't explode; it knocked the Earth out of its normal orbit.

4. Kinetic energy can be measured in:

A. watts

B. calories

C. grams

D. amperes

5. Which of the following statements is true?

A. Energy is measured in joules, and power is measured in calories.

B. Power is energy divided by time.

C. Batteries release energy, but TNT releases power.

D. Power signifies a very large value of energy.

E. All of the above.

6. For each of these, mark whether it is a unit of energy (E) or power (P):

A. horsepower

B. kilowatt-hour

C. watt

D. calorie

7. Hybrid vehicles run on:

A. electric and solar power

B. solar power and gasoline

C. electric power and gasoline

D. nuclear power and gasoline

8. What is the main reason that hydrogen-driven automobiles have not replaced gasoline ones?

A. Hydrogen is too expensive.

B. Hydrogen is too difficult to store in an automobile.

C. Hydrogen is radioactive, and the public fears it.

D. Hydrogen mixed with air is explosive.

9. Compared to an equal weight of gasoline, U-235 can deliver energy that is greater by a factor of (choose the closest value)

A. 2200

B. 25,000

C. one million

D. one billion

10. Which of the following contains the most energy per gram?

A. TNT

B. chocolate chip cookies

C. battery

D. uranium

11. Compare the energy in a kilogram of gasoline to that in a kilogram of flashlight batteries:
    A. The gasoline has about 400 times as much energy.
    B. The gasoline has about 10 times as much energy.
    C. The gasoline has about 70 times less energy.
    D. They cannot be honestly compared, since one stores power and the other stores energy.

12. Which is least expensive—for the same energy delivered?
    A. coal
    B. gasoline
    C. natural gas
    D. AAA batteries

13. The kinetic energy of a typical 1-gram meteor is approximately equal to the energy of
    A. 10 grams of TNT
    B. 150 grams of TNT
    C. 1/100 grams of TNT
    D. 10 grams of gasoline

14. Coal reserves in the United States are expected to last for
    A. hundreds of years
    B. three or four decades
    C. 72 years
    D. less than a decade

15. A limitation for all electric automobiles is:
    A. Low energy density per battery.
    B. Batteries explode more readily than gasoline.
    C. Electric energy is not useful for autos.
    D. Electric motors are less efficient than gasoline motors.

16. Solar power is about (choose all that are correct)
    A. 1 watt per square meter
    B. 1 kW per square meter
    C. 1 megawatt per square km
    D. 1 gigawatt per square km

17. The efficiency of inexpensive solar cells is closest to
    A. 1%
    B. 12%
    C. 65%
    D. 100%

18. A human, running up stairs, can briefly use power of approximately
    A. 0.01 horsepower
    B. 0.1 horsepower
    C. 0.2 horsepower
    D. 1 horsepower

19. A 12-oz can of soft drink (not the "diet" or "lite" kind) contains about
   A. 10 Calories
   B. 50 Calories
   C. 150 Calories
   D. 2000 Calories

20. A large nuclear power plant delivers energy of about
   A. 1 megawatt
   B. 1 gigawatt
   C. 100 gigawatts
   D. 1000 gigawatts

21. Electricity from a AAA battery costs the consumer about:
   A. 1¢ per kilowatt-hour
   B. 10¢ per kilowatt-hour
   C. $1 per kilowatt-hour
   D. $1000 per kilowatt-hour

22. Electricity from a wall plug costs the consumer about:
   A. 1¢ per kilowatt-hour
   B. 10¢ per kilowatt-hour
   C. $1 per kilowatt-hour
   D. $1000 per kilowatt-hour

23. The energy per gallon (not per pound) of liquid hydrogen, compared to gasoline, is about
   A. 3× less
   B. about the same
   C. 3× more
   D. 12× more

24. Most of the hydrogen we use in the United States comes from
   A. pockets of hydrogen gas found underground
   B. hydrogen gas extracted from the atmosphere
   C. hydrogen produced in nuclear reactors
   D. It is manufactured from fossil fuels and/or water.

25. You have 10 tungsten bulbs, and each uses 100 watts. You leave them all on for an hour. The energy used is
   A. 10 kilowatt-hours
   B. 1 kilowatt-hour
   C. 10 kilowatts
   D. 1000 watts

26. The kinetic energy of a bullet, per gram, is (within a factor of 2)
   A. about the same as the energy released from 1 gram of TNT
   B. about the same as the kinetic energy in a typical 1-gram meteor
   C. about the same as the energy released by 1 gram of chocolate chip cookies
   D. none of the above

2

CHAPTER

Atoms and Heat

Quandaries

When the asteroid hit the Earth 65 million years ago, it had a kinetic energy equivalent of 100 times its own weight in TNT. In the impact, virtually all that energy was turned into heat. The temperature of the rock (turned into vapor) was over a million degrees Fahrenheit, over a hundred times hotter than the surface of the Sun.

Why? How does kinetic energy turn into heat? What is heat? How did this lead to an explosion?

All objects sitting in a room should reach the same temperature. Yet if you pick up a cup made of glass, it feels cooler than a cup made of plastic. Many people unconsciously recognize plastic by its relative "warm feel."

How can two objects be the same temperature and yet one feels cooler? What mistaken assumption are we making?

Many scientists are worried that the Earth is warming. Some models predict that the continued dumping of carbon dioxide into the atmosphere (from the burning of fossil fuels) could soon warm the Earth by about 5 degrees F. If this happens, we expect the oceans to rise by a foot or more—even if no ice melts. Some low islands could be swamped.

Why should the sea level rise if no ice melts?

When we heat our homes by burning fuel, we are wasting energy. We could do a much better job by pumping heat from the cold outdoors into our homes.

Pump heat from the cold outdoors? This sounds like nonsense. Isn't the burning of fuel 100% efficient, since all the energy goes into heat? How could we possibly do better than that?

Atoms and Molecules and the Meaning of Heat

Press your hands together hard and rub them vigorously for about 15 seconds. (It is actually a good idea to do this right now, before you read further, if nobody is watching.) Your hands feel warmer. The temperature of the skin has risen. You turned kinetic energy (energy of motion) into heat.

In fact, heat *is* kinetic energy, the kinetic energy of molecules.[1] Your hands feel warmer because, after rubbing, the molecules are shaking back and forth faster than they were prior to your rubbing. That's what heat really is: the shaking of atoms and molecules, rapid in speed, but microscopic in distance.

This is a good time to discuss the makeup of matter. All substances are made of atoms, and there are only about[2] 92 different kinds of these: hydrogen, oxygen, carbon, iron, etc. A complete list appears in a chart known as a *periodic table*, shown in figure 2.1.

Each of the atoms in the periodic table has a number associated with it called the *atomic number*. This represents the number of protons in the atom; it is also the number of electrons in the atom (usually). The atomic number of hydrogen is 1; for helium, the atomic number is 2; for carbon, it's 6; for oxygen, it's 8; and for uranium, it's 92.

Molecules are combinations of atoms that stay clumped together. A molecule of water is written H_2O, meaning that it is made of two atoms of hydrogen (that's the H_2) and one atom of oxygen (that's the O). Helium molecules contain only one atom (He), and hydrogen gas molecules contain only two attached atoms of hydrogen (H_2). But molecules can be very large. The molecule known as DNA, which carries our genetic information, can contain billions of atoms.[3] When molecules break apart or come together, it's called a *chemical reaction*.

In all materials, the molecules are constantly shaking. The more vigorously they shake, the hotter the material is. When you rub your hands together, you make the molecules in your hands shake faster. How fast do they shake? The answer is startling: the typical velocity of shaking is about the same as the speed of sound, about 700 miles per hour, 1000 feet per second, or 330 meters per second. That's fast. Yet the particles (at least in a solid) can't travel very far. They

---

[1] Molecules are collections of atoms stuck to each other; an example is O_2 = oxygen gas, with two oxygen atoms making one oxygen molecule.

[2] Why do I say "about"? Some of the known elements are very rare or absent in nature because they are radioactive and have decayed away. Two such elements are technetium (element 43) and plutonium (element 94). If we count only the stable elements, the number is 91. If we count the radioactive ones, the number is over 100.

[3] And these atoms can be combined in different ways. That's how DNA encodes your genetic information. DNA molecules for different animals have different lengths.

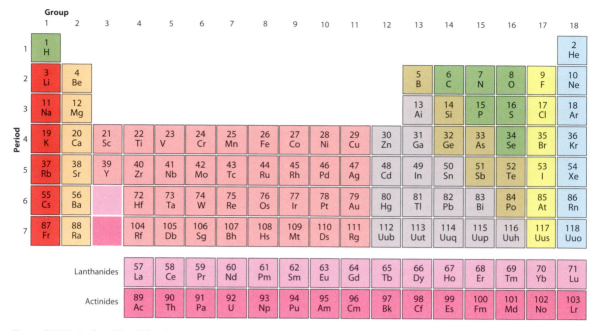

Figure 2.1 Periodic table of the elements.

bump into their neighbors and bounce back. They move fast but, like a runner on a circular track, their average position doesn't change.

Atoms are too small to be observed with an ordinary microscope. Their size is about 2×10^{-8} cm $= 2 \times 10^{-4}$ microns.[4] If you move across the diameter of a human hair (typically 25 microns), you will encounter 125,000 atoms from one side to the other. A red blood cell (8 microns across) has about 40,000 atoms spanning its diameter. Some molecules are so large (such as DNA) that they can be seen under a microscope, although the individual atoms in the molecules can't be resolved.

Even though you can't see atoms, you can see the effect that their shaking has on small, visible particles. With a microscope, you can see the shaking of tiny bits of floating dust (1 micron in diameter). This phenomenon is known as *Brownian motion*.[5] The shaking comes from the dust being hit on all sides by air molecules, and if the dust is sufficiently small, this bombardment does not average out.[6]

---

[4] A micron (μ or μm) is another name for a micrometer. It is 10^{-6} m $= 10^{-4}$ cm.

[5] This shaking of small particles was first observed on pollen grains in water by an English botanist, Robert Brown. Since he didn't know about atoms hitting the dust, the most reasonable interpretation at the time was that the movement indicated the small particles were alive! A detailed explanation, including predictions of the amount of shaking versus particle size, was deduced by Albert Einstein in 1905. Based on his work, most scientists were finally persuaded to believe the atomic theory.

[6] For a wonderful simulation of this effect, see the Web site at http://ephysics.physics.ucla.edu/ntnujava/gas2D/ebrownian_motion.htm. There is a similar site at Galileo.phys.Virginia.EDU/classes/109N/more_stuff/Applets/brownian/brownian.html.

The Speed of Sound and the Speed of Light

Is it a coincidence that the speed of molecules is approximately the speed of sound? No—sound travels through air by molecules bumping into each other. So the speed of sound is determined by the speed of molecular motion. Sound traveling through a gas cannot move faster than the velocity of the gas molecules.[7]

You can easily measure the speed of sound yourself. One way is to watch someone hit a golf ball, chop wood, or hit a baseball. Notice that you see the event before you hear the noise. That's because the light gets to you very quickly, and then you have to wait for the sound. Estimate your distance to the person, and estimate how long it takes for the sound to reach you. If the distance is 1000 feet, the delay should be about 1 second. (If you do this at a baseball game, it is helpful to sit as far from home plate as possible.) The velocity is the distance divided by the time.

When I was a child, and afraid of thunder and lightning, my parents taught me a way to tell how far away the sound and light was coming from. For every 5 seconds between the lightning flash and the thunder, they said, the lightning was 1 mile away. If there was a 10-second delay, then the lightning strike was 2 miles away. To me at that age, a mile was about the same as infinity, and so that put me at ease. The rule works because light travels so quickly that it covers a mile in a tiny fraction of a second. In other words, the light arrives virtually instantly. But thunder, since it is a sound, travels at the slower speed of sound: 330 meters per second = 1000 feet per second = 1 mile per 5 seconds (approximately) = 700 miles per hour (mph).

Knowing the speed of sound can be useful to measure distances. In 2003, I found myself on a boat some distance from a glacier that was dropping huge pieces of ice into the water. I measured that the sound took 12.5 seconds to reach me. That way I knew that the edge of the glacier was 2.5 miles away (1 mile for every 5 seconds). Until I did this, I had thought I was much closer; the huge size of the glacier had misled me.

The speed of light is much greater: 186,000 miles per second, or 3×10^8 meters per second. Although that sounds super fast, we can express it in a way that makes it sound much slower. Modern computers take about 1 billionth of a second (a *nanosecond*, or ns) to do a calculation. (Many go faster, but you should know that a nanosecond is typical.) In that billionth of a second, light travels only about 1 foot (30 centimeters). That's why computers must be small. Computers must often retrieve information to do a calculation, and if the information is too far away, it has to waste several cycles to get it.[8] If the computer speed is 3 GHz, then the light goes only 4 inches in one cycle.

Remember: The speed of light is about 1 foot in 1 computer cycle (1 ns).

---

[7] In a solid, the sound can travel faster than the molecules, since the molecules are effectively touching. They don't have to move in order to transmit a force to the next molecule.

[8] That was a fundamental oversight in the classic movie *2001: A Space Odyssey* (1968). A computer (named "Hal") was portrayed as being large enough for a human to walk into. Incidentally, the next letter after H is I, after A is B, and after L is M. So the following letters spell out IBM. Arthur C. Clarke, who wrote the story, insisted that was not his intention.

The Enormous Energy in Heat

The average speed of the molecules that compose this book is the speed of sound, but they are all moving in random directions. Suppose that I made them all move in the same direction. Then the entire book would be moving at the speed of sound, 720 miles per hour. Yet the total energy would be exactly the same.

This example illustrates the enormous energy that is contained in the heat of ordinary objects. Unfortunately, it is often not possible to extract that energy for useful work. We'll discuss this further when we get to the section on heat engines. There is no good way to change the directions of the shaking so that all the molecules move together. Yet we can do the opposite. When an asteroid hit the Earth 65 million years ago, all the molecules were initially moving at 30 kilometers per second in the same direction. After the impact, the directions were all different.

When kinetic energy is turned into heat, we can think of this process as coherent, regular motion becoming randomized. The molecular energy changes from being neatly "ordered" (all molecules moving in the same direction) to being "disordered." The term *disorder* is very popular in physics. The amount of disorder can be quantified, and that value is given the name *entropy*. When an object is heated, its entropy (the randomness of its molecular motion) increases. I'll discuss entropy further at the end of this chapter.

Hiss and Snow: Electronic Noise

Radios, when tuned between stations, sometimes give a hissing sound. What is the origin of that hiss? Old TV sets, when there is no station present, show white spots jumping on the screen that reminded people of snow. What is the snow?

The surprising answer is that the snow and the hiss are due to the same thing: electrons jumping around in the electronics of your set. They are in constant motion due to heat, and when there is no other signal present, you get to watch (or listen) to them move. Even though they are not molecules, they share the energy of shaking.

Lowering the temperature can reduce such noise, and high-sensitivity electronics often have to be cooled to reduce the hiss and the snow. In chapter 9, I'll talk about a device for seeing in very low light that had such a cooling system attached. But too much cooling can cause the device to cease operation, since a transistor (discussed in chapter 11) actually depends on the fact that room-temperature electrons have some kinetic energy. Without that kinetic energy, the electrons become trapped and electricity doesn't flow. If you cool a transistor, and remove that energy, the transistor no longer functions.

Now that we have described heat as the kinetic energy of the molecules (and sometimes of the electrons too), we can address a trickier question: what is temperature?

Temperature

Temperature is closely related to heat. Stop for a moment and think about it. When it is 100°F outside, it is hot. When it is below 32°F, water freezes. But it is very tricky to state exactly what temperature is. Temperature is what you read with thermometers. But what does it measure? The answer is surprisingly simple:

> Temperature is a measure of the hidden kinetic energy of the molecules.

By the *hidden kinetic energy*, I mean the usually unobserved energy of fast (speed of sound) but microscopic (in distance moved) shaking. When we get to the section on temperature scales, I'll give the equation that allows you to calculate the kinetic energy from the temperature.

The temperature increases when the average shaking energy of its molecules is greater. (We use the word *average* because at any given instant, some of the molecules may be moving faster than others, and some slower, just like dancers on a dance floor.) If two objects have the same temperature, then their molecules have the same kinetic energy of vibration.

Here is a surprising consequence of what I just said. Suppose that two bars, one made of iron and the other of copper, have the same temperature. Then their molecules must have the same kinetic energy, on average. Will the iron molecules and the copper molecules have the same average speed? The surprising answer is *no*. The iron molecules, which are lighter (see figure 2.1), will be shaking faster, on average.

In chapter 1, I said that kinetic energy is given by $KE = 1/2\ mv^2$. Copper and iron have different molecular masses m. So the heavier copper molecule must have a smaller velocity v in order to have the same kinetic energy KE. See why temperature was once even more of a mystery than heat?

> **Remember:** At the same temperature, lighter molecules move faster (on average) than heavier ones.

The Zeroth Law of Thermodynamics

The key discovery that makes temperature a really useful idea is the simple fact that two things that touch each other tend to reach the same temperature. That is why a thermometer gives you the temperature of the air—because it is in contact with the air, so it gets to the same temperature. The fact that objects in contact tend to reach the same temperature was such an important observation that it was given a fancy name: the *zeroth law of thermodynamics*.[9]

---

[9] The *first* law of thermodynamics, as you may remember from chapter 1, is the fact that energy is conserved. We'll state the second and third laws later in this chapter. The zeroth law was added only after the other laws were articulated, and apparently everybody thought it should go first, so it got the number zero.

Touch a hot iron object to a cold copper one. Because they are touching, the fast molecules in iron now bang into the slower ones in the copper. The iron molecules lose energy, and the copper ones gain. The temperature of the iron will drop, and that of the copper will rise. Only when the temperatures are the same does the transfer of energy stop. The "flow" of heat is actually the sharing of kinetic energy. Heat (kinetic energy) is given up by the hot material to the cold one. The flow stops only when both materials have the same temperature.

This means that if you put a bunch of things in the same room and wait, eventually they will all reach the same temperature. Of course, that doesn't work if one of the objects is a source of energy, such as a burning log. But if no energy is going in or out of the room, all objects will eventually reach the same temperature.

WHERE IS OUR HYDROGEN?

The element hydrogen is, by far, the most abundant element in the Universe. Hydrogen atoms make up 90% of the atoms in the Sun. The same is true for the large planets of Jupiter and Saturn. Yet in the atmosphere of the Earth, hydrogen gas is virtually absent. Why? Where is our hydrogen?

There is a remarkably simple answer, and it comes from the zeroth law of thermodynamics. The Earth once had lots of hydrogen, but we lost it to space. Hydrogen in the atmosphere of the Earth would have the same temperature as the nitrogen and oxygen. Therefore, the molecules of hydrogen have the same kinetic energy, on average. But since hydrogen is the lightest element (its atomic weight is only 1/16 that of oxygen), it must have a higher velocity. Since energy depends on the square of the velocity, the velocity must be a factor of 4 larger (so the square is 16). This high average velocity turns out to be enough for the hydrogen to escape from the Earth like a rocket![10] The Sun and Jupiter have much stronger gravity than the Earth, so they kept their hydrogen. We'll discuss escape velocity in more detail in chapter 3. The Earth lost its hydrogen gas because our gravity is too weak.

THE COLD DEATH

Stars are very hot, and molecules in space are very cold. Eventually, the stars will stop burning, and eventually everything in the Universe may reach the same temperature. By keeping track of everything, we can calculate what that temperature is. If we ignore the expansion of the Universe (see chapter 13), then the average temperature of the Universe turns out to be –270°C.[11]

---

[10] The average velocity of the hydrogen molecules is not sufficient for them to escape, but some hydrogen molecules have well above the average, and those are the ones we lose. Some nitrogen and oxygen molecules are lost this way too, but since their average velocity is so much lower than that for hydrogen, their loss is negligible.

[11] Most of the particles in the Universe are invisible, very low temperature particles of light (called the *cosmic microwave background*) and similarly low temperature neutrinos. The cold death occurs when all the energy is shared equally, including these numerous very cold particles.

Because the Universe is expanding, the eventual temperature may be even lower. Philosophers have called this the "cold death" of the Universe, and the thought of it gets some people depressed. But being cold doesn't necessarily mean life will be uninteresting. A detailed analysis made by physicist Freeman Dyson showed that even as the Universe gets very cold, life can continue, and the complexity of organized thoughts could get greater and greater. That might take additional evolution, but we have hundreds of billions of years for that.

What would life be like in such a Universe? What would the descendants of humans look like? Some people estimate that, because of the extreme cold, in order to remain complex, active creatures, they would have to be very large, perhaps as large as planets are now, and maybe even bigger.

Temperature Scales

The concept of temperature was invented long before it was understood. It was measured using devices called *thermometers*. People could make thermometers that would always agree, more or less because (as stated in the zeroth law) it doesn't matter what material the thermometer is made of. So temperature became a standard idea. We'll talk about how thermometers work later in this chapter.

There are two common temperature scales: the Fahrenheit scale and the Centigrade scale. Centigrade has recently been renamed *Celsius*.[12] Celsius is also abbreviated C, just like Centigrade, and Fahrenheit is abbreviated F. The scales are defined such that the freezing point and melting of water is 32°F and 0°C, and the boiling and condensation point of water is 212°F and 100°C.[13]

We can convert between Fahrenheit and Celsius by the following rules. Let T_C be the temperature expressed in the Celsius scale, and T_F be the temperature in the Fahrenheit scale. Then

$$T_C = (T_F - 32)(5/9)$$
$$T_F = (9/5)T_C + 32$$

Examples (try the equations yourself):

Freezing of water: $T_F = 32$ gives $T_C = 0$
Boiling of water: $T_C = 100$ gives $T_F = 212$
"Room temperature": $T_C = 20$ gives $T_F = 68$

---

[12] The name of the Centigrade scale was changed to Celsius to honor Anders Celsius, a professor of astronomy who built some of the world's best thermometers in the 1700s. But the name change was made only in the 1970s.

[13] An amusing historical detail is that Celsius set up his original temperature scale to put 0 at the boiling point of water and 100 at the freezing point—exactly backward from the way we use it today. Higher temperature was colder! It is interesting to think that it wasn't originally obvious that higher temperature should be warmer. It is just a convention.

DEGREES

Until recently, it was common to refer to temperature in *degrees*. A temperature of $T_F = 65$ was read as "65 degrees Fahrenheit" and written 65°F. However, the word *degree* doesn't add any meaning, and some people were confused by it. (It has nothing to do with angles, which are also measured in degrees.) So scientists are now adopting a new convention of dropping the degree symbol. Thus, 32°F is usually shortened to 32 F. You'll see it both ways. There is no physics in this; it is just notation. I'll sometimes use the traditional terminology, just because of the fact that that is how you will hear it used most, and because it sometimes makes it clear that we are talking about temperature.

Note that Celsius degrees are bigger than Fahrenheit degrees. A change of 1 C is a change of 9/5°F = 1.8°F ≈ 2 F. As an approximation for *changes* in temperature, remember:

$$1°C \approx 2°F$$

Digression: Which is metric, C or F? The original Fahrenheit scale was designed to make 0 F the coldest temperature that could easily be reached in a laboratory. That was done by mixing ice and salt, and that is what is called 0 F. The temperature of 100 F was originally chosen to be body temperature. (They made a slight mistake, and average body temperature is actually about 98.6 F.) On this scale, water freezes at 32 F and boils at 212 F. When the Centigrade scale was officially adopted (by the French, under Napoleon), they decided that the two standard points should be the freezing and boiling points of water. So on the Centigrade scale, water freezes at 0 C and boils at 100 C. Some people think that the Centigrade scale was more "metric" than the Fahrenheit scale, and that is nonsense. Both scales were based on standard points 100 degrees apart; they just chose different standard points.

ABSOLUTE ZERO

What happens if the molecules actually come to a stop, and have zero kinetic energy? When all motion of the molecules stop, we say the temperature of the material is at "absolute zero." This happens at –273 C = –459 F.[14]

Using this fact, we can define a new temperature scale called the *absolute* or *Kelvin* scale (named after William Thompson[15]). Physicists find the Kelvin scale to be very convenient because it simplifies equations. For example, if we use the Kelvin scale, then the average kinetic energy E per molecule is given by a very simple equation:

$$E = 2 \times 10^{-23} \, T_K$$

---

[14] Don't confuse *–459 F* with *Fahrenheit 451*. The latter is the title of a Ray Bradbury science fiction book and is meant to be the temperature at which books burn.

[15] Thompson was appointed to the nobility by Queen Victoria in 1892 and given the title Baron Kelvin of Largs.

where T_K is the temperature in Kelvin (or degrees Kelvin). The constant, given in the equation as 2×10^{-23}, is very small only because atoms are so small. *Don't bother learning this number.* It is not important to know the numerical value for the kinetic energy of the particles. It is important to know their velocity (1000 feet per second, about the speed of sound) and that if you double the temperature (on the Kelvin scale), then you double the kinetic energy.

The most remarkable fact about this equation is that it doesn't depend on the kind of material. That's just the zeroth law again. I find that to be an amazing and surprisingly simple law of physics. Ponder it for a few moments. Temperature is just the hidden kinetic energy. At room temperature, the kinetic energy of the atoms in the air is identical to the kinetic energy of the atoms in this book. That fact eluded scientists for hundreds of years. The only really tricky part is that the energy must be measured per molecule. This equation begins to illustrate what physicists sometimes refer to as the "beauty" of physics. It isn't really beauty in the traditional sense. It is just an insight, a simplicity that is missed by people who don't study physics.

You can convert from the Kelvin scale to the Celsius scale by subtracting 273:

$$T_C = T_K - 273$$

Thus, for example, $T_K = 273$ is the same as $T_C = 0$. Put another way, 273 K = 0 C.

The Space Shuttle Columbia *Tragedy*

On 1 February 2003, the *Columbia* space shuttle broke apart in flames as it reentered the atmosphere, killing all seven astronauts on board.

The space shuttle always generates enormous heat when it reenters the thicker parts of the Earth's atmosphere. That's because it has very large kinetic energy, and to slow down (so that it can land), it must get rid of that energy.

To calculate the energy per gram, we need to know the velocity. When the space shuttle orbits, it travels the Earth's circumference of 24,000 miles in 1.5 hours, so its velocity is 24,000/1.5 mi/h = 16,000 mi/h = 7000 m/sec = 22 times the speed of sound. At the time that it began to fall apart, the shuttle had slowed to 18.3 times the speed of sound. That is known as Mach 18.3. We'll show why it has to move so fast in chapter 3.

In the following optional calculation, we show that if the kinetic energy of the space shuttle were all turned to heating it, its temperature would rise to

The Mach rule:
$$T = 300 \, M^2$$

where M is the Mach number. This is a useful equation that you will not find in any other textbook. For $M = 18.3$, this gives $T = 100,000$ K. That is 17 times as hot as the surface of the Sun. This is why the pieces of the shuttle glowed so brightly—friction with the air made them very, very hot.

There is no way to avoid this turning of kinetic energy to heat on reentry.[16] The space shuttle is designed to have heat-resistant ceramic "tiles" on the bottom surface. During reentry, these tiles face the onrushing air and glow with a temperature of thousands of degrees. They can lose this heat by conduction with the air and by radiation. They cool off by the time that the shuttle lands.

The shuttle contains little fuel and no explosives. It was the kinetic energy of motion, turned into heat, that destroyed the vehicle.

> **High temperatures:** Here's a little trick you might find helpful. Suppose that an object (such as a meteor, or the interior of the Sun) has a temperature of 100,000 C. How hot is it in K? The answer is 100,273 K. That looks pretty close to 100,000. They differ by only 0.27%. Here is a useful rule: When temperatures are really high, then the temperature in C is approximately the temperature in K.

OPTIONAL: CALCULATION

Let's derive the Mach number equation. Now here is a trick that can allow you to get the answer very quickly. We know that at room temperature (300 K), the molecules in the shuttle are moving at about the speed of sound—i.e., at Mach 1. Suppose that all the kinetic energy of the orbiting shuttle was randomized—i.e., turned into heat. Then the molecules would be moving at Mach 18.3 (since that is how fast the shuttle was moving). So as the energy of orbit turns into the energy of heat, the molecules' hidden motion speeds up by a factor of 18.3.

What does that do to their hidden kinetic energy—i.e., to the temperature? Remember that the kinetic energy is $E = (1/2)mv^2$. So if you increase v by a factor of 18.3, you increase the kinetic energy by a factor of $(18.3)^2 = 335$. That means that you increase the temperature by a factor of 335, from 300 K to 335×300 K = 100,000 K.

Put another way, if you move at Mach number $M = 18.3$ and turn your kinetic energy into heat, your temperature will rise to a temperature $T = 300\ M^2$. This equation can be used for any Mach number M, and it gives the temperature in Kelvin.

Thermal Expansion: Sidewalk Cracks, Highway Gaps, New Orleans Levees, and Shattering Glass

When the atoms in a solid heat up (i.e., they move faster; i.e., their velocity increases; i.e., they get more kinetic energy), they tend to push their neighbor atoms farther away. The effect is small, but important—most solids expand a little bit when heated. A typical number, worth remembering, is that a 1 C temperature rise makes many substances expand by somewhere between 1 part in 1000 and 1 part in 100,000.

---

[16] In principle, the shuttle could have "retrorockets" that would slow it down in the same manner that rockets sped it up. To do this, however, would take large rocket engines, stages, and fuel just as big as those used in the launch. Someday, if technology developments allow engines and fuel that are much smaller, this might prove possible.

These sound like small numbers, but the span of the Verrazano-Narrows Bridge in New York City is 4260 feet. When the temperature changes from 20 F to 92 F (a typical seasonal change in New York City), then the length of the bridge changes by about 2 feet.[17]

Another effect of change in temperature is the change in shape of the bridge. Because the suspension cables get shorter in the winter cold, the height of the middle of the suspension is 12 feet higher in winter than in summer. Why is this more than the 2 feet we calculated for the span? The answer is in the geometry: the cables shorten by only 2 feet, but because of the shallow way in which they hang, that makes the center rise 12 feet. Try this with a horizontal string. If you hold it tight, it is straight. But loosen it just a bit, just a centimeter, and it will sag a lot more than a centimeter.

That expansion means that the molecules are not as close to each other, and that weakens the force of attraction. As a result, hot metals are not as strong as cool metals. It was the heating of the metal columns in the World Trade Center that weakened them and led to the collapse of the building.

Sidewalk cement is typically laid with grooves between squares about 5 feet = 60 inches on a side. In a 1 C temperature change, its length would change by 35 parts in a million—i.e., by 60 inches × 35 × 10^{-6} = 0.002 inch. For a 40 C change, that is 0.08 inch, almost a tenth of an inch. That may not sound like very much, but if there were no grooves, the concrete would be compressed and might buckle, causing random cracks. (Just as with the bridge, and the string, a small expansion can cause a big buckling.) The small groove, placed by the person who paved the cement, gives room to expand and prevents the cracking. (Or, rather, it puts *neat* cracks in ahead of time, instead of letting ugly, random cracks form.)

Large confined pieces of cement or concrete will crack if they are exposed to changes in temperature, unless those cracks are introduced. That creates serious design and engineering problems. Imagine, for example, that you were asked to build levees to protect the city of New Orleans from flood. (Much of the city is below sea level.) You can't surround it with solid concrete levees, because those would crack when the temperature changed. You would have to make the levees out of individual pieces, with spaces between. Those would have to be filled with sliding joints or some other flexible material. But if not done properly, those connecting regions might turn out to be the weakest part of the levee.

In fact, that is exactly what happened. Figure 2.2 shows part of the levee in New Orleans that failed after Hurricane Katrina. The levee is clearly made out of rectangular segments—that was done to leave room for expansion. But it was the expansion joints that failed, not the concrete itself. The expansion joints didn't break from heat—that's what they were designed for. But they were weaker than the reinforced concrete, so when the pressure of the flood put a great force on the levee, that's where they broke—at the weakest points.

---

[17] To calculate how much, we take the temperature difference to be 72 F = 40 C. If we look up the thermal expansion for steel, we find that the amount is 12 parts per million *for each* C, so multiply the expansion by the temperature change of 40 C to get 480 parts per million. That sounds small, but the bridge is 4260 feet long. Multiply 480 parts per million (480E–6) by 4260 to get a change in length of 2 feet.

Figure 2.2 Levees in New Orleans, shortly after Hurricane Katrina, broken at their thermal expansion joints. They didn't break from heat, but from the pressure of the flood. However, the expansion joints were the weakest part of the levee. (Courtesy of U.S. Army Corps of Engineers.)

SHATTERING GLASS

If you heat a glass pan in the oven, then put it in cool water, it will often crack or even shatter. A few decades ago, a special glass was developed that didn't crack; it was trademarked as Pyrex, and it is very popular for kitchen glass— e.g., measuring cups and pans. What makes Pyrex special? Why do sudden temperature changes cause some materials to crack and not others?

The glass cracks because the outside cools more rapidly than the inside, making it a different size. It starts to bend, like a bimetallic strip, but glass is brittle, and it breaks. Pyrex is a special glass that expands much less than ordinary glass; that is why it usually doesn't break when cooled.

Why doesn't the glass crack when initially heated in the oven? The answer is that when heated slowly, the heat passes through the glass, and all of the glass is at approximately the same temperature. It is the difference in temperature between the inside and the outside of the gas that causes the different expansion, and leads to the cracking.

TIGHT LIDS

Tight lids on jars are such a common problem that I own several special devices to help open them, mostly large wrenches that get a good grip on the lid. But my mom taught me a different way to do it: put the lid under hot water for a few seconds. The expansion of the lid, although it is tiny, is often enough to loosen the lid so it can be opened. (I'd use a rag to hold the hot lid.) This works only if the metal expands more than the glass. That happens if the expansion coefficient is greater, or if the lid gets hotter than does the glass.

Global Warming and the Rise of Sea Level

Many climate experts believe that the temperature of the Earth is rising because of the carbon dioxide being dumped into the atmosphere by the burning of fossil fuel. (We'll discuss this in detail in chapter 10.) The predicted warming over the next 30 years is between 1.5 and 5 C, depending on which model turns out to be more accurate. For the moment, assume that we will warm by 5 C = 9 F.

One of the most surprising effects of the warming is the rise of sea level—not because ice melts (although that does contribute), but simply because water expands so much. The volume expansion of water is 2×10^{-4} per degree C. For 2.5 degrees C, that amounts to $2.5 \times 2 \times 10^{-4} = 5 \times 10^{-4} = 0.0005$. The average ocean depth is about 12,000 feet. When the oceans expand, they will rise by 0.0005 of this—i.e., by about 6 feet. That would flood much of the coastal areas of the world, including much of Bangladesh and much of the populated area of Florida.[18]

This scenario is scary enough that people have become seriously interested in doing the calculation carefully. More detailed calculations have been done that take into account the fact that it will be primarily the surface waters that are warmed, and the variability of the expansion of water. (When the temperature of water is below 4 C, it actually shrinks when heated. Much of the deep ocean is close to 4 C.) The 1996 report of the Intergovernmental Panel on Climate Change (IPCC) estimates that, taking all these things into account plus melting of glaciers, the effect would be a rise in sea level between 15 and 95 centimeters—i.e., between 6 inches and 3 feet.

Thermometers

Most thermometers make use of the small expansion in order to measure temperature. They typically fill a small glass bulb with fluid and attach a tube with a tiny long hole (figure 2.3). When the temperature is raised, the fluid expands and moves up the tube. Markings on the tube indicate the temperature.

In a real thermometer, the diameter of the bulb (holding most of the liquid) is much greater than the diameter of the tube. Note that the thermometer would not work if *both* the glass and the fluid expanded the same amount. Thermometers take advantage of the fact that fluids can be found (e.g., mercury and alcohol) that expand more than glass. Alcohol, died a red color, is commonly used because its expansion is particularly large. Most of the alcohol is in the bulb at the bottom. When it expands, it forces fluid into the tube. Without the bulb, the expansion would not be enough to be visible. A large amount of fluid (in the bulb) expands; it has nowhere to go (because the glass holding it does not

---

[18] Pieter Tans once told me that of those who live on the coast, only the Dutch would be unaffected. "We know how to build dams," he said.

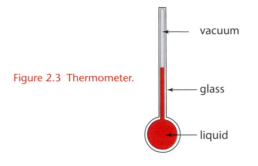

Figure 2.3 Thermometer.

vacuum

glass

liquid

expand) except up the tube. The tube usually has a vacuum in it so that air pressure does not impede the flow.

TEMPERATURE IN THE SHADE VERSUS IN THE SUN

Why do meteorologists measure temperature in the shade, rather than in the Sun? Aren't people more interested in the temperature in the Sun? Why don't they report it?

It turns out that there is a good reason. Thermometers are supposed to measure air temperature. When you place them in a room, they eventually reach the same temperature as the air; that's the zeroth law of thermodynamics. However, if you put a thermometer in direct sunlight, the red-colored alcohol absorbs more sunlight than does the transparent air. That makes the thermometer hotter than the air. Of course, heat will flow from the thermometer into the air, but if the Sun keeps shining on the thermometer, the thermometer will always be hotter. So a thermometer in the Sun does not measure the air temperature. On the other hand, the temperature of the air in the shade is usually the same as that in the Sun.[19] So if you really want to know the temperature of the sunlit air, measure it in the shade.

What happens if another object sits in the Sun? It too can get hotter than the air. You've probably had the experience of walking on hot sand, or of touching a car that has been sitting out in the Sun. Because these objects absorb sunlight readily, they are often hotter than the air. It is an old tradition in New York City (where I grew up) of publishing newspaper photos on a hot day showing someone frying eggs on an automobile hood. The hood is hotter than the air. It is much hotter. That's because the Sun is shining on it.

As a consequence, the "temperature in the Sun" is not a well-defined concept. Different objects will have different temperatures. Air close to a hot auto will be hotter than air close to a mound of snow, even if they are only a few feet apart. In fact, even the temperature of an object in the Sun is not well-defined, since its surface (exposed to sunlight) will usually be hotter than its interior.

[19] Some people would say that the term "temperature of the air in the Sun" is not correct, since even air does absorb some power from sunlight. But that effect is very small, so that the average motion of air molecules in sunlight is actually very similar to the motion in nearby shade.

Another type of thermometer works on the principle that different metals will expand by different amounts. If you take two bands of different metals and bind them to each other, you get a *bimetallic strip*. As one side expands more than the other, the strip bends. The amount of bending will be very large for even a small amount of expansion. The bending metal can pull a lever that moves an indicator over a temperature scale. Thermometers using bimetallic strips are used in oven thermometers and in old thermostats.

Yet a third type, called a *digital thermometer* (often used in medicine) takes advantage of the fact that the electrical properties of certain materials change when the temperature changes. A small circuit with a battery can measure these changes and put the result on a digital display.

DOES EVERYTHING CONTRACT WHEN COOLED?

No. Cold water (below 4 C ≈ 39 F but not frozen) expands when it is further cooled. As it freezes into ice, water expands even more. This is a strange behavior, and it happens because water molecules start arranging themselves into mini-structures, even while in a liquid state.

Without this peculiar behavior of water, life on the Earth might not have endured. In oceans and lakes, once the water gets colder than 4 C, the freezing water expands, and with its low density it floats on top of the other water. When it freezes, it expands even more, and so ice forms on the surface. This ice and layer of cold water insulates the water below, and keeps it from getting colder.

If cold water were denser than warm water, then in winter the cold surface water would sink to the bottom, and the warm water would rise to the top, where it would be chilled by contact with the cold air. If water contracted when it froze into ice, then even the ice would sink to the bottom. Some people speculate that the entire ocean would eventually reach the freezing point and turn into ice, and whatever life there was in the water would be killed.

SR-71 SPY PLANES

SR-71 planes flew so fast that friction from the air heated the outer surface to over a thousand degrees C. The thermal expansion was so great that if the wings were made in the usual way, they would crack. According to the designers (see the book *The Skunk Works* by Ben Rich), they solved this problem by making the fittings of the plane loose—almost like the cracks placed in concrete. A good tight fit was obtained only when the metal expanded, at high speed. A tricky consequence of this was the fact that the planes leaked fuel (through these loose fittings) until the outside heated up sufficiently. (I know, this is hard to believe, but it's true.) An image of the SR-71 is shown in figure 2.4.

Conduction

When two objects come in contact, the touching (collisions of surface molecules) allows them to share kinetic energy. The zeroth law implies that the hotter object (greater kinetic energy per molecule) will lose some of its kinetic energy and the cooler object will gain some. Eventually, they will be at the same

Figure 2.4 SR-71 spy plane. (Photo courtesy of NASA.)

temperature. But this doesn't happen instantly. Moreover, the rate is different for different materials. We say that different materials "conduct heat" at different rates.

Let's look at one of the "quandaries" at the beginning of this chapter. Even though both are at room temperature, a plastic cup and one made of glass feel different. The glass one seems cooler. (If you've never noticed this, find two such cups now, and do the experiment.) But why should that be? If both objects were sitting together in the room, they were at the same temperature, right?

Yes, the plastic and the glass were at the same temperature. But plastic and glass conduct heat at different rates. Your finger is warmer than room temperature, because you are generating heat in your body at an average rate of about 100 watts. When you touch the glass, it conducts the heat away rapidly, and so the temperature of your fingertip drops slightly. That is what your nerves sense: not the temperature of the glass, but the temperature of your skin. When you touch plastic, the heat is not conducted away as rapidly, so your skin doesn't cool as much. You think (incorrectly) that the glass is cooler than the plastic. In fact, they are the same temperature. The glass, however, cools your warm skin faster than the plastic does.

Solid, Liquid, Gas, and Plasma

Aristotle said there were only four elements: air, earth, water, and fire. In retrospect, that sounds silly—unless what he was really referring to was what we now call the *states* of matter. Air is the most common gas, earth the solid, water the liquid, and fire the most common plasma.

At low temperatures, the shaking of molecules in a substance is low, and the molecules tend to stick together in a rigid form we call a *solid*. When the substance gets hotter, the molecular motion increases to the point that the bonds to nearby molecules are weakened. The molecules still stick, but they can now slip past each other. When they reach this point, we say we have a *liquid*.

The most remarkable thing about this change is that it happens so abruptly. Water at 31 F is a solid, and water at 33 F is a liquid. The change from solid to liquid is called a change in *phase*.

As we continue to heat the water, the molecular shaking increases, but until the temperature reaches 212 F (= 100 C), the molecules slip but they still stick. At 212 F, the shaking is finally enough to overcome the attractive forces between the molecules, and they break apart. This is the phenomenon called *boiling*, and the escaped molecules are now a *gas*.

Even below 212 F, some molecules will have sufficient energy to break away. This happens because not all the molecules have the same energy; some are shaking faster, and some slower. The faster ones are the ones that can break away. When they do that, and leave the surface, then the molecules left behind are the slower, colder ones. That's what we call *evaporation*. Now you can see why evaporation makes the liquid cooler—it's just because the hotter molecules are leaving.

At even hotter temperatures, collisions between the molecules are sufficient to break them apart into individual atoms. If the atoms are themselves broken apart, so that electrons are knocked off their surfaces, then we call the gas a *plasma*.[20] A plasma consists of electrons with negative electric charge. The remaining atom fragment, which has a net positive charge, is called an *ion*. A plasma has no net electric charge because it is a mixture of negatively charged electrons and positively charged ions. We will discuss positive and negative charges in much more detail in chapter 6.

Here's an important fact: The temperature at which a solid melts (e.g., 32 F for ice) is the same as the temperature at which the liquid (water, in this case) freezes. Similarly, water boils at 212 F. If you have a hot gas, and cool it, it will begin to condense—that is, turn into a liquid—when you lower the temperature to 212. This symmetric behavior seems obvious to some people, and surprising to others.

Solids, liquids, and gases are commonplace, but many people think plasmas are exotic. They are more common than you might guess. If gases are hot enough that the collisions knock electrons off the molecules, then the result is a plasma. A candle flame is a plasma. The gas inside a fluorescent lightbulb is a plasma. The surface of the Sun is a plasma. A bolt of lightning is largely plasma.

EXPLODING TNT

Let's think again about what happens when TNT (trinitrotoluene) explodes. According to table 1.1 in chapter 1, the chemical energy that is released is 0.65

---

[20] The word *plasma* was originally used in biology and was appropriated for physics by Nobel Laureate Irving Langmuir. If you are interested, see L. Tonks, "The Birth of 'Plasma,'" *American Journal of Physics* 35: 857 (1967).

Calories per gram of TNT. When TNT explodes, it suddenly converts 0.65 Cal/g to heat. This new thermal energy is much greater than its prior thermal energy, which amounted to only 0.004 Cal/g.[21] In other words, after the explosion, the internal kinetic energy increases by a factor of 167. If the molecules didn't break apart (they do—and that complicates it a little), the absolute temperature would suddenly become 167 times greater than the prior temperature (300 K). That makes the temperature $167 \times 300 = 50,000$ K. Note that if we convert back to Celsius, C = 50,000 –273 ≈ 50,000 C (rounding to the nearest 1000).

Of course, 50,000 C is very hot, much hotter than the surface of the Sun (which is about 6000 C) Nothing is a solid at 50,000 C. The forces between the molecules are not strong enough to hold them together. That means that our gram of TNT is suddenly converted to a very hot gas, perhaps even into a plasma.

What will that hot gas do? Even at normal room temperatures, gases take typically 1000 times the volume of a solid. So just the fact that it turns into a gas makes it expand by a factor of 1000. But since it is hot, it expands even more—by another factor of 167 (the ratio of temperatures before and after). We'll discuss the extra factor of 167 in the next section, but for now, accept that figure. Put that factor on top of the factor of 1000, and we get a total expansion in volume of 167,000. (This is only a rough estimate.)

To summarize, here is our picture of what happens when TNT explodes: The solid material is suddenly converted to a hot gas. The hot gas expands rapidly until its volume goes up by a factor of 167,000. The expanding gas pushes everything out of the way. Any nearby material picks up the velocity of the gas. Terrorists typically surround the explosive with a pipe, or pieces of metal (e.g., nails). When the metal fragments fly out at high speed, they are what do the most harm.[22]

Gas Temperature and Pressure: The "Ideal Gas Law"

Why did the heated gas in the last section expand by an additional factor of 167? It helps to understand the difference between a solid and a gas. In a solid, the atoms bounce back and forth but never leave their relative positions. As the solid gets hot, this added bouncing makes the solid expand. But when the energy of the molecules becomes sufficiently great, the atoms push their way out. At high temperature, the molecules no longer stay in the same place but

---

[21] For room temperature, we take $T_K = 300$. (That's 81 F, which is a hot room, but it gives a nice round number in degrees Kelvin.) The energy per molecule is given by the equation we already discussed: $E = 2 \times 10^{-23}$ K. Putting in the numbers, this equation gives the heat energy per molecule = 2×10^{-23} K $\times 300 = 600 \times 10^{-23}$ J = 1.4×10^{-24} Cal. TNT has 2.6×10^{21} molecules in 1 g. So the thermal energy in 1 g of TNT is the energy per molecule multiplied by the number of molecules: $E_{TNT} = (1.4 \times 10^{-24}) (2.6 \times 10^{21}) = 0.004$ Cal/g. So the thermal energy at room temperature is much less than the chemical energy released in the explosion.

[22] The military has built "fragmentation bombs" and "fragmentation grenades" based on the same principle. The colloquialism "to frag" originally meant to attack someone with a fragmentation grenade.

move much more freely. They bump into other molecules, and they bounce against any walls in the containers that they are in. The bouncing tends to push the walls outward. A force must be applied to the walls to keep them from moving.

The pressure of a gas is defined as the force it exerts on one square meter of area. The key result is:

$$P = \text{constant} \times T_K$$

This equation is part of the "ideal gas law." It is called *ideal* because most real gases deviate from it a little bit, yet it is usually a good approximation.[23]

The importance of this law is as follows: If you double the absolute temperature, you double the pressure of the gas. If you raise the absolute temperature by a factor of 167 (as in our TNT example), then the pressure increases by a factor of 167. That's why hot gases exert so much pressure.

AUTOMOBILE AIRBAGS

The airbags that are used to protect you during an automobile crash are balloons that inflate very rapidly—in a thousandth of a second—between the time that the crash is detected by the automobile electronics and the time that your head would smash into the windshield. How can you fill a balloon that rapidly? The answer is, naturally, with an explosion. Airbags contain about 50 to 200 grams of an explosive called sodium azide. Its molecules consist of 1 atom of sodium and 3 of nitrogen; it has the chemical formula NaN_3. When triggered by an electric pulse, it explodes into sodium metal and nitrogen gas. The released gas inflates the balloon.

LEIDENFROST LAYERS, SAUTÉING, AND FIREWALKING

Have you ever seen a drop of water land on a hot saucepan? It seems to float above the surface and move about as if there were no friction. If you have never seen this, try it. Put on a pair of glasses to protect your eyes. You'll see the drop sizzle and then float just above the surface of the pan.

This happens because the rapid heating of the water turns it into a gas and pushes the drop away from contact with the pan. The gas has very little friction, and so the droplet moves over the surface. The gas also conducts heat very poorly, since it is a thousand times less dense than the water (so there are 1000 times fewer molecules present to carry the kinetic energy from the sauce pan to the water).

The thin layer of gas that insulates the drop of water is called a *Leidenfrost layer*, after Josef Leidenfrost, the scientist who, in the sixteenth century, was the first to understand why water droplets floated on hot pans.

For a class demonstration, this effect is easily demonstrated with liquid nitrogen. Nitrogen is a gas that composes approximately 79% of air. It turns to

---

[23] In many physics and chemistry texts, the ideal gas law is written $P = nkT_K$, where n is the number of molecules per unit volume, and k is Boltzmann's constant. Another way to write it is $PV = NkT$, where N is the total number of molecules.

liquid when cooled to $77\,K = -196\,C = -321\,F$. Pour some on a table, and watch the little droplets of liquid nitrogen scoot over the tabletop, suspended on thin layers of nitrogen gas.[24]

Some people believe that the Leidenfrost effect can explain "firewalking," the ability of people to walk barefoot over hot coals without burning their feet. If the skin of your foot is moist (e.g., from sweat) and you step on a hot coal, the water is very rapidly boiled into a thin layer of gas. The water vapor from the sweat has a temperature of 100 C; it penetrates into the hot coals and prevents the much hotter gases from the interior from reaching the feet. Although the water vapor is hot, it is also a poor conductor of heat, so it doesn't heat the foot very quickly.

Look up firewalking on the Internet; you will find lots of commercial organizations that will lead you through a firewalking ritual as part of a self-improvement and confidence-building program. (If you can walk on fire without being burned, you can do anything. . . .) But I don't recommend that you try walking on hot coals without professional supervision. The professionals who run these firewalking clinics use additional tricks. They may make sure that your feet are adequately damp (e.g., from walking on moist sand near the sea); they also use special coals that have a thick cool insulting layer of ash outside the hot burning core. Here is something you can try with relative safety: Next time you are at the beach on a hot day, and the sand is too hot to walk on comfortably, wet your feet, and try again. You'll discover that you can walk a few tens of meters before the sand becomes unbearably hot. Of course, the temperature of the sand didn't change, just the flow of energy into your feet. And be careful, even hot sand can scald your feet. If you never get to leave the city, you can try the same thing on a hot sidewalk. But carry some sandals with you in case your feet begin to burn.

AUTOMOBILE: EXPLOSIONS UNDER THE HOOD

We've talked about turning energy (e.g., energy of motion) into heat, but can we do the opposite? There is a huge amount of energy hidden as heat. Can it be turned into useful energy?

Yes. Exploding TNT turns chemical energy into heat, the heat causes the material to turn into hot gas, and the expanding hot gas can rip apart rock. That counts as useful work.

We can also tame this process to do some more gentle work, like running an automobile. Gasoline and air are injected into a chamber called a *cylinder* (because of its shape) making an explosive mixture. A spark (from the spark plug)

---

[24] I know a professor, Howard Shugart, who pours a little liquid nitrogen in his mouth and then forcefully spits it out. A huge plume of mist shoots out (mostly water droplets condensing from the cooled air), and the class cheers. I have never had the courage (or been foolish enough) to try this. I've heard of other professors who gargle with liquid nitrogen, their throats protected by the Leidenfrost layer. I'll leave this to others. But I do pour liquid nitrogen over my hand, keeping the surface tilted so that the cold liquid runs off. There is little sensation of cold. But don't cup your hands and catch a little bit in one location. The sudden freezing of the skin will leave an effect very similar to that of a severe burn.

ignites the mixture, it explodes,[25] and the mixture turns into a hot gas. The high pressure from this gas pushes a piston, which in turn pushes a series of gears that turn the wheels.

The explosions in an automobile are generally kept small so that they won't rip the engine apart. Your car probably has 4 to 8 cylinders, and these are run in sequence to provide a fast series of bursts that approximate continuous power. If you would really like to get a sense of how a gasoline engine works, to see the animated image take a look at http://auto.howstuffworks.com/engine1.htm. When run by a Web browser, it shows the cycles of the engine. At the top of this Web image is a spark plug. Gas and air are introduced through the valves; the spark plug ignites the mixture, forcing the piston down in the cylinder. At the end of its motion, another valve is opened; the cylinder moves upward (carried by the momentum of an attached flywheel) to expel the burned gases, and the cycle repeats.

HEAT ENGINES

Any engine that runs by turning heat into mechanic motion is called a *heat engine*. An automobile engine is a heat engine; so is a locomotive steam engine, and a diesel. Nuclear submarines and nuclear ships (some of our aircraft carriers are nuclear) are also run by a heat engine. Nuclear power is used to heat water to steam, and the steam is run through a turbine (a fancy fan) to make it spin. The spinning motion is conveyed to the propeller to push the submarine (or ship) forward. We'll talk more about creating heat from nuclear energy in chapter 5.

What kind of engine is *not* a heat engine? Think for a moment and see what you can think of. I'll put some examples in the footnote to make it easy to *not* peek at the answer until you've thought about it.[26]

WASTED ENERGY

In an automobile engine, the chemical energy from the gasoline and air mixture is turned into heat, and the pressure from the hot gas pushes the piston. But not all of the energy turns into this useful work. Some of the heat is conducted away to the outside air and is "wasted." For typical automobiles, only about 10% to 30% of the chemical energy is converted to useful propulsion.[27]

---

[25] As I noted in chapter 1, engineers sometimes like to distinguish between an *explosion* and a *deflagration*. In an explosion, the surface of burning moves faster than the speed of sound. In this precise terminology, the burning of gasoline in an automobile is a deflagration, not an explosion. But I won't use this fine distinction.

[26] An electric motor, used (for example) in an electric car. The sail of a boat. A windmill (used to grind flour). A windup toy. The muscles in your body.

[27] Assume that an automobile gets 30 mpg as it travels 50 miles per hour over a level road. Although the auto has a peak power of 150 horsepower, assume that it uses only about 25 horsepower under these conditions. The density of gasoline is about 6.2 pounds per gallon. Using these numbers, you can show that the rate of gasoline use is about 10,000 grams per hour. With 10 Calories per gram, the energy in the gasoline being used is about 30 Cal/sec = 123 kW. But the energy that the engine actually delivers is only, typically, 25 hp = 18 kW. Based on these assumptions, the energy efficiency is $18/123 = 0.15 = 15\%$.

The rest is wasted—in the form of heat that escapes or that has to be removed. In fact, gasoline engines waste so much energy that special cooling systems are built in to get rid of the wasted heat. That is what the *radiator* in the front of the car does—it cools water by letting air blow by it, then uses the cool water to remove waste heat from the engine (so that it doesn't "overheat"), and then sends the hot water back to the radiator to cool off.[28] Much of the heat also leaves in the exhaust of the car.

It is possible to use the energy more efficiently, but there are surprising limits. As you will see in the next section, there are limits to how well a heat engine can perform.

LIMITED EFFICIENCY OF HEAT ENGINES

Here is a puzzle: The thermal energy of water at room temperature is about 0.04 Calories per gram. That is small, but it is 5 times as much as in a battery. And water is cheap. Why not use the thermal energy in water as a fuel?

It turns out that there is a very fundamental theorem that limits how much of such heat can be turned into useful energy (e.g., kinetic or potential energy). This theorem was one of the greatest achievements of the theory of heat. To understand this theorem, you first have to realize that heat can be extracted (turned into useful energy) only when it is flowing from a hot region to a cold region. For example, when gasoline burns, it is hotter than the surrounding air, and that allows it to expand and push against a piston. If the surrounding air were just as hot as the exploded gasoline, it would have a similar pressure and the piston would not move. Heat engines depend on such a temperature difference to do their work.

Let the hot temperature (e.g., of the exploded gasoline) be T_{HOT} (in degrees Kelvin) and the temperature of the gas after it has been cooled be T_{COLD}. The amazing theorem is that the efficiency of the engine will be given by the following equation:

$$\text{Efficiency is less than or equal to}$$
$$1 - (T_{COLD}/T_{HOT})$$

Perfect efficiency is 1 (i.e., 100%). Thus, for example, if the gasoline explodes at 1000 K, and is cooled to 500 K before it exhausts from the cylinder, then the efficiency of the engine will be less than or equal to $1 - (500/1000) = 0.5 = 50\%$.

This is a remarkably simple rule, and it is always true when trying to extract energy from heat. It is not relevant for batteries or solar cells that extract energy directly from chemicals or from light. But it shows why heat engines, to be efficient, must be hot.

Let's go back to the puzzle: Why not extract heat energy from room-temperature water? Imagine a boat that scoops water out of the sea, extracts

---

[28] If the radiator stops working, the engine gets very hot (it "overheats"), the lubricating oil decomposes, and without lubrication, the metal pistons no longer slide smoothly in the metal cylinders—they scrape and eventually bind. We use an ironic word for the process of the metal binding to other metal: we say that the engine "freezes"—even though it all happened because of the high temperature.

the heat, and uses it to run the propeller, turning the water into ice. The ice can then be thrown overboard. This would be very nice. Let's calculate the efficiency for such an engine. Since the boat is at room temperature, and that is the same (we assume) as the temperature of the water, then T_C and T_H are equal. Then the efficiency is less than or equal to $1 - (300/300) = 0$. So the efficiency is zero.

You need a temperature difference in order to extract any useful energy from heat. You cannot extract heat from a single object and turn it into useful energy unless there is something colder that you can use. This fact is so important that it has been given another fancy name: the *second law of thermodynamics*.

It is not necessary for you to memorize the efficiency equation. But you should know that to have high efficiency, you have to have large temperature differences (e.g., between the hot exploded gasoline and the cool outdoors). If temperature differences are small, then you cannot extract very much useful energy from heat.

VOLKSWAGEN "BUG" AND THE EFFICIENCY EQUATION

In the 1960s, the Volkswagen company introduced the car that was commonly known as the "Beetle" or the "Bug." At a time when other cars averaged averaged 6 to 15 miles per gallon (mpg), it got 30 mpg. That was, in part, because it was little. But it also ran its engine at a higher temperature, a temperature at which higher efficiency was obtained, according to the efficiency equation. If T_H gets very large, then the ratio T_C/T_H gets very small, and the efficiency $1 - (T_C/T_H)$ gets close to 1—i.e., close to 100% efficiency.

When I bought my first Beetle, in 1966, there was another advantage: the car was believed to produce very little air pollution. That's because at the high temperature of the engine, virtually all the carbon particles in the exhaust were burned into carbon dioxide. The result was the virtual total absence of "smoke" at the back exhaust. But a few decades later, people began to worry about other kinds of pollution—in particular, nitrous oxides, NO and NO_2. These two gases, referred collectively as NOx, were not considered pollution in 1966! It turns out that at high temperatures, ordinary air (N_2 and O_2) react chemically to form nitrous oxide, and nitrous oxide is more important in the formation of smog than carbon particles. The Beetle produced huge amounts of nitrous oxides because of its high engine temperature. The nitrous oxide production could not be reduced without reducing the temperature of the engine, and if they did that, then the efficiency of the engine would go down. When new legislation limited the nitrous oxides allowed from new cars, the old Bug was phased out. The "new" Volkswagen Beetle (no longer in manufacture) uses a water-cooled engine that operates at low temperatures to avoid making nitrous oxides, but as a result, it is not as efficient as it could be.

Refrigerators and Heat Pumps

A heat engine requires a temperature difference, something hot (to provide the energy) and something cold (for the heat to flow to). In an automobile engine, this is created by burning (exploding) gasoline. As the hot gas expands, it does useful work (i.e., it turns the wheels of the car). It is possible to reverse this process: to take mechanical motion and use it to create a temperature difference. The device that does this is called a *refrigerator* or a *heat pump*.

A typical refrigerator works by using a mechanical force to reduce the pressure in a chamber. Then the gas law equation, $P = \text{constant} \times T$, implies that the temperature of the gas will decrease. That cool gas can be used to freeze ice, or just to cool a room. That is how refrigerators and air conditioners work.

The mechanical force that reduces the pressure must push the piston against the room air pressure. This motion slightly heats the air. So in a refrigerator, not only is one side being cooled, but the other side is being heated. Energy is conserved, so any heat that leaves the refrigerator must result in energy transferred elsewhere, usually to the surrounding air in the room. Thus, refrigerators heat the rooms that they are in. Air conditioners are designed to cool a room and put the extra heat outside. That's why air conditioners must be placed in windows or other locations with access to the outside. You can think of an air conditioner as a device that uses mechanical motion (usually from an electric motor) to pump heat from inside the room (where it is warm) to the outside (where it is cold).

The reverse also works. On a winter day, you can take an air conditioner, install it backward, and use it to pump heat energy from the cold outdoors into a warm room. That means it takes some of the thermal energy out of the cold outside air—making it even colder—and brings that energy indoors to make the indoors warmer. When used in this fashion, the device is usually called a *heat pump*. Heat pumps are widely used in cold parts of the United States. It is exactly the same as using an air conditioner in reverse, to make the outside colder and the inside warmer.

Here is a puzzle with a surprising answer: Suppose that you have a gallon of fuel, and a cold house. What is the best way to heat your home? You can burn the fuel, and use the heat produced. But here is a much better way: use the fuel in a heat engine, and use the mechanical motion that is produced to run a heat pump. The heat pump will extract heat from the cold outdoors, and pump it into the room. It turns out that the heat pump method will put typically three to six times more heat into your room than if you were to just burn the fuel and use the heat emitted. This factor over burning is called the *coefficient of performance*, or COP. Of course, the heat engine is not 100% efficient, so some of the energy is still turned into heat; use this to supplement the heat pump.

Does that mean that we are wasting fuel when we heat our homes by burning fuel (gas, coal, or wood) instead of using that fuel to run a heat engine/heat pump system? The surprising answer is *yes*. But an engine/pump system is more complicated and its costs more. It generally isn't used unless the outside temperature is very cold, since otherwise it is cheaper to buy more fuel than it is to buy the expensive engine/pump system. But as we run out of fossil fuel, and it becomes more expensive, we can expect to see much wider use of engine/pump heating systems.

Look back now at the fourth "quandary" listed at the start of this chapter.

Laws of Thermodynamics

Here is a complete list of the laws of thermodynamics:

Zeroth law:	Objects in contact tend to reach the same temperature.
First law:	Energy is conserved (if you consider all the forms, including heat).
Second law:	You can't extract heat energy without a temperature difference.
Third law:	Nothing can reach the temperature of absolute zero.

The second law can also be understood as the fact that all objects in contact with each other tend toward *equilibrium*—i.e., they all tend toward the same temperature. A famous consequence of the second law is that whenever heat flows, the total "disorder" in the Universe tends to increase. The third law is plausible, since it is hard to remove heat from an object without having something that is colder, and so it is difficult to remove heat from any object that is close to absolute zero.

It is not necessary that you memorize this numbered list of laws. It is far more important that you know the facts—i.e., that objects in contact tend to reach the same temperature, that energy is conserved, etc.

Heat Flow: Conduction, Convection, and Radiation

Heat energy moves from one place to another in three ways, called *conduction*, *convection*, and *radiation*:

- **Conduction:** Energy flows by contact. We discussed this earlier when we talked about touching glass versus plastic. Hot molecules transfer energy to cooler ones by direct contact. A good conductor is something that transfers heat rapidly from one molecule to the next. Metals are usually good conductors, as is glass. Plastic is a poor conductor. If you want to insulate something from heat, you use a poor conductor. If you want a pan that is heated at one point to distribute that heat over its whole surface, you make it out of a good conductor (e.g., aluminum or copper).
- **Convection:** Energy is carried by a moving material, usually a gas or a liquid. When the hot material reaches some cold things, it usually transfers its energy by contact—i.e., by conduction. Example: An electric heater in your room warms the nearby air (by conduction). That air then moves throughout the room (convection), warming things that it touches (by conduction). A fan can help convection. Hot air also tends to rise (see chapter 3), and that starts the air in a room circulating on its own; that's called *natural convection*. A convection oven uses circulating hot air to heat food.
- **Radiation:** Energy moves through empty space, carried by (possibly invisible) light. When you stand in sunlight, you are warmed by

radiation from the Sun. When you stand in front of an infrared heat lamp, you are warmed by the invisible infrared radiation. (We'll discuss such invisible light in much more detail in chapter 9.) A microwave oven cooks by radiation. Microwaves penetrate into food, so they will cook the insides of some foods as rapidly as they cook the outsides.

The word *radiation* is used for virtually any energy that flows through space. This includes nuclear radiation (which can cause cancer; see chapter 4), visible light, ultraviolet light (which can cause sunburn; see chapter 9), and microwaves.

OPTIONAL: ENTROPY AND DISORDER

I mentioned earlier that we can quantify the concept of *disorder* into a number called the *entropy*, and that when heat flows, the net entropy of the Universe tends to increase. This subject receives a great deal of attention from philosophers, and so it is worth a bit of further discussion.

When entropy changes due to heat flow, the calculation is simple: When heat flows into an object, its numerical increase in entropy is Q/T, where Q is the amount of heat (usually measured in joules) and T is the temperature. When heat leaves an object, the entropy of that object decreases by Q/T.

When heat flows from a hot object (with temperature T_H) to a cold object (which has temperature T_C), the total change in entropy is:

$$\text{Total change in entropy} = Q/T_C - Q/T_H$$

The first term will always be bigger than the second one (since T_C is smaller than T_H), and so the total entropy will increase. This is the deep meaning of the fact that the entropy of the Universe is increasing. The Universe is becoming more and more disordered.

Disorder can also increase without heat flow. For example, if you burst a balloon, then the atoms are no longer confined into a small region but have spread out through the atmosphere. This kind of disorder can be included in with the other.

It is important to realize that the entropy of an object can go up or down; it is the total entropy of the Universe that is always increasing. My goal with this book is to decrease the entropy in your brain. That is a fancy way of saying that I hope you learn something. As you learn, you will radiate heat, and that will increase the entropy of the world around you. The net entropy of the Universe goes up, but I hope that your own entropy goes down.

When an object cools off, its entropy goes down, but the heating of the surroundings more than makes up for that, so the total entropy of the Universe increases. The entropy of the Earth is decreasing with time, as is the entropy of the Sun. The Sun is emitting visible light; the Earth is emitting infrared light (see chapter 9), and as a result, the total entropy of the Universe (when you include the entropy of the light) goes up.

Some philosophers (and some physicists) have argued that the increase in entropy of the Universe is what determines the direction of time—i.e., why we remember the past and not the future. (That really is a deep question, not at trivial as it sounds.) But it can also be argued that it is the local decrease in entropy—i.e., when we learn things—that gives us the sense of time.

Have fun thinking about these ideas. There have been several popular books devoted to the subject. The second and third laws of thermodynamics can be reformulated to read as follows:

Second law: The entropy of the Universe tends to increase.
Third law: The entropy of an object goes to zero at $T = 0$ K.

Understanding how these reformulations are equivalent to the original statements is part of the advanced study of thermodynamics.

Chapter Review

Atoms, the basic constituents of matter, are about 10^{-8} cm in diameter. There are about 50,000 of them in the diameter of a red blood cell, which is about the smallest object visible with visible light. Heat is the shaking of molecules, the fact that they have kinetic energy. The velocity of shaking is comparable to the velocity of sound, about 1000 feet per second = 330 meters per second. For solids, the atoms remain in the same location despite this violent shaking. The effects of shaking can be observed in Brownian motion. Brownian motion also makes itself evident in electronic noise such as hiss.

If two objects have the same temperature, then the average kinetic energy of the molecules in the two objects is the same. However, the speed of the molecules is not equal. If two molecules with different masses have the same kinetic energy, the lighter molecule will have the higher velocity. The lightest molecules of all, those of hydrogen gas (H_2), move so fast that most of them have escaped the gravity of the Earth and are no longer present in the atmosphere.

Temperature can be measured on the Fahrenheit scale or the Celsius (Centigrade) scale. But more useful for physics is the absolute or Kelvin scale, for which 0 K corresponds to a kinetic energy per molecule of zero. A *change* of 1 K is equal to a *change* of 1 C is equal to a *change* of 9/5 F $\approx$ 2 F.

Most objects expand when they get warm, an amount typically by a part per 1000 to a part per 100,000 for each degree C. This fact is used for thermometers, but it also results in sidewalk cracks, and it could cause substantial sea level rise if global warming warms the oceans.

Heat can be transferred by conduction, since atoms in contact with others can share their kinetic energy. Gases expand when they get hot. A good approximation is the "ideal gas law," which says that the gas pressure is proportional to the absolute temperature. Heat can also flow from radiation and from convection.

An explosion occurs when an object gains so much energy that it becomes a very hot gas. It is the high pressure of the gas and the resulting rapid expansion that make up an explosion. Such explosions also occur in the cylinders of internal combustion engines, and we use them to supply energy to our automobiles.

Energy converted to heat cannot always be converted back efficiently. The limit is the efficiency equation, efficiency $\leq 1 - T_C/T_H$. Such heat is often considered wasted, and may result in the ultimate "cold death" of the Universe.

Entropy is a measure of the disorder in molecules. Warm objects, by their shaking, have more disorder, and therefore higher entropy. Whenever heat flows, the entropy of the Universe increases, although the entropy of an object (such as your brain) can decrease. Indeed it does, when you learn something.

The four major laws of thermodynamics are: (0) objects in contact tend to reach the same temperature, (1) energy is conserved, (2) extracting useful energy from heat requires a temperature difference, and (3) the entropy of the Universe is always increasing.

Energy can be used to extract heat from an object. That is the basic principle of the refrigerator, the air conditioner, and the heat pump.

Discussion Topic

If you go to a high altitude, the temperature of the air is usually lower. What do you think that does to the sound velocity? (This issue will turn out to be important when we discuss UFOs, in chapter 7.)

Internet Research Topics

1. Look up "firewalking" on the Internet. See if you can find organizations that offer firewalking training, and also see if you can find sites that explain how firewalking is possible. Describe what you find.
2. What are the most common elements in the human body? Compare them to those in the Earth's crust. Which do you consider the biggest surprises?
3. Look up "heat engines" on the Internet. There will be a lot of technical discussion meant for engineers. Can you find any novel heat engines? Look up "nitinol engine." Can you find heat engines that claim to work on small temperature differences? Do they discuss the poor efficiency that you get with small temperature differences (from the efficiency equation)?
4. Look up "heat pump" on the Internet. What can you find out about the coefficient of performance (COP)? What is the cost? Would a heat pump be a good investment in the area where you live? (Don't guess. Work out some numbers.)
5. Why is it warmer in the summer than in the winter? Is the Earth farther from the Sun? (No.) Does the intensity of the Sun change? (No.) Why does Australia have winter when we have summer?

Essay Questions

1. Read an article that involves physics or technology that appeared in the last week or two. (You can usually find one in the *New York Times* in the Tuesday Science section.) Describe the article in one to three paragraphs, with emphasis on the technological aspects (not on business or political aspects). If you don't understand the article, then

you can get full credit by listing the things that you don't understand. For each of these items, state whether you think the writer understood them.

2. Describe in a page what aspects of this chapter you think are most important. What would you tell your friends, parents, or children are the key points that are important for future presidents or just good citizens?

3. When objects are heated, they usually expand. (There are some exceptions.) Explain why, and give examples that illustrate how this behavior causes problems and how it can be usefully applied.

4. Describe what is meant by an *explosion*, and explain what is going on in terms of atoms and molecules.

5. Give examples of "small" explosions that have useful purposes, particularly explosions that the average person doesn't even realize are explosions.

6. Estimate how many atoms there are in a sheet of paper. (This question is deliberately vague. Take the paper size to be anything reasonable. Use the average size of an atom as described in the text.)

7. Discuss the efficiency equation. What are its implications for automobile engines?

8. If you double the temperature of a gas, what happens to the velocities? Do they double? (The answer is no.) Work out how much the velocity increases.

Multiple-Choice Questions

1. If you double the energy content of a kilogram of gas, the temperature of the gas (measured on the absolute K scale)
    A. is unchanged
    B. increases by the square-root of 2
    C. doubles
    D. is multiplied by 4

2. About how fast are molecules in air moving?
    A. 1000 ft/sec
    B. the speed of light
    C. 9.8 m/sec
    D. 9.8 cm/sec

3. Temperature is the measure of
    A. average momentum
    B. average kinetic energy
    C. average velocity
    D. average total energy

4. In a bucket of water, the instantaneous speed of the molecules is closest to
    A. zero, since they don't move
    B. whatever speed the wood is moving
    C. approximately 1.7 meters per second

D. approximately 1000 feet per second

E. approximately 186,000 miles per second

5. A refrigerator operating in a room
    A. warms the room
    B. cools the room
    C. has no effect on the room
    D. removes water vapor from the room

6. A refrigerator *with its door open* is operating in a room. It
    A. warms the room
    B. cools the room
    C. has no effect on the room
    D. removes water vapor from the room

7. A table has the same temperature as the air above it. That means that the molecules in the air and in the table have
    A. the same average velocity
    B. the same average energy
    C. the same average acceleration
    D. the same average mass

8. Ice melts at
    A. 0 K
    B. 0 C
    C. 0 F
    D. 100 C

9. A gas heater warms a room mostly through
    A. convection
    B. conduction
    C. radiation
    D. depletion

10. The velocity of sound
    A. increases when the temperature increases
    B. decreases when the temperature increases
    C. depends on the pressure of the air
    D. is constant, independent of the temperature

11. If a gasoline engine produces a hotter explosion, then the efficiency of the engine should
    A. be the same
    B. go up
    C. go down

12. A cup full of water and made of plastic feels warmer than one made of glass because
    A. plastic is warmed by water
    B. plastic conducts heat less than glass
    C. plastic conducts heat better than glass
    D. plastic dissolves in water

13. The atom with the fewest number of protons is
   A. helium
   B. carbon
   C. hydrogen
   D. oxygen

14. How many atoms are there in the width of a human hair? About
   A. 25
   B. 125,000
   C. 273,000,000,000
   D. 6×10^{23}

15. The speed of sound is approximately
   A. 1 foot per nanosecond
   B. 1 foot per second
   C. 1000 feet per second
   D. 186,000 miles per hour

16. Put a hot glass in cool water. The glass shatters because
   A. it heats the surface water to boiling
   B. the outer surface of the glass contracts rapidly, but the inner part doesn't
   C. the outer surface of the glass expands rapidly
   D. the rapid conduction of heat triggers strong vibrations

17. Sea level is rising from global warming. The main cause is
   A. expanding sea water
   B. melting glaciers
   C. expanding rock under the sea
   D. contracting earth (while the water stays constant)

18. The temperature warms by 2 C. That is approximately the same as
   A. 1 F
   B. 2 F
   C. 4 F
   D. 1/2 F

19. Heat a room using the least energy by
   A. burning natural gas (methane)
   B. using natural gas to run a heat pump
   C. burning coal
   D. burning gasoline

20. There is almost no hydrogen gas in the atmosphere because
   A. it escapes the Earth's gravity
   B. there is very little hydrogen in the oceans or land
   C. hydrogen molecules move slower than oxygen and nitrogen
   D. hydrogen has all sunk to the Earth's core

21. At Mach 10, if all the energy of a meteorite went into heating the meteorite, its temperature would be about
   A. 300 C
   B. 6000 C

C. 30,000 C
D. 100,000 C

22. The New Orleans levees failed because
    A. thermal expansion broke them
    B. they were made of continuous concrete that could not stand the pressure
    C. they had leaks at the thermal expansion joints
    D. they were weaker at the thermal expansion joints

23. If temperature rises by 5 C = 9 F, then the rise in sea level will be about
    A. 2 inches
    B. 2 feet
    C. 12 feet
    D. 97 feet

24. For a typical auto, the fraction of the gasoline energy wasted as heat is about
    (careful, this may be a trick question)
    A. 1%
    B. 10%
    C. 20%
    D. 80%

25. The temperature in the sun
    A. is always hotter than in the shade
    B. is always the same as in the shade
    C. is sometimes cooler than in the shade
    D. is not well defined

26. When a material is cooled
    A. it stays the same size
    B. it contracts
    C. it expands
    D. Some materials contract and some expand.

27. The melting temperature of an object is usually
    A. equal to its freezing point
    B. higher than the freezing point
    C. lower than the freezing point
    D. equal to the boiling point

28. Absolute zero is the temperature of
    A. frozen water
    B. the Universe
    C. liquid helium
    D. nothing

29. When a liquid boils, the increase in volume (liquid to gas) is typically a factor of
    A. 10
    B. 100
    C. 1000
    D. 1,000,000

30. If the temperature of a gas in a container goes from 0 C to 300 C,
    the pressure will
    A. stay the same
    B. double
    C. become 300 times greater
    D. become infinite

31. Firewalking is similar to
    A. water on a saucepan
    B. the space shuttle reentry
    C. an automobile air bag
    D. water skiing

32. For gasoline to explode, it needs
    A. to be mixed with oxygen
    B. to be mixed with nitrogen
    C. to be mixed with carbon dioxide
    D. No mixing needed; just a spark.

33. To be more efficient, the temperature difference (between the ignited hot fuel
    and the cool part of the engine) should be
    A. as small as possible
    B. as large as possible
    C. It doesn't matter.

34. Heat flow through empty space (no atoms present)
    A. is impossible
    B. occurs through conduction
    C. occurs through convection
    D. occurs through radiation

35. Entropy (optional topic) measures
    A. heat
    B. temperature
    C. disorder
    D. energy

36. Molecular motion stops at
    A. 0 K
    B. 0 C
    C. 0 F
    D. 32 F

Gravity, Force, and Space

Being in orbit is like being infatuated—you are constantly falling, but you aren't getting closer.

Gravity Surprises

When an astronaut is orbiting the Earth, the head of the astronaut is "weightless." When he sneezes, his head will snap back . . .

. . . exactly as much as it does when he is sitting on the Earth's surface.

What? Aren't astronauts weightless?

At an altitude of 100 miles, you are virtually at the edge of space, since more than 99.999% of the Earth's atmosphere is below you. At this altitude, the force of gravity is less than at the Earth's surface. It's strength, compared to surface gravity, is only . . .

. . . 95% as great.

That's not much of a change. It's almost as strong as on the Earth's surface. So why doesn't the astronaut "feel" this gravity?

The $10 million X Prize was awarded in 2004 to the first private company to send a rocket to an altitude of 100 km. Some think that this is the beginning of private exploitation of space. But getting into orbit would take more energy. How much more?

About 30 times more.

Does that mean that the X-Prize winner was nowhere near getting into orbit?

If the Sun suddenly turned into a black hole, but its mass didn't change (it is presently about 300,000 times the mass of the Earth), then the orbit of the Earth . . .

. . . wouldn't change.

But don't black holes suck up everything near them? No.

These facts surprise most people. That's because they misunderstand several important concepts, including weightlessness, orbits, and the behavior of gravity.

The Force of Gravity

Any two objects that have mass attract each other with a force we call *gravity*. You probably never noticed this for small objects because the force is so weak. But the Earth has lots of mass, and so it exerts a big gravitational force on you. We call that force your *weight*. The fact that gravity is actually a force of attraction is not obvious. Prior to the work of Isaac Newton, it was assumed that gravity was simply the natural tendency of objects to move downward.

If you weigh 150 pounds and are sitting about 1 meter (3.3 feet) from another person of similar weight, then the gravitational force of attraction between the two of you is 10^{-7} lb. This seems small, but such forces can be measured; it is about the same as the weight of a flea.

You weigh less when you stand on the Moon, because the Moon doesn't put as big a force on you. If you weigh 150 lb on the Earth, you would weigh only 25 lb on the Moon. You haven't changed (you are made up of the same atoms), but the force exerted on you is different. Physicists like to say that your *mass* hasn't changed, only your weight. Think of mass as the amount of material, and weight as the force of attraction of gravity.

Mass is commonly measured in kilograms. If you put a kilogram of material on the surface of the Earth, the pull of gravity will be a force of 2.2 lb. So a good definition of a kilogram is an amount of material that weighs 2.2 lb when placed on the surface of the Earth. That number is worth remembering, since kilograms are commonly used around the world.[1]

Suppose that you weigh 150 lb on the Earth. Then your mass is about 68 kg (i.e., it is 150/2.2). Go to the surface of Jupiter, and you will weigh nearly 400 lb. On the surface of the Sun, you will weigh about 2 tons (T), at least for the brief moment before you are fried to a crisp. But in all cases, your mass will be 68 kg.

The equation that describes the pull of gravity between two objects was discovered by Isaac Newton and is called *Newton's law of gravity*. It says that the

---

[1] A more accurate value is that there are 2.205 lb in 1 kg and 0.454 kg in 1 lb, but don't bother memorizing these more precise numbers.

force of attraction is proportional to the mass—double the mass, and the force doubles. The force also depends on a special relationship with the distance called an *inverse square law*. The law is inverse because when the distance gets larger, the force gets smaller. The law is square because if you triple the distance, the force decreases by nine; if you make the distance increase by 4, then the force goes down by 16, etc. This law is usually written in the following way:

$$F = G\frac{Mm}{r^2}$$

M is the mass of the pulling object, m is the mass of the object being pulled, and r is the distance between them. G is a constant that makes the units come out right. *You don't have to memorize this equation.* I often put in equations that you aren't required to learn. They are there so that you can see how they are used (I'll use this equation several times in this chapter), but although you should try to follow the calculations, I don't expect you to be able to do them yourself.

For large objects like the Earth, some of the mass may be very close (right under your feet) and some can be far away (on the other side of the Earth). It turns out that to get the right answer for a spherical object, you can just use the distance to the center and you'll get the right answer. If you are standing on the surface of the Earth, then the right value of r to use will be the Earth's radius, about 4000 miles (6000 km). If you are in a satellite, then the distance to the center is the height (distance above the surface) plus the radius. I will not ask you to make such calculations, but I explain this in case you want to understand some of the calculations I'll do in this chapter, such as those for a black hole.

As you get farther away from the surface, r increases, and the gravity decreases. If you are 4000 miles above the surface of the Earth, your distance to the center has doubled, and the force decreases by a factor of 4.

I'll now use that fact to explain one of the surprises listed at the beginning of this chapter. At an altitude of 100 miles, the force of gravity is weaker. How much weaker? The only thing that changes in the gravity equation is the value of $1/r^2$. Here are the values for the two distances:

Surface of the Earth: $r = 4000$ mi. $1/r^2 = 1/16,000,000$
Satellite: $r = 4100$ mi. $1/r^2 = 1/16,810,000$

The force when you are on the surface is greater than when you are in orbit, since the denominator is smaller. Let's compare those two numbers. Let's take the ratio of the forces. It is easy to do with a pocket calculator:

Ratio $= 16,000,000/16,810,000 = 0.95 = 95\%$

The forces are different, but not by much. The weight in space is 95% of the weight on the ground. It is only 5% less.

Huh? Wait a minute! Isn't something is space "weightless"? Why are we concluding that it weighs 95% as much? We need to talk about the meaning of *weightlessness*. We'll get to that soon.

Push Accelerates—Newton's Third Law

Here is something that might surprise you: If you weigh 150 lb, not only is the Earth attracting you with a force of 150 lb, but you are attracting the Earth with a force of 150 lb too. This is an example of Newton's third law, which states: If an object exerts a force on you, then you exert the same force back on it. In my mind, this law is so fundamental that it should have been Newton's first law. (Newton's first law is that an object in motion tends to stay in motion unless there is an outside force. Newton's second law, $F = ma$, is one that we'll get to soon.)

But if you are so small, how can you exert such a large force on the Earth? The answer is that, even though you are small, your mass exerts a force on every piece of the Earth, simultaneously. When you add all those forces together, the sum is 150 lb. So you are pulling up on the Earth exactly as much as the Earth is pulling down on you.

Think of it in the following way: If you push on someone else's hand, that person feels your force. But you feel the force too. You push on the other person; the other person pushes back on you. The same thing works with gravity. The Earth pulls on you; you pull on the Earth.

Orbiting the Earth, and Weightlessness

Imagine an astronaut in orbit in a capsule 100 miles above the surface of the Earth. If the astronaut weighed 150 lb on the Earth, we showed that at the higher altitude, the astronaut weighs only 95% of that—that is, 142.5 lb, i.e., he is 7.5 lb lighter.

But aren't orbiting astronauts weightless? Movies show them floating around inside spaceships. How can they do that if they weigh almost as much as they do when they are on the surface of the Earth?

To understand the answer to this paradox, we have to think about what it means to be "weightless." Suppose that you are in an elevator, and the cable suddenly breaks. The elevator and you fall together. During those few seconds before you crash into the ground, you will feel weightless. You will float around inside the elevator. You will feel no force on your feet, and your shoulders will not feel the weight of your head. (Your head falls with your chest at the same rate, so the muscles in your neck aren't needed to keep your head above your chest.) In those brief seconds, you have the same "weightless" experience as the astronauts. All the time, of course, the Earth is pulling you rapidly toward it.[2] You have weight, but you feel weightless. A movie made of you in the elevator would show you floating around, apparently without weight, while you and the elevator fell together. You would look just like the astronauts floating around in the International Space Station.

[2] There are rides at some amusement parks that allow you to fall for long distances and experience weightlessness, at least for a small number of seconds. We'll calculate how many seconds in a later section.

Now imagine that instead of falling, the elevator is shot out of a gun with you inside and it flies 100 miles before hitting the ground. During that trip, you will again feel weightless. That's because you are in motion along with the elevator. You and it fly in the same arc.[3] Your head and chest are both moving in that arc together; there is no force between them and your neck muscles can be completely relaxed. Your head will seem to have no weight, at least to you. To an outside observer, you are falling.

Prior to sending them into orbit, potential astronauts were flown in airplanes following such arcs in order to see how they responded to the sensation of weightlessness and to get them used to it.

When you and the elevator are moving together under the force of gravity (either falling or shot in an arc), there seems to be no gravity. From that alone, you might think that you were far out in space, far away from the gravity of any planet, star, or moon. From inside the elevator, you can't tell the difference.

Now imagine that at the top of a very tall tower (200 kilometers high) is a large gun, pointing horizontally. Out of this gun, we are going to shoot the elevator, with you inside. If we pick a low velocity (e.g., 1 mile per second), you and the elevator will curve toward the Earth, and crash into it, as in path *A* in figure 3.1. But if instead we pick a higher velocity like 5 miles per second, you and the elevator will follow path *B*: you will curve toward the Earth, but because the Earth curves too, you will miss the edge of the Earth. You will keep on curving downward, but you will never hit. You are in orbit. The force of gravity makes the path of the elevator—let's call it a space capsule now—curve downward. But if that curvature matches the curvature of the Earth, then it misses the surface and stays at a constant height.[4]

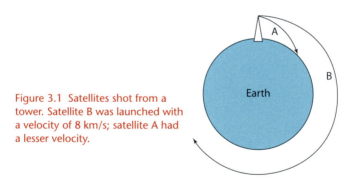

Figure 3.1 Satellites shot from a tower. Satellite B was launched with a velocity of 8 km/s; satellite A had a lesser velocity.

This may seem preposterous, but it is reasonable to think of an astronaut in orbit around the Earth as being in a state of perpetual falling. That's why he feels weightless.

You can think of the Moon as doing the same thing. It is attracted to the Earth by gravity, but it has high sideways motion. Even though it is falling toward the Earth, it always misses.

---

[3] Your path could be traced by a geometric curve known as a *parabola*.

[4] If your velocity is not exactly horizontal, or if your velocity is a little low or high, then the orbit will not be a circle but an ellipse.

The Velocity for Low-Earth Orbit (LEO)

To stay in a circular orbit just a few hundred miles above the surface, the velocity of a satellite must be about 8 km/s, which is about 18,000 mi/h, or about 5 miles per second. (The actual value depends slightly on the altitude; we'll derive this number from a calculation later in this chapter.) At this velocity, the satellite orbits the 24,000-mile circumference of the Earth in about 1.5 hours.

> **Remember:** A satellite orbiting in a LEO (low-earth orbit) goes 5 miles every second and takes 90 minutes, 1.5 hours, to go around the Earth.

If the astronaut inside a satellite wants to land, he does not fire his rockets in a direction away from the Earth; he fires his rockets toward the direction he is headed—in the forward direction! The force of the rocket exhaust slows down the satellite, so it is no longer going fast enough to miss the edge of the Earth. If he fires the rockets enough to stop the satellite completely, then the satellite will simply fall straight downward. Gravity brings the satellite back to the Earth. If the satellite moves faster than 5 miles per second, it will leave the circular orbit and head out into space. At about 7 miles per second, it will have sufficient velocity to reach the Moon and beyond. This velocity is called the *escape velocity*. We'll discuss this concept further later in this chapter.

Analogy with a Rock and Sling

There is another way to think about Earth satellites. Forget gravity for a moment. Imagine that you have a rock tied at the end of a string, and you are spinning it in a circle above your head. The string provides the force that keeps the rock from flying away, and this force also keeps the rock in circular motion. If the string breaks, the rock flies off in a straight line. Gravity does the same thing for an Earth satellite: it provides the force that keeps the satellite in a circular orbit.

An old weapon called the *sling* is based on this principle. A rock is held by a leather strap and spun in circles over the head. Arm motion helps it pick up circular speed. It is the strap that keeps the rock in circular motion. When the strap is released, the rock flies in a straight line toward its target. Such a sling was the weapon that, according to the Bible, David used to kill the giant Goliath.

In a similar manner, if we could suddenly "turn off" the force of gravity, the Moon would leave its circular orbit and head off in a straight line. It would not continue in a circle. The same is true for all the satellites now in orbit around the Earth. And with the Sun's gravity turned off, the Earth would head out into space too, at its previous orbital speed (around the Sun) of 20 miles per second.

Geosynchronous Satellites

Weather satellites and TV satellites have a very special orbit: they are geostationary. This means that they stay above the same location of the Earth at all times. It also means that the same weather satellite will be able to watch the development of a storm, or of a heat wave, continuously. And if you are receiving a signal for your TV, you never have to re-point the antenna because the satellite remains in the same direction above your house at all times.

How can this be, since satellites must orbit the Earth to avoid crashing back down? The answer is elegant: The satellites orbit the Earth at such a high altitude, 22,000 miles, where the gravity is weak, that they orbit at a relatively low velocity, and take 24 hours to go around the longer distance. Since the Earth rotates once in that period, if they orbit above the equator, they will stay above the same location. Both are moving—your home with the TV dish and the satellite—but their angle with respect to each other doesn't change.

There is a catch. If the satellite is to stay in the same location in our sky relative to the ground, it must orbit over the equator. Can you see why that is true? A geostationary satellite moves in a circle around the center of the Earth. If the satellite is not in an equatorial orbit, then it will spend half of its orbit in the Northern Hemisphere and half in the Southern. Only if it orbits above the equator can it stay precisely above the same Earth location at all times.

As a result, all geostationary satellites are right above the equator. If you look at them up in the sky, they all line up in a narrow arc. That creates a problem. If the satellites are too close to each other, their radio signals can interfere. So international treaties are required to divide up the space.

Sometimes you will hear these satellites referred to as *geosynchronous* satellites. NASA uses the term *geosynchronous* to refer to any satellite that orbits the Earth in 24 hours, including ones in polar orbits. If a satellite is above the equator, then it is called geostationary. In this book, I won't bother with that fine distinction. NASA engineers need to know the difference, but future presidents don't.

The problem with geosynchronous satellites is that they are so far away. 22,000 miles is over 5 times the radius of the Earth. From way out there, the Earth looks small. That's good, if you are trying to watch weather patterns. One photo can show an entire ocean. Moreover, you can have your weather satellite devoted to a special region, such as the East Coast of the United States, or Europe. But recall, it can't be directly above the United States or Europe, since it has to be above the equator. Look at weather satellite photos and see if you can tell where the satellite is located. You'll see that it isn't looking straight down, but at an angle from its location above the equator.

Geosynchronous satellites are also useful for TV broadcasts. That's because the antenna on the ground, which is usually a microwave dish, has to be pointed only once. Set it and forget it. The disadvantage is that the satellite transmitter must be extra powerful to send a signal 22,000 miles. For commercial satellite broadcasting, the pointing advantage has won out over the power disadvantage.

Satellite TV is even more popular in developing nations than in the United States. That's because it can be set up by an international company without having to build elaborate infrastructure within the country. It can broadcast to nearly half the world at the same time. I recall being really surprised when I visited the ancient and almost unchanged city of Fez in Morocco. The old part of the town has streets and buildings that have not changed for a thousand years—except, that is, for all the satellite dishes on the roofs.

Here is an unusual application for geosynchronous satellites: If you are kidnapped, and you don't know where you have been taken, try to spot a satellite dish. If the dish is pointing straight up, then you know you are on the equator. If it is pointing horizontally, then you know you are at the North or South Pole. But be careful. Even at the equator, the satellite doesn't have to be overhead. The satellite could be above the Congo, and you could be in Brazil. So you really have to determine the direction of north to make good use of the satellite dish information.

Spy Satellites

Spy satellites carry telescopes to look down on the surface of the Earth and see what is going on. They were once used exclusively by the military, to see the secrets of adversaries, but now they are widely used by government and industry to look at everything from flooding and fires to the health of food crops.

The ideal spy satellite would stay above the same location all the time. But to do that, it must be geosynchronous, and that means that its altitude is 22,000 miles. At those distances, even the best telescopes can't resolve things smaller than about 20 feet. (We'll derive that number when we discuss light.) That means that such a spy satellite could see a football stadium, but it couldn't tell if a game were being played. Such satellites are good enough to watch hurricanes and other weather phenomena, but they are not useful for fine details, such as finding a particular terrorist.

Thus, to be useful, spy satellites must be much closer to the Earth. That means that they must be in low-Earth orbit (LEO), no more than a few hundred miles above the surface. But if they are in LEO, then they are not geosynchronous. In LEO, they zip around the 24,000-mile circumference of the Earth in 1.5 hours; that gives them a velocity relative to the surface of 16,000 miles per hour. At this velocity, they will be above a particular location (within 100 miles of it) for only about a minute.[5] This is a very short time to spy. In fact, many countries that want to hide secret operations from the United States keep track of the positions of our spy satellites and make sure that their operations are covered over or hidden during the brief times that the spy satellite is close enough to take a photo.

LEO satellites cannot hover. If they lose their velocity, they fall to the Earth. If you want to have continuous coverage of a particular location, you must use an airplane, a balloon, or something else that can stay close to one location.

The military and intelligence agencies are developing quiet, high-flying drones (airplanes without a pilot) to do the most critical kind of spying. But these vehicles can be shot down more easily than a satellite.

GPS—a Medium-Earth Orbit (MEO) Satellite

One of the wonders of the last decade is the Global Positioning System (GPS). A small GPS receiver costs under $50, and it will tell you your exact position

---

[5] At 16,000 mi/h = 4.4 mi/sec, it will go 200 miles in 45 seconds.

on the Earth within a few meters. I've used such a receiver in the wilderness of Yosemite, in the souks of Fez, and in the deserts of Nevada. You can buy a car with a built-in GPS receiver that will automatically display a map on your dashboard showing precisely where you are. The military uses GPS to steer its smart bombs to land at just the location they want, within a few meters. Many cell phones now include a GPS system.

A GPS receiver picks up signals from several of the 24 to 32 GPS satellites orbiting the Earth. It is able to determine the distance to each satellite by measuring the time it took for a signal to go from the satellite to the receiver. Once the receiver has measured the distance to three satellites, it can then calculate precisely where on Earth it is located.

To understand how the GPS receiver calculates its position, consider the following puzzle: A person is in a U.S. city. He is 800 miles from New York City, 900 miles from New Orleans, and 2200 miles from San Francisco. What city is he in?

Look on a map. There is only one city that has those distances, and that is Chicago. Knowing three distances uniquely locates the position. GPS works in a similar manner, but instead of measuring distances to cities, it measures distances to satellites. And even though the satellites are moving, their locations when they broadcast their signals are known. They tell the computer in your GPS receiver where they are, so the GPS receiver can figure out where it is.

You might expect that the GPS satellites would be in geosynchronous orbits. But they are not, primarily because such a large distance would require that their radio transmitters have much more power to reach the Earth. They were not put in low orbit (LEO) because they would often be hidden from your receiver by the horizon. (Each one would be in view for only a few minutes if it passed 100 miles overhead.) So they were placed in a medium-Earth orbit (MEO) about 12,000 miles high. They orbit the Earth every 12 hours.

Using Gravity to Search for Oil

I said earlier that every object exerts a small gravitational force on every other object. Remarkably, measurement of such small forces has important practical applications. If you are standing over an oil field, you feel slightly less gravity than the gravity you feel standing over solid rock. Oil is less dense than rock; it has less mass (per cubic kilometer), and so its gravity isn't as strong.[6] Such small gravity changes can even be measured from airplanes flying above the ground. An instrument can make a *gravity map* that shows the density of the material under the ground. Maps of the strength of gravity, taken by flying airplanes, are commonly used by businesses to search for oil and other natural resources.

A more surprising use of such gravity measurements was to make a map of the buried Chicxulub crater on the Yucatan peninsula, the crater left behind

---

[6] Optional note: The gravity of a spherical object acts as if it all originates from the center of the object. But this is true only if the sphere's mass is uniformly distributed. If we treat the Earth as such a sphere, then you can mathematically think of the mass of an oil field as a sum of a uniform Earth and a little bit of "negative" mass that cancels out some of the gravity. If you are close to the oil field, you will sense the reduced gravity because this little bit of negative mass will not attract you as much as if it were denser rock.

when an asteroid killed the dinosaurs. The crater was filled in by sedimentary rock that was lighter than the original rock, so even though it is filled, it shows a gravity "anomaly"—i.e., a difference from what you would get if the rock were uniform. An airplane flying back and forth over this region made sensitive measurements of the strength of gravity, and they are represented in a map generated by a computer.[7]

In this map, the tall regions are regions in which the gravity is stronger than average, and the low regions are locations where the gravity is slightly weaker. The crater shows several concentric circles, with the largest over 100 kilometers in diameter. The inner rings probably formed when material from under the huge crater was forced upward, partially filling it.

Manufacturing Objects in Space

When the space program began, many people thought that there would be significant advantages to being in a weightless environment. In a satellite, things wouldn't sag under their own weight. It might be possible to make better (rounder) ball bearings or to grow more perfect crystals (used in computers and other electronics) in a weightless environment.

This promise has been largely unfulfilled. The additional cost of doing the work in a satellite has not turned out to be worth it. It costs about $10,000 to launch a kilogram of anything into orbit. In the near future, commercial companies hope to reduce that price to $1,000/kg. It's hard to make a profitable factory in space when it costs that much just to get there. There is no reason in principle why getting to space must be expensive; we'll show in a later section that the energy required is only 15 Cal/g. Some time in the future, if travel to space becomes as cheap as an airplane ride, then the idea of factories in orbit may be more likely.

Escape to Infinity

Suppose that you want to completely leave the Earth. Just going into orbit isn't enough; you want to get far away, maybe take a trip to the Moon, Mars, or a distant star. That takes more energy than just going into orbit. How much more? The answer is surprisingly simple: exactly twice as much. You can get that much kinetic energy by going 1.414 times faster than orbital speed. (That's because energy depends on the square of the velocity, and $1.414^2 = 2$.) Since orbital velocity is 5 miles per second, the escape velocity is $5 \times 1.414 \approx 7$ miles per second $\times 11$ km/sec.

These are good approximate numbers to know:

> Orbital velocity: 5 miles per second = 8 km/sec (for LEO)
> Escape velocity: 7 miles per second = 11 km/sec

Just like orbital velocity, the escape velocity doesn't depend on the mass of the object.

---

[7] The map can be seen at: http://media.skyandtelescope.com/images/Chicxulub_Crater_l.jpg).

How much energy is there in an object moving that fast? That can be calculated from the kinetic energy equation $E = 1/2\ mv^2$. The answer is interesting: for each gram, the kinetic energy is about 14 Calories.[8] That's only 40% more than the energy in a gram of gasoline. So in principle, you could send a gram into space with the fuel of 1.4 grams of gasoline, assuming that you didn't have to accelerate the gasoline (but used it, for example, in a large gun). As you'll see when we discuss rockets, they do accelerate most of their fuel, and that wastes 97% of the energy in the fuel they use.

Science Fiction Planets

One of the most common "errors" in science fiction movies is the implicit assumption that all planets in all solar systems have a gravity about equal to the Earth's gravity. There is no reason why that should be so. Pick a random planet, and you are just as likely to be a factor of 6 times lighter (and bouncing around like astronauts on the Moon) or 6 times heavier and unable to move because of your limited strength. Imagine a person who weighs 150 pounds on the Earth trying to move on a planet where he weighs 900 pounds.

Falling—and the g Factor

You'll note that the force of gravity depends on the mass of the object. A heavier object has a greater force on it. So why do all objects fall at the same speed?

The answer is that heavier objects have a bigger force but that they *need* a bigger force to accelerate. It takes more force to push a big mass than a small one. The precise statement of this principle is called Newton's second law: the acceleration of an object is given by

$$a = F/m$$

This law is usually written: $F = ma$. I just solved for a. For gravity, the F is proportional to the mass m. So if you double the mass of the object, the force is doubled, so the acceleration is the same! In fact, even if you triple or quadruple the mass, the acceleration will still be the same. All objects are accelerated together.

At the surface of the Earth, acceleration of everything is called the *acceleration of gravity* and is given the symbol g. The actual value of this number is:

The acceleration of gravity:
g = 22 mph increase in velocity every second
= 32 feet per second increase each second
= 10 meters per second increase per second

That means that when you are falling, your speed increases by 22 mph every second. Fall for 1 second, and your speed is 22 mph. Fall for 2 seconds, and your speed is 44 mph. Fall for 3 seconds, and it is 66 mph.

[8] To use the equation, we first convert everything to kg and m/sec. The mass of 1 gram is 0.001 kg. Escape velocity is 11,000 m/sec. $E = 1/2\ (.001)(11,000)^2 = 60,000$ joules = 14.5 Calories.

What about running? A world-class male sprinter can do the 40-meter dash in 4.4 seconds. His average acceleration is 4.1 meters per second (mps) every second, or about 9.4 mph every second. That's about 40% of a g.

How far does an object fall? Here is the equation: $H = 1/2 \, gt^2$. Don't bother memorizing this, but I will use this equation in the next section.

The X Prize

As mentioned at the start of this chapter, the $10 million X Prize was awarded in 2004 to the first private company to send a rocket to an altitude of 100 km. Here is an interesting question: What velocity would be needed to send the rocket that far?

Here is a nice way to solve the problem. The velocity needed is exactly the same as the velocity you would get from an object falling 100 km. I calculate that in the optional section that follows. The answer is 0.86 miles per second.

OPTIONAL: CALCULATION: VELOCITY OF A 100-KILOMETER FALL

What is the speed to reach 100 km? I use the equation $H = 1/2 \, gt^2$ from the previous section and solve for t, to get $t = sqrt(2H/g)$. Putting $H = 100$ km = 100,000 m, $g = 10$ mps per sec, I get $t = $ sqrt(2 × 100,000/10) = sqrt(20,000) = 141 seconds. A falling object gains a velocity of 22 mph every second, so the falling object will go 141 × 22 = 3102 mph. (We ignore air friction.) That sounds like a lot, but only because it is given in miles per hour. Convert it to miles per second by dividing by the number of seconds in an hour, 3600, and you get that it is only 0.86 miles per second.

Compare 0.86 mps to the speed you need to get into orbit: 5 mps. So the X Prize winner would have needed to go 5/0.86 = 5.8 times faster to get into orbit.

How much more energy does that take? Remember that the energy depends on the square of the velocity. So a velocity that is 5.8 times greater would have an energy that is $5.8^2 = 34$ times greater. The X Prize–winning rocket had 34 times too little energy to get into orbit. If shot from a gun, it would have re-quired 34 times more fuel. If you use a rocket, then you might need more than 34 times as much fuel, since the rocket must carry much of the fuel with it during the initial acceleration. We'll discuss this later in this chapter, when we talk about rockets. In the end, to get into orbit, you wind up needing a very big and expensive rocket. The size of the space shuttle rocket was not determined by government stupidity or extravagance, but by physics.

There are potentially cheaper ways to get to space than using rockets. We'll talk about some of them later in this chapter too.

Air Resistance and Fuel Efficiency

A moving car feels the force of air on its front, and that tends to slow the car down. To keep going at the same velocity, the engine must make up the lost en-ergy. We'll show that much of the gasoline used by the car is to overcome the force of this air resistance. That is an important fact to know. At a velocity of 30 m/sec = 67 mi/h, the force of air on the front of the car will be about 500 lb!

To keep the car from being slowed down by this force, the engine must exert an equal and opposite force. That takes a lot of gasoline. At high velocities, more than 50% of the gasoline is used to overcome this air resistance. As a result, car designers have worked hard to "streamline" the shape of automobiles. If the front surface of the car is tilted (rather than flat, as in the old autos from the 1920s), then the force of the air resistance is reduced. On a tilted surface, air molecules can bounce off obliquely instead of hitting the front and bouncing straight back. In such a car, the force can be as low as 100 lb.

Many truck drivers are in business for themselves, and they have to pay for the extra gasoline used to overcome air resistance. Maybe you've noticed the smooth curves that some truck drivers have added to the cabs of their trucks to minimize air resistance. Reducing this force can save substantial money on gasoline. The top of the cab has had a contoured shape added to it (called a *cab fairing*) to make the air bounce off smoothly, at an angle, instead of hitting the flat face of the truck head on (figure 3.2). This type of alteration is sometimes called *aerodynamic smoothing*, and it saves gasoline. Instead of air bouncing off a surface aimed right at it, it is only deflected to the side, and that reduces the force of resistance.

Note also that you save even more gasoline by driving slower. At 1/2 the speed, the force is 1/4 as much. That means that the vehicle uses 1/4 as much

Figure 3.2 Truck with cab fairing to reduce air resistance. (Photo courtesy of NASA.)

gasoline to overcome air resistance. Likewise for ordinary cars: drive slower, and you'll get more miles per gallon, because even though you travel through the same amount of air, you exert less force on the air.

The "g-Rule"

There is a very good reason to think of acceleration in terms of g: it enables you to solve important physics problems in your head. Suppose that I told you that I am going to accelerate you in a horizontal direction by $10\,g$. How much force will that take? The answer is simple: 10 times your weight! When astronauts are accelerated by $3\ g$ (the maximum acceleration of the space shuttle), the accelerating force must be 3 times their weight. I call this the "g-rule":

The force to accelerate an object
= the number of g's times the weight

To get the number of g's, just calculate the acceleration in m/s per sec, and divide by 10. (Some books use the more precise value of 9.8.) So for example, we write that an acceleration of $a = 20$ m/s per sec $= 2\ g$. This is an acceleration of $2\ g$. To accelerate an object by 20 m/s each second (so that its velocity is 20 m/s after the first second, 40 m/s after the 2nd second, etc.) takes a force equal to twice the weight of that object.

The g-rule is actually just a different way to write Newton's second law. I mentioned earlier that this law reads: $F = ma$, where F is the force in a unit called the Newton (about the weight of an apple), m is the mass measured in kilograms, and a is the acceleration measured in mps per second. But this equation isn't very useful unless you get accustomed to using Newtons instead of pounds as a measure of force. In my g-rule version, you can use any units you want.

Optional: Can you see why the g-rule is identical to $F = ma$?

RAIL GUN

To go into orbit around the Earth, a satellite must have a velocity of 8 km/s. Why not give it this velocity in a gun? Could we literally "shoot" a satellite or an astronaut into space?

The answer is: You might be able to do this, but the astronaut would be killed by the force required to accelerate him. Let's assume that we have a very long gun, an entire kilometer long. If we calculate the required acceleration to shoot the astronaut into space, we get $a = 3200\ g$—i.e., the acceleration is 3200 times the acceleration of gravity.[9]

That's a lot of g's. Remember the g-rule: the force that it will take to accelerate you by $3200\ g$ will be 3200 times your weight. So if you weigh 150 lb, the

---

[9] Optional calculation: The distance $D = 1$ km $= 1000$ m, and the velocity $v = 8$ km/s $= 8000$ m/s. The relationship between distance, acceleration, and velocity is $v^2 = 2aD$. (We didn't cover that equation in this text, but you'll find it in other physics books.) Put in v and D, and solving for a gives $a = 3200$ m/s per sec. To convert this to g's, divide by $g = 10$ to get $a = 32000/10 = 3200\ g$.

force that must be applied to you is $3200 \times 150 = 480{,}000$ lb $= 240$ tons. That is enough to crush your bones.

Suppose that we want to experience no more than $3\ g$. How long would the gun have to be to get you going at $v = 8$ km/s? The gun would have to be 1000 km long! That is, of course, ridiculously impractical.

Of course, acceleration over such long distances is not impractical—it's what the space shuttle does. It takes off, it accelerates at $3\ g$, and it must go over 1000 km to reach orbital speed. So the space shuttle accelerates as if it were in a very long gun, but without a barrel. This process actually takes the space shuttle farther than 1000 km since $3\ g$ is only the peak acceleration; for most of the flight, the acceleration is less.

ACCELERATION DURING AIRPLANE TAKE-OFF

The take-off speed for a commercial airplane is about 160 miles per hour. What acceleration is needed to achieve this speed on a 1-kilometer runway? Let me make that question a little more personal. You are sitting in such an airplane. In a few seconds, you will be at the other end of the runway, and you will be moving (along with the airplane) at 160 miles per hour. What force does the seat have to apply to your back to accelerate you?

I show in the footnote[10] that this requires the airplane to accelerate at 2.6 m/s per sec. We can convert this number to g units by dividing by 10, to get that the acceleration is $2.6/10 = 0.26\ g$. So the force that the airplane must push on you to get you going that fast is 0.26 times your weight, i.e., about a quarter of your weight. If you weigh 150 lb, then the push you feel on your back will be about 39 lb. Think of this the next time you are in an airplane taking off. Does this number feel about right?

Circular Acceleration

Physicists like to define velocity as having *magnitude* and *direction*. If your velocity changes, we call it *acceleration*. But suppose that you change only your direction and not the actual number of meters per second? We still call that acceleration because many of the equations we've been using still work.[11]

The most important example of this kind of "acceleration" is when you go in a circle. If the magnitude of your speed is v, and this doesn't change (you keep going the same number of meters per second or miles per hour), and the circle has radius R, then we say that the circular acceleration is

$$a = v^2/R$$

---

[10] I convert 160 miles per hour to meters per second by dividing by 2.2. So the airplane's final speed is $160/2.2 = 73$ m/s. I use the same equation that I used in the prior footnote, $v^2 = 2aD$, so $a = v^2/(2D)$. Put in $v = 73$, and $D = 1000$, for $a = 73^2/2000 = 2.6$ m/s^2.

[11] Optional for people who have studied vectors: Velocity is defined in physics as a vector. If the velocity at time t_1 is $\mathbf{v}_1$, and at t_2 is $\mathbf{v}_2$, then the acceleration vector is defined as $(\mathbf{v}_2 - \mathbf{v}_1)/(t_2 - t_1)$. Even if the velocity is changing only in direction, the difference vector $(\mathbf{v}_2 - \mathbf{v}_1)$ is not zero. For circular motion, its magnitude is given by the equation in the text.

This kind of acceleration is very important for the fighter pilot who is trying to change his direction rapidly. For example, if he is moving at velocity of 1000 mi/h and is turning in a circle of radius $R = 2$ km, then his acceleration turns out to be 10 g. That's about as much as a fighter pilot or an astronaut can tolerate. We do the calculation in an optional footnote.[12] The problem is that at 10 g, a human's blood pressure is not great enough to keep the blood in the brain, and that causes the pilot to faint. Perhaps you've seen a movie of a pilot being spun in a huge cylinder as a test of what accelerations the pilot can stand before fainting.

High g, from circular acceleration, is also a method used to separate the components of uranium for making a nuclear weapon. Such a device is called a *centrifuge*. The heavier parts of the uranium feel a greater force, and they are pulled more strongly toward the outer parts of the spinning cylinder. Such centrifuges are mentioned often in the news. In 2004, Libya disclosed the fact that it had purchased large centrifuge plants for the production of nuclear weapons material. Parts of such centrifuges have been found hidden in Iraq. We'll talk more about these centrifuges in chapter 5.

Gravity in Space—According to Science Fiction

Science fiction movies, such as *2001: A Space Odyssey* (1968), sometimes show a large rotating satellite. There is then "artificial gravity" from the rotation, and astronauts can walk around on the outside. This actually makes sense. The astronauts will feel a force on their feet (which point away from the center of rotation) that will appear to them indistinguishable from gravity. I'll show in an optional footnote that the edge of a satellite with a 200-meter radius must be moving at about 44 m/s.[13] That means that it will rotate once every 40 s. This matches the rate seen in the *2001* movie.

Many science fiction movies show space voyagers in spaceships walking around as if on the Earth, even when the spaceship is not rotating. Is that nonsense? Where does that gravity come from?

I can make sense out of it by assuming that the spaceships are not moving at constant speed. If the ship engine accelerates the ship by $a = 10$ m/s each second $= g$, then the ship will put a force on any astronauts inside to accelerate them too. A person of mass m will feel a force $F = ma$. But since the engines have set $a \approx g$, the force on an astronaut will be $F = mg$. That is exactly his Earth weight. If he places his feet on the back surface of the ship (opposite in direction to the path the ship is taking), then he will feel like he is standing on the Earth.

---

[12] Optional calculation: Acceleration of fighter pilot. First we convert to metric units. We convert miles per hour to meters per second by dividing by 2.24, giving $v = 1000/2.24 = 446$ m/s. We also convert R to meters: $R = 2000$ m. Plugging in these values gives $a = (446)^2/2000 = 100$ m/s^2. We convert this to g's by dividing by 10 to get $a = 100/10 = 10$ g.

[13] Optional calculation: If the satellite has radius R and is rotating with a rim velocity v, then the force on the astronauts' feet will be $F = ma$. If we use the formula for circular acceleration, $a = v^2/R$, then this becomes: $F = mv^2R$. To make this equal to the astronauts' weight, we set $F = mg$. Solving for v, we get $v = $ sqrt(gR). For a satellite with a radius of $R = 200$ meters, the rim velocity must be $v = $ sqrt$(10 \times 200) = 31$ m/s.

In fact, he can't even tell the difference; the acceleration serves as a "virtual" gravity.[14] He can even stand on the side of the ship, if the ship is being accelerated sideways by 1 g.

Here is an interesting number: the distance a spaceship would travel in a year if it had constant acceleration of g. I do the calculation in an optional footnote.[15] The answer is 5×10^{15} m, which is about a half light-year (the distance that light travels in a year). The distance to the nearest star (not counting the Sun) is about 4 light years.

Black Holes

Big planets have high escape velocities. For Jupiter, the escape velocity is 61 km/s. For the Sun, it is 617 km/s. Are there any objects for which the escape velocity is higher than the speed of light, 3×10^5 km/s $= 3 \times 10^8$ m/s? The surprising answer is yes. We call such objects *black holes*.[16] They get their name from the fact that even light cannot escape, so we can never see their surface—they are black. I'll show in chapter 12 that no ordinary object (made out of mass as we know it) can go faster than the speed of light. That means that nothing could escape a black hole.

If the black hole is invisible, how do we know it's there? The answer is that even though it can't be seen, we can see the effects of its very strong gravity. Even when there is nothing visible, the gravity of the black hole is so strong that we know there must be something of great mass present, so we deduce it must be a black hole.

To be a black hole, you have to do one of two things: either have a lot of mass, or pack a moderate amount of mass into a very small radius. Several black holes are known to exist: even though we can't see them (no light leaves the surface), we know that they are there because of the strong gravitational force they exert. The known black holes are all as massive as a star or greater.

---

[14] Movies sometimes show the virtual gravity as being off to the side (as when the astronauts are standing on the side of the ship and looking forward out a window). This could be accomplished by having special sideways thrusters. Every hour or so, the thrusters could rotate the ship around so that it doesn't wind up going too far sideways.

[15] Optional calculation: The distance that you travel at acceleration g is given by the following distance equation: $D = 1/2\ gt^2$. We put in $g = 10$. For standard physics units (MKS, or meters, kilograms, and seconds), we need to have t in seconds. We can calculate the number of seconds in a year as follows: 1 year = 365 days = 365×24 hours = $365 \times 24 \times 60$ minutes = $365 \times 24 \times 60 \times 60$ seconds = 3.16×10^7 seconds. Plugging in this value, we finally get $D = (1/2) \times 10 \times (3.16 \times 10^7)^2 = 5 \times 10^{15}$ m. One light-year is the speed of light (3×10^8 m/s) multiplied by the number of seconds in a year (which we just showed was 3.16×10^7 s). That gives 9.5×10^{15} meters.

[16] Why is the physics of black holes included in a text titled *Physics and Technology for Future Presidents*? Is there any practical use for this knowledge? The answer is no, not really. It is included, as are a few other things in this book, just because most people have heard about them and are curious. But you never know. Knowing the size of a black hole once won the author of this book a free guide to Paris. Outside of Shakespeare Books on the west bank of the Seine was a sign offering this prize if anyone could answer the question, "What size would the Earth have to be for it to be a black hole?"

But if you were to take the mass of the Earth and pack it inside a golf ball (in principle, this is possible), then it would be a black hole. The mass would be the same, but the radius would be so small that the gravitational force on the surface of the golf ball would be enormous.

The Sun is more massive, so you don't have to pack it so tightly. It would be a black hole if you packed it inside a sphere with a radius of 2 miles.

As noted earlier, the force of gravity on an object of mass m (e.g., you) exerted by a spherical object of mass M (e.g., the Sun) is given by $F = GMm/r^2$. (No, you still don't need to know this equation.) Notice that the size of the object (the Sun) doesn't enter the equation. That means that even if the Sun were compressed into a black hole, its gravity at the distance of the Earth wouldn't change, because its mass M didn't change.

Of course, if the Sun became a black hole, it would have a tiny radius of only 2 miles. The gravity on the surface of the black-hole Sun would be much greater than the gravity on the surface of the present sun, because r is so much smaller. That's why black holes have a deserved reputation for strong gravity: they are so small that you can get very close to a large amount of mass.

There are several black holes that were created, we believe, when the inner part of a star collapsed from its own weight into a very small radius. The object in the sky known as Cygnus X-1 is thought to be a black hole from such a collapse. If you have spare time, look up Cygnus X-1 on the Internet and see what you find.

Many people now believe that large black holes exist at the center of the large collections of billions of stars known as galaxies, such as the Milky Way. We presume that these were created when the Milky Way formed, but we know almost none of the details about how this happened.

Even more remarkably, the Universe itself may be a black hole. That's because the black-hole radius for the Universe is about 15 billion light-years, and that is approximately the size of the observable Universe. In other words, the Universe appears to satisfy the black-hole equation. But you probably won't be surprised at one inescapable[17] conclusion: we can never escape from the Universe.

Momentum

If you shoot a powerful rifle, then the rifle puts a large force on the bullet sending it forward. But the bullet puts a backward force on the rifle, and that is what causes the "kick." The rifle can suddenly go backward so rapidly that it can hurt your shoulder. If you don't have your feet firmly planted on the ground, you will be thrown backward.

Recall that we mentioned Newton's third law earlier—the fact that if you push on something, it pushes back on you. Newton stated this as "for every action, there is an equal and opposite reaction." But we no longer use this old terminology of *action* and *reaction*. Instead, we say that if you push on an object (such as a bullet), then the bullet also pushes back on you with the same

---

[17] Pun intended.

force for exactly the same amount of time. Of course, the bullet is lighter, so it is accelerated much more than the gun is. That's $a = F/m$ again. Same force on the gun and the bullet, but the bullet is lighter, so the acceleration will be greater.

Based on the fact that the bullet pushes on the rifle for the same time that the rifle pushes on the bullet, we can derive an extremely important equation, sometimes called the *conservation of momentum*.[18] For the rifle (subscripts R) and the bullet (subscripts B), the equations are

$$m_B \, v_B = m_R \, v_R$$

The equation for the rifle recoil is simple: the mass of the bullet times its velocity is the same as the mass of the gun times its recoil velocity. Of course, the velocities are opposite; this is sometimes indicated by putting a minus sign in front of one of the velocities. (I didn't bother to do that here.) When the gun is stopped by your shoulder, then you recoil too—but less, because you have more mass.

The product mv is the quantity called the *momentum*. As I mentioned earlier, one of the most useful laws of physics is the conservation of momentum. Before the rifle was fired, the bullet and gun were at rest; they had no momentum. After the rifle fired, the bullet and the rifle were moving in opposite directions, with exactly opposite momenta (the plural of *momentum*), so the total momentum was still zero.

Here's another way to say that: When you fire a gun, the bullet gets momentum. You and the rifle you are holding get an equal and opposite momentum. If you are braced on the ground, then it is the Earth that recoils with that momentum. Because the Earth has large mass, its recoil velocity is tiny and difficult to measure.

If the objects are in motion before the force acts, then conservation of momentum means that the *changes* in momentum must be equal and opposite. Let's apply that to the comet that crashed into the Earth and killed the dinosaurs. To make the calculations easy, assume that before the collision the comet with mass m_C was moving at v_s = 30 km/s (a typical velocity for objects moving around the Sun). Assume that the Earth was at rest. After the collision, the total momentum would be the same. The Earth would have mass m_E (that now included the mass of the comet) and have velocity v_E. So we can write

$$m_C \, v_C = m_E \, v_E$$

Solving for v_E, we get

---

[18] Optional derivation: If an object is at rest, and it experiences a force F for a short time t, then its velocity will increase by $v = at = (F/m)t$. Write this as $mv = Ft$. If there are two objects, and the forces are equal and opposite, and the time is exactly the same, then Ft is equal and opposite for the two objects, and that means that the quantity mv must also be equal and opposite for the two objects. So every little bit of mv given to one object will be balanced by an equal mv given the other. The total change in mv will be the same for both.

$$v_E = m_C v_C / m_E$$

The mass of the comet[19] is about 10^{19} kg. Everything else is known, so we can plug into this equation and get the v_E, the velocity of recoil of the Earth:

$$v_E = (10^{19})(30,000)/(6 \times 10^{24})$$
$$= 0.05 \text{ m/sec}$$
$$= 5 \text{ cm/sec}$$
$$= 2 \text{ in/sec}$$

That's not much of a recoil, at least when compared to the prior velocity of the Earth, which is 30 km/s $\approx$ 1,000,000 inches per second. So the Earth was hardly deflected. It's orbit changed, but only by two parts in a million.

How much does a truck recoil when hit by a mosquito? Not much. In the following optional section, I calculate that the mosquito slows the truck by a fraction of a micron per second.

OPTIONAL: CALCULATION: MOSQUITO IMPACT ON A TRUCK

Let me estimate how much a truck recoils when it is hit by a mosquito. Assume, from the point of view of the truck, that a mosquito with mass $m = 2.6$ mg = 2.6E–5 kg, is moving at 60 mi/h = 27 m/s. Assume that the truck weighs 5 metric tons = 5000 kg. I use the same equation, except I let the 2.6-milligram mosquito represent the comet:

$$v(\text{truck}) = m(\text{mosquito}) \times v(\text{mosquito})/m(\text{truck})$$
$$= 2.6E–5 \times 27/5000 \text{ m/sec}$$
$$= 1.4E–7 \text{ m/sec}$$
$$= 0.14 \text{ micrometers per second}$$

Although the conservation of momentum is one of the most important laws of physics, it is violated in many action movies. For example, if the hero in *The Matrix* (1999) punches the villain, and the villain goes flying across the room, then the hero should go flying backward (unless he is braced on something big and massive). Likewise, small bullets, when they hit a person, seem able to impart very large velocities to the person that they hit, so the person goes flying backward.[20] In reality, any such momentous bullet would simply rip a hole in the target and come out the other side.

[19] A comet with a radius $R_C = 100$ km = 10^5 m has a volume of $V_C = 4/3\pi R_C^3$ (for a sphere) = 4.2×10^{15} cubic meters (m³). Assuming that the comet is made mostly of rock and ice, the density is probably about 2500 kg/m³, so the mass is $2,500 \times 4.2 \times 10^{15} = 10^{19}$ kg.

[20] As a fan of this movie, I explain to myself that according to the script, "reality" is just a computer program called *the matrix*. Therefore I can assume that whoever programmed *the matrix* simply forgot to put in conservation of momentum. (Or, possibly, Neo and the others altered the program to violate the laws of physics.)

Rockets

Imagine trying to get into space by pointing a gun downward and firing bullets so rapidly that the recoil pushes you upward. Sound ridiculous? Yet that is exactly how rockets work (figure 3.3).

Figure 3.3 Rocket. (Courtesy of NASA.)

Rockets fly by pushing burned fuel downward. If the fuel has mass m_F and is pushed down with a velocity v_F, then the rocket (which has mass m_R) will gain an extra upward velocity v_R given by the same kind of equation we used for the rifle:

$$v_R = v_F \, m_F / m_R$$

Compare this to the rifle equation and to the comet/Earth collision equation.

Because the rocket weighs so much more than the fuel that is expelled every second (i.e., m_F/m_R is tiny), the amount of velocity gained by the rocket is much less than the fuel velocity. As a consequence, rockets are a very inefficient way to gain velocity. We use them to go into space only because in space there is nothing to push against except expelled fuel. Another way to think of this is as follows: rockets are inefficient because so much of the energy goes into the kinetic energy and heat of the expelled fuel, rather than into the kinetic energy of the rocket.

The preceding equation gives the velocity *change* when a small amount of fuel is burned and expelled. To get the total velocity given the rocket, you have to add up a large number of such expulsions. Meanwhile, the mass of the rocket (which is carrying the unused fuel) is changing as fuel is used up. A typical result is that the rocket must carry huge amounts of fuel. The mass of fuel used is usually 25 to 50 times larger than the payload put into orbit.

For a long time, this huge fuel-to-payload ratio led people to believe that shooting rockets into space was impossible—after all, how could you even hold the fuel if it weighed 24 times as much as the rocket? The problem was solved by using rockets with multiple stages, so that the heavy containers that held the initial fuel never had to be accelerated to the final orbital velocity. For example, the space shuttle has a final payload (including orbiter weight) of 68,000 kg = 68 tons, but the boosters and the fuel weigh 1,931 tons, a factor of 28 times larger.[21] Of course, the booster never gets into orbit, only the much smaller shuttle.

Balloons and Astronaut Sneezes

When you inflate a balloon and then release the end, the balloon goes whizzing around the room, driven by the air being pushed out the end. What is happening is nearly identical to the way a rocket flies. Before you release the balloon opening, the balloon and the air have zero total momentum. When you let the air come out, it rushes out with high speed, pushed by the compressed air in the balloon, and it pushes back. It pushes back on that air, and that air pushes on the balloon. The released air goes one way, and the balloon containing its remaining air goes the other way, and the two momenta cancel.

When you sneeze, the sudden rush of air outward likewise can push your head backward. In the opening of this chapter, there was a puzzle: how much does the astronaut's head snap back? I read in a newspaper once that the head would snap back at a dangerous speed because the head was weightless in space. That's not true, of course. The writer of the article had confused weight with mass. The astronaut's head has no weight, but it has every bit as much mass as it did on Earth. The force of the sneeze will accelerate the head backward by an amount given by $F = ma$, but since m is the same as on the Earth, the acceleration a will be no greater.

Skyhook

Ponder the space shuttle. To put a 1-gram payload into orbit requires 28 grams of extra weight (fuel + container + rocket). For a rocket that is going at escape velocity, the efficiency is even worse.[22] But let's assume this number, and compare it to the energy it would take to lift an object to the same altitude. Suppose that we build a tower with an elevator that goes all the

---

[21] The external tank holds 751 tons of fuel, and there are two solid rocket boosters that weigh 590 tons each, for a total of 1931 tons.

[22] That's because the rocket has to go much faster than its spent fuel is ejected, so that this fuel has high velocity in the upward direction. Its kinetic energy is wasted.

way up to space.[23] How much energy would it take to haul the gram up to the top, using the elevator? According to the earlier section "Escape to Infinity," the energy required to take a gram of material to infinity is 15 Calories. That's the energy in 1.5 grams of gasoline (not including the air). So using a rocket takes 28/1.5 = 19 times more fuel using a rocket rather than using an elevator.

Many people have pondered the fuel waste from rockets. Although a tower to space seems impossible, it may make sense to hang a cable down from a geosynchronous satellite and use it to haul payload up, an idea once referred to as "Project Skyhook." The recent discovery of very strong carbon nanotubes has revived the idea. Arthur C. Clarke used this idea in his 1977 science fiction novel *Fountains of Paradise*.

A more likely idea is to "fly" to space on an airplane. Airplanes have two attractive features: they use oxygen from the atmosphere as part of their fuel (so they don't have to carry it all, as do rockets), and they can push against the air, instead of having to push against their own exhaust. Although it is possible in principle, the technology to achieve 8 to 12 km/s with airplanes does not yet exist. We'll talk more about airplanes in a moment.

Earlier in this chapter, I mentioned the possibility of using rail guns. These are long devices that achieve their high projectile velocities by using electric and magnetic forces to push on the projectile. But recall their limitation: they must be very long in order to avoid huge accelerations and the huge *g*-forces that would result. Even trained fighter pilots can endure only about 10 *g*'s, and they can endure that only for a few seconds.

Ion Rockets

The inefficiency of rockets comes from the fact that typical chemical fuels only have enough energy to give their atoms a velocity of 2 to 3 kilometers per second. Rockets could possibly overcome this limitation by shooting out *ions*, a name for atoms that have an electric charge. You can find out a lot about ions on the Internet. Like the rail gun, the ions can be given their high velocity through electric forces, so they are not limited to the velocity of 2 to 3 km/s that is typical of rocket fuel. For example, a proton accelerated to 100,000 volts (I'll explain what this means in chapter 6) has a velocity of 4400 km/s. This makes ion rockets potentially much more efficient than chemical rockets, but so far nobody has figured out how to make the mass of the expelled ions sufficiently great to be able to launch a rocket from the Earth. They are more useful when a low thrust is needed for an extended period of time. NASA is using ion propulsion on the Dawn Mission to the asteroids Vesta and Ceres, which was launched on 27 September 2007 and should reach Vesta on 25 August 2011.

---

[23] In the Bible, such a tower was attempted in ancient Babylon, and it is also referred to as the Tower of Babel. Its purpose was to reach heaven. To prevent the Babylonians from succeeding, God made all the workers speak different languages. Thus, according to the Bible, this is the origin of the multitude of languages spoken by humans. It is also the origin of the verb "to babble."

Airplanes, Helicopters, and Fans

Airplanes fly by pushing air downward.[24] Every second, the airplane tends to pick up downward velocity from the Earth's gravity. It stays at the same altitude by pushing enough air downward that it overcomes this velocity.

The fact that wings push air down is most readily observed in a rotary-wing aircraft, otherwise known as a helicopter. (Helicopter pilots call the ordinary airplane a "fixed-wing" aircraft.) In fact, the helicopter blades are designed to have a shape identical to wings, and air is pushed past them when they spin. They push the air down, and that is what pushes the helicopter upward. If you stand under a helicopter rotor when it is spinning, you can feel the air being forced downward.

It is probably more convenient, however, to observe how wing-shaped blades push air by standing in front of a fan. Fans work the same way as helicopter blades, and as airplane wings. The blade's movement through the air pushes air perpendicular to the direction of motion of the blade.

For the airplane and the rocket, the v_R needed is the velocity to overcome the pull of gravity. In one second of falling, gravity will give any object a velocity of

$$v = gt \approx 10 \text{ m/s}$$

This falling velocity must be cancelled by accelerating upward, and this is done in an airplane by pushing air downward. Air is typically a thousand times less dense than the airplane (1.25 kg/m^3), so to get enough air (i.e., to make the mass of the air large), the wings must deflect a large amount of air downward.

The wake of a large airplane consists of this downward flowing air, often in turbulent motion. It can be very dangerous for a second plane that encounters this wake, since the amount of air flowing downward is large.

Hot Air and Helium Balloons

The first way that humans "flew" was in hot air balloons, in 1783 above Paris. These make use of the fact that hot air expands and it takes more volume compared to an equal mass of cool air. Another way to say this is that the density of hot air (the mass per volume) is less than that of cool air.

In a liquid or gas, things that are less dense tend to float. That's why wood floats on water. (If it has a density less than 1 g/cm^3; some wood sinks.) The heavier fluid "falls" and flows under the less dense object, pushing it upward.

---

[24] In most physics books, the lift on the airplane wing is explained by use of a principle called *Bernoulli's law*. The "derivation" is done using a diagram that typically shows the air trailing the wing as if it is completely undisturbed. The astute student will be bothered by this. How can the air put a force on the wing but the wing not put a force on the air? A careful analysis (done in advanced aerodynamics books) shows that to establish a flow with higher velocity above the wing than below, the air velocity distribution far from the plane is not undisturbed; in fact, it has air deflected downward with a momentum rate equal to the upward force on the wings, as it must to satisfy momentum conservation.

Boats float only if their average density (hull plus empty space inside) is less than that of water. That's why a boat will sink if it fills with water.

If you fill a balloon with hot air, it will rise until it reaches an altitude at which the density of the surrounding air matches that of the balloon. (Of course, you have to include the mass of the balloon, and any weight that it carries, along with the mass of the air inside.)

Better than hot air are light gases such as hydrogen and helium. Air weighs about 1.25 kg/m³ (at sea level). The same volume of hydrogen weighs about 14 times less—i.e., about 0.089 kg = 89 g.[25] If we fill a 1-cubic-meter balloon with hydrogen, it will tend to float. Calculate how much 1 m³ of air weighs and how much 1 m³ of hydrogen weighs. The *lift* (upward force) of the balloon is just equal to the difference between these two weights. Putting this into numbers, this means that a cubic-meter balloon will have a lift of $1.25 - 0.089 = 1.16$ kg. That means that if you hang an object from it that weighs less than one kilogram (including the mass of the balloon skin), the balloon will still go upward.

Helium gas isn't quite as light as hydrogen gas, so if the balloon is filled with helium, the lift isn't quite as much. Here's the calculation: helium gas weighs about 0.178 kilograms per cubic meter, so the lift on a helium balloon would be $1.25 - 0.178 = 1.07$ kg. Note that even though the helium is twice as dense as hydrogen, the lift is almost as good.

But 1.16 kg of lift for 1 m³ of balloon is not really much. That's why, despite the cartoons you may have watched on TV, even a large packet of balloons are not enough to lift a 25-kg child. (If you have some spare time and want to be amused, look up "Lawn-chair Larry" on the Internet.)

Hot air balloons have even less lift. If you heat the air to 300 C, then its temperature in absolute scale is 600 K. That's twice its normal temperature, so its density is half of its usual density. The lift of a cubic meter would be $1.25 - 1.25/2 = 0.62$ kg/m³. Notice that the lift is significantly worse than hydrogen or helium. To lift a person who weighs 100 kg (including basket, balloon skin, and cables) would take $100/0.62 = 161$ m³ of hot air. If the balloon were shaped like a cube, the side of such a balloon would be $\sqrt[3]{161} = 5.4$ meters = 18 feet. That's why hot air balloons have to be so big, and why they don't lift much.

Floating on Water

The same principles apply to floating on water. Saltwater is denser than freshwater, so the density difference between it and you is greater; that's why you float higher in saltwater. Even strong swimmers take advantage of their buoyancy (the fact that they are less dense, on average, than water). If water becomes filled with bubbles, its average density can become less than yours, and you will sink. The *New York Times* on 14 August 2003 had an article describing how a group of boys drowned because they were in bubbly water. Here is a quote from that article:

---

[25] Nitrogen has an atomic weight (number of neutrons plus protons) of 14. Hydrogen has an atomic weight of 1. The number of atoms in a cubic meter is the same for both gases, so the factor of 14 simply reflects the larger nitrogen nucleus.

To four teenagers from the suburbs, Split Rock Falls was a magical place—cool water rushing between the granite walls of a mountain ravine, forming pools for hours of lazy summertime swimming.

On Tuesday afternoon the four men—Adam Cohen, 19; Jonah Richman, 18; Jordan Satin, 19; and David Altschuler, 18—returned to their favorite childhood summer haunt to find it engorged by a summer of heavy rain. By the end of the day, all four men, each an experienced swimmer, was dead, drowned in the waters they knew well.

In what officials here described as one of the worst drowning accidents ever in the Adirondack State Park, all four died after Mr. Altschuler slipped off a narrow granite ledge into a foaming pool of water whipped into a frenzy by a tumbling waterfall. In a final act of friendship, Mr. Richman, Mr. Cohen and Mr. Satin, who had grown up together on Long Island, jumped after him to try to save his life, police and officials said. The laws of physics were against them, though.

"They call it a drowning machine," said Lt. Fred J. Larow, a forest ranger with the State Department of Environmental Conservation, who helped recover the bodies here, about 20 miles east of Lake Placid. "The water was so turbulent and aerated that there was no way they could stay above water. Even the strongest swimmer in the world couldn't have survived it."

Undersea volcanic eruptions have also led to bubbly water in the oceans, and in such water even big ships will sink.

Submarines can adjust their depth below the ocean surface by using ballast tanks to take in or expel water. When they take in water, the air in the tanks is replaced by the heavier water, and the average density of the submarine increases. This makes the submarine sink. The only thing that will stop the sinking is the expulsion of water from the tanks. If the submarine sinks too far, then it is crushed by the weight of the water above it; that makes it even more dense, and so it sinks faster. That is called the *hull crush depth*. In the movie *Crimson Tide*, the hull crush depth was 1800 ft, about 1/3 mi.

The submarine in that movie (the USS *Alabama*) was able to save itself by getting its engine running and using its short "wings" to get lift in a way similar to the way that an airplane does—by moving forward and pushing water downward. A submarine can also push compressed air into its ballast tanks, driving out the water, and decreasing its average density.

Sperm whales are said to be able to dive as deep as 2 mi. Diving deep is easy, since as the whale goes deeper, any air in its lungs (remember, a whale is a mammal) is compressed, and that makes the whale less buoyant. So once the whale is denser than water, it will sink. Coming up is the hard part. Whales save enough energy in their effortless dives to be able to swim back up to the surface.

Air Pressure on Mountains and Outside Airplanes and Satellites

Air pressure is simply the weight of the air above you. In any fluid or gas, the pressure is evenly distributed, so that the air at sea level will push an equal amount up, down, and sideways. As you go higher, there is less air above you, so the pressure decreases. At an altitude of 18,000 feet (3.4 miles, or 5.5 kilometers), the pressure is half of what it is at sea level. That is the altitude of Mount Kilimanjaro in Africa, and many people walk to the top every year. They go above half the atmosphere! It is hard to breathe up there, and nowhere on the Earth do people live continuously at that altitude. But if you spend only a day or two going up and down, you can take it. Go up another 18,000 ft to an altitude of 36,000 ft, and the pressure is reduced, by another factor of 2, to one-quarter the pressure at sea level. That's a typical altitude for jet airplanes. And that rule continues; for every additional 18,000 ft, the pressure (and the density of the air[26]) drops by another factor of 2. You cannot live at such a low density of air, and that is why airplanes are pressurized. You can live at that pressure if the "air" you are given is pure oxygen (rather than 20%), and that is what you would get from the emergency face masks that drop down in airline seats if the cabin ever loses pressure.

Want to know the decrease in air pressure? Divide the altitude by 18,000 ft (or by 5.5 km, if that's how you've measured it). The number you get is the number of halvings you have to do. Suppose that your altitude (e.g., in an airplane) is 40,000 ft. Divide that by 18,000 to get about 2 halvings. That means that the pressure is reduced by a factor of $(1/2) \times (1/2) = 1/4$. So the pressure (and the air density) is 4 times less than at sea level.

Now consider a "low-Earth orbit" satellite, at $H = 200$ km above sea level. Let's do all our calculations in kilometers. So we use the half-altitude distance, 5.5 km. Then $H/5.5 = 200/5.5 \approx 36$. Now multiply 1/2 by itself 36 times to get:

$$P = (1/2)^{36} = 1.45 \times 10^{-11} = 0.00000000001$$

That pressure is 10 trillionths times as small as the pressure at sea level.[27] Satellites need this low pressure to avoid being slowed down by collisions with air. A typical altitude for LEO (remember? low-Earth orbit) is 200 kilometers.

Because the density of air decreases with altitude, a helium balloon will not rise forever. Eventually, it reaches an altitude at which the outside air has the same density as the helium (with the weight of the balloon averaged in), and then it stops rising. I noticed this as a child, when I was disappointed to see that the helium-filled balloon I had released did not go all the way to space, but went high up and then stopped rising.

That's why balloons are not a possible way to get to space.

Convection—Thunderstorms and Heaters

When air is heated near the ground, its density is reduced and it tends to rise, just like a hot air balloon. Unconstrained by any balloon, the air expands as it rises, and an interesting result is that it will remain less dense than the surrounding air until it reaches the *tropopause*, the level at which ozone absorbs sunlight, causing the surrounding air to be warmer. When the hot air reaches the tropopause,

---

[26] The air density doesn't drop quite that much because the air up there is cooler.

[27] The equation implicitly assumed that the temperature of the air continues to cool at higher and higher altitude. But above the tropopause, the temperature rises, and that makes this calculation interesting, but not really accurate.

its density is no longer less than that of the surrounding air, so the air stops rising.

On a summer day, when thunderstorms are growing, it is easy to spot the tropopause. It is the altitude at which the thunderstorms stop rising and begin to spread out laterally. The tropopause is a very important layer in the atmosphere. It is the location of the ozone layer, which protects us from cancer-producing ultraviolet light. We'll talk more about this layer in chapter 7 because of the important effect it has on sound and in chapter 9 when we discuss ultraviolet radiation and its effects.

Convection is the name we give to the process of hot air expanding and rising. When you have a heater in a room near the floor, the rising hot air forces other air out of the way; that results in a circulation of the air. This is a very effective way to warm a room—much faster than heat conduction through the air. But it invariably results in the warmest air being near the top of the room. If you put the heater near the top, the warm air just stays up there. On a cold day, in a room with a heater near the floor, stand up on a stepladder and feel how much hotter it is near the ceiling.

Hurricanes and Storm Surge

Hurricanes are created when tropical water is very warm and heats the air above it. They begin like a thunderstorm, but they grow fierce because of the huge energy in the warm water. The key to predicting the strength of a hurricane season is to look at satellite maps of the *sea-surface temperature*. When the Caribbean sea is very warm, especially near the southern coast of the United States, then great storms are likely to be generated.

The hot air rises, and its low density reduces the weight of air above the water; that is the same as reducing the air pressure (the weight per square meter). As a result, the water in the center of the hurricane rises. This rise is called the *storm surge*. It often does more damage than the high winds of the hurricane. It results in a higher ocean, so if you live on the coast, the ocean may be in your living room. The surge rises higher if it happens to occur at high tide. In addition, if the winds of the hurricane happen to be toward the coast, then that pushes the water up even higher. There are many areas on the Barrier Islands off North Carolina and in the Florida Keys where the storm surge can completely cover an island.

The winds of the hurricane are a result of the rapidly rising hot air. The circular motion of the hurricane winds comes about because these winds are created on a spinning Earth.

It is the storm surge that takes boats and puts them inland. It is the storm surge that overturns automobiles. Winds alone will not do that. The pounding surf of a raging hurricane-driven ocean will demolish many of the strongest buildings. Water is a thousand times denser than air.

Angular Momentum and Torque

In addition to ordinary momentum (mass times velocity), there is another kind of momentum that physicists and engineers find enormously useful in their calculations called *angular momentum.* Angular momentum is similar to ordinary momentum but it is most useful for motion that is circular—i.e., rotation. Think of it as the tendency for spinning objects to keep spinning. It is called *angular* momentum because a spinning object keeps changing its angle of rotation.

If an object of mass M is spinning in a circle of radius R, moving at velocity v, then its angular momentum L is

$$L = MvR$$

What makes angular momentum so useful is that, like ordinary momentum, when there are no external forces on an object, it is "conserved"—i.e., its value doesn't change. Have you ever spun on ice skates? Actually, ice skates are not necessary—just stand in a spot and start spinning with your arms stretched out. If you've never done this, then I strongly recommend you try it right now. As you spin, rapidly pull your arms in. (You'll never forget the experience, and you can have great fun entertaining children with it.) To the surprise of most people, pulling in your arms while you are spinning will cause you to suddenly spin much faster. You can predict that from the angular momentum equation. If the angular momentum L is the same before and after the arms are pulled in, and the mass of the arms M is the same, then vR must be the same. If R gets smaller, then v must get bigger.

Angular momentum conservation also explains why water leaving a tub through a narrow drain begins to spin. In fact, it is very unlikely that the water in the tub wasn't already spinning, at least a little bit. But when the distance to the drain (R, in the equation) gets small, the v in the equation gets very big. A similar effect occurs in hurricanes and tornadoes. Air being sucked into the center (where there is low pressure due to air moving upward) spins faster and faster, and that's what gives the high velocities of the air in these storms. The air in hurricanes gets its initial spin from the spin of the Earth, and amplifies by the angular momentum effect. So hurricanes really do spin in different directions in the Southern and Northern Hemispheres. The tendency of objects to pick up spin from the Earth as they move is called the *Coriolis effect.*}

It is not true that sinks or tubs drain differently in the Northern and Southern Hemispheres. Their direction depends on the small residual rotation in the water left over from the filling of the tub, or from a person getting out, because the Coriolis effect is much smaller than such random spin.

Conservation of angular momentum can be used to understand how a cat, dropped from an upside-down position, can still land on his feet. (Don't try this one at home! I never have, but I've seen a movie . . .) If the cat spins his legs in a circle, his body will move in the opposite direction, keeping his total angular momentum equal to zero. That way he can bring his legs underneath him. Astronauts do this trick when they want to reorient themselves in space. Spin an arm in a circle, and your body will move in the opposite direction. You can try that trick on ice skates too.

The conservation of angular momentum has other useful applications. It helps keep a bicycle wheel from falling over, when the wheel is spinning. It can also cause a difficulty: if kinetic energy is stored in a spinning wheel (usually called a *flywheel*), then the angular momentum makes it difficult to change the direction that the wheel is spinning in. This makes the flywheel's use for energy storage in moving vehicles, such as buses, problematical. It is often addressed by having two flywheels spinning in opposite directions, so although energy is stored, the total angular momentum is zero.

Angular momentum can be changed by a suitable application of an outside force. The required geometry is that the force must act obliquely, at a distance. We define *torque* as the tangential component of the force times the distance to the center. So for example, to start a bicycle wheel spinning, you can't just push on the rim in a radial manner. You have to push tangentially. That's torque. The law relating torque and angular momentum is as follows: the rate of change of angular momentum is numerically equal to the torque.

You can probably see why mastery of the equations of angular momentum is very useful to engineers and physicists in simplifying their calculations. But you don't have to learn them. If you need a calculation someday, just hire a physicist.

Chapter Review

Weight is the force of gravity acting on mass. The force of gravity obeys an inverse square law, so when the distance increases by (for example) a factor of 3, the force becomes 9 times weaker. It is this force that keeps the Moon in orbit around the Earth, and the Earth in orbit around the Sun. If gravity were turned off, satellites would move in straight lines rather than in circles. Even at great distances, the force of gravity never goes completely to zero. The sensation of weightlessness felt by astronauts is really the sensation of continuous falling.

All satellites must keep moving, or they fall to the Earth; they cannot hover. In low-Earth orbit (LEO), the satellite moves at 8 kilometers per second and orbits the Earth in 1.5 hours. LEO satellites are useful for Earth observations, including spying. Geosynchronous satellites are useful for applications where the position of the satellite must stay fixed with respect to the ground. A medium-Earth orbit (MEO) is between these. GPS satellites are in MEO. A GPS receiver determines its location by measuring the distance to three or more of these satellites.

Close to the surface of the Earth (where we live), objects fall with constant acceleration until they are slowed down by air resistance. Air resistance also limits the fuel efficiency of automobiles. Satellites must fly high (>200 km) to avoid air resistance.

Acceleration can also be measured in units of g, the acceleration of gravity. The g-rule says that to accelerate an object to $10\,g$ requires a force 10 times as great as that object's weight on the Earth's surface. This is known as Newton's second law. Humans cannot endure more than about $10\,g$. The space shuttle never accelerates more than $3\,g$. Circular motion can also be considered acceleration, with Newton's second law applying, even if the magnitude of the speed doesn't change. Based on this, we can calculate the velocity that a satellite must have to stay in a circular orbit at different altitudes above the Earth.

The surface gravity on other planets and on asteroids is very different than it is on the Earth, science fiction movies notwithstanding.

To escape to space completely requires an energy of about 15 Cal/g. This is enough energy to lift you up a tall-enough elevator, if one could be built (Project

Skyhook). If you have a velocity of 11.2 km/s, then your kinetic energy is sufficient to escape. Black holes are objects whose escape velocity exceeds the speed of light.

Gravity measurements have practical applications. Since oil is lighter than rock, it has a weaker gravity, and that fact has been used to locate it. Gravity measurements give us the best image of the Chicxulub crater.

When a gun is fired, the bullet goes forward and the gun goes backward. This is an example of the conservation of momentum. Other examples: rockets go forward (very inefficiently) by shooting burnt fuel backward; airplanes and helicopters fly by pushing air downward.

Objects float when their density is less than that of the fluid or gas they are in. That includes boats and balloons. Hot air rises because it is less dense than the surrounding air, and that happens for hot air balloons and thunderstorms. The density and pressure of air decreases with altitude according to a halving rule. The pressure drops by a factor of 2 every time your increase your altitude by 5.5 kilometers.

Angular momentum (a momentum that applies to circular motion) is also conserved, and that causes contracting objects to speed up. Examples include sink drains, hurricanes, and tornadoes.

Discussion Topics

1. Could gravity measurements be used to discover secret tunnels underground?
2. Are there any science fiction movies that seem to violate the laws of physics discussed in this chapter? Can you come up with "reasonable" explanations that make the physics in the movie OK—for example, artificial gravity from a constantly accelerating spaceship?
3. A spinning wheel has angular momentum. What if you change the direction of the spin? Do you think that changes the angular momentum?
4. What do you think it should take to qualify being called "an astronaut"?

Internet Research Topics

See what you can find on the Internet about the following topics:

1. Manufacturing objects in space
2. Spy satellites
3. Black holes
4. GPS
5. Ion engine
6. Project Skyhook
7. Rail guns
8. Ion propulsion
9. Lawn-chair Larry

Essay Questions

1. Automobiles and trucks are often designed with a tapered ("aerodynamic") rather than a blunt front surface. Why is this? What does the tapered front accomplish? How important is it?
2. Artificial-Earth satellites fly at different altitudes, depending on what they need to do. Describe the differences between low-Earth satellites, medium-Earth satellites, and geosynchronous satellites. For which applications would you use each?
3. Which would fall faster, a high-diver doing a "swan dive" (in which he spreads out his arms and falls chest-first), or a diver with his arms pointed above his head going straight down? Why? Explain the relevant physics.
4. When you dive into the water, having your head down is much less painful than a "belly-flop." Discuss this in terms of the "area" that hits the water.
5. Other than keeping us on the ground, does gravity have any practical applications? What can we learn about the Earth by remote measurement of gravity?

Multiple-Choice Questions

1. At high velocities, most of the fuel used in an automobile is used to overcome
   A. gravity
   B. momentum
   C. air resistance
   D. buoyancy

2. Balloons rise until
   A. the surrounding air becomes too dense
   B. the surrounding air becomes too thin (not dense)
   C. the surrounding air becomes too cold
   D. they reach space

3. At an altitude of 200 km, the downward force of gravity on an Earth satellite is
   A. the same as on the surface of the Earth
   B. a little bit weaker
   C. zero
   D. about half as strong as on the surface

4. Airplanes fly (maintain their altitude) by
   A. pushing fuel downward
   B. pushing fuel backward
   C. pushing fuel upward
   D. pushing air downward

5. Rockets fly by
   A. using antigravity
   B. pushing air downward

C. pushing fuel downward

D. being lighter than air

6. The storm surge in a hurricane comes from
   A. the strong waves
   B. increased gravity
   C. the force of winds on buildings
   D. the low pressure

7. When an astronaut in space sneezes, his head is snapped back with higher velocity than on Earth, because
   A. momentum is conserved
   B. his head is massless
   C. there is no air in space
   D. Wrong! His head is not snapped back with higher velocity.

8. A rail gun
   A. accelerates rails
   B. can shoot bullets faster than ordinary guns
   C. doesn't really accelerate things
   D. uses ions for propulsion

9. The Sun would be a black hole if it were squeezed into a radius of
   A. 2 miles
   B. 2,000 miles
   C. 2 million miles
   D. It already is a black hole.

10. The method that would take the least energy to get something to space is
    A. lifting it by elevator (if the elevator existed)
    B. using an ion rocket
    C. flying it in a balloon
    D. launching it in a three-stage rocket

11. If an artificial satellite orbited the Earth at 240,000 miles (the distance to the Moon), it would orbit the Earth in a period of
    A. 90 minutes
    B. 1 day
    C. 1 week
    D. 1 month

12. The time it takes a geosynchronous satellite to orbit the Earth is
    A. 90 minutes
    B. 1 day
    C. 1 week
    D. 1 month

13. The force of gravity between two people standing next to each other is
    A. zero
    B. too small to measure
    C. small but measurable
    D. greater than 1 lb

14. You can see where the tropopause is by
    A. looking at Earth satellites
    B. seeing how high birds fly
    C. looking at the tops of thunderstorms
    D. seeing where lightning comes from

15. If satellites fly too low, they will crash because
    A. of stronger gravity
    B. of weaker gravity
    C. of lower air density
    D. of air resistance

16. Ice-skaters spin faster by pulling in their arms. This illustrates that
    A. momentum is conserved
    B. angular momentum is conserved
    C. energy is conserved
    D. angular energy is conserved

17. Someone firing a rifle is pushed back by the rifle. This illustrates that
    A. momentum is conserved
    B. angular momentum is conserved
    C. energy is conserved
    D. angular energy is conserved

18. A problem with rail guns is that
    A. they use too much energy to be practical
    B. they violate momentum conservation
    C. the high accelerations can cause damage
    D. they are limited to about the speed of sound

19. The maximum acceleration of the space shuttle is about
    A. 1 g
    B. 3 g's
    C. 10 g's
    D. 18 g's

20. The lowest practical altitude for a satellite orbiting the Earth is about
    A. 10 km
    B. 200 km
    C. 600 miles
    D. 24,000 miles

21. For very high velocities, the best kind of rocket to use is
    A. ion rocket
    B. hot air rocket
    C. hydrogen/oxygen
    D. TNT

22. The altitude of a geosynchronous satellite is closest to
    A. 200 miles
    B. 22,000 miles
    C. 2,000,000 miles
    D. 93,000,000 miles

23. Skyhook, if built, is
A. a way of getting to very high orbits
B. a way to get into space using less energy
C. a safer way to get into space
D. a faster way to get into space

24. If you fall for one second, your speed will be approximately
A. 8 mph
B. 22 mph
C. 32 mph
D. 45 mph

25. Artificial gravity can be created in a spaceship by
A. accelerating the ship
B. moving at high velocity
C. using ion engines
D. There is no such thing.

26. Orbital velocity is about
A. 5 miles per sec
B. 7 miles per second
C. 8 miles per second
D. 11 miles per second

27. The Sun is mostly hydrogen, yet the Earth has very little hydrogen gas. That's because:
A. it escaped to space
B. it combined with oxygen to make water
C. it combined with carbon to make hydrocarbons
D. it was burned by fusion when the early Earth was hot

28. The orbit of GPS is
A. LEO
B. MEO
C. GEO (same as HEO)
D. TEO

29. A geosynchronous satellite orbits the Earth
A. once every 90 minutes
B. once every 6 hours
C. once every 24 hours
D. once every month

30. A typical spy satellite can continuously observe a location on the ground for about
A. 10 seconds
B. one minute
C. 90 minutes
D. 12 hours

31. Compared to the weight of the payload, the fuel used in a rocket to orbit typically weighs
A. about the same
B. less than half as much

C. about twice as much
D. over 20 times more

32. A cubic meter balloon can lift
A. about one gram
B. about 2 pounds
C. about 50 pounds
D. about 150 pounds

33. Spy satellites usually fly at
A. LEO
B. MEO
C. HEO
D. escape velocity

4

CHAPTER

Nuclei and Radioactivity

1. This book is radioactive.
2. You are radioactive too, unless you have been dead for a long time.
3. The United States Bureau of Alcohol, Tobacco, and Firearms tests wine, gin, whisky, and vodka for radioactivity. If the product does not have sufficient radioactivity, it may not be legally sold in the United States.
4. Of those killed by the Hiroshima atomic bomb, the best estimate is that fewer than 2% died of radiation-induced cancer.
5. Biofuels (e.g., ethanol made from corn or sugar cane) are radioactive. Fossil fuels are not.

These statements are all true, and yet they surprise most people. That reflects the confusion and misinformation that pervades the public discussion of radioactivity. I hope that when you finish this chapter, you can come back and read these five statements and say, "Of course."

Radioactivity

Radioactivity is the explosion of the nucleus of the atom. What makes this explosion so important and fascinating is the enormous energy released, typically a million times greater than in chemical explosions for the same number of atoms.

Atoms are small but not completely invisible. A device called a scanning tunneling microscope (called an STM by experts) can pass over individual atoms, feel their shape, and then present that on a computer screen in the form of an image. A similar device can pick up individual atoms, carry them, and place them at new locations. Figure 4.1 shows 35 xenon atoms arranged to form the letters *IBM* on the surface of a nickel crystal.

It is this ability to manipulate individual atoms that has led to the excitement about the new field called *nanotechnology*. The name comes from the

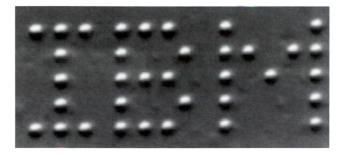

Figure 4.1 "Visible" atoms. The letters *IBM* were written by arranging individual xenon atoms on the surface of a nickel crystal. The atoms were manipulated and photographed using a scanning-tunneling microscope. This work was done by a team led by Donald Eigler. Guess which company they worked for. (Image used with permission from IBM. Copyright IBM.)

fact that an atom is about 1/10 of a nanometer (a billionth of a meter) in diameter.

To put the size of an atom in perspective, consider the following examples: a human hair has a thickness of about 125,000 atoms and a human red-blood cell has a diameter of about 40,000 atoms. These numbers are big, but not huge. I didn't have to use scientific notation. So atoms are small, but they are not infinitesimally small.

Each atom consists of a cloud of electrons with a tiny nucleus in the center. The nucleus has a radius of about 10^{-13} cm, which means that it is 100,000 times smaller than the atom itself. To visualize this ratio, imagine that an atom were enlarged until it was the size of a baseball or football stadium (300 m). Then the nucleus, similarly expanded, would only be the size of a mosquito (3 mm). Since its linear size is 10^{-5} times the size of the atom, its volume is 10^{-15} times the volume of the atom (because you calculate volume by taking the cube of the linear size). That's like the volume of the stadium compared to the volume of the mosquito. Imagine trying to fill an entire stadium with mosquitoes. It would take 10^{15} of them. That's a quadrillion.

This enormous disparity often gives rise to the statement that the atom is mostly "empty space." Some could argue, however, that the space isn't really empty; it is filled with the electron wave. We'll talk more about that in chapter 11. Yet, even though it has only 10^{-15} of the volume of the atom, the nucleus contains more than 99.9% of the mass of the atom. The nucleus is very small, but very massive. That was not predicted; try to imagine the surprise and disbelief of scientists in 1911 when Ernest Rutherford discovered this incredible fact. It seems completely implausible. But it is true.

Within 20 years of Rutherford's discovery, we learned that the nucleus itself was made up of even smaller pieces. The most important of these are protons and neutrons:

- **Protons** weigh almost 2000 times as much as electrons and have the same magnitude of electric charge, but they are opposite in sign. (We'll discuss the sign of the charge in chapter 6. Electrons, by convention, have negative charge, and protons have positive charge.)

- **Neutrons** are similar in mass to protons (they are actually about 0.3% heavier), but they have no electric charge—i.e., they are "neutral," hence their name.

So here is the basic picture of the atom: It has a very small nucleus made of protons and neutrons. Surrounding this is a relatively large volume occupied by electrons. But most of the mass is in the tiny nucleus. The nucleus of an atom weighs almost exactly the same as the entire atom itself since the electrons are so light.

Scientists love to deconstruct. So it is natural for them to wonder whether protons and neutrons are made of smaller objects. The answer was uncovered in the last few decades of the twentieth century: protons and neutrons are made of particles called *quarks*[1] and a variable number of lightweight *gluons* that hold the quarks together (they are named after glue). We'll discuss these further in an optional section at the end of this chapter. What are quarks made of? According to the unproven string theory, they (as well as electrons) are made of something called *strings*. These strings are not really like ordinary strings; they are very short and exist in multiple dimensions. The key idea in string theory is that everything is made up of the same kind of object. To summarize:

- Matter is made of molecules (e.g., water is made of H_2O).
- Molecules are made of atoms (e.g., H_2O = hydrogen and oxygen).
- Atoms are made of electrons orbiting a nucleus.
- Nuclei are made of protons, neutrons, and other light particles (e.g., gluons).
- Protons and neutrons are made of quarks and gluons.
- Quarks and electrons may be made of strings (if string theory is right).

Elements and Isotopes

The number of protons in the nucleus is called the *atomic number*. This number also indicates the number of electrons orbiting the nucleus. An atom of hydrogen has 1 proton in the nucleus (and 1 electron in orbit), and we say that it has atomic number 1. An atom of helium has 2 protons in the nucleus and 2 electrons in orbit. It has atomic number 2. An atom of uranium has 92 protons in the nucleus and 92 electrons in orbit. We say it has atomic number 92. Each element has a different atomic number. Table 4.1 lists some of the atomic numbers of elements that we will be discussing in this chapter:

As mentioned earlier, the nucleus consists primarily of protons and neutrons. Neutrons don't have electric charge, so they don't change the behavior of the atom (at least, not much). But they do make the nucleus heavier. Atoms of an element with different numbers of neutrons are called different *isotopes* of that element.

---

[1] The name *quark* was proposed by Cal Tech physicist Murray Gell-Mann, who found the term used in James Joyce's novel *Finnegan's Wake*.

Table 4.1 Atomic Number of Selected Elements

Element	Atomic number (N_p)
Hydrogen	1
Helium	2
Lithium	3
Carbon	6
Nitrogen	7
Uranium	92
Plutonium	94

For example, the nucleus of ordinary hydrogen (the abundant kind) always contains one proton and no neutrons. But about 1 in every 6,000 hydrogen atoms has a nucleus that contains an extra neutron. That kind of hydrogen is called *deuterium*, or *heavy hydrogen*. Water made from heavy hydrogen weighs more; it is called *heavy water*. Heavy water was very important during World War II in the development of the nuclear reactor. In fact, Hitler had a special plant to purify deuterium (useful to make a nuclear reactor), and the Allies sent a team to blow that plant up.

About 10^{-18}, or a billionth of a billionth, of ordinary hydrogen has 2 neutrons in the nucleus. This kind of extra-heavy hydrogen is called *tritium*. Tritium is the only kind of hydrogen that is radioactive. It is used in medicine and in hydrogen bombs.

We'll be talking a lot about deuterium and tritium, especially when we get to nuclear reactors and bombs. So learn those terms. Here are useful memory tricks:

- In *deuterium*, the proton and neutron in the nucleus form a *duo*.
- In *tritium*, the proton and neutrons in the nucleus form a *trio*.

Over 99% of uranium found in the Earth has a nucleus with 92 protons and 146 neutrons, making up a total of 92 + 146 = 238 particles in the nucleus. This is called U-238. But about 0.7% of the uranium has only 143 neutrons in the nucleus instead of 146. This is a different isotope of uranium called U-235. It is very important, because U-235 plays a key role in the atomic bomb and nuclear reactors.

Both U-238 and U-235 have 92 protons. That means that they both have 92 electrons. Since it is the electrons that play the major role in ordinary chemistry, both isotopes react very similarly with other elements, such as oxygen and water. That's why they are both called uranium. But when we are interested in the properties of the nucleus, particularly in nuclear explosions, the difference in neutrons becomes extremely important.

Radiation and Rays

Now let's return to the radioactivity, the explosion of the nucleus. A common chemical explosion (e.g., TNT) takes place when a large molecule suddenly breaks up into smaller molecules. In a similar way, a radioactive explosion takes place when a nucleus breaks up into smaller parts.

We'll begin with the most common type of radioactivity, in which a relatively small particle is thrown out from a big nucleus. It flies out like a bullet, at very high speed, sometimes approaching the speed of light. When this process was first discovered by Henri Becquerel in 1896, nobody knew what was coming out. The projectiles couldn't be seen directly, but they passed through matter and could expose photographic film. The projectiles were called *rays*, probably because they travel in nearly straight lines. They had properties similar to x-rays, which Wilhelm Conrad Roentgen had discovered a few years earlier in 1895. Different kinds of rays were found, with somewhat different properties, and they were named after the letters of the Greek alphabet.[2] Some rays (e.g., from uranium) could be stopped by a piece of paper; these were called *alpha rays*. Rays with more penetration were called *beta rays*. The most penetrating of all were called *gamma rays*. (There were *delta rays* too—but those turned out to be the same as low-energy beta rays, so the term is little used.)

The old terminology has changed; instead of saying rays, we now say *radiation*. It is worthwhile to learn this formal terminology:

- **Radioactivity** refers to the explosions of atomic nuclei.
- **Radiation** consists of the pieces that get thrown out in the explosion.

Each ray (or particle) is like a tiny bullet, so tiny that you don't feel it if it hits your body. Alpha rays and beta rays bounce off many atoms before they stop; with each bounce, they can knock apart a molecule or mutate a gene. The slowing bullet leaves a trail of damaged molecules along its wake. The damage is small but, if you are hit by a large number of particles, the total effect can make you ill or even kill you. Gamma rays tend to be absorbed by a single atom; however, they frequently break up the atom or even its nucleus, so secondary radiation is emitted. It is that secondary radiation that often does the most damage.

"SEEING" RADIATION—THE CLOUD CHAMBER

When alpha or beta rays pass through a gas, they knock electrons off atoms, creating a trail of charged particles called *ions*. If the gas has a lot of water vapor or alcohol vapor mixed in, and the gas is cool, then the water or alcohol tends to form little droplets on these ions. In essence, they form clouds along the path of the radiation. White tracks suddenly appear when an alpha or a beta ray passes through a chamber set up in this way. Figure 4.2 shows some images taken in the original cloud chamber, invented by Charles Wilson (for which he got the Nobel Prize in 1927). The streaks are the cloud particles along the path of the radiation.

The tracks are very similar in nature to the vapor trails left behind when a jet airplane engine passes overhead. The track doesn't actually show the airplane itself, but it shows where it has passed.

---

[2] The first few letters of the Greek alphabet, printed in lower case, are α (alpha), β (beta), γ (gamma), δ (delta). The last (24th) letter is ω (omega).

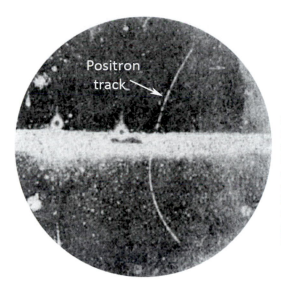

Positron track

Figure 4.2 Cloud chamber showing a track made by antimatter. The thin curved line is a trail of cloud particles left behind by a positron. This image was the first antimatter ever observed, and it won Carl Anderson a Nobel Prize. The horizontal broad white region was a divider that the positron passed through. (Photo courtesy of U.S. Department of Energy.)

A cloud chamber is a marvelous thing to watch. Radiation from a radioactive source will show short white lines of cloud particles suddenly forming along a path. The droplets are heavy, so they drift to the bottom of the chamber. Meanwhile, new paths are suddenly appearing above them. Every once in a while, radiation all the way from space will cause a long path to appear in the cloud chamber. This radiation is known as *cosmic radiation.*

Radiation and Death: The rem

The biological damage done to cells hit by radiation is measured in a unit called the *rem.* I can give you a rough idea of how big a rem is from the following example.[3] Suppose that each square centimeter of your body is penetrated by 2 billion gamma rays. If that happens, then the radiation dose to that part of your body is approximately 1 rem. A rem usually refers to the amount of damage to each gram of your body. If your whole body is exposed, then we say you got a whole body dose of 1 rem, which means that each gram of your whole body suffered the same damage.

Two billion gamma rays sounds like a lot of radiation, so it might make you think that a rem is a huge amount of damage. But remember, the nucleus is very small. When it emits energy, the energy is big only in a relative sense. For example, the gamma rays entering your body will deposit their energy, and that will cause your body to warm up. But the amount of warming can be calculated[4]—it is less than 1 billionth of a degree C. The radiation does damage individual molecules, and that is the source of all the real danger. Most of

---

[3] For this example, I assumed that the gammas had an energy of 1.6×10^{-13} joules each, a value defined to be 1 million electron volts, or 1 MeV. That is the energy of an electron from a wire with a voltage of 1 million volts.

[4] The calculated value is 2×10^{-11} C.

the time, the damage can be repaired by the cells in your body. But your body is not always successful. Many people estimate that the exposure to this much radiation (1 rem to every cell in your body—i.e., 1 rem whole body) will induce cancer with a probability of 0.0004, or 0.04%. We'll talk more about that shortly.[5]

The term *rem* was originally an acronym.[6] Physicists will never miss a chance to honor one of their own, so it was inevitable that a new unit would be introduced, the Sievert.[7] The conversion is simple: there are 100 rem in 1 Sievert. If you look at modern textbooks, you'll see Sieverts used more and more often. But most public reports still persist in using the rem, so we will too. (**Memory trick:** The *Sievert* is capitalized, and it is the *big* unit. It consists of 100 of the smaller, uncapitalized *rem*.)

RADIATION POISONING

If every cubic centimeter in your body is exposed to 1 rem, then we say you have received a whole-body dose of 1 rem. If the whole-body dose is more than 100 rem, the damage done to the molecules of the cells is enough to disrupt the metabolism of the body, and the victim becomes sick. This is called *radiation poisoning*. If you know someone who has undergone radiation therapy (to kill a cancer), then you know what the symptoms of mild radiation sickness are: nausea, listlessness (popularly referred to as "loss of energy"), and loss of hair. The severity of the illness depends very sensitively on the dose, as table 4.2 shows.

Table 4.2   Dose and Radiation Illness

Whole-body dose	Resulting radiation illness
Below 100 rem (below 1 Sievert)	No short-term illness
100 to 200 rem (1 to 2 Sieverts)	Slight or no short-term illness; nausea, loss of hair; rarely fatal
300 rem (3 Sieverts)	50% chance of death if untreated within 60 days
More than 1000 rem (more than 10 Sieverts)	Incapacitation within 1 or 2 hours; survival unlikely

The radiation illness gets much worse as the dose increases. At about 300 rem, the victim has about a 50% chance of dying from the illness over the next few weeks, if not treated. In medical terminology, 300 rem is said to be the

[5] In this calculation, I am assuming the linear hypothesis. We'll discuss that later in this chapter.

[6] Short for "Roentgen equivalent in man," where the unit of Roentgen is the amount of radiation it takes to release a certain number of electrons per gram of material.

[7] Sieverts are named after Rolf Sievert, the former chairman of the International Commission on Radiation Protection (ICRP).

LD50—which stands for the lethal dose that will kill 50% of those exposed. Memorize the following:

LD50 for radiation is 300 rem = 3 Sieverts

You will frequently hear the term *millirem* being used when people talk about radiation leakage. One millirem is 1 thousandth of a rem. LD50 for radiation is 300,000 millirem. The reason that I want you to memorize these numbers is that you will very likely encounter radiation in your life, maybe at a doctor's office, maybe if there is a leakage of radioactivity somewhere. You'll probably hear the term millirem, rather than rem. A dental x-ray gives, typically, a few millirem to your jaw (not to your whole body). It is useful to understand how small a millirem really is.

NUCLEAR ASSASSINATION

Alexander Litvinenko was an officer of the Soviet KGB and later the Russian Federal Security Service (FSB), but in 1998 he accused them of ordering assassinations and other nefarious activities. He fled to London, where he wrote two books full of allegations that implicated then Russian President Vladimir Putin. In November 2006, he fell ill and died three weeks later. An autopsy revealed that he had suffered over 1000 rem of radiation exposure; he died from radiation illness. His blood contained 0.05 curies of the radioactive isotope polonium-210. (One *curie* is the number of decays per second in a gram of radium: 3.7×10^{10}.) Nobody knows how the polonium was administered to him, but some speculate that it was put into Litvinenko's food.

Naturally, many people suspected Putin, or at least the FSB. Newspapers at the time reported that only an advanced nuclear laboratory could have produced that much polonium. However, I knew that wasn't the case. Polonium-210 is also used in antistatic brushes—the alpha particles emitted ionize the air (just as radiation ionizes the air in a cloud chamber) and that allows the static charge on film to leak off, so it no longer attracts dust.

To see how hard it would be to get polonium-210, on the Internet I ordered a photographer's antistatic brush. It cost me $8.95, and it contained 250 microcuries of Po-210. To get the dose given to Litvinenko, I would have had to order 200 of these, and that would have cost $1800. (Unless I got a discount for the bulk order.) Expensive, but it didn't require the resources of the FSB to obtain.

Does that mean that polonium-210 is an unrecognized danger, a new murder weapon, and that we have now entered a new era of nuclear terrorism? Not really. It is actually quite tricky to extract the polonium from the ceramic pellets that hold it in the antistatic brush, although that is only chemistry—an advanced nuclear laboratory is hardly necessary. And it is far cheaper and easier to assassinate someone with arsenic. I suspect that whoever murdered Litvinenko may have wanted to do it in a dramatic way to make a point.

Radiation and Cancer

Get ready for a paradox: the average dose that it takes to induce cancer is a whole-body dose of approximately 2500 rem.[8] But a smaller dose of 1000 rem will kill you within a few hours from radiation illness. So how can anyone get cancer from radiation? You might (incorrectly) think that the victim would always die first. The solution to this paradox is that even with a small dose, there is a proportionate chance that cancer will be induced. This phenomenon is called the *linear effect*.

Suppose, for example, that someone gets 25 rem, which amounts to 1% of a 2500-rem cancer dose. There will be no radiation illness. But the fact that it is 1% of a cancer dose means that there will be a 1% chance that it will induce cancer. If a billion people (or any other number) each get 25 rem, then 1% of them will get this extra cancer. (I say *extra* cancer, because in addition to this 1% who will get radiation-induced cancer, 20% of them will die from "natural" cancer.) The data used to show this are plotted in figure 4.3.

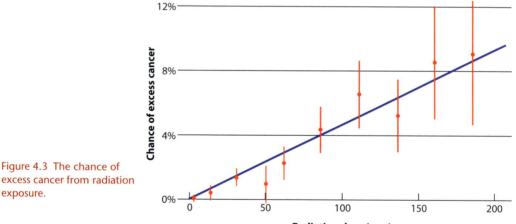

Figure 4.3 The chance of excess cancer from radiation exposure.

The dots indicate the probability of excess cancers from the doses shown on the horizontal axis. The vertical lines above and below each data point show the uncertainty in the measurement. A straight line passes through most of the data—that's why it's called the linear effect. Look at the number for 100 rem. That's 4% of the 2500-rem cancer dose, and it gives a 4% chance of excess cancer. 25 rem is 1% of a cancer dose, and (according to the line) that gives about 1% chance of excess cancer.

Let me now explain why we consider 2500 rem to be a cancer dose. Suppose that we take 2500 rem and break it into 100 parts, and expose 100 people to 25 rem each. None will get radiation illness. Each person has a 1% chance of ex-

---

[8] It was once thought that it took 4 times more than this, and some old books will still show the value of 10,000 rem. But 2500 rem is the current best estimate.

cess cancer. So, on average, for 100 people, we expect 1 excess cancer total from our 2500 rem. Of course, to get the one cancer, we had to spread the dose out among many people, to avoid killing anyone outright from radiation illness.

THE LINEAR HYPOTHESIS

Is the linear effect valid even for very low doses? That turns out to be a very important question. The answer will affect many public issues: the levels of allowed radiation for the public; whether we need to evacuate a contaminated area; how carefully we need to store radioactive waste; and even whether we can use x-rays to detect smuggling (since people might be inadvertently exposed). The assumption that the line in the plot correctly predicts cancer effects at very low levels is called the linear hypothesis.

Is the linear hypothesis true? Unfortunately, we don't know the answer. Suppose that you are exposed to 2.5 rem—that is, 1/1000 of a cancer dose. Does that exposure give you a 0.1% (i.e., 1/1000 chance) of extra cancer? If you believe the linear hypothesis, then the answer is yes. But look at the data point all the way on the left, the point that shows the measured cancer rate for an exposure of about 2.5 rem. Notice the vertical position of the point: it actually lies at zero excess cancers. But the uncertainty is large enough that the error bar passes through the linear hypothesis line, and so maybe 2.5 rem does give a 1/1000 chance of excess cancer. But maybe it doesn't. Given the uncertainty, we don't know.

The assumption that the line represents the true chance of cancer even at very low doses is called the *linear hypothesis*. The basis for belief in the linear hypothesis is the fact that cells are being damaged all the time, by chemicals, by invasive germs, by stress, and by aging. If you damage them a little more, even if just a tiny bit more, you increase that damage, and therefore increase the cancer probability.

Yet not all experts believe the linear hypothesis. They argue that the cell can repair minor damage, so if the damage is spread out enough, every body will be able to recover. We know, for example, that the linear hypothesis doesn't work for radiation *sickness*. Although 1,000 rem will cause fatal radiation illness, 1 rem per person will cause no radiation illness whatsoever, not even with low probability. Moreover, the linear hypothesis doesn't work for most other kinds of poisons, such as arsenic. Many chemicals (including some vitamins) are essential to human life in low doses but fatal in high doses.

The proponents of the linear hypothesis argue that cancer is very different from radiation sickness. Radiation illness occurs when the radiation damage overwhelms the body's ability to recover. It is analogous to poisons such as arsenic, which also have thresholds. Cancer appears to be a much more probabilistic illness. You develop cancer when you get, by chance, exactly the worst possible kinds of mutations, creating cells that grow and divide and will not be turned off by normal bodily controls.

Why don't we just do the scientific study and find out if the linear effect is valid at low doses? The reason is that cancer is a common disease, and that makes small increases difficult to observe. Even without radiation, 20% of people die from cancer. So with 2500 people, you expect 500 cancers anyway, even with no exposure to radiation. Now let's look at what will happen if each of the 2500 people is exposed to 1 rem. According to linear hypothesis, we expect one additional cancer in the group—i.e., 501 cancers total. Statistical fluctuations make such a small effect virtually impossible to verify. Even with large numbers of people exposed (e.g., in the Chernobyl nuclear reactor accident; see the next section), the effect tends to be obscured by statistical fluctuations and systematic uncertainties.

But a premature death from cancer is a tragedy for anyone, even if it doesn't appear in the statistics. One additional cancer is *significant* (especially to the person affected), even if it is not *statistically significant*. That's why many people think we should assume the linear hypothesis, even if it is experimentally unverified, and use it as a basis for public policy. Since the linear hypothesis forms the basis for much public discussion, it is important for future presidents to know what it implies. It is equally important to know that it may not be correct.

THE CHERNOBYL DISASTER

In 1986, the Chernobyl nuclear power plant in Ukraine had a violent accident. There was an explosion in the vessel that contained the radioactive fuel, and a huge amount of radioactivity was released into the atmosphere by the subsequent fire (figure 4.4). We'll discuss the innards of nuclear reactors in the next chapter; for now, all you need to know is that a vast amount of radioactive ma-

Figure 4.4 Chernobyl nuclear power plant, after the fire that released the radiation.

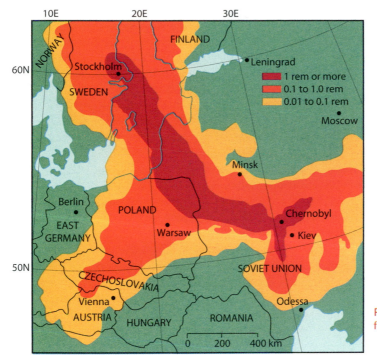

Figure 4.5 Map of radiation from the Chernobyl disaster.

terial was released. Dozens of the firefighters who put out the Chernobyl blaze died from radiation sickness. Radioactivity from the plant was carried by wind over populated areas. This was one of the biggest news items of the 1980s; everybody who was an adult at that time remembers it. There was fear around the world as the radioactive plume drifted. Some radiation from this event was detected in the United States.

The Chernobyl event was one of the most famous and important events of the entire decade. It was in the newspapers for months. It is cited today by many people, those who oppose the further use of radioactive processes (such as nuclear power) and those who are in favor of nuclear power.[9] People use Chernobyl data to support their point of view, regardless of what that point of view is. So it is important to understand this event. It would be worthwhile to read more about the Chernobyl accident. You will find many links to the Chernobyl disaster on the Internet.

Even though a lot of effort went into mapping the spread of the radiation, most of the damage was done in the first few days. So it is hard to know the total radiation exposure to humans. The firefighters who died received doses of several hundreds of rem. Initial estimates were that between 25,000 and 40,000 people had received a median dose of 45 rem. The government decided to evacuate all regions in which a person would receive a lifetime dose of 35 rem or more. The cloud of debris blew over 1,000 miles, toward Stockholm, Sweden, giving most people even in that distant city a dose exceeding 1 rem. A map of the areas most severely exposed is shown in the figure 4.5. An estimate by the

---

[9] Those in favor say that this event shows the very *worst* that can happen.

International Atomic Energy Agency is that the total dose to the world population was 60,000,000 rem. This number is obtained by estimating the number of rem received by each person on the Earth and then adding those numbers together.

If we assume that the linear hypothesis is correct, we can now calculate the number of cancers induced in a remarkably simple way: just divide the number of rem by 2500 to get 60,000,000/2500 = 24,000 cancers worldwide.

Several groups hotly disputed this value. There was even a counterestimate made by an antinuclear group saying that the number of deaths was actually 500,000. I believe that this estimate was made by looking at death rates in the Chernobyl region and attributing all excess deaths to the accident. That is not an accurate way to do it. A movie titled *Chernobyl Heart* was based on these very large estimates.

Who is right? That requires a careful look at the way the data were analyzed. Some people will tend to believe the International Atomic Energy Agency, because they have established a good reputation with past scientific work. Others will distrust them, because the people who did the work are scientists, many of whom have worked for organizations such as the Department of Energy (DOE, formerly the Atomic Energy Commission, or AEC), and are therefore suspected of pronuke bias. But they made the best scientific estimate, so I think the 24,000 figure is probably most accurate.

Now let's calculate the expected cancers for the people near the nuclear plant. Based on the numbers given earlier, assume that 40,000 people received 45 rem on average.[10] Multiply these to get the total of 1.8 million rem. Divide by 2500 to find the total number of cancers expected, equal to 720. (These people were already counted in the 24,000 number.)

This is clearly a great tragedy. And yet, a surprising fact for many people is that even if this calculation is correct, and the number of cancers predicted is accurate, it will be difficult to identify the people who are being killed by the Chernobyl accident. There are too many other cancers.

Let's consider the nearby victims. Among 40,000 people, we would expect 20% to die of cancer anyway. (This is probably a low estimate for the Chernobyl vicinity because people in this region also have significant rates of illness from heavy smoking and alcoholism.) That means 8000 cancer deaths that are *not* a result of the radiation. On top of these, we predict another 720 deaths from the radiation exposure. So the total cancers, instead of being the "normal" 8000, will rise to 8720. Of these 8720, only 8% were caused by the Chernobyl accident. It is difficult to know who is dying from "ordinary" cancer and who is dying from the radiation effects, unless the kinds of cancer induced by radiation are different. (There are some differences: thyroid cancer is more prevalent, for example, from radiation—although it rarely leads to death.)

In the population of 100 million people who received measurable radiation, we expect there to be approximately 20 million cancers from other causes. Let's pretend for a moment that this number is exact: 20,000,000 "ordinary" cancers. Because of Chernobyl, this number will be increased to 20 million + 24 thousand = 20,024,000. Put another way, the probability of anybody getting

---

[10] 45 rem was the median, not the mean, so this is a rough approximation.

cancer, if exposed, has been increased from 20% to 20.024%. Given the natural variations in cancer, most people think that we will not be able to see the increase statistically.

Yet those 24,000 people are all individuals who otherwise would not have died from cancer. There is a strange paradox here: tragedy is occurring, and yet is almost invisible among the much larger tragedy of cancer from other causes.

CANCER FROM THE NUCLEAR BOMBING OF HIROSHIMA

I stated at the start of this chapter that, in Hiroshima, fewer than 2% of the deaths caused by the atomic bomb were from cancer. The reason is simple: unless you were far from the center of the blast, you were killed by the blast or fire. Still, some who survived the fireball received enough radiation to die of radiation sickness. The best estimate is that there were about 52,000 survivors (not killed by those other effects) who received 0.5 rem or more; the average dose for these people was 20 rem. That means that the total dose was $52,000 \times 20 = 1,040,000$ rem. Divide this by 2500 rem per cancer to get the total expected number of cancers: $1,040,000/2500 = 416$. That is 0.8% of the 52,000 people.

Estimates of the total number of people killed by the Hiroshima bomb vary between 50,000 and 150,000. (It is hard to know how many people were there when the destruction was so total in the core of the city.) But you can see now why the cancer deaths were less than 2% of this.

Cancer deaths could have been higher if the bomb had been exploded when it was close to the ground, rather than at high altitude, because that would have increased the amount of radioactive *fallout*. We'll discuss fallout further in chapter 5.

HIGH RADIATION IN DENVER

Denver is located in a geologic area that emits (from the ground) higher-than-average amounts of the radioactive gas *radon*. A reasonable estimate[11] is that the average yearly *excess* in Denver (compared to the U.S. average) is about 0.1 rem per person per year. For 2.4 million people living in Denver for 50 years, this excess amounts to $(0.1)(2.4 \times 10^6)(50) = 1.2 \times 10^7$ rem, which should cause 4800 excess cancers.

So here is another paradox: the actual cancer rate in Denver is lower than the average in the United States. How can that be? Does it mean that the linear hypothesis is wrong? Maybe exposure to low levels of excess radiation helps immunize you from cancer! Or could there be other effects, even more important than radioactivity, that cause cancer? Can you guess what they are? Do you expect that there would be differences in lifestyle, eating habits, exposure to sunlight and ultraviolet radiation, or genes? Anything else? Does that mean that someone who moves to Denver lowers his risk of cancer?

---

[11] At an EPA Web site (www.epa.gov/rpdweb00/understand/calculate.html), you can calculate the radiation dose you would get in different locations around the country. To convert their answer from mrem to rem, divide by 1000.

In case you are hoping that I can give you the answer, I can't. Nobody knows the answer. But there are lots of possible explanations. In the meantime, I would not consider the excess radiation of Denver to be an important factor in a decision to live there.

However, before you buy a house in the Denver area, you might want to measure the radon level to make sure that you are not buying one of the highly radioactive ones. This is a serious comment—some houses have been measured to have dangerous levels.[12] The Environmental Protection Agency (EPA) offers guidelines for making such measurements, and commercial devices are available.

TOOTH AND CHEST X-RAYS

You are exposed to radiation every time you get an x-ray photograph taken of you. Such photographs are now called *x-rays* themselves, but that is just short for "x-ray photograph" or "x-ray image." You may have noticed that the person who takes the image leaves the room when the x-ray machine is turned on. For a tooth x-ray, a lead shield may be placed over you body to protect your "vital organs." This frightens many people. What is the danger?

Next time an x-ray is taken, ask the technician for the dose in rem. (Odds are, the technician won't know, but will simply assure that you it is safe.) A typical dose for a dental x-ray is less than 1 millirem (i.e., 0.001 rem); for the purpose of the calculation, let's assume that the dose is 1 millirem = 1 mrem = 10^{-3} rem.

The numbers that we gave for cancer assumed a whole-body dose. That means that they gave the cancer rate assuming that every part of your body got the same dose. But when your tooth and jaw are exposed, it is probably not more than a pound of flesh that is exposed. Let's assume that it is 1% of your body. Then, by the linear hypothesis, such an exposure is only 1% as dangerous as a whole-body dose. That makes it equivalent to 1% of a millirem whole-body—i.e., 10^{-5} rem. From the linear hypothesis, it takes 2500 rem per cancer. That would require $2500/10^{-5} = 250{,}000{,}000 = 250$ million tooth x-rays. Put another way, one tooth x-ray will induce cancer with the chance of 1 in 250 million = $1/250{,}000{,}000 = 4 \times 10^{-9}$. The risk of you dying from an abscessed tooth is probably much higher.

Let's now calculate the risk of cancer from a chest x-ray. Modern chest x-rays are about 25 millirem to about 50 lb of your body. That is a much greater dose than you get from a tooth x-ray. It is 25 times more millirem to 50 times more body, which means that the dose is $25 \times 50 = 1250$ times greater. The number of cancers per chest x-ray should also be 1250 times greater than for the tooth x-ray. That number comes out to be $1250 \times 4 \times 10^{-9} = 5 \times 10^{-6}$. Let's assume that without a chest x-ray, your chance of cancer is exactly 20%. Then after the x-ray, your chance has risen to 20.000005%.

---

[12] Ironically, the highest level ever measured was in a demonstration house built by the Department of Energy. To save energy, the house was built with recirculating air, and as a result, the radon that leaked into the house from the ground could not easily escape.

X-RAYS AND PREGNANCY

Radiation can be particularly dangerous to the fetus. Mutation in one of the stem cells (cells that can turn into other cells) can lead to mental retardation, malformed growth, or cancer. If the mother has a tooth x-ray, or one to her ankle, then the danger is tiny.

As with other low-dose effects, our actual knowledge of the effects of x-rays is limited; it is based primarily on high-level exposures (e.g., in accidents and during World War II) and the linear hypothesis. The United Nations Scientific Committee on the Effects of Atomic Radiation (UNSCEAR) has studied this problem and concluded that the risk to the fetus is about 3% for each rem of exposure.

If a dental x-ray were applied directly to the fetus, the linear hypothesis says that the 1-mrem dose would result in a risk 1000 times smaller—i.e., 0.003% for each mrem. If the radiation is delivered to the tooth, and the only radiation reaching the fetus is 100 times smaller, then the risk goes down by a similar factor. It is possible that the dangers of the mother's untreated tooth can be greater to the fetus than these small amounts of radiation.

ULTRASOUND

Many pregnant women are exposed to ultrasound. Although technically ultrasound is a form of "radiation" (as is sound itself), it does not have any of the ability to cause mutation that x-rays or beta rays have. Ultrasound is high-frequency sound, and it does not have the capability of delivering enough energy to individual cells to cause mutations.

Ultrasound could have other negative effects, but ultrasound radiation should not be confused with nuclear radiation. We'll talk more about ultrasound in chapter 9.

Radiation to Cure Cancer

This may seem paradoxical, but one of the most effective methods to cure cancer uses radiation. Cancers are not healthy cells; they have been mutated to absorb nourishment and divide rapidly. Because they are specialized for division rather than long life, many cancer cells are more vulnerable to radiation poisoning than are healthy cells. So a common way to treat cancer is to hit it with high levels of radiation.

This is typically done by aiming the radiation right at the cancer. The radiation can enter the body from many directions, but it is focused on the cancer. The focusing is designed to make sure that the cancer cells get a higher dose than do the surrounding cells. Cells in the vicinity of the cancer also have their metabolism disrupted, and that can induce radiation illness. The goal is to kill the less resistant cancer cells; other cells get sick but recover. The difficulty, as with all cancer treatment, is that it is necessary to kill virtually all the cancer cells in order to prevent the cancer from returning.

Radiation treatment is often combined with chemotherapy (cancer cells are also less resistant to poisons), so you cannot always tell which effects are due to which treatments. Maybe for this reason, radiation treatment is often confused with chemotherapy. Don't confuse them.

Some people avoid radiation treatment because they fear the side effects. Others avoid it because they fear that the radiation will induce cancer. That argument does not make numerical sense. Compare the tiny risk of an additional induced cancer to the extreme danger of the cancer that is already present. But the fear of radiation is so great that some doctors cannot convince the cancer patient to get this treatment.

Dirty Bombs

Can a terrorist explode a tank of radioactive material in the middle of a city and make the city uninhabitable for the foreseeable future? A device that releases large amounts of radioactivity is called a *dirty bomb* or a *radiological weapon*. This scenario has received a lot of attention in the news. Yet it is harder to do this than you might think.

Assume, for the moment, that you are a terrorist and you are designing such a weapon. You want to spread radioactive material over a square kilometer. Your goal is to make the area so dangerous that people would get radiation illness if they were exposed for 1 hr. Let's say that the bomb will be roughly 1 m in size and will be delivered by a truck. To do this, you will have to concentrate the radioactive material by a factor of 1 million (since 1 km^2 contains 1 million m^2). That's a real problem for the terrorist. He is transporting something that is a million time more intense in its radioactivity than when it is spread out. That means that someone within 1 meter of the bomb will get a lethal dose in 1 millionth of an hour rather than in an hour—i.e., in 3.6 milliseconds. To avoid that, the person might try to stay 10 ft (3 m) from the bomb. But that distance would still give an LD50 dose in 36 milliseconds.[13]

To avoid this, the terrorist might try shielding the radiation with lead. A half-centimeter shield of lead will reduce the radiation by a factor of 3—that's not very much. A full-centimeter shield will reduce it by 9, and a 2-centimeter shield will reduce it by 81. Let's try that. The weight of a 1-meter cube with lead walls 2 cm thick is about 1.6 tons. You could carry that in a truck. It would still give an LD50 dose, at a distance of 10 meters, in 3 seconds. That doesn't mean that a radiological weapon is impossible. It just means that it is much harder than many people suppose.

José Padilla was a Chicago street thug recruited by al Qaeda to build and detonate a radiological bomb in the United States. (I recommend a Google search on his name.) He was captured by the police before he made any definite plans. But in the deposition for his trial, we learned that al Qaeda had told him to abandon plans for a radiological weapon and instead use natural gas to

---

[13] The radioactivity spreads out over a larger area. The area at a distance of 3 m is about a factor of 9 greater, so the dose is 9 times lower.

blow up some apartment buildings. I think it is likely that al Qaeda realized that a working radiological weapon was essentially beyond the means of terrorists and instead are likely planning to use again the method that devastated the World Trade Center—exploding fossil fuel.[14]

What Are the Rays, Really?

It took many years to figure out what the rays really were. Here are the answers:

- **Alpha rays** are chunks consisting of 2 protons and 2 neutrons moving at a high velocity. This is identical to the nucleus of a helium atom. Because we know these rays are actually particles, they are sometimes referred to as *alpha particles*, or simply *alphas*. When alpha rays finally slow down, they usually attract 2 electrons (either free electrons or ones that were weakly attached to atoms) and form a helium atom.
- **Beta rays** are energetic electrons. They are much lighter than particles like alphas, yet they move so fast that they have energy comparable to that of the slower alpha particles. When betas finally stop (after numerous collisions), they usually attach themselves to an atom.
- **Gamma rays** are packets of very energetic light. Gammas travel at the same speed as light (3×10^8 m/sec), but they typically carry a million times as much energy as a single packet of visible light. (We'll talk about light packets, also called *photons*, in chapter 11.)
- **Neutrons** are massive particles like protons, but with no electric charge. Neutron emission is very important in the chain reaction, which we will discuss in chapter 5. The neutron bomb is a nuclear weapon that emits a large number of neutrons. Its purpose is to induce radiation illness in people, while doing relatively little damage to buildings. The ethics of such weapons were hotly debated in the 1970s.
- **X-rays** are the most famous of the radiations, since they have become so important in medicine and dentistry. They are packets of light, like gamma rays, but a factor of 10 to 100 times less energetic. They pass through many materials, such as water and carbon, but are quickly stopped by elements with a high atomic number, such as calcium and lead. Because x-rays are stopped by calcium, this allows them to be used to look at tooth decay and broken bones. An x-ray photograph is actually a shadow of the calcium projected on a piece of film. The character Superman could not use his x-ray vision to look through lead, since it is one of the heaviest elements and therefore absorbs x-rays.[15] Lead is used to protect from x-rays. As I said earlier,

---

[14] For an essay I wrote on radiological weapons, see "The Dirty Bomb Distraction" at www.muller.lbl.gov/TRessays/29-Dirty_Bombs.htm.

[15] He would also not be able to look through uranium or plutonium, but I don't think that ever made it into the stories. For some unexplained reason, he was highly allergic to kryptonite, which

when you get a tooth x-ray, the dentist will likely put a lead-filled apron over other parts of your body to make sure that scattered x-rays don't reach your vital organs.

- **Cosmic rays** are any radiation that comes from space. Cosmic rays consist of protons, electrons, gamma rays, x-rays, and muons— unusual particles that can pass through a hundred meters of rock, and are themselves radioactive. Because of their penetrating power, they have been used to x-ray the Egyptian pyramids.

- **Fission fragments** are a particularly dangerous kind of radiation that is emitted when a nucleus undergoes fission—i.e., it splits into two or more pieces. Fission fragments are chunks containing large numbers of protons and neutrons, and they are themselves usually highly radioactive. Their real danger comes from when they stop and then redecay. These are the radioactive particles that make fallout from nuclear bombs so dangerous. Fission fragments also make up the most radioactive part of nuclear reactor waste.

- **Cathode rays** were discovered emanating from hot metal that had a high voltage applied. It wasn't known that the rays are actually electrons; the electron hadn't been discovered yet. That had to wait for the 1897 work of J. J. Thomson. Most people still call the device that uses these beams *cathode ray tubes*, or CRTs. As of the early twenty-first century, many TV and computer screens are still made of CRTs, but they are quickly disappearing. If you're young, your children may never hear this term, since in the near future all CRTs will likely be replaced with thin screens.[16]

- **Neutrinos** are the most mysterious of all the rays. They are usually emitted at the same time that beta rays (electrons) are emitted from a nucleus. Neutrinos have such small mass that, even when they have only moderate energy, they move nearly at the speed of light. They do not "feel" electric or nuclear forces, so they pass through the entire Earth with only a small chance of hitting anything. They are so strongly emitted by the Sun that over 10^{10} of them are passing through every square centimeter of your body every second. Nevertheless, they are the least dangerous radiation you are exposed to. Because they pass through matter so easily, neutrinos are sometimes referred to as *ghost particles*.

- **Radiation from cell phones** occurs in the form of microwaves, which are very low energy packets of light, even lower energy than those of visible light. Microwaves deposit their energy in the form of heat; that's why they can be used in microwave ovens. Microwaves do not break DNA molecules in the body, and therefore they pose no risk of causing cancer in the way that alpha, beta, and gamma rays and even sunlight can. Much of the fear of microwaves comes from the fact that they share the name *radiation* with the other, far more dangerous forms such as gamma radiation.

I presume is a compound containing the element krypton. Maybe that came from too much childhood exposure.

[16] For more on the history, see: www.aip.org/history/electron/jjrays.htm.

You Are Radioactive

A typical human body contains approximately 40 grams of potassium. Most of this is the stable, nonradioactive isotope potassium-39. Each nucleus of potassium-39 contains 19 protons and 20 neutrons, totaling 39 (that's why it is called potassium-39). But about 0.01% of the potassium atoms have an extra neutron in their nucleus; these are called potassium-40. Potassium-40 is radioactive. This means that your body contains 40/10,000 = 0.004 g = 4 mg of a radioactive cancer-producing isotope. The number of radioactive potassium-40 atoms in your body is 6×10^{19}. This is not an artificial radioactivity, but it is left over from the formation of potassium in the supernova that gave birth to our Solar System (more on this in chapter 13).

Potassium-40 is often abbreviated as *K-40*. The *K* comes from the Latin name "kalium" for pot ashes—the original source of *pot*assium. Parts of the word kalium also survive in the word "alkali."

Approximately 1000 atoms of K-40 (read this aloud as "potassium-40") explode in your body every second. Your body is radioactive. About 90% of the explosions produce an energetic electron (beta ray); most of the rest produce an energetic gamma ray. So there are about 1000 self-inflicted radiations per second from your own body. This radioactivity within your body produces a dose of approximately 0.016 rem = 16 millirem over a 50-year period. If the linear hypothesis is correct, we calculate the cancer induced by dividing the rem by 2500. Your chance of having a self-induced cancer is $0.016/2500 = 6.4 \times 10^{-6}$, i.e., about 6 chances in a million. That's small, although it is higher than your chances of winning a typical grand lottery.

The results are more interesting if you think about the consequences for a large population. There are about 300 million people in the United States. Multiply 300 million by 6 millionths of a cancer per person, and you find that $300 \times 6 = 1800$ people will die of cancer over the next 50 years in the United States, induced by their own radioactivity. That averages to 36 per year in the United States. If you sleep near to somebody, then their radioactivity can affect you (see the discussion topic at the end of this chapter).

A second source of radioactivity in our bodies comes from carbon-14, also called *radiocarbon*, and abbreviated as *C-14*. The C-14 nucleus is similar to that of the ordinary C-12 nucleus, except that it has two extra neutrons (increasing the atomic weight from 12 to 14). But, it turns out, those extra neutrons make carbon-14 radioactive. In carbon-14, one of the neutrons will explode, emitting an electron and a neutrino. When the electron and neutrino are emitted, the neutron turns into a proton, so the nucleus has one more charge—that changes it from carbon to nitrogen. On average, half of the carbon-14 atoms in your body will explode in 5730 years. That period of 5730 years is called the *half-life* of C-14.

Every gram of carbon in your body has 12 atoms of carbon-14 exploding every minute. That is equivalent to 1 explosion every 5 seconds, on average for every gram . In an average body, there are about 3,000 such radioactive explosions every second.[17] This is in addition to the 1,000 K-40 decays mentioned

---

[17] This assumes that you weigh 150 lbs and are 23% carbon (typical for humans).

earlier. Rather than state it this way, many scientists will say that the radio-activity is about 4000 Becquerels. That way, they can honor Henri Becquerel (who discovered radioactivity), impress the nonexpert (but they won't impress you), and totally confuse people who don't know that the number of Becquerels just means the number per second.

Now here is the really fascinating thing about C-14: we can use it to measure how long things have been dead. To see how this works, we have to understand a very strange phenomenon in radioactive decay that is called the *half-life rule*.

The Mysterious Half-Life Rule

As mentioned earlier, half the atoms of C-14 decay in 5730 years. It is natural to assume that the remainder would decay in the next 5730 years, but that is not what happens. Only half of them decay in the subsequent 5730 years. And then in the following 5730 years, only half of the remainder will explode. The remaining fractions after various years are shown in table 4.3.

Table 4.3   Radioactive Decay of C-14

Age in years	Number of half-lives	Fraction remaining
5730	1	1/2
11,460	2	1/4
17,190	3	1/8
22,920	4	1/16
$5730 \times N$	N	$1/2^N$

The value of the half-life is different for different radioactive isotopes, but the behavior is similar (table 4.4). For K-40, the half-life is 1.25 billion years. That means that in 1.25 billion years, half of the K-40 decays. In the next 1.25 billion years, half of the remaining atoms decay. There have been nearly four half-lives since the Earth was formed about 4.6 billion years ago. That's why there is so much K-40 left; since the formation of the Earth, there hasn't been enough time for all the K-40 to decay.

By the way, did you notice how similar this rule is to the rule we used to calculate the density of air at different altitudes? Recall that at an altitude of 18,000 ft, the density of air is half of what it is at sea level. Go up another 18,000 ft, and the density is reduced by another factor of 1/2, and so forth.[18] At 180,000 feet (10 such steps) the density is $(1/2)^{10}$. The math is identical to the formula for radioactive half-lives!

---

[18] These numbers are approximate because the temperature also changes with altitude and that also affects the density.

Table 4.4   Half-Lives of Some Important Isotopes

Polonium-215—0.0018 second	Tritium (H-3)—12.4 years
Polonium-216—0.16 second	Strontium-90—29.9 years
Bismuth-212—60.6 minutes	Cesium-137—30.1 years
Sodium-24—15.0 hours	Radium-226—1,620 years
Iodine-131—8.14 days	Carbon- 14—5,730 years
Phosphorus-32—14.3 days	Plutonium-239—24000 years
Iron-59—6.6 weeks	Chlorine-36—400,000 years
Polonium-210—20 weeks	Uranium-235—710 million years
Cobalt-60—5.26 years	Uranium-238—4.5 billion years

RADIOACTIVE DECAY

If you have a large number of radioactive atoms, half of them will explode in one half-life. (It may not be exactly half, since the rule is based on probabilities.) Thus, one half-life later, there are only half as many radioactive atoms. That means that the number of radioactive explosions per second will only be half as great as it was in the beginning. This reduction of radiation was originally called *radioactive decay*. But now the word *decay* is applied much more universally. Physicists typically talk about a single nucleus undergoing radioactive decay. In this context, that word is used far more commonly than the word *explosion*.

NUCLEI DIE, BUT THEY DO NOT AGE

The half-life rule is true for all the known kinds of radioactivity. But the more you think about it, the more mysterious it is. People don't die following the half-life rule. When we are born (at least in the United States), we are expected to live about 80 years. If we make it to age 80, we don't expect to live another 80 years—yet that is the way it would work if our physical aging followed the half-life rule.[19] Please appreciate how strange the half-life behavior is! It is as if the atom does not age. The old carbon-14 is identical to the young one. No matter how old it is, its expected half-life is still 5730 years.

We don't really understand this phenomenon, but physicists sometimes "explain" it by saying that radioactive decay is determined by the laws of quantum mechanics, which are probabilistic laws. For K-40, we say that the probability of decaying in 1.25 billion years is 50%. This probability doesn't change, so no matter how old the atom is, its probability of decay remains 50% for the next 1.25 billion years. Of course, we haven't really explained anything, because we don't know *why* the laws of physics should be probability laws.

[19] In fact, at age 80 years we expect to live, on average, another 9 years.

RTGs: Power from Radioactivity

On 20 January 2006, the United States launched its *New Horizons* satellite to go all the way to Pluto. It should arrive in 2016. It will send back data and photos about that tiny planet. (I still call it a planet, even though the International Astronomical Union thinks it isn't one.) Where will it get the energy for its transmissions?

Solar power? No. Since Pluto is about 30 times farther from the Sun than is the Earth, the solar power is reduced by a factor of $30 \times 30 = 900$, to about one watt per square meter. Good solar cells convert about 41% of that to electricity (see chapter 1), but 0.41 watts per square meter is still too weak to be used.

Other sources were considered, including batteries and fuel cells. But they cannot provide continuous power for 10 years, and they are heavy. Weight is an important consideration, since only a light satellite could be launched at a high enough speed to reach Pluto in a decade.

For these reasons, NASA chose to use radioactivity to supply the power. The satellite contains 11 kg (24 lb) of plutonium-238 (Pu-238). Its radioactivity creates heat at the rate of about 600 watts/kg, for about 6.6 kW total. A thermoelectric generator (wires consisting of different metals in contact) converts about 7% of this heat to electricity, giving about 460 watts of electric power. The combination is called a radioisotope thermoelectric generator, or RTG.

NASA used Pu-238, which is 1 neutron lighter than the chain-reaction version Pu-239, because Pu-238 decays with a half-life of 87 years. This rate is fast enough for a large number of nuclei to decay over the 10-year mission and produces enough high power at low weight. Yet, the rate is low enough that the atoms are only partially used up during the 10-year mission, so the power doesn't decrease too much over that period.

RTGs have been used for many years. The *Voyager* spacecraft used an early version of an RTG in 1977 for the same reason as *New Horizons*: it was to be the first satellite to go beyond the planets into "deep space."[20]

Some people oppose the use of RTGs because it involves plutonium. They argue that a failed launch could cause the plutonium to fall to the Earth and cause environmental damage. For the past few years, we have had no facility to produce Pu-238 in the United States, and so we have been buying this material from Russia. There are now proposals to create new facilities in the United States to make it.

OPTIONAL: HOW DO WE MAKE PU-238 FOR RTGS?

You don't have to know how Pu-238 is made, but you might find it interesting. It is worth reading through this section quickly, just to get a sense of how they do it. You don't have to remember the details. In a nuclear reactor (discussed in the next chapter), neutrons are absorbed on U-238, making U-239, which then beta decays into neptunium-239 (Np-239), and then Pu-239. That's our source of Pu-239, used in bombs. But some of the Pu-239 absorbs a neutron, becoming Pu-240. That (as we'll see in the next chapter) pollutes the bomb

---

[20] *Star Trek: The Motion Picture* (1979) features this satellite as a key element of its plot.

material. As time goes on, some of the Pu-240 absorbs another neutron and becomes Pu-241. Pu-241 is radioactive, and it emits an electron (beta decay, with a half-life of 14 years) to become americium-241 (Am-241). The americium decays to make Np-237.

At this point, the Np-237 is separated into a special container and put back into the reactor. In the nuclear reactor, the Np-237 can absorb another neutron to become Np-238. This is radioactive with a half-life of 2 days. It emits an electron and becomes Pu-238. The container is then removed, and the plutonium is separated from any neptunium.

Sound complicated? It is. This is radioactivity high-tech. There are hundreds of other processes that produce radioactive materials for special uses. Many of these are for medicine. The medical technician who uses specialty isotopes usually has no need to know the complex technology that is used to produce them.

SMOKE DETECTORS

The most common smoke detectors have a small radioactive source underneath the cover. It is typically an alpha particle emitter, with alpha particles that travel only about 1 centimeter in the air before they stop. These alpha particles knock electrons off the air molecules, and that makes the air electrically conductive. This conductivity is measured with a battery. If the air is conductive, then the alarm does not sound.

However, if smoke drifts under the cover, then the electrons tend to stick to the smoke particles. That means that they are no longer free to move, and the electric current stops. When the electronics detects that electric current is no longer flowing, it sounds the alarm, usually a piercing screech.

Obviously, it is important that the battery works. Smoke alarms measure the battery strength; if it gets weak, the electronics emits short beeps to alert you.[21]

MEASURING AGE FROM RADIOACTIVITY

If a rock has minerals in it that contain potassium, then it is often possible to tell when the rock was first made. That's because all potassium on the Earth contains about 0.01% of the isotope K-40, and K-40 has a nice property: when it undergoes a radioactive decay, it turns into argon gas. This gas is then unable to escape from the solid rock and it accumulates there. The gas can escape only if the rock is melted.

To see how that can be used to measure the age of a rock, let's consider a specific example. Suppose that we find a rock formed from lava—i.e., it was once liquid. We would like to know when it turned solid—i.e., when it became a rock. We look and see whether the rock has potassium in it. If it does, then we know part of that potassium is turning into argon gas. We can see how much argon gas is trapped in the rock and use that information to determine how

---

[21] The beeps, in my experience, appear to be on a musical note that is very hard to locate. Moreover, the beep typically comes every 10 minutes so that the weak battery isn't completely drained. That makes it much much harder to locate.

long the rock has been a solid. This technique, called *potassium-argon dating*, is enormously useful in geology for measuring the age of rocks and of ancient volcano flows.[22]

Archeologists can use the radioactive isotope of carbon, C-14, to measure the age of fossils. This is called *radiocarbon dating*. C-14 is produced in the atmosphere by cosmic rays. It is absorbed into plants when they create carbohydrates out of atmospheric carbon (by "breathing" carbon dioxide). We eat the plants, or animals that ate the plants, or animals that ate the animals that ate the plants . . . and so we get C-14 into our bodies. Since the path from the atmosphere to our bodies happens so fast (typically, less than a year), our carbon has nearly the same radioactivity as the atmospheric carbon: 12 decays per minute for each gram of carbon.

When we die, and no longer eat food, the carbon-14 decays, and it is not replaced. If an archeologist finds a fossil and measures that each gram of carbon in it is has only 6 decays per minute of C-14 rather than 12, then he knows that creature died one half-life ago—i.e., 5730 years ago. This method is the primary means for measuring ages in archeology.

Suppose that the archeologist measures 3 decays per minute. (Remember, 12 decays/minute indicates an age of 0.) What is the age of the fossil? Careful—this is potentially a trick question. Try it, and then check the footnote[23] for the answer.

After 10 half-lives, the radioactivity has been reduced by a factor of $(1/2)(1/2)(1/2)(1/2)(1/2)(1/2)(1/2)(1/2)(1/2)(1/2) = 1/2^{10} = 1/1024 = 0.001$. So instead of 12 decays per minute, the archeologist will measure only 12 decays per 1024 minutes. Such low rates are very difficult to measure, so C-14 is useful only to ages of about 10 half-lives—i.e., about 57,300 years. Beyond that, the rate is too low.[24]

> **Puzzle:** Why doesn't the carbon-14 in the atmosphere decay away as it does in our bodies?
>
> **Answer:** It does, but it is constantly replenished by new cosmic rays. The level in the atmospheric carbon is set by the level at which the decay exactly balances the production. That turns out to be 1 C-14 for every 10^{12} ordinary carbon atoms. And at that density, 1 g of carbon (which will have 10^{-12} g of C-14) will have 12 decays every minute.

[22] I have used potassium-argon dating to measure the ages of craters on the moon. When an asteroid or a comet hits the moon, it melts some of the rock. Any old argon in the rock is released. In a few seconds, the rock solidifies again and begins to accumulate argon from decaying K-40. When we melt the samples in our laboratory, we can measure the amounts of argon and potassium, and from those numbers we can deduce the age.

[23] If the age were one half-life, we would expect the rate to be 6 decays per minute. After an additional half-life, the rate should be cut to 3 per minute. This matches the observation. So the fossil is 2 half-lives old = 11,460 years.

[24] A better way is to count the remaining C-14 atoms, rather than the decays. Even when the decay rate is low, there is a very large number of atoms left. This can be done with an instrument known as an accelerator mass spectrometers (AMS). The first person to succeed in doing such a measurement is the author of this book.

RADIOACTIVE ALCOHOL

Let's return now to another of the surprising examples that appeared at the beginning of this chapter. In the United States, alcohol for consumption must be made from fruits, grain, or other plants. It is illegal to make it from petroleum. (I don't know why, but it is the law.) Of course, any alcohol made by the fermentation of plant matter contains recent, radioactive carbon-14. In contrast, petroleum was created by decaying vegetable matter that got buried 300 million years ago. A half-life is only 5730 years, so the petroleum was formed from living matter that died over 50,000 C-14 half lives ago. There is no detectable C-14 left in it. This absence provides an easy way for the U.S. government to test to see if alcohol was produced from petroleum. The U.S. Bureau of Alcohol, Tobacco, and Firearms tests alcoholic beverages for C-14. If the expected level of radioactivity from C-14 is present, then the beverage is fit for human consumption. If the alcohol is not radioactive, then it is deemed unfit for human consumption.

For the same reason, biofuels (made from living plants) are radioactive, but fossil fuels are not. This is not a reason to fear biofuels! The radioactivity is not much greater than in the human body. (It is a little bit bigger, per pound, since fuel contains a larger fraction of carbon than does the body.)

Environmental Radioactivity

Is all cancer caused by radioactivity in the environment? No. If you live in a typical city, you are exposed to about 0.2 rem of radiation every year. Most of this comes from radon gas seeping up from the rocks in the ground and cosmic radiation coming from space, and some of it comes from medical x-rays, if you get them. In 50 years, the typical American is exposed to a total of about 15 rem of mostly natural radioactivity. To calculate the expected cancer, you follow our rule: Take the total number of rem (rem per year multiplied by number of years) and divide by 2500. This gives 15/2500 = 0.004 = 0.4%. But 20% of people die from cancer, not 0.4%, so the cancer must be coming from something else.

What else? Some people think it is food, or pollutants, or something else that we might be able to eliminate. But if we add up all known carcinogens, we still can't account for the bulk of cancer. So there is some other cause. It could be something as simple as the natural exposure to the highly reactive chemical known as oxygen[25]—which we can't eliminate unless we give up breathing. Nobody knows.

VOLCANIC HEAT AND HELIUM BALLOONS

The rock in the Earth is radioactive, largely from potassium, uranium, and thorium in the ground. If you have ever been in a deep mine, you know it is very warm—not cold, as it is in a shallow mine or cave. The reason is that heat

---

[25] This is not a joke, but a serious scientific hypothesis put forth by eminent biochemist Bruce Ames.

is gradually seeping up from inside the Earth. Uranium and thorium decay underground and produce a large number of alpha particles. The underground heat is the energy lost by the alpha particles as they slow down in collisions with other atoms. When the alpha particles finally stop, they pick up electrons and turn into helium gas. This gas collects underground along with natural gas (methane) and is extracted along with the methane. It is what we use to fill helium balloons.

The total power generated inside the Earth from radioactivity is about 2×10^{13} watts. That sounds like a lot, but the sunlight hitting the Earth has 2×10^{17} watts, a factor of 10,000 greater.[26] So the energy coming down is much greater than the energy coming up.

The heat generated underground by radioactivity is responsible for volcanoes, thermal springs, and geysers. Under a large glacier (during the ice ages, glaciers were several kilometers thick), the heat from the Earth is enough to melt the ice at the bottom of the glacier, and that keeps it slipping.

The Earth is 6370 km thick (to the center), so it is surprising that 20% of the radioactive component of heat comes from radioactivity near the surface, in the "thin" shell of rock known as the *crust*. The crust is only 30 km thick on average, but it contains a much higher density of the radioactive uranium, thorium, and potassium. At the bottom of the crust, 30 km down, the temperature of the rock is about 1000 C, enough to glow red-hot.

WHY AREN'T MOST ATOMS RADIOACTIVE?

This might sound like a silly question, until you learn the answer. We believe that, in the early Solar System, most atoms *were* radioactive. There are intensely radioactive isotopes of hydrogen, oxygen, nitrogen, calcium, and all the other atoms of which we are made. These were once abundant. But most of them had short half-lives, from fractions of a second to millions of years, and as a result, most of them decayed away.[27] Now, 4.6 billion years later, we are left with only three kinds of atoms: those that are not radioactive (e.g., C-12), those that have very long half-lives (e.g., K-40 and uranium), and those that have been produced in the recent past (e.g., C-14).

OPTIONAL: THE CAUSE OF RADIOACTIVITY—THE "WEAK FORCE" AND TUNNELING

Chemical explosives must be triggered—e.g., gunpowder is exploded by the impact of the hammer of the gun, and TNT is usually detonated with the help of an electrical signal. What triggers the nuclear explosion of radioactivity? Alpha decay and fission (to be discussed soon) come from a quantum-mechanical phenomenon called *tunneling*. In tunneling, a particle that is tightly bound to another one by nuclear glue (the gluons) makes a *quantum leap* away, to a location where the force is weak. At this distance, the force on the particle is

---

[26] It is interesting to note that for the planet Jupiter, the heat coming from the interior is greater than the heat coming from the Sun, although this heat is probably generated by the gravitational contraction of the planet rather than from radioactivity.

[27] In the context of the Earth's age, even a period of millions of years is *short*.

dominated by its electric charge, and it is strongly repelled. That repulsion is what gives the particle its high energy.

For beta (electron) radiation, the answer is totally different. For many years, it was hypothesized that there was a new "force" in the nucleus that was responsible, a force that was so weak that didn't do anything else except, once in a while, trigger radioactive decay. We now know that this force is real, that it is related to the electric force, and that it operates in such a way that it triggers the explosion. Because of its history, it is called the *weak force*. The fact that the force is weak means that it has a very low probability of acting in any one second. For example, it takes 5730 years for a C-14 atom to have a 50% chance of decay.

We now know that the weak force can do more than just cause beta decay. It can also put a force on the particle. The mysterious particle called the *neutrino* has no electric charge. It does not feel electric forces; it feels only the weak force and gravity. Despite its name, the weak force of the neutrino is stronger than the force of gravity. When a neutrino goes through the Earth, there is a small but nonzero chance that it will collide with one of the atoms in the Earth, because it can feel their weak force.

There are particles that don't feel the weak force at all. The most important one is the particle called the *photon*. A photon is a particle of light, sometimes called a *wave packet* of light. (We'll discuss these further in chapter 11.) X-rays and gamma rays are also photons. Photons don't feel the weak force, but they do feel gravity. They pick up energy when falling in a gravity field, and their paths are deflected when traveling close to a massive body like the Sun or the Earth. Another particle that we think does not feel the weak force is the *graviton*, a particle consisting of an oscillating packet of gravitational field.

Is Radioactivity Contagious?

By this, I mean that if you are exposed to something that is radioactive, do you become more radioactive yourself? Do you "catch" it, like you catch a cold? In the world of science fiction, the answer is yes. People exposed to atomic bombs come away glowing in the dark. But in the real world, the answer is no, at least most of the time, for most kinds of radioactivity.

There are two ways that you can become more radioactive by being exposed to radiation. The first is if you actually get some radioactive material stuck on you, or if you breath it in. You don't really become radioactive, you just become dirty with radioactive dirt. This could happen if radioactive debris from a bomb lands on you or if you touch radioactive dust while taking a tour of the inner parts of a nuclear reactor. (When I did this recently, I had to wear special clothing, and I was instructed to touch nothing.)

But there is a kind of radioactivity that really can make you radioactive: neutrons. Some forms of radioactive explosions emit neutrons, and when these hit your body, they can attach themselves to the nuclei of your atoms. For example, if you add 2 neutrons, you can turn your nonradioactive C-12 nucleus into a radioactive C-14 nucleus. In reality, to do this would require so many neutrons that you would be dead from radiation illness. But objects exposed to intense neutrons do become radioactive.

FORENSIC RADIOACTIVITY: "NEUTRON ACTIVATION"

Radioactivity is unique in its ability to detect a tiny number of atoms hidden among a vast number of others. In your body, only one carbon out of 10^{12}, i.e., only one per trillion, is radioactive. Yet we can count the number from the decays, because they emit such energetic particles. Suppose that you wanted to detect an atom that is not radioactive? Then the trick is to make it radioactive by hitting it with neutrons. If it makes a unique radioactive isotope, then its presence can be measured. This clever technique is called *neutron activation*, and it is a very useful way to detect elements and their isotopes that are present at tiny (parts per billion and less) amounts. Such rare constituents can sometimes be used as "fingerprints" to identify the factory in which an object was made, or the part of the world where it came from.

To activate a sample, the sample is placed in a nuclear reactor, which bombards it with a large number of neutrons. The sample is removed, and its radioactivity is measured to look for the characteristic rays for the desired element.

In 1977, this method was used by Luis Alvarez and his team to search for the rare element iridium. They found enough iridium to conclude that it must have come from an extraterrestrial impact (since meteors, comets, and asteroids contain abundant iridium). From this discovery, they concluded that a large impact had taken place 65 million years ago—at the very time that the dinosaurs went extinct.

THE GLOW OF RADIOACTIVITY

Objects that are intensely radioactive can cause the surrounding air to glow. That's in part because the radiation can knock electrons off the air molecules, and when those electrons reattach, they emit light. Likewise, high-energy electrons can cause water to glow. But such a glow is usually seen only at levels of radiation that are extremely high, such as inside a nuclear reactor.

Even weak radiation can cause strong light emission if it hits special materials called *phosphors*. In a phosphor, even weak rays have their energy absorbed by the molecules of the phosphor. A short time later, the atoms release this energy in the form of ordinary light.

The front of an old-fashioned CRT screen is made of red, blue, and green phosphors. Take a magnifying glass up to such a CRT screen and look at it. You will not see any white phosphors—only red, green, and blue. (These phosphors, despite their name, are not made of the element phosphorus, but of other materials.) When the phosphors are hit by electrons (cathode rays), they emit light, and that is what creates the image. If you put a radioactive material near your TV screen, the rays emitted from it would also make the TV screen glow—provided that they can get through the thick glass.

RADIOACTIVE WATCHES AND LUMINOUS DIALS

Before the dangers of radioactivity were completely recognized, watches had their dials painted with a combination of phosphor and the radioactive element radium. They glowed brightly in the dark. They also gave people cancer, particularly the workers who painted the dials (and licked their brushes to straighten the fibers). It is not legal to make watches with radium dials anymore, although you can sometimes find one at a flea market.

Not all radioactivity is equally dangerous. A watch with a radium dial emits enough gamma rays that it could double your yearly exposure to radioactivity from natural sources. Many people would conclude that such an exposure is not a big risk, but there is greater risk if the radium leaks from the watch.

I do wear a watch that contains tritium (H-3), the radioactive isotope of hydrogen that has a half-life of 12 years. The tritium emits beta rays, and when these hit the phosphor on the dial of the watch, the phosphor glows. In 12 years, unfortunately, my watch will glow only half as brightly. The reason I feel safe wearing this watch is that the low-energy electron (beta particle) emitted when tritium decays has so little energy that it stops within a few thousandths of a centimeter. So it never gets out of the watch. It goes only far enough to make the phosphor emit light.

Someday, we may have computer screens illuminated with tritium. Right now, it is too expensive; most people mistakenly fear anything that is radioactive, so people haven't built factories to make tritium cheaply. But a tritium screen would be on all the time, with no battery needed. Of course, after one half-life (about 12 years), the screen would be only half as bright, assuming that you still owned the same computer 12 years later.[28]

Plutonium

Small particles of radioactive material, such as plutonium, can be very dangerous. Even a tiny dust mote can contain 10^{14} plutonium atoms. If you were to breathe in a large number of such particles, and they got lodged in your lungs, then one small region of your lungs could receive large doses by having billions of nuclei decay at the same spot. This is why people worry about plutonium. It has been called "the most poisonous substance known to man." That is incorrect, and a bad exaggeration, but when you hear it said (and you will), you will know that the origin of the fear is the possibility of small particles breathed in and absorbed in your blood. Plutonium is less toxic than, for example, botulism toxins. We'll discuss this further in the next chapter. Large pieces of plutonium are much less dangerous—unless they are used to make a nuclear bomb.

The author's mentor Luis Alvarez used to keep a piece of plutonium on his desk, as a paperweight. (That was when Alvarez worked at Los Alamos, on the atomic bomb project.) Why didn't the plutonium cause cancer in his hand? The reason is that the radiation from plutonium consists of alpha particles that slow down rapidly when traveling through matter. They will be stopped by piece of paper. So although the alpha particles enter the skin, they enter only the outer layers of skin, which are either dead already or soon to be shed. (The primary way that skin protects itself from carcinogens is by constantly shedding its outer layers and replacing them with fresh skin from underneath.) In contrast, in the lungs the live cells are in contact with the

---

[28] Optional: If you want to learn more about the risks of radium and tritium watches, look at www.orau.org/ptp/collection/radioluminescent/radioluminescentinfo.htm.

atmosphere. That is also why lungs are so vulnerable to cancer from smoke, whereas the skin is not.

Some people have suggested that terrorists might make a plutonium bomb, not a nuclear explosion, but just one that will take a piece of plutonium and blow it into small particles that could get into the lungs. That is possible, in principle, although getting a metal into particles of the right size is very difficult to do. A terrorist might kill more people by making a botulism-toxin bomb instead of a plutonium one. And botulism toxin is much easier to obtain. It often appears in home-made mayonnaise if the mayonnaise is not kept properly cool. And it is the active ingredient in Botox—used by some people (including at least one former presidential candidate) to reduce skin wrinkles and (they think) make them look more attractive.

Fission

Fission refers to a special kind of radioactivity, the sudden splitting of a nucleus into two or more large parts. It was named in analogy to the fission of biological cells. Fission occurs in two forms: *spontaneous* and *induced*.

In spontaneous fission, the nucleus behaves like other radioactive nuclei—i.e., it sits around unchanged for typically one half-life, and then at some random time it decays by suddenly flying into pieces. Spontaneous fission is almost nonexistent in nature.[29] It does occur in some artificially produced isotopes.

A second kind of fission is induced fission. This kind is far more important for this book. Induced fission can occur if the right kind of nucleus is hit by a neutron. The neutron is absorbed, and the resulting nucleus, even with just this one tiny addition, becomes unstable, and fissions. This kind of fission is the fundamental basis of both nuclear reactors and nuclear weapons (e.g., the "atomic bomb"). We'll discuss induced fission in more detail in chapter 5.

When a nucleus fissions, the bulk of the mass is usually divided into two unequal pieces called (appropriately) fission fragments. These pieces are usually radioactive with relatively short half-lives (from seconds to years), and they can be very dangerous to humans. They are the primary source of the residual radioactivity after a nuclear weapon has been exploded, and the main danger in radioactive fallout from such a weapon.

Fusion

Fusion is the source of energy for the Sun, and as a consequence, it is the ultimate source of energy for virtually all life on Earth.[30]

---

[29] The radioactivity of natural uranium is primarily alpha emission. However, one in about 20,000 of the decays is spontaneous fission.

[30] We now know that there are colonies of life under the sea that live off the energy of deep sea "vents." These derive their energy from the radioactivity of the Earth's crust, and so there is some life on Earth that survives independently of the Sun.

When Charles Darwin published his book *The Origin of Species* (1859), he had a lot to say that went beyond evolution. He found fossil seashells high up in the Andes mountain ranges and concluded that these regions had once been under the sea. He estimated that the Earth must be at least 300 million years old, based on his calculation that it would have taken that long for erosion to create some of the great valleys in England. That was adequate time, he concluded, for natural selection to account for changes in species.

But when his book was first published, it received the criticism of the great physicist William Thompson (later knighted as Lord Kelvin and after whom the Kelvin temperature scale is named). He told Darwin that the Earth could not possibly be that old, or the Sun would have completely burned out. Even if the Sun were made completely out of coal, it would have used up all its energy long ago. A more energetic source, Kelvin suggested, was meteors! (Kelvin knew that meteors carried much more energy than coal.) Kelvin estimated that meteoric heat could account for a solar lifetime as long as 30 million years. But 300 million years was not possible, at least according to all the physics known at their time.

Darwin had no answer. In the second and later editions of his book, he removed his thesis that the Earth was hundreds of millions of years old.[31]

Of course, we now know that Darwin's original estimate of the minimum age of the Earth was correct; Kelvin was wrong. The Earth is about 4,500 million years old. The Sun has been burning for longer than that. It isn't coal that provides the energy, or meteors, but nuclear reactions. But the source is not radioactivity, or even fission. It is a kind of nuclear reaction that we have not yet discussed: *fusion.*

Fusion refers to the coming together of particles, in contrast to *fission*, which is the separation of particles. It may seem strange that you can get energy by bringing particles together, but it is true, if you pick the particles correctly. The basic source of fusion in the Sun comes from bringing together 4 hydrogen nuclei to make helium. In the process, other particles are also created. In a typical solar fusion, in addition to the helium we will get 5 gamma rays (for which we use the symbol γ), 2 neutrinos (ν), and 2 positrons (e^+). (A *positron* is just like an electron, but while the electron has a negative electric charge, the positron has a positive charge.) In symbols, we write solar fusion as:

$$4\,H \rightarrow He + 2\,e^+ + 2\,\gamma + 2\,\nu$$

You can read this formula as: 4 hydrogens fuse together and then break apart into (that's the arrow) 1 helium, 2 positrons, 2 gammas, and 2 neutrinos. Most of the energy is in the positrons, gammas, and neutrinos, not in the helium.

Isn't that fission? Notice that there are more particles after the reaction than before. So why is this called fusion, rather than fission? There is no really good reason for this name. It is based on the convention that a new element has been

[31] Amusing digression: There is no universal agreement in professional publications on how to refer to "the Earth." Some people point out that we don't refer to "the Mars" or "the Jupiter" and that Earth is a proper name, just like Mars and Jupiter. So some people always say "Earth" without "the," as in the following sentence: "The age of Earth is about 4.5 billion years."

created, helium, that is heavier than any of the nuclei that we started with (which were all hydrogen). In fission, the elements created (the fission fragments) weigh less than the original uranium or plutonium.

It is all the light particles (e^+, γ, ν) that carry most of the kinetic energy. The neutrino escapes out of the Sun, but the other particles collide with other atoms (mostly hydrogen), share their energy with these atoms, and heat the Sun. It is this radioactively induced heat that makes the Sun shine. The energy released in one fusion reaction is typically 25 MeV.

What Is an MeV?

Let's begin with the unit called the *electron volt*, abbreviated *eV* (small *e*, capital *V*).[32] The eV is useful when talking about individual atoms, since chemical reactions typically have an energy between 0.1 and 10 eV. An MeV is a million eV. Here are some conversions that I recommend you need not bother to memorize: 1 eV = 1.6×10^{-19} joules = 3.8×10^{-23} Calories. One *mole* of material has 6×10^{23} atoms. In chemical reactions, if each atom releases 1 eV of energy, then the energy released is $3.8 \times 10^{-23} \times 6 \times 10^{23} = 23$ Calories per mole.

To compare this with previous work: Table 1.1 in chapter 1 shows that methane, burned in air, releases about 13 Cal of heat per gram. One mole of methane is 16 g, so methane releases $13 \times 16 = 208$ Calories per mole. That's about 9 eV per molecule.

Why the World Is Interesting

The Sun is a star, and the fusion that takes place in the Sun is very similar to the fusion that takes place in stars. If the only fusion that took place was the combining of 4 hydrogen atoms into 1 helium atom, then the world would be a very dull place. The reason is that complex life, as we know it, requires heavier atoms such as carbon and oxygen. Not much of interest (life, intelligence) can happen when all you have is hydrogen and helium. The only molecules you can get are H_2 molecules. Carbon makes very complicated molecules (such as DNA), and that makes interesting life possible.

We believe that the heavier atoms such as carbon and oxygen are created inside stars. Carbon is formed from the fusion of three helium atoms. All the carbon in your body and all the oxygen in the atmosphere was once buried deep in a star, where it was created. Fortunately for those of us who like an interesting world, that star eventually exploded, spewing its debris into space. Eventually, the material clumped up to form a new star (which we call the Sun) and a bunch of planets (Earth, Venus, Mars, etc.). With all the carbon and oxygen on the Earth, life (as we know it) was possible. We are literally made of the ashes of an exploded star. The Sun is a secondary star, created from such debris of an earlier star.

---

[32] The *V* is often capitalized, presumably because it arises from a person's name, Volta. (Yet the word *volt* is often not capitalized, which goes to show that there are no absolute rules followed.)

OPTIONAL: THE DETAILS OF FUSION

You don't have to know the contents of this section unless you are interested. The fusion reaction that I gave earlier, with 4 hydrogens turning into 1 helium and a bunch of other things, does not usually take place in just one step. If the star is a primary star, made of only hydrogen and helium, then the first step is that 2 hydrogens will combine to form a positron, a neutrino, and a deuteron (a particle with a nucleus containing one proton and one neutron). In symbols, the reaction looks like this:

$$H + H \rightarrow d + e^+ + \nu$$

Even though this formula shows hydrogen as H, it normally takes place in such a hot place that its electron has been knocked off—i.e., it is a plasma. Thus, the hydrogen in the equation is really just a proton. The electron, moving in the plasma, doesn't take part in the reaction.

Next, the deuteron combines with a hydrogen atom to create an isotope of helium called *helium-3* and written as ^{3}He:

$$d + H \rightarrow {}^3He + \gamma$$

Last, two helium-3 atoms combine to form ordinary helium:

$$^3He + {}^3He \rightarrow He + 2H + \gamma$$

Note that in the end, the original hydrogen atoms have been transformed into helium plus a few other things.

Why Isn't the Earth a Star?

Why doesn't fusion take place on the surface of the Earth? Or inside a tank of hydrogen gas? The simple reason is that they aren't hot enough. Why does that matter?

The nuclei of all the elements have protons and neutrons in them. The protons give them a positive charge. Positive charges repel, as we will discuss in more detail in chapter 6. In ordinary matter, two nuclei will never get very close because of this repulsion.

To overcome the repulsion, you have to give the nuclei enough energy that they are not stopped by the electric force. Energetic atoms are hot atoms. So if you heat the atoms enough, they will have enough kinetic energy to allow their nuclei to overcome the electric repulsion and touch. The temperature that you need depends on the specific fusion reaction, and how often you need to have the fusion take place. The temperature in the center of the Sun is believed to be about 15 million C. That's hot, but not too hot. Even at 15 million C, what takes place is a relatively slow fusion—most of the hydrogen fuel has not yet burned. Some stars, which are hotter, completely burn out in a few million years, and that doesn't give planets around them enough time for interesting life to evolve.

How Did the Sun Get Hot Enough to Ignite Fusion?

Of course, the Sun is very hot now, because of the fusion that is taking place. But how did the Sun get hot in the first place? We believe that it was simply the gravitational attraction of the matter that made up the Sun. It was all that debris, the meteors, falling into each other—the same mechanism that Kelvin thought was the total source of heat for the Sun. When all that matter got pulled together, the mutual gravity gave it kinetic energy; when it all settled, that energy was converted to heat. This will happen only if the object is big enough so that the temperature rises over a million C. If the mass isn't that large, then the fusion isn't ignited, and the object never turns into a star. In fact, the definition of a star used by most astronomers is an object that is large enough that the heat at the core ignites fusion.

The outer surface of the Sun is not that hot. Its temperature is about 6000 C. That is not hot enough for fusion. All of the fusion takes place at the core of the sun. The heat works its way out, and the surface then glows at a much lower temperature.

Jupiter is not quite big enough to become a star. Its mass is about 0.1% of the Sun's. It is believed that to become a star (i.e., to ignite fusion), an object must be substantially larger, about 8% of the mass of the sun, almost 100 times more massive than Jupiter.

Fusion for Power

One possible fuel for fusion is hydrogen. There is lots of this in the oceans, since they are made of H_2O. Moreover, when you burn hydrogen in fusion, all you get is helium, a harmless gas, useful for balloons and (as we will discuss later) to cool superconductors. It sounds like the ideal energy source, to replace the burning of fossil fuels and nuclear reactors.

Fusion for power has been a goal of scientists ever since the 1950s. Alas, it has proven difficult to achieve. The problem is simple: Most schemes to produce fusion require temperatures hotter than in the center of the sun—millions of degrees C. If we are going to use fusion to generate electricity, then we need hotter temperatures so that the fuel will burn quickly. The plans for controlled thermonuclear reactors (CTRs, in the jargon of the energy business) have temperatures of 100 million C. If you want the reactions to take place in a millionth of a second (to make a hydrogen bomb), then the temperature must be even greater.

Anything that hot explodes. That's the fundamental problem. But there are possible solutions. One of these is called *magnetic confinement* and uses a magnetic field to hold the hot hydrogen. We'll talk about some of the technical details of the devices to do this, including the large international device known as Tokamaks and ITER, in chapters 5 and 10. Another is to fuse just little bits of hydrogen and let it explode. Why not? Isn't that what we do with gasoline and air in our automobiles? That is an approach being developed too. Last, there are schemes for *cold fusion*. Most of these have been shown not to work, but there are some new ideas that show some promise. I'll discuss these further in the next chapter.

Back to the Beginning

Look at the surprising examples that opened this chapter. Do they seem less surprising now? If they were surprising when you first read them, why?

Chapter Review

Radioactivity is the explosion of the atomic nucleus, the small core of the atom that has only 10^{-5} of the atomic radius and 10^{-15} of its volume, but over 99% of its mass. The pieces that shoot out from this explosion are called *rays* or *radiation*. Important kinds of radiation include alpha particles (consisting of two protons and two neutrons), beta particles (electrons), and gamma rays (bundles of energetic light).

The nucleus has over 99% of the mass of the atom, but only 10^{-5} of the radius and 10^{-15} of the volume. It is made primarily of protons and neutrons, which are in turn made of quarks. The name of the element is related to the number of protons. Different isotopes of the same element have different numbers of neutrons.

When radiation passes through matter, it breaks apart molecules and knocks electrons off atoms, leaving charged particles called ions. The track of ions can be made visible in a cloud chamber, or (if the material is a phosphor) by emitted light.

The path of damage done by radiation can cause radiation sickness, if it is sufficiently intense. The LD50 for this is 300 rem, and each rem would take about 2 billion gamma rays per square centimeter. At lower levels, the main damage is to DNA and has the possibility of inducing cancer. The linear hypothesis states that the number of cancers depends only on the rem dose, and not on the number of people who share this dose. In this scenario, about 2500 rem cause one cancer. The linear hypothesis has not been tested for low doses. If it is correct, then the Chernobyl nuclear accident will kill about 24,000 people from cancer. Those people will not be easily identified, since about 20% of people die of cancer from other causes.

Most natural material, including your body, is radioactive from K-40, C-14, and other natural materials. This fact can be used to estimate the ages of rocks and bones.

Radioactive decay follows the half-life rule: for every additional half-life (5730 years for C-14), another half of the nuclei will decay. Nuclei die in this way, but until they die, they do not age. This is one of the mysterious behaviors that comes from quantum mechanics.

Natural radioactivity in rock is the source of heat within the Earth, including the heat responsible for volcanoes. Alpha particles emitted by K-14, uranium, and thorium, slow down and attract electrons and become helium atoms—used in toy balloons.

Plutonium-238, made in nuclear reactors, produces enough heat from its radioactivity that it is used to provide power for satellite missions to deep space.

We believe that originally, when the Sun was formed, most elements were made of radioactive isotopes. The radioactive ones have now mostly decayed, so we are left primarily with those that are not radioactive.

The two main causes of radioactive decay are tunneling (a kind of quantum leap), and the weak force. The ghost-like particles called neutrinos are sensitive only to the weak force and to gravity. They are so strongly produced by the Sun that over 10 billion of them pass through each square centimeter of your body every second.

Most radiation does not make the material it hits radioactive, with the exception of neutrons. The radioactivity that neutrons induce can be used to find trace amounts of rare materials such as iridium (used to understand the dinosaur extinctions).

Plutonium is a human-made radioactive element. The main isotope Pu-239 has a half-life of 24,000 years. It is used in nuclear bombs and reactors. It is toxic, but not quite as much as other chemicals including botulism toxin.

Fission refers to the breakup of a nucleus into two or more large parts. Although it happens naturally, the usual method is to induce fission with neutrons. Fission is used in nuclear reactors and in uranium and plutonium bombs. Fusion refers to the coming together of nuclei to make a larger nucleus. Fusion usually produces many small pieces too. Fusion of hydrogen into helium produces the energy that drives the Sun. Most of the heavier elements (including carbon, nitrogen, and oxygen) are the elements that make life "interesting" and were created by fusion within a star. Fusion usually takes high temperatures to get started, but then it produces enough heat to keep the process going.

Discussion Topics

1. What do you think about the linear hypothesis? If the deaths aren't observable, should they be counted, even if the linear hypothesis has a good basis in theory?

2. "If a tree falls in the woods, and nobody ever observes it (or its consequences), did it happen?" That is a favorite quandary of philosophy students. Some people argue that deaths from very small levels of radioactivity cannot be detected statistically, and therefore they might as well be ignored. Do you agree?

3. Of course, we can't get away from our own radioactivity. But think about double beds. If we spend a lot of time very close to someone else, we are exposed to their radioactivity as well as our own. Suppose we spend 1/3 of a day in close proximity to another person. Then we might expect to *increase* U.S. potassium-induced cancers by about 1/3, or about 13 per year. Actually, only their gamma rays will reach you; the beta rays (electrons) will stop in their own bodies. The gamma rays are 10% of the radiation, so the potassium-induced cancers will be only about 1 per year. So if everyone in the United States slept in double beds, the number of additional cancers expected in the next 50 years is 50.

   That may sound silly. But remember, those are 50 people who otherwise would not die of cancer. And that's just in the United States. The current world population is about 6 billion people, 20 times larger than that of the United States alone. So over the world, we expect not 50 deaths from double beds, but 1000.

Do you think we should do something about the double bed crisis? In the 1950s, many married couples slept in twin beds. (See a replay of any 1950s situation comedy for evidence. If they ever show the bedroom, it will have twin beds.)

Should we return to twin beds? We could instead install shields between people that protect them from each other's gamma rays.

At what point is it silly to worry? What level of "additional cancer" would you consider to be negligible? Would you accept a different level of risk for yourself than you would allow for the entire United States? A risk of 0.000006 sounds negligible for me, but 50 preventable cancers in the United States sounds substantial—yet they represent the same numbers! This is at the heart of a fundamental paradox for public policy, and I don't have the answer. (It is also a great opportunity for a demagogue looking for something to use to scare people.)

4. If I wanted to scare you, I would quote 1000 deaths from double beds. If I wanted to make you feel that the risk was negligible, I would tell you that the chance of you getting cancer from sleeping in a double bed is 0.000002. Both numbers are correct. So be wary when you hear frightening statistics; see if there is a way of stating them that might lead to a different conclusion.

5. The cancer rate in Denver is lower than in Berkeley. Some people have argued that this shows that radiation protects you from cancer. Is this valid? Why not? Could the result be true, even if the argument is not valid?

6. Think about a person who lived so near the Chernobyl accident that he received a dose of 100 rem and then contracted cancer. A rem dose of 100 gives a $1/25 = 4\%$ chance of cancer. Therefore his chance of getting cancer has been increased from 20% to 24%. But what is the public perception? Will the friends and relatives believe that the cancer was induced by the accident, even though only 1 out of 6 (i.e., 4 out of 24) of such people actually had their cancer triggered by the radioactivity rather than by the usual causes?

7. Khidhir Hamza was the chief Iraqi bomb designer, before he defected to the United States. He wrote a book about his experience, titled *Saddam's Bombmaker*. In it, he says:

> The Russian reactor was also targeted. Somebody must have closely studied its layout, because the reactor was destroyed, though not the fuel in the core. (Later, Saddam would claim to UN inspectors that it was destroyed.) The next day, stupidly, tragically, senior scientists and engineers were called in to salvage the computers and electronic equipment, and found themselves wading through radioactive water. Many of them, including Basil al-Qaisy, one of the best electrical engineers in Iraq, would eventually come down with cancers.
> —from Khidhir Hamza and Jeff Stein, *Saddam's Bombmaker* (New York: Scribner, 2000), p. 246

Discuss this paragraph in view of what you know. Can you attribute the cancers to the radioactivity? Did Saddam's chief bombmaker make a mistake?

8. Suppose that we had developed the neutron bomb in World War II instead of the ordinary atomic bomb. The neutron bomb kills people but does much less damage to buildings. But also suppose that years later, the military invented a bomb that did both—the atomic bomb. And suppose that President Jimmy Carter decided that replacing our stockpiled neutron bombs with atomic bombs because they would do more damage would be useful to deter war in Europe. Would the public have decided that Carter did the right thing, that replacing the neutron bomb with one that did more destruction, was more humane? Some people would argue that this conclusion is the logical consequence of the criticism of the neutron bomb. What do you think?

Internet Research Topics

1. What are the applications of *neutron activation*? How is it used in science? By the police? In medicine?
2. Do all scientists accept the linear hypothesis? See what you can find out. Why do some reject it? What are the implications if the linear hypothesis is not true?
3. Are RTGs in use for any current satellites or other space missions? Do some members of the public oppose their use? Why? What are the arguments on both sides? Have other countries used RTGs? Do some people think that RTGs are nuclear reactors?
4. Look up discussions of radioactivity on the Web, and see what you can sense about the way people think about it. Are there people who believe that radioactivity is all human-caused and that we can eliminate it from the environment?
5. How much does natural radioactivity vary from location to location on the Earth? Are there places other than Denver with high natural levels? Can you find records of the radioactivity for the place you live?

Essay Questions

1. Describe the two ways in which radiation can hurt people. What are the prejudices and other mistaken ideas that people think are true but aren't? Give examples from two historical events. Give relevant numbers if you remember them; if you don't, make good guesses (and state that they are guesses). Is radiation ever administered to people on purpose? Explain.
2. Describe what is meant by the *linear hypothesis* for nuclear radiation effects. What does it say about a *threshold*? Is the linear hypothesis true for radiation illness? Why is there a debate about the linear hypothesis—can't the issues be answered scientifically? Give an example to show how the linear hypothesis affects public discussion of radioactivity.
3. Suppose that it is announced that a laboratory in your city accidentally released some radioactive gases into the environment and that this would cause an exposure to people who worked nearby of about

1 millirem (0.001 rem) over the next year for each person who works there. Assume that 1000 people work there. The mayor asks you to speak to the public to alert them to the risks they face. What would you say? Give numbers when appropriate.

4. Describe the different ways that radioactivity is used to measure the ages of different objects. Give as much detail as you can. Include such things as what is meant by the "age" of an object.

5. "The trouble with most folks isn't so much their ignorance; it's knowing so many things that aren't so."—Josh Billings. Give three examples of common scientific misperceptions that could cause people to reach incorrect conclusions on important public issues. Describe what you would tell them to correct their misinformation.

6. The public has a fear of many things that relate to radioactivity. Give some reasons why people fear radioactivity and explain whether this fear is rational or irrational. What is the worst that can perceivably and in reality happen with a nuclear power plant? What has happened, historically?

7. What is the difference between fission and fusion? Where do they occur in the Universe? (Include human-made events and devices.) What elements are typical for fission? For fusion?

Multiple-Choice Questions

1. Energy in the Sun is produced primarily by
   A. fossil fuels
   B. neutrinos
   C. fission
   D. fusion

2. The mass of the nucleus is closest to
   A. 99% of that in the entire atom
   B. 1% of that in the atom
   C. 10^{-5} of that in the atom
   D. 10^{-15} of that in the atom

3. Radioactivity in the Earth leads to
   A. energy for volcanoes
   B. helium for toy balloons
   C. heat for geysers
   D. all of the above

4. LD50 refers to a
   A. lethal dose
   B. legal dose
   C. large dose
   D. lowest dose

5. A major reason that your body is radioactive is that
   A. it is slightly contaminated by debris from nuclear tests
   B. it is made radioactive by medical x-rays

C. you eat radioactive carbon in your food
D. you are hit by neutrinos from the Sun

6. The primary cause of death from the Hiroshima bomb was
A. radiation-induced cancer
B. the blast from the bomb
C. fallout from the bomb
D. smoking induced by fear of the bomb

7. Fission fragments are
A. harmless particles emitted in radioactivity
B. among the most dangerous kind of radiation
C. ghost-like particles that pass through the Earth
D. the source of most energy in the Sun

8. Wine made from fossil fuels is not radioactive because
A. the radioactivity has decayed away
B. there is no carbon in such wine
C. ancient plants never were radioactive
D. the half-life of the key elements is too long

9. If the half-life of an element is big, that means that
A. it decays faster than one with a small half-life
B. it decays slower than one with a small half-life
C. it belongs to a very heavy element
D. it emits lots of neutrons when it decays

10. Muller wears a wristwatch with tritium in the watch hands because
A. he is not worried about radioactivity
B. tritium is not radioactive
C. the levels of radiation that emerge are safe
D. no radiation emerges from the watch hands

11. After three half-lives, the fraction of nuclei remaining is
A. 1/2
B. 1/3
C. 1/4
D. 1/8

12. Most of the elements found on the Earth were created
A. in the first few million years of the Earth's existence
B. within a star
C. in the Big Bang
D. in a supernova explosion

13. The fraction of the U.S. population that dies from cancer is typically
A. about 1 in 10,000
B. about 1 in 100
C. about 1 in 5
D. Most of us die from cancer.

14. Watch dials containing radium glow because
A. all radioactivity creates light
B. batteries supply energy when a button is pushed

C. the radiation hits a phosphor

D. They don't glow. Only tritium glows.

15. Choose all the items that derive from radioactivity:

    A. helium for children's balloons

    B. volcanic lava

    C. warmth in deep mines

    D. most cancer in the United States

16. After four half-lives, the surviving tritium atoms are removed. Compared to completely new tritium atoms, they are expected to live

    A. 1/8 as long

    B. 1/4 as long

    C. 1/2 has long

    D. exactly as long

17. One rem of radiation dose takes how many gammas per square centimeter?

    A. about 1

    B. about 5000

    C. about a million

    D. about 2,000,000,000

18. Which requires the larger dose?

    A. radiation poisoning

    B. cancer

    C. They are very close in amount needed.

19. The linear hypothesis is (choose all that are correct)

    A. known to be true

    B. widely used even though it is not proven

    C. known to be false

    D. generally ignored in analysis of radiation deaths

20. A Sievert is how many rem?

    A. 1

    B. 0.01

    C. 100

    D. 1000

21. Cancer is lower in Denver because

    A. the radiation is lower

    B. the radiation is higher

    C. there are more gamma rays even though there are fewer betas

    D. We don't know why.

22. The number of deaths from the Chernobyl accident is about

    A. less than 1

    B. 24,000

    C. 124,000

    D. over a million

23. Radioactivity is used in

    A. flashlights

    B. TV screens

    C. fluorescent lights
    D. smoke detectors

24. Neutron activation is used
    A. to treat cancer
    B. to search for rare atoms
    C. to provide energy
    D. to create light

25. Dirty bombs may be less of a threat than people fear, because
    A. once spread out, the rem level drops below the radiation illness threshold
    B. dirty bombs require plutonium, and that is hard for terrorists to obtain
    C. the radioactivity is too small to induce cancer
    D. the radiation does not actually get out of the bomb case, even when exploded

26. Radioactivity in the Earth is responsible for
    A. helium we use in balloons
    B. warmth to keep the oceans liquid
    C. creation of the oxygen we breathe
    D. navigation of birds

27. Volcanic heat comes from
    A. radioactive explosions
    B. the weight of rock
    C. accumulated sunlight
    D. fusion deep in the Earth

28. The best way to measure the age of an ancient bone is
    A. potassium-argon dating
    B. neutron activation
    C. controlled thermonuclear fusion
    D. radiocarbon dating

29. An RTG (radioisotope thermoelectric generator) is
    A. a type of bomb
    B. used in hospitals for power
    C. carried in pacemakers to supply power
    D. carried in satellites for power

30. Uranium is found on the Earth, even though it is radioactive. That's because
    A. it is constantly made by cosmic rays
    B. it has a very long half-life
    C. it is created in fusion in the sun
    D. it is created by fusion in the Earth

31. Carbon-14, compared to carbon-12,
    A. has more protons
    B. has more neutrons
    C. has more electrons
    D. has a shorter half-life

32. Alexander Litvinenko was assassinated with
  A. plutonium
  B. tritium
  C. radium
  D. polonium

33. The radiation from cell phones
  A. does not cause cancer
  B. causes cancer, but not enough to worry
  C. causes over 1% of the cancer in those who use them
  D. protects users from cosmic rays

34. The threshold for radiation illness is about
  A. 1 rem
  B. 100 rem
  C. 500 rem
  D. 2500 rem

35. The natural radioactivity (rate of nuclear decays) in a typical human body is closest to
  A. 0.2 per hour
  B. 12 per hour
  C. 12 per minute
  D. 4000 per second

36. The source of energy in the Sun is
  A. uranium chain reaction
  B. radioactivity of tritium
  C. fusion of hydrogen
  D. radioactivity of potassium-40

37. Which of the following is most toxic per gram?
  A. arsenic
  B. Botox
  C. water
  D. plutonium

38. The radioactivity from your own body
  A. does not cause cancer because it is natural
  B. is responsible for more than half of the cancer in humans
  C. does cause cancer, but at a very low rate
  D. helps reduce natural cancer, just as radiation therapy does

39. Drinking alcohol is required to be radioactive because
  A. it proves that it is not made of petroleum
  B. it is healthier
  C. it shows that the alcohol was "aged"
  D. It is not required; just the opposite, alcohol for consumption must not be radioactive.

5

CHAPTER

Chain Reactions, Nuclear Reactors, and Atomic Bombs

A Multitude of Chain Reactions

Nuclear explosions, cancer, the population bomb, lightning, the spread of viruses (both biological and computer), and snow and rock avalanches all have something in common: they are based on the principle of the chain reaction. Other related phenomena are Moore's law, which has governed the computer revolution; compound interest; and the polymerase chain reaction (PCR), which was used to prove that many people on death row were, in fact, innocent. You'll see in this chapter how understanding any one of these phenomena gives insight into understanding all the others. We begin with the story of the game of chess.

Chess

According to legend, the game of chess was invented by Grand Vizier Sissa Ben Dahir, who presented the game to King Shirham of India as a gift.[1] In gratitude, the king offered the Grand Vizier any reward requested—provided, of course, that it sounded reasonable. The Grand Vizier asked for only the following:

[1] In fact, the ancient game of chess was quite different from the modern game. Until about 500 years ago, the queen could move only one square at a time, and the bishop could jump over pieces in the manner of a knight.

One grain of wheat, representing the first square of a chessboard. Two grains for the second square. Four grains for the next. Then eight, sixteen, thirty-two . . . doubling for each successive square until all sixty-four squares were counted.

The king was impressed with the apparent modesty of the request, and he immediately granted it. He took his chessboard, removed the pieces, and asked for a bag of wheat to be brought in. But, to his surprise, the bag was emptied by the 20th square. The king had another bag brought in, but then realized that the entire second bag was needed for just the next square. In fact, in 20 more squares, he would need as many bags as there were grains of wheat in the first bag! And that was only up to square 40 (figure 5.1). Legend does not record what the king then did to his Grand Vizier.

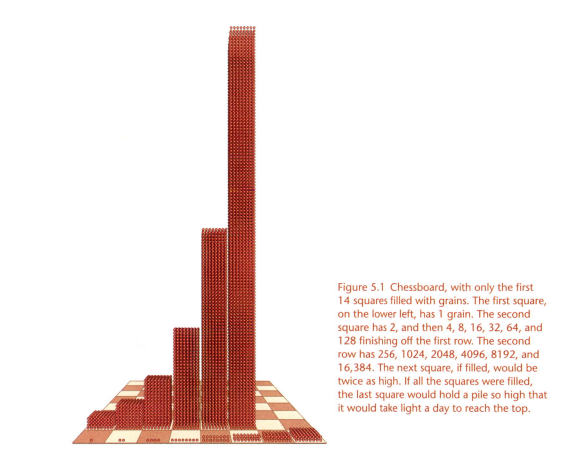

Figure 5.1 Chessboard, with only the first 14 squares filled with grains. The first square, on the lower left, has 1 grain. The second square has 2, and then 4, 8, 16, 32, 64, and 128 finishing off the first row. The second row has 256, 1024, 2048, 4096, 8192, and 16,384. The next square, if filled, would be twice as high. If all the squares were filled, the last square would hold a pile so high that it would take light a day to reach the top.

The number of grains for the last square can be calculated by multiplying 2 times itself 63 times. That doesn't take very long on a pocket calculator. But it is faster if you use the power-law key, marked y^x on a scientific calculator. If you multiply 2 by itself 63 times it is equal to 2^{63}. So put in 63 as x, and 2 as y. The answer is $9223372036854775808 \approx 0.922 \times 10^{19} \approx 10^{19}$. If you include the grains on the first 63 squares, the sum is about twice[2] as large, $2 \times 10^{19} = 2^{64}$. The calculation can also be done easily on a spreadsheet.[3]

If all the grains for the last square were stacked into a cube, the cube would have about 2 million grains on each edge.[4] If each grain were 1 mm in size, then each edge of the giant cube would be 2,000,000 mm = 2,000 m ≈ 2 km. The cube would be 2 km long, 2 km wide, and 2 km high. It would take a very large chessboard to hold this cube. If the grains had to fit on a square that was 2 cm ≈ 2 cm, its height would be 20 billion km. At that height, it would take almost a day for light from the top of the pile to reach the bottom. This amount of wheat on the board would exceed the total world production for over a thousand years.

The amazing feature of this problem is that with just 63 steps, each one quite modest (you are only doubling), you get a really huge number. This type of rapid growth is called *exponential growth*. It is called that because you raise a number (in this case, 2) to a power (63) called the *exponent*. Exponential growth is the secret behind all the phenomena we discuss in this section.

We had a very similar phenomenon in the chapter on radioactivity (chapter 4), except it worked backward. For each half-life, the number of remaining atoms was cut in half. That is the opposite of exponential growth, and is called exponential decay.

Nuclear Bomb

When a neutron hits a uranium-235 (U-235)[5] nucleus, it has a high probability of triggering the breakup of that nucleus into two large pieces. This process is called fission (named in analogy to the fission of a biological cell). The two large pieces are called fission fragments (we discussed these in chapter 4), and they are arguably the most dangerous kind of radiation that comes from nuclear bombs. In addition to the two large fragments, two neutrons are usually released. These neutrons make the chain reaction possible. If there are other U-235 nuclei around, then these neutrons might hit them and cause additional fissions. That doubling process can continue until very large numbers of nuclei are split.

---

[2] But the sum is *only* twice as large. Do you see why? Try it with smaller chessboards, maybe one with only 4 squares. You'll see that you always put one more on the last square than on all the preceding squares added together. If you are mathematically inclined, you might try to prove this as a theorem.

[3] For example, in Microsoft Excel, put the number 2 in location A1. In location A2, put the equation "=A1*2". Then use the spreading function to extend this to cells down to cell 64. The cell A20 will then contain the formula A19*2 and the value of 2^{20}; the cell A64 will contain the value of 2^{64}.

There is an even simpler way to do this in Excel. Just put the symbols "=2^20" into any box. The carat ^ means "to the power of," so 2^20 = 2^{20}. Excel will then evaluate this value.

[4] That's because 2 million cubed = (2 million)^3 = $(2 \times 10^6)^3$ = $8 \times 10^{18} \times 10^{19}$.

[5] Uranium is the heaviest atom that is found in relatively high abundance on the natural Earth. Uranium has atomic number 92—i.e., it has 92 positively charged protons and 92 negatively charged electrons. U-235 also has 143 neutrons. The atomic "weight" is the sum of 92 protons + 143 neutrons = 235 heavy particles in the nucleus.

In the first fission (the first "generation"), 1 atom is split into 2. In the second generation, 2 are split—then 4, 8, etc. In the 64th generation, the number of atoms fissioned will be 10^{19}, the same number as we found for the chess problem. The total number of atoms fissioned (including the ones in prior generations) is about twice that amount: 2×10^{19}.

How many atoms are there in, say, 10 kg of uranium? (I pick this number because the International Atomic Energy Agency, or IAEA, calls that a "significant amount"—an amount that could be used in a nuclear weapon.) The answer[6] is 2.6×10^{25}, much larger than the number we fissioned in 64 generations. How many generations will it take to reach this number? To find the answer, you could keep on multiplying 2's until you get there. It won't take that long. In an additional 20 generations (84 total), you'll reach 2×10^{25}. In other words, after 84 successive doublings, every atom in 10 kg of uranium will be fissioned.

How much energy is released? Each fission of a uranium nucleus releases about 30 million times as much energy as a molecule of TNT. So the 10 kg will be equivalent to about 30 million $\times$ 10 kg of TNT = 300 million kg = 300 kilotons of TNT. That is the fundamental idea behind an atomic bomb. (Many people prefer to use the term nuclear bomb today, since it is the nucleus of the atoms that fissions, and it is the nuclear energy that is released.) The first nuclear bomb released the energy of about 20 kilotons of TNT, less than what we just calculated, and that shows that not every atom was fissioned before the bomb blew itself apart.

Neutron-induced fission also takes place with plutonium-239, abbreviated Pu-239.[7] But in plutonium, typically three neutrons are released per stage. How many stages does it take to reach 2×10^{25}? To find the answer, multiply 3 by itself until you get this number. The answer (so that you can check yourself) is in the footnote.[8] We'll talk more about nuclear chain reactions later in this chapter.

I used a lot of numbers in this section, so let me summarize the important ones: 10 kg of U-235 is enough for a nuclear weapon. If the chain reaction proceeded by doubling, it would all be split in only 84 generations. Plutonium takes fewer generations since each fission releases three neutrons instead of two.

The image in figure 5.2 shows the nuclear explosion from the bomb dropped during World War II at Nagasaki. Estimates of the deaths caused are uncertain, but there were probably between 50,000 and 150,000. The "mushroom-shaped cloud" is not a result of the nuclear aspect of the bomb, but comes from any large explosion. As the hot gas from the explosion rises, it tends to make that shape. Ordinary conventional explosions also have the mushroom shape.

[6] If you have studied chemistry, then you may know how to calculate this number. Since the atomic weight of U-235 is 235, that implies that 235 g will contain 1 mole. A mole is equivalent to 6×10^{23} atoms. (That number is known as Avogadro's number.) In 10 kg of U-235, there are $10,000/235 = 42.6$ times as many atoms as 235 g—i.e., it will contain $(42.6)(6 \times 10^{23}) = 2.6 \times 10^{25}$ atoms.

[7] Plutonium is an artificial element, created in nuclear reactors. It has atomic number 94—i.e., it has 94 protons in the nucleus and 94 electrons orbiting the nucleus. The Pu-239 nucleus also contains 145 neutrons. The atomic weight 239 is the sum of 94 protons and 145 neutrons.

[8] The answer is 53 generations. That's even less then the number of squares on a chessboard!

Figure 5.2 Mushroom cloud of the Nagasaki bomb. (Photo courtesy of U.S. Department of Energy.)

The cooler part of the cloud spread out when it reached the tropopause (about halfway up in the photo), but the really hot center continued to rise.

The Fetus: A Chain Reaction in the Womb

You began life as a single cell, the result of the fusion of your father's sperm and your mother's egg. That cell then divided, in a process called fission. Then the two resulting pairs divided again. How many times did the cells have to divide to make up a complete human body? There are about 10^{11} cells in a human. As you can check, $10^{11} = 2^{37}$, so the answer is 37 stages of doubling. Even if each division stage took a day, the whole process would be over in 37 days.

So why does it take 9 months to be born? The answer is that the cells can't keep dividing that rapidly. They have to grow between divisions, and that takes nutrients. Soon after the process begins, the growth is limited by the body's ability to bring nutrition to the dividing cells. If you did keep doubling in size with each successive generation, then the baby would double its weight in the last day. My wife claims that is exactly what being pregnant felt like.

Cancer: An Unwanted Chain Reaction

When a baby reaches full size, the body "turns off" the chain reaction that is responsible for growth. It appears to have several mechanisms to do this. Many cells maintain the ability to turn the chain reaction back on, if it is needed. For example, if you are wounded or your skin is cut, then the local cells start reproducing again. They can fill the wound with remarkable speed, since they follow the doubling rule of the chain reaction.

The potential danger of unrestricted splitting is so great, however, that our cells have several mechanisms to prevent this from happening when it isn't needed. If all else fails, the cell can be signaled to kill itself, a process known as apoptosis.

If your cells are unlucky enough to receive several specific mutations, they lose this ability to commit suicide. If that happens, then the cells can grow unregulated, continuing to double and double and double. When that happens, we say that the cells have become a cancer. If the cells stay in one region of the body, their reproduction may eventually be limited by their ability to get nutrients; such a collection of local cells is a benign cancer. However, if the cells are the type that can break off and drift in the blood or other body fluids to a different part of the body, then the cancer is termed malignant. Cancers that can spread in this way reach areas where nutrition is abundant, and they continue to grow in an unlimited chain reaction. Eventually, their growth can interfere with vital body functions, and the victim dies.

The reason that cancer can be so devastating, and that people can die from it so quickly, is because it takes advantage of the chain reaction to grow with great rapidity.

The Population Bomb

In 1798, Thomas Rohr Malthus wrote his "Essay on the Principle of Population."[9] It is hard to find a more influential essay in the history of public thought. He argued that population growth obeyed the rules of a chain reaction.[10] If the average parent has 2 children (per parent), then there will be a doubling every generation. If the average parent has 1.4 children, then there will be a doubling

---

[9] You can get a copy at www.esp.org/books/malthus/population/malthus.pdf.

[10] He said that population would grow by "geometric ratio." By the terminology of the time (still in used in some math classes), that means that the population grows by the same factor with each generation. If the factor is 2, then we call it the doubling rule. But any factor greater than 1—e.g., 1.4—will still result in an unlimited chain reaction. With the factor 1.4, the population still doubles; it just takes 2 generations ($1.4^2 \approx 2$).

every two generations (since $1.4 \times 1.4 \approx 2$). The doubling would take longer, but it would still lead to a chain reaction.

Malthus argued that the available food supply will not show a similar exponential growth. It will grow much more slowly because it is limited by available land, water, and other resources. As a result, the growth of population will always outrun the growth of food. Malthus thought that the only thing that would stop this population bomb was disease and famine. Based on this observation, others argued that famine was not only inevitable, but it served an important purpose. That was a very bleak outlook on life. Malthus' analysis has been so influential that some call economics "the dismal science" based on this pessimistic outlook.

Many people today still think that this population bomb is the ultimate disaster working on the human population. But there is an alternative to starvation: birth control. It is interesting to note that Malthus thought that the bomb was imminent in 1798. We have managed quite a few doublings since then, and we are still well fed.[11] In 1968, Paul Erlich wrote The Population Bomb, in which he predicted that massive starvation would hit the world in the 1970s. Recently, he maintained that it would happen in the 2000s. I expect him to issue a new warning soon, saying that the devastation will take place by the end of the 2010s.

But there is reason to be more optimistic. According to recent studies, world population is departing from exponential growth. If you want, you can read the paper "The End of World Population Growth" in the scientific magazine *Nature*.[12] The United Nations estimated in 2003 that the world population growth will slow and that the total population will not exceed 10 or 12 billion before it starts to decline. The reason for the slowing growth is not understood, but it is possible that humans tend to have fewer children when they become well-fed and secure. If this is true, then the secret to the end of the population bomb would then be a truly delightful one: make everyone in the world wealthy. To avoid polluting the world (when everyone can afford SUVs), it may be necessary to increase energy conservation to match the wealth. This appears to be possible.[13]

Mass Extinction Recovery

Sixty-five million years ago, the dinosaurs were destroyed, yet the mammals survived. Why? Well, the extinctions were not as simple as many people think. It wasn't true that all the mammals survived. In fact, it was likely that that 99.99% of all mammals were killed at the same time. For recovery, all it takes is for one or more "breeding pairs" to survive. Imagine, for example, that two rats made it

---

[11] As of today, there appears to be enough food to feed everybody alive today. Starvation and hunger are present in the world, not because of lack of food, but because available food is not reaching people who need it.

[12] The reference is Wolfgang Lutz, Warren Sanderson, and Sergei Scherbov, *Nature London*, **412**(68466846): 543–545 (2001).

[13] If you are interested, see my essay "The Conservation Bomb" at www.muller.lbl.gov/TRessays/05_Conservation_Bomb.htm.

through the bad period. Assume that it takes about a year for rats to grow and to breed. This would allow the number of rats to double every year. After just 56 years (an instant in geologic time), there would be so many rats that they would completely cover the surface of the Earth like a carpet.[14]

Of course, that kind of massive rat bomb never happened. The growth of rats was limited by the availability of food, by disease, and by competition with other animals. But the example shows that it is difficult to spot great catastrophes in geologic history unless the extinctions are so great that they actually eliminate entire species—i.e., they leave no breeding pairs. The rats recovered; the dinosaurs didn't. It turns out that 65 million years ago, all large animals disappeared. That may be because such animals are rare and require lots of territory to stay alive. When 99.99% are killed, they are less likely to be able to find mates than are the little guys.

A population explosion sometimes occurs when a new species is introduced into an environment in which there are no natural predators. Twenty-four rabbits were released in Australia in 1859. Seven years later, 14,253 rabbits were shot for sport on property owned by Thomas Austin, the person who released the original two dozen. By 1869, Austin had killed over 2 million rabbits on his property, and he realized what a major mistake he had made. Wild rabbits are still a major pest throughout Australia.

Nobody knows what led to the plague of rats that gave rise to the story of the Pied Piper of Hamelin.[15]

DNA "Fingerprinting": The Polymerase Chain Reaction (PCR)

In every cell of your body, there is a collection of molecules called DNA[16] that contain the information that runs your body. For different people, the molecules are virtually identical. The DNA contains the genetic code that tells the cells how to reproduce, how to respire, how to function. But there is a tiny component that is different for different individuals, such as the parts that determine eye color. Half of your DNA came from your father and half from your mother, so your DNA is very similar to theirs, but not identical. (If it were identical to that of another person, you would be either an identical twin or a clone.)

DNA fingerprinting consists of looking at the parts of the DNA that vary among different people. If you pick enough of these regions, you have a unique identification. For close blood relatives, more of these regions will be identical than for people who are not related.

A potential difficulty in DNA fingerprinting is that methods to read the code—that is, to determine the exact sequence of molecules in the relevant

---

[14] The area of the Earth is 5×10^{18} cm^2. After 56 generations, the number of rats would be $2^{56} = 7 \times 10^{16}$. That allows a 10 cm × 10 cm area for each rat, including the oceans.

[15] If you aren't familiar with Robert Browning's poem, I recommend it highly. It is online at www.indiana.edu/~librcsd/etext/piper/text.html.

[16] DNA stands for deoxyribonucleic acid, if that helps. Try a Google search on it.

regions of the DNA—do not work with just a few molecules. DNA fingerprinting methods require billions of copies of the DNA molecules.

That's where the chain reaction comes in. It takes advantage of the fact that DNA is a molecule that is designed to duplicate itself. Just before a cell splits into two cells, the DNA molecule makes a copy of itself so that the same DNA can go in each half. Kary Mullis, a San Diego biologist and surfer (water, not Web) realized the potential value of this fact, during a drive in the California mountains. His invention, called PCR (for polymerase chain reaction), is a method that has transformed biology, and won him a Nobel Prize in chemistry in 1993.[17]

Mullis realized that if he had even one DNA molecule, he could use the chain reaction to get billions of copies. The procedure involves using chemicals that will trigger duplication of a segment of the DNA molecule—a segment that is known to be different for different people. The stages in the chain reaction are achieved by temperature cycling of the fluid containing the DNA. When the temperature is cool, the desired part of the DNA makes what is called a complementary strand, which stays attached to the original DNA. The mixture is then heated (to near boiling), and the two strands separate. When cooled, both the original DNA and the complementary strand duplicate; these are separated by heating, and the cycle is repeated. After 35 cycles (taking less than an hour) there will be $2^{35} = 3.4 \times 10^{10}$ copies, i.e., 34 billion. This gives the scientist enough material to determine the exact genetic code of the fragments.

PCR APPLICATIONS: INNOCENT AND GUILTY; THOMAS JEFFERSON AND SALLY HEMMINGS

DNA fingerprinting is a method that can be used to identify people based on only a few cells from their body. The method was used to identify the remains of people after the World Trade Center attack and the space shuttle *Columbia* disaster. It was used to free innocent people who had been awaiting execution for crimes they didn't commit. By 2007, DNA fingerprinting had shown that over two hundred prisoners scheduled for execution were innocent of the crimes for which they had been convicted. Back in 2003, Governor George Ryan of Illinois became concerned that innocent people might be executed, and so he commuted the sentence of every Illinois prisoner scheduled to be executed—156 of them. Bits of blood left at the scene of the crime had matched their blood type, but this far more sensitive method might be able to show that the blood was not theirs.

PCR can be used to convict criminals too. It is used to identify rapists, beyond a reasonable doubt. It is a very reliable way to prove paternity, by the match between the father and the child. It has even been used 200 years after the father was dead—to provide evidence that descendants of Sally Hemmings, a slave owned by President Thomas Jefferson, were also the descendants of Jefferson. In this case, DNA of Jefferson's known descendants was compared to DNA of descendants of Sally Hemmings. The match was reported to be weak, but still much stronger than would have been found between unrelated people.

---

[17] Mullis's Nobel Lecture is available at nobelprize.org/nobel_prizes/chemistry/laureates/1993/mullis-lecture.html.

Illness and Epidemics: Chain Reactions of Viruses and Bacteria

A virus or bacteria duplicating in your body uses the chain reaction to reach enormous numbers. If your body has to devote major parts of its resources to killing the germ, then you feel sick. If it can't succeed in stopping the exponential growth, you die.

The math of a chain reaction also describes the spread of an epidemic. Consider a single person infected by the smallpox virus. That person can spread the virus to someone else by contact or by saliva droplets spread from the breath. If one person infects two people, and they infect an additional two people, and that pattern continues, then it takes only 33 such stages to infect the entire world (since $2^{33} = 8.6$ billion > World population). Worse, suppose that the first person infects 10 people, and they each infect 10 others. Then, after only 10 such stages, the number infected will be $10 \times 10 \times 10 \times 10 \times 10 \times 10 \times 10 \times 10 \times 10 \times 10 = 10^{10}$, and that is greater than the entire world population. Such spread, in the past, has been limited by the fact that people didn't travel widely, and so an epidemic could be localized. But today, with airplane travel, an infected person could infect thousands.

Note that not all illnesses are chain reactions. When humans are infected with anthrax (as happened during the terrorist attacks in 2001), the disease was not spread from one person to another. Those infected became ill, and some died, but the disease did not spread like a chain reaction.

Computer Viruses: Electronic Chain Reactions

Computer viruses obey the same laws as other chain reactions. A virus in your computer system can spread to other systems by the copying or sharing of infected programs. E-mail can spread like a chain reaction if it allows a message to be automatically forwarded—for example, to everyone on your mailing list—or if it contains an attached program that fools you into opening it.

Such computer viruses can spread with remarkable speed, in part because the multiplication number at each stage can be large. If, for example, an infected computer infects 100 others, then the virus could spread around the word in fewer than 4 stages (since $100^4 = 10^8$, and that is greater than the total number of computers in the world), at least in principle. Of course, the infection would not spread to people who were on nobody's e-mail list, or who run antivirus programs that intercept and "kill" the infection.

Urban Legends

Stories, jokes, and rumors can spread in a chain reaction. If you hear something that fascinates you, and you tell two people, and each of them tells two people, then the doubling law enables the story to spread like an explosion. One of the most intriguing of these is called an urban legend. You hear that baby alligators, given as presents to children and then flushed down the toilet when they grew too large, are now living in the sewers of the city. You tell some friends. The story spreads, even though it isn't true. Other famous urban legends include razor blades put in apples on Halloween and cell phones that cause explosions at gasoline stations.

A characteristic of such urban legends is that you often tell someone, only to have them respond, "Oh, I know that. Everyone knows that," even though the story turns out to be untrue.

The Internet allows the extra-rapid spreading of urban legends because it allows them to jump long distances. You don't have to be in close proximity to tell the story. At the same time, urban legends can now be debunked easily too. There is now a Web site devoted to urban legends, their history, and their truth: www.snopes.com. Look at it and see if you can find things you "know to be true" that actually aren't.

Avalanches: Chain Reactions of Rock or Snow

A stone falls from a ledge, and knocks out 2 more stones. Each of them knocks out 2, and so on. This is an avalanche. The doubling rule applies.

If each rock knocks out less than 1 additional rock, then the avalanche will fade away and die. Suppose, for example, that we reach a part of the slope for which each rock knocks out, on the average, 0.5 additional rocks, and then stops. Assume that the avalanche reaches this region of the slope with 64 rocks tumbling down. Then after 4 stages, the number of rocks being knocked out is equal 64 multiplied by $(0.5)^4 = 64 \times 1/16 = 4$. This typically happens when the avalanche reaches a part of the slope that is no longer steep, so the rocks are more strongly planted in the ground and harder to knock loose.

A snow avalanche can be similar, although the snow doesn't usually exist in discrete objects like rocks. But, like a rock avalanche, it will die when the slope is no longer steep.

Lightning: An Electron Avalanche

Sparks, and their bigger relatives called lightning, are also examples of chain reactions. In fact, they are very similar to rock avalanches. Sparks occur when an electron has such a high electric voltage (see chapter 6), that it breaks off whatever holds it and accelerates through air. If it picks up enough energy (by its repulsion from other electrons left behind), then it can break another electron off a molecule of air, doubling the number of moving electrons. Now we have 2, and that can double to give 4, then 8, then 16. The number of electrons increases exponentially, and that is the spark (or the lightning). In lightning, collisions of the electrons with the air molecules heat the air, causing it to expand rapidly (making thunder) and to glow (making the visible lightning stroke).

Compound Interest: Seen as a Chain Reaction

Compound interest refers to the fact that you can earn interest on your interest. If you invest money at an annual rate of, say, 5%, then after a year, you have 1.05 times as much as you started with. After 2 years, the amount is $(1.05) \times (1.05) = (1.05)^2$ times greater, and after 14 years, the amount is $(1.05)^{14} \approx 2$ times the original amount. Your money will continue to double every additional 14 years. After 28 years, it will have grown by a factor of 4 times, and after three doublings (42 years), it will have grown by a factor of 8.

Compound interest is a form of a chain reaction. The doubling creates two amounts, each equal to the original, and each of these will continue to double. That's why the math is the same.

Suppose that you start with $1000 and would like to become a billionaire. All it takes is a factor of a million $\approx 2^{20}$. From this math, you can see that a factor of a million takes 20 doublings. At 14 years per doubling period, it would take 280 years, and your billion dollars would be worth a lot only if there was negligible inflation. To really become a self-made billionaire, you have to have a doubling time of no more than 1 or 2 years.

Moore's Law of Computers: Exponential Growth

The same doubling rule that we see in chain reactions occurs in other phenomena. One of the most famous is in computer technology. In 1965, Gordon Moore, one of the founders of the integrated circuit industry, noticed that the number of basic components that could be put on a chip had doubled every year for the previous six years. From what he knew of the technology, he expected the trend to continue, at least until 1975. By that time, instead of 50 components per chip, he predicted there would be 65,000!

Moore's prediction sounded so ludicrous that cartoonists made fun of it by taking his prediction to its extreme—it implied that someday consumers would buy their own hand-held computers, and even be able to buy them in a department store. Such a cartoon appeared in the original paper (figure 5.3).[18] These days, when the cartoon has proven to be true, it is hard to imagine that in 1965, this was supposed to be a funny but ridiculous extrapolation.

Figure 5.3 Prediction of ridiculous future if Moore's law holds (drawn in 1965).

[18] Gordon E. Moore, Cramming more components onto integrated circuits, *Electronics* 38(8): 114–17 (1965).

As Moore's predictions began to come true, the newspapers picked up on it, and called the phenomena Moore's law. It seemed to apply to other aspects of computers besides the density of components, including processor speed and magnetic disk memory. The average doubling period, spread out over the last 35 years of the twentieth century, turned out to be about 18 months. Thus, the explosion of computers that has taken place in the last 35 years is really analogous to a nuclear explosion. No wonder it has dazzled so many people. A plot of this growth is shown in figure 5.4.

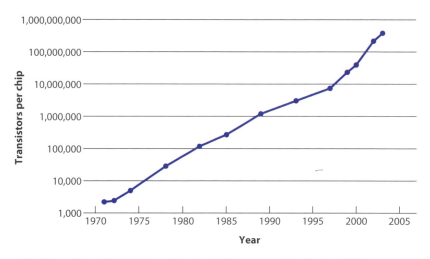

Figure 5.4 Moore's law. Note that each horizontal line represents a factor of 10 increase over the previous line. In the last 40 years, the capability of computers has increased by roughly a factor of a million.

The plot shows the number of transistors (individual switches function like the old vacuum tubes of the earliest computers) on the most advanced commercial Intel chip, plotted versus year. Notice that the scale on the left is logarithmic—which means that each horizontal line means a factor of 10 increase over the lower line. When the doubling law is plotted on a logarithmic plot, it becomes a straight line!

This increase even seems more amazing to me, because when I was growing up a portable radio had fewer than 10 transistors. (People now use the term transistor to refer to a "transistor radio." But for this plot, we mean the number of separate transistor components.)

The Nobel Prize in physics in 2000 was given to Jack Kilby, who along with Robert Noyce invented the integrated circuit that allowed this expansion to take place. But we don't really understand the reason for Moore's law. Every year, for the past two decades, there have been articles in magazines explaining why Moore's law is soon going to fail. So far, these articles have always had "good" reasons, and they have always turned out to be wrong. I believe that Moore's law will continue to hold for at least another decade, but I can't predict beyond that. We are about to reach the limit of smallness (since a circuit cannot be smaller than the size of an atom), but we have yet to truly exploit the third dimension (i.e., putting circuits not only alongside each other, but also on top of each other.)

Folding Paper

A particularly easy way to study the doubling rule is by folding a sheet of newspaper. Suppose that you take a sheet and fold it in half. There are then two layers. Fold again for four layers, and again for eight layers.

It is an old trick to bet someone that they can't fold a newspaper sheet eight times. Let's look at what happens when you try that. (And I encourage you to actually try it!) Let's look at the paper after seven folds. There would be $2^7 = 128$ layers. To see how thick that is, measure 128 pages from a book. When I do that, I find it is about 1/4 inch thick = 0.25 inch.

Notice that each time you fold the paper, the width is halved. The New York Times, when laid out flat, is 27 inches wide. After seven folds, it should be 1/128 times narrower—i.e., 27/128 = 0.21 inch wide.

Now, for your next and eighth fold, you are supposed to fold something that is 0.25 inch thick and only 0.21 inch wide. You are trying to fold something that is thicker than it is wide! That's why it can't be done—unless you use a very long sheet of paper instead of a newssheet.

Tree Branching

Here is one last example illustrating how the doubling rule can lead, in a small number of steps, to a large number of objects. Suppose that a tree has a trunk that divides into three large branches, and each of these divides into three more branches. Suppose that the branches continue to divide another six times, and then you arrive at three leaves at each end. How many leaves are there on the tree?[19] Do you suppose that Nature uses a trick like this to simplify the code required in the design of trees? Suppose, in addition to the doubling rule, it put in a random process. So, for example, the probability of creating two branches might be 50%, while the probability of three new branches might be 30%, and the probability of dividing into four is 20%. That would make for a more interesting tree. Take a look at actual trees and see what you think.

Nuclear Weapons Basics

As soon as it was discovered that a neutron-induced fission creates more neutrons, it was clear that there was a potential method for releasing enormous nuclear energy. The concept of the nuclear chain reaction had actually been patented in England by the nuclear physicist Leo Szilard in 1932. The first actual

---

[19] $3^9 = 19,683$. If each leaf is 10 cm × 10 cm with area 100 square cm, the total area of leaf surface is 196,830 square centimeters = 19.68 square meters.

chain reaction was achieved by a team led by Enrico Fermi at the University of Chicago in 1942.

As discussed earlier in this chapter, the chain reaction makes use of the fact that more than one neutron comes out per uranium fission. If those neutrons can be made to hit other uranium nuclei, then soon the doubling rule will result in the fission of nearly every nucleus. It takes only 80 doublings. The key to achieving this is the concept of *critical mass*.

Critical Mass

If the uranium chain reaction is to work, there must be enough material so that the emitted neutron hits another uranium nucleus, instead of escaping between the nuclei and out of the bomb. If enough uranium surrounds the initial fission, so that the neutrons will not escape, then we say we have a *critical mass* of uranium. For many years, the value of this critical mass was highly classified. This was because many people thought it was larger than it turned out to be. The critical mass for a bomb based on uranium fission is different than that for plutonium fission. Part of this is due to the fact that more neutrons are emitted when plutonium fissions.

To make a critical mass, there must be enough material so that after each fission, more than one of the neutrons that is emitted will hit another nucleus to keep the chain reaction going. A simple calculation[20] indicates that this requires a sphere of uranium 13.5 cm in radius, weighing 200 kg = 440 lb. There was no hope during World War II that so much U-235 could be obtained, and that may be the reason why the Germans (under the direction of the famous physicist Werner Heisenberg) abandoned the effort. But the United States (under J. Robert Oppenheimer) invented ways to reduce the amount needed. According to the book *The Los Alamos Primer*, written by R. Serber during the U.S. effort, the most important of these was to add a neutron reflector at the surface. According to Serber, the critical mass can be reduced to 15 kg for U-235 and to 5 kg for Pu-239. That much plutonium would fit in a cup.[21]

The term *critical mass* has worked its way from physics into our everyday language as a metaphor. One or two people, working on a problem, may not be enough. But if you assemble a critical mass of people, the progress can be explosive.

Uranium Bomb

The nuclear bomb that destroyed Hiroshima was a "gun"-type bomb that obtained its energy from the fission of U-235. By *gun*, I mean that a piece of U-235 was shot by a cannon at another piece of U-235; the combination was above the critical mass, and so a fission chain reaction began that released the

---

[20] This calculation is performed in the book *The Los Alamos Primer*, by Robert Serber (University of California Press, 1992); see p. 28.

[21] The density of plutonium is 20 g/cm^3, so 5 kg = 5000 g would fit in 250 cm^3, which is about the volume of a standard cup.

enormous nuclear energy causing the explosion. The entire bomb, including cannon, weighed 4 tons. The energy released from the fission chain reaction was 13 kilotons of TNT equivalent. The day after Hiroshima was destroyed, President Harry Truman mistakenly announced that the yield was 20 kilotons. This was the first uranium device ever exploded. It had not been tested. (A prior test on U.S. land at Alamogordo, New Mexico, was of a plutonium bomb.) The design was so simple that a test was decided to be a waste of uranium. After the bomb was dropped, there was not yet enough new uranium to make a new one, although the Oak Ridge plants were producing enough that a new bomb could be ready soon.

A photo of the Hiroshima bomb is shown in figure 5.5. The cylindrical shape indicates the presence of the gun (more like a cannon) in the interior.

Figure 5.5 The uranium "gun" bomb dropped by the United States on Hiroshima. The shape reflects the presence of a cannon inside. (Photo courtesy of U.S. Department of Energy.)

Plutonium bombs are more difficult (see the next section). For that reason, a bomb that uses uranium is the material of choice for a terrorist, since the design is so simple. But such a bomb requires highly enriched U-235, and that is not easy to make. When you dig uranium from the ground, it is 99.3% U-238, and only 0.7% U-235. It is only the rare isotope U-235 that can be used for a bomb. Separating this isotope from its more common form is extremely difficult to do.

Saddam's Bomb

When the United States defeated Iraq in 1991, one of the conditions that Saddam Hussein agreed to was U.N. inspections of his nuclear facilities. The U.N.

Figure 5.6 Iraqi Calutron. This field shows the pieces left after the International Atomic Energy Agency destroyed Saddam's Calutron. This Calutron had enriched uranium to 35% U-235; that was not yet enough to use for a bomb, but only a few more steps would have brought the enrichment to 90%. (Photo courtesy of U.S. Department of Energy.)

discovered that he had developed devices to separate U-235 from natural uranium. But these devices (figure 5.6), instead of being the modern centrifuge or laser systems that we had anticipated, were Calutrons (short for "CALifornia University Tron"). The Calutron was the slow but sure method invented by Ernest Lawrence (the person Lawrence Berkeley Laboratory is named after). Lawrence had invented this method during World War II (figure 5.7), and his system had separated virtually all the U-235 that was used in the attack on Hiroshima.

Gas Centrifuge Uranium Enrichment

The most modern and efficient method for separating U-235 is to use a gas centrifuge. Uranium is combined with fluorine to make a gas, uranium hexafluoride. This is then placed in rapidly rotating cylinders. The gas with the heavier U-238 tends to concentrate on the outer part of the cylinder, leaving the lighter U-235 near the center. These are then piped out, as shown in figure 5.8. In fact, the enrichment in one centrifuge is small, and the gas must be pumped through thousands of centrifuges in order to get enough enrichment for nuclear power plants or nuclear weapons.

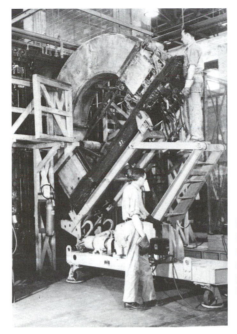

Figure 5.7 The original Calutron, built at the University of California at Berkeley. The uranium ions moved in a semicircular path along the C shape. They were bent into this path by strong magnets. (Photo courtesy of U.S. Department of Energy.)

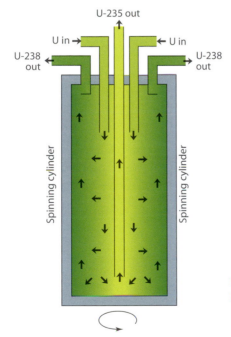

Figure 5.8 Centrifuge design for U-235 purification.

Centrifuges and Nuclear Proliferation

Modern centrifuges can be made to be efficient and relatively small. Because they spin so fast, they must be made of very strong materials to keep from breaking apart. One of the key new materials is called maraging steel, used primarily for uranium centrifuges, for rocket bodies, and for high-performance golf clubs. (Look it up on the Web.) U.S. intelligence services become very suspicious of countries that begin importing or making substantial quantities of maraging steel, unless they are major manufacturers of golf clubs.

A typical centrifuge plant would have several thousand centrifuges (figure 5.9), but the entire collection could fit inside a large class lecture hall. Such a system could produce enough enriched uranium for several nuclear bombs per year. Hidden centrifuge plants are very difficult for intelligence agencies to locate. They don't require large amounts of power, and they are very quiet (thanks to the exquisite balancing of the cylinders that must be achieved to keep them from spinning themselves apart).

Figure 5.9 Centrifuge cascade in Ohio. (Photo courtesy of U.S. Department of Energy.)

Gas centrifuges have been the method of choice for recent proliferation. Centrifuges were developed by A. Q. Khan of Pakistan, and the technology was shared with other countries, including North Korea and Libya. We found out about the Pakistani program in 2003, when Libya decided to abandon their centrifuge effort, and to cooperate with nonproliferation efforts. Components of the Libyan centrifuge are shown in figure 5.10).

Plutonium Fission Bomb

Both the bomb tested at Alamogordo and the one dropped on Nagasaki were plutonium bombs, using Pu-239. Plutonium is relatively easy to get; it is

Figure 5.10 President George W. Bush examining cylinders from the Libyan centrifuge plant. (Photo courtesy of U.S. Department of Defense.)

produced in most nuclear reactors, including those intended to produce electric power, and then it can be separated using chemistry. However, it normally has a high fraction of Pu-240, which is highly radioactive. This radioactivity tends to predetonate the bomb—i.e., make it explode before the chain reaction is complete. As a result, a special design has to be used: *implosion*. This is extremely difficult to design and engineer and build, and probably could not be built by a small organization such as a terrorist group. The resources of a full country (Pakistan, North Korea) are probably necessary.

The bomb dropped on Nagasaki yielded 18 kilotons of explosion. It used only 6 kg of plutonium (about 13.5 lb). That much plutonium could easily fit in a coffee mug. The higher yield per gram (compared to uranium) results from the fact that plutonium emits more neutrons in fission than does uranium, so the reaction goes faster, and we get a more complete chain reaction before the plutonium is blown apart.

If 6 kg of plutonium completely fissions, it releases the energy equivalent of about 100 kilotons of TNT. But the explosion is so great that the bomb is thrown apart before all of the chain reaction has been completed. The real challenge of the bomb project during World War II was to get the plutonium so compressed that the chain reaction would "go to completion." In fact, the 18-kiloton yield of the first bomb, exploded in a test at Alamogordo, shows that 18% of the nuclei fissioned. In the North Korea test of 2006, they got a yield of 400 tons (0.4 kiloton) showing that they got less than 1/2% fission. (They could not use less than a critical mass.) That's why most people think that their test was a "dud." Their second test, in 2009, was larger; one good estimate was 1.6 kilotons. But even in that test, the fission was only 1.6%.

The plutonium is often arranged as a hollow shell, with explosives on the outside. The explosives drive the shell into a little blob and compress it (even though it is solid). The compression pushes the atoms close enough together that neutrons produced in the chain reaction are unlikely to be able to leak between them. Thus, compressed plutonium has a smaller critical mass than uncompressed plutonium.

Figure 5.11 The plutonium implosion bomb that the United States dropped on Nagasaki. From the shape, you might guess that it contained a sphere of explosives. (Photo courtesy of U.S. Department of Energy.)

A photo of this bomb is shown in figure 5.11. Notice that the bomb is more spherical than the Hiroshima uranium bomb. That reflects the spherical shell of explosives for the implosion.

You should look hard at this photo and think about the enormous destruction done by such a small device. That reflects the factor of a million between chemical and nuclear energy.

The explosives often use a special kind of explosive "lens," a special shape in the explosive that tends to make the explosion converge on a point.

A U.S.–trained physicist named Khidhir Hamza, who worked as the chief nuclear weapons designer for Saddam Hussein, says the Iraqi bomb was not going to be a gun-style design. Instead, they were going to use uranium but reduce the critical mass using an implosion.[22]

Thermonuclear Weapon or "Hydrogen Bomb"

A hydrogen bomb is also called a *thermonuclear weapon* because it uses the heat from a plutonium or uranium fission bomb to fuse molecules of deuterium and tritium[23] (see the section "Fusion" in chapter 4). The process takes place in three stages. First, the explosion of a fission bomb creates an intense heat. Second, this heat causes the deuterium and tritium to reach energies that are sufficient to overcome their natural repulsion to each other (the nuclei of both are positively charged) and fuse. Third, this fusion releases energy and neutrons; the high-energy neutrons cause fission in a uranium container (made of U-238) that surrounds everything else, and that releases even more energy.[24] The biggest hydrogen bomb ever tested (they have never been used in war) released an energy equal to over 50 million tons of TNT. That is million, not thousand!

---

[22] Hamza eventually defected and now lives in the United States. He told his story in the book, *Saddam's Bombmaker* (Scribner, 2001).

[23] Deuterium is a hydrogen atom with a neutron in the nucleus, in addition to the usual proton. Tritium is a hydrogen atom with two neutrons in the nucleus and one proton.

[24] Even though U-235 must be used to keep a chair reaction going, if there are high energy neutrons made in the fusion reaction, these will split the U-238 and release energy. But U-238 will not by itself sustain a chain reaction. Thus, the U-238 serves a purpose only when added to a fusion bomb.

The "secret" of the hydrogen bomb, kept highly classified until just a few years ago, is that a plutonium fission bomb emits enough x-rays that they can be used, after bouncing off the uranium cases, to compress and ignite the tritium/deuterium combination. There is a second secret, although this has been public for a longer period. Instead of using tritium, the bomb can contain a stable (not radioactive) isotope of lithium called Li-6. This is a solid, which means that the material is stored at high density. The neutrons from the fission weapon break up the Li-6 to make the tritium. Thus, the fuel is created in the same microsecond that the bomb is exploding. The fusion fuel is usually lithium combined with deuterium, called lithium deuteride.

BOOSTED FISSION WEAPON

You can increase the energy of a fission bomb by adding a small container with tritium/deuterium gas. In the heat of the explosion, the tritium and deuterium fuse, releasing more energy and more neutrons. The additional neutrons mean that the chain reaction in the fission bomb chain becomes more complete, and that increases the yield of the bomb. A boosted fission weapon uses fusion, but it uses the fusion for neutrons to split plutonium, rather than for energy production, so it usually isn't considered a fusion bomb.

TERMINOLOGY: ATOMIC BOMBS, HYDROGEN BOMBS, AND ALL THAT

A bomb that uses the energy of the nucleus to release energy can safely be called a *nuclear bomb*. Some prominent people (starting with President Eisenhower, Edward Teller, and lastly President George W. Bush) pronounced nuclear as "nukular"—but most academics say that is incorrect. Yet many nuclear engineers and bomb designers continue to use the "nukular" pronunciation, making it into a kind of tradition. President Harry Truman referred to the bombs dropped over Japan as "atomic bombs." This name is still used too. The argument for this is that previous bombs were really "molecular bombs" that took advantage of the chemical reactions of molecules. The *atomic bomb* is the first one that releases enormous energy from within the atom. I've also heard that the bomb designers worried that the word *nuclear* would make people think it was a biological weapon, since prior to World War II, the term nucleus was commonly associated with the core of a biological cell.

The bomb based on fusion of hydrogen is often referred to as a *hydrogen bomb*. A name typically used by scientists is *thermonuclear bomb*. The word *thermonuclear* refers to the fact that the fusion takes place because of the high temperature (that's the *thermo* part). The material that fuses consists of tritium and deuterium, two isotopes of hydrogen.

Also used are the abbreviations *A-bomb* and *H-bomb*.

Terrorist Nukes

There is great fear among people in the United States that a group of terrorists could make a simple nuclear bomb, smuggle it into a U.S. port, and cause damage far greater than they did in 2001.

If they wanted to make a uranium bomb, the problem would be obtaining the purified U-235. That requires a multibillion-dollar research and development program building Calutrons or centrifuges. The program would be very fragile and easily destroyed unless well-hidden. So the biggest fear is that some other large organization (Pakistan? North Korea?) could give or sell them the purified U-235.

If the terrorists wanted to make a plutonium bomb, they would have to master the exceedingly difficult art of implosion. That requires precision explosives, machined to high tolerance. It requires extensive testing, with images taken of the implosions to see what to fix.

I don't think it likely that terrorist groups could accomplish this. It takes the resources and privacy of a country, such as Afghanistan, Iran, or North Korea. More likely, the terrorists would buy or otherwise obtain the weapon from such a country. Not all of the nuclear weapons of the former Soviet Union have been accounted for. That doesn't mean that they were stolen; it could just be poor accounting. But it is conceivable that someone has such a weapon and is willing to sell it. Such weapons are secured by special codes.

How much damage could a smuggled nuke do? Figure 5.12 shows series of images showing the damage from various bombs if detonated in central San Francisco. The red circle indicates regions of widespread fire; the blue circle shows areas where most homes and buildings are destroyed by the blast wave, and the yellow circles shows regions of moderate risk to people from flying debris.[25]

Fallout

Much of the danger from large (megaton) nuclear weapons comes from the nuclear fallout. This consists of the fission fragments from the uranium or plutonium in the bomb. Fallout is particularly bad if the bomb is exploded near the ground. (The Hiroshima and Nagasaki bombs were exploded high in the air, to maximize the blast to as much of the city as possible.) If the bomb is exploded near the ground, then a lot of dirt and other materials are caught up in the fireball of the explosion. This rises in the air. Ordinarily, much of the radioactivity would occur high in the air, where it hurts nobody. But if there is a lot of dirt mixed in, then the fission fragments tend to fall with the heavy dirt, and bring the radioactivity to the ground. This is a major problem for the larger bombs.

A large fraction (over 5%) of the fission fragments are the isotope strontium-90, a highly radioactive material that has a half-life of 29 years, which gets into the food supply. Back in the 1950s, when many people were worried about the long-term effects of nuclear testing, the term *strontium-90* was well known by the general public. It falls on grass, is eaten by cattle, passed on in their milk to children, and gets concentrated in the bones (since its chemistry is similar to that of calcium).

[25] These were calculated using the nuclear effects calculator at the Web site of the Federation of American Scientists, www.fas.org/programs/ssp/nukes/nuclear_weapon_effects/nuclearwpneffct calc.html.

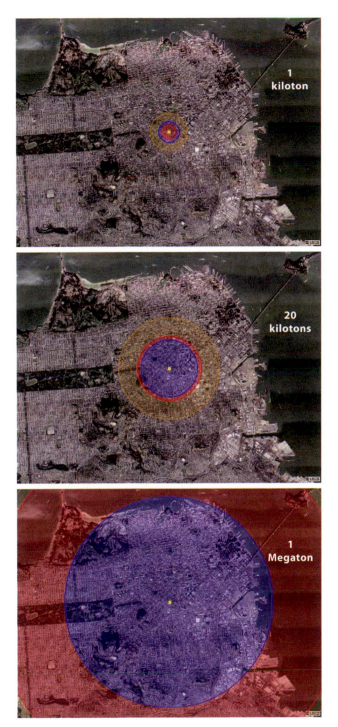

Figure 5.12 The expected effects of a nuclear explosion at ground level in San Francisco. The circled region is destroyed by blast. In the bottom plot, the outlying regions are likely destroyed by fire. Top: 1 kiloton is comparable to the North Korea 2006 test. Middle: 20 kilotons is comparable to the bombs dropped during World War II. Bottom: 1 megaton is comparable to the bomb size carried by U.S. B-52 airplanes. (Calculated using the FAS online computer at www.fas.org/nuke/intro/nuke/effects.htm)

Present Stockpile of Nuclear Weapons

The United States currently has about 12,500 nuclear weapons, although not all are in active use.[26] There are about 10 different "designs," but most involve both fusion and fission. Russia has a similar stockpile. Why so many? During the Cold War, the United States feared a surprise attack from Russia (replay of Pearl Harbor?) and assumed that most of its own weapons would be destroyed in such an attack. The United States wanted to make sure that even if only 1% of its weapons survived, that would be enough to destroy Russia. The assumption was that if Russia knew this, they would never attack. See the movie *Dr. Strangelove* for an ironic account of the possible consequences of this strategy.

The big issue for the stockpile now is reduction (through treaties) and *stockpile stewardship*. This refers to the fact that as our weapons get old, some people argue that they may fail. In olden days, we would assure their functionality through periodic testing, but we have now entered an era when we have decided to end all testing. (This is largely an attempt to keep other nations from developing nuclear weapons.) So there is a large program at Livermore and Los Alamos to try to develop methods of testing the reliability of the weapons without having to set off any of them. It is a big technical challenge.

Nuclear Reactors

A nuclear reactor is a device in which a *sustained* chain reaction takes place. It doesn't involve doubling; instead, from each fission, only one of the emitted neutrons (on average) hits another nucleus to cause another fission. It is as if every couple had, on the average, two children. Then the population would not grow. The power output from a sustained nuclear reaction doesn't grow but is constant.

The power comes out in the form of heat, just as it does when burning coal or gasoline. Frequently, the heat is used to boil water into steam. This steam is then used to run a turbine. (A turbine is really just a fan; as the steam expands through it, it makes the fan turn.) Think about this: a super high-tech nuclear submarine really just uses uranium to boil water!

For their fuel, commercial nuclear reactors use primarily U-235, just as in a nuclear bomb. But the uranium is not enriched to bomb quality. Recall that natural uranium has only 0.7% U-235; the rest is U-238. For use in a bomb, the U-235 has to be enriched to about 80%. But for a nuclear reactor, it has to be enriched only to about 3%. (An exception is the Canadian reactor, called *Candu*. We'll discuss this in a moment.)

Why can a reactor use less enriched fuel? There are two reasons: The first is that they don't require that both neutrons hit U-235; only one. So if one of the two neutrons is absorbed, that's OK—for the reactor, not the bomb. Having a lot of U-238 around isn't so bad.

---

[26] For recent information, I recommend the Web site maintained by the Federation of American Scientists: http://nuclearweaponarchive.org.

But there is a more important reason: a nuclear reactor uses a *moderator*. A moderator is a chemical mixed in with the fuel that tends to slow down the neutrons without absorbing them. The most popular moderators are ordinary water (H_2O), heavy water (deuterium oxide: D_2O),[27] and graphite (which is nearly pure carbon). The moderators consist of nuclei that are light and don't absorb neutrons. The neutrons hit the moderator and bounce off, but in the process they lose a little energy. After enough such bounces, the neutrons are no faster than expected from their temperature. They are called *thermal neutrons* to reflect the fact that they have slowed down to such velocities.

In the commercial nuclear reactor, the fast neutrons emitted in fission bounce off the moderator and become thermal (slow) neutrons. These neutrons are more readily absorbed on other U-235 nuclei, so the enrichment (concentration) of the U-235 need not be 80%, but only 3%.

Can a Reactor Explode Like an Atomic Bomb?

An atomic bomb requires fast neutrons (not moderated) in order to have the entire 80 generations over with before the bomb blows itself apart. After 80 generations, the temperature reaches many millions of degrees. The only reason the bomb doesn't blow apart at that point is that there isn't enough time! With moderated neutrons, the chain reaction is much slower, since the neutrons are slower.

This is an important fact: *Commercial nuclear reactors depend on using slow neutrons.* The reason this is important is that if the nuclear reactor begins to "run away"—i.e., if the operator makes a mistake[28] and the chain reaction begins to grow exponentially (doubling)—then the slowness of the neutrons limits the size of the explosion. Once the temperature rises to a few thousand degrees K, the atoms are moving faster than the neutrons, and so the neutrons can't catch up to them them; the chain reaction stops. The energy released will blow up the reactor, but that energy will be about the same that you would get from TNT. It's an explosion, but it is a million times smaller than a nuclear bomb.

A chain reaction that *depends* on slow neutrons cannot give rise to a nuclear explosion. For that reason, a commercial nuclear reactor cannot blow up like a nuclear bomb. It is important to know this and to be able to explain the logic to the public, since this fact is not widely known.

There are real dangers from nuclear reactors (see the section "The China Syndrome," later in this chapter). Blowing up like a nuclear bomb is not one of them.

OPTIONAL: SLOW NEUTRONS AND U-235

Why are slow neutrons more likely to be absorbed on U-235? The physical reason is simple: a slowly moving neutron feels the nuclear force for a

---

[27] Recall that deuterium is a hydrogen atom with both a proton and a neutron in the nucleus.

[28] Unfortunately for public confidence, the best-known commercial nuclear reactor safety officer is Homer Simpson.

longer time and is more readily pulled by the nuclear force toward the U-235 nucleus.

Of course, the slow neutrons are also more strongly attracted to U-238. But the effect turns out to be much stronger for U-235. So having slow neutrons means that you can use 3% enriched U-235 instead of 80%.

Canadian nuclear reactors use D_2O, heavy water, as a moderator. This is more expensive, but heavy water is more effective in slowing neutrons without absorbing them. As a result, they can use natural unenriched uranium, which is only 0.7% U-235. Their reactor is called a *Candu* reactor, after Canada and deuterium.

Plutonium Production

In a nuclear reactor, only one of the neutrons from uranium fission is used to produce another fission. The other is absorbed. This can be done by using *control rods* made of material that absorbs neutrons without releasing energy. Some of the neutrons are absorbed on U-238, which typically makes up 97% of the uranium in the reactor. When U-238 absorbs a neutron, it becomes U-239. This is radioactive and decays (it emits an electron and a neutrino and has a half-life of about 23 minutes) to an isotope of neptunium, Np-239. This isotope of neptunium is also radioactive. It emits an electron and a neutrino, with a half-life of 2.3 days, to turn into the very famous isotope of plutonium, the one that can be used for a nuclear weapon, Pu-239.

That is how we manufacture plutonium. We make it from U-238 by hitting it with neutrons in a nuclear reactor. The plutonium is a different chemical element from uranium, so when the fuel is removed, the plutonium can be chemically separated. That is not hard to do. The extraction of plutonium is called *uranium reprocessing*. When the United States gives nuclear power plants to developing countries, the United States does not allow them to do their own reprocessing, for fear that they would get a supply of plutonium in this way. Of course, the United States does give them nuclear fuel to run the reactors—but that is a mixture of U-235 and U-238, with too large of a fraction of U-238 for it to work as a bomb.

BREEDER REACTORS

The Pu-239 is usually not considered nuclear waste, because it can be used itself to run a nuclear reactor. It is nuclear fuel. Moreover, if you put it in a nuclear reactor, you get three neutrons per fission instead of two. In a reactor, operating at constant (not exponentially growing) power, you want only one neutron per fission to produce another fission. What do you do with the extra two neutrons? Answer: Put U-238 in the reactor, and make more plutonium.

Thus, a reactor can make (out of U-238) more Pu-239 fuel than it consumes! Such a reactor is called a *breeder reactor*. It has the potential of turning all uranium, not just 0.7% of it, into nuclear fuel, and thereby increasing the available fission fuel by a factor of 140. The time to double fuel in a breeder reactor is about 10 years.

There has been public opposition to breeder reactors. The two most common objections are:

- **The "plutonium economy"**: Breeder reactors would allow much greater use of nuclear power, but that means that plutonium would be widespread. Besides the fact that plutonium is radioactive, and therefore dangerous, some might be diverted to terrorists to make nuclear bombs. Proponents respond that the dangers of plutonium have been greatly exaggerated and that terrorists would not be able to make plutonium bombs because it is extremely difficult to get the required implosion to work adequately.
- **Reactor explosion:** The most efficient kind of breeder reactor would use fast, not slow, neutrons. This is called a *fast breeder*. But if fast neutrons are used, then the main safety aspect of the ordinary reactor is lost. In a fast breeder, the chain reaction could spread uncontrollably, and instead of just a meltdown, the reactor really could explode like an atomic bomb. Proponents respond that they would put in lots of other safety systems that would prevent this from happening.

Dangers of Plutonium

Plutonium has been called "the most toxic material known to man." There is widespread fear of plutonium and of the potential plutonium economy. Because plutonium is so important in public discussion, it is worthwhile giving some of the physics facts.

Here are the key facts[29]: Plutonium is toxic both because of its chemical effects and because of its radioactivity. The chemical toxicity is similar to that of other "heavy metals" and is not the cause for the widespread fear. So instead, let's consider only the dangers from the radioactivity.

Plutonium-239 is radioactive with a half-life of 24,000 years. The radiation from the decay is an alpha particle. It does not have enough energy to penetrate the dead layer of your skin, so it causes harm only if it gets into your body. This happens if you eat it or if you breathe it into your lungs.

For acute radiation poisoning, the lethal dose is estimated to be 500 mg—i.e., about half a gram. A common poison, cyanide, requires a dose 5 times smaller to cause death: 100 mg. Thus, for ingestion, plutonium is very toxic, but 5 times less toxic than cyanide. The primary risk from ingesting plutonium comes from the danger of inducing cancer.

For inhalation, the plutonium can cause death within a month (from pulmonary fibrosis or pulmonary edema); that requires 20 mg inhaled. To cause cancer with high probability, the amount that must be inhaled is 0.08 mg = 80 micrograms. The lethal dose for botulism toxin (the active ingredient in Botox, a widely advertised chemical used by people to reduce skin wrinkles) is

---

[29] For a detailed technical analysis of the toxicity of plutonium, see www.muller.lbl.gov/papers/PlutoniumToxicity.pdf.

estimated to be about 0.070 micrograms = 70 nanograms (ng).[30] Thus, botulism toxin is over a thousand times more toxic. The statement that plutonium is the most dangerous material known to humans is false—it is an urban legend. But it is very dangerous, at least in dust form.

How easy is it to breathe in 0.08 mg = 80 micrograms? To get to the critical part of the lungs, the particle must be no larger than about 3 microns. A particle of that size has a mass of about 0.140 micrograms. To get to a dose of 80 micrograms requires 80/0.14 = 560 particles. In contrast, the lethal dose for anthrax is estimated to be 10,000 particles of a similar size. Thus, plutonium dust, if spread in the air, is more dangerous than anthrax—although the effects are not as immediate.

How easy is it to turn plutonium into dust and spread it into the air? Most people believe it is very difficult to do so. But others argue that if you vaporize the plutonium, it might form small droplets of just the right size. These droplets would have to stay separate from each other, and not coalesce (like raindrops do) into larger particles. Experiments done with vaporized plutonium indicate that it does not form particles of the critical size. But it is hard to know what will happen in all circumstances.

Plutonium metal, in chunk form, is not very dangerous, but it does get warm from the energy released every second from the radioactive alpha decays. Only the alphas emitted on the surface of the plutonium actually get out of the metal, and those do not have enough energy to penetrate the dead layer of your skin.

Depleted Uranium

When U-235 is enriched, there is some U-238 left over. This is called *depleted uranium*. It is about half as radioactive as ordinary uranium since the U-235 and the radioactive isotope U-234 are gone.[31] The remaining U-238 does decay by emitting an alpha particle, with a half-life of 4.5 billon years, roughly the age of the Earth. That's why there is so much left on the Earth—only half of the original U-238 has decayed.

In contrast, U-235 has a half-life of 0.7 billion years. In the 4.5-billion-year age of the Earth, it has gone through 4.5/0.7 = 6.5 half-lives. That has reduced its abundance by a factor of $2^{6.5}$ = 90. That's why there is so little left.

Depleted uranium is used by the military for certain kinds of weapons, particularly shells that are used to attack tanks and other armored vehicles. Depleted uranium is not used because of its radioactivity, but because of two other inherent features: (1) It is very dense; with a density of 19 grams per cubic centimeter, it is almost twice as dense as lead. That is important for penetration. (2) When it hits a metal shield, it tends to form highly concentrated streams, instead of spreading out and splattering. This also helps it to penetrate armor.

---

[30] The toxicity of chemicals such as botulism toxin is not well known, since we don't do experiments on humans, and many people feel that experiments on animals are also improper. Some people estimate that the LD50 for botulism may be as low as 3 ng (rather than 70).

[31] Although U-238 itself is less radioactive than natural uranium, I've been told that impurities bring the radioactivity of depleted uranium back up to the level of natural uranium.

People oppose the use of depleted uranium because it leaves radioactive material on the battlefield. Proponents say that the danger of radioactivity is small compared to the damage done by war, and that the alternative (lead) is also highly poisonous.

Gabon: A Nuclear Reactor 1.7 Billion Years Ago in Africa

In 1972, the French discovered that the uranium they were mining in Gabon, Africa (at a location known as Oklo), did not have 0.7% U-235, but closer to 0.4%! At first they were worried that someone had been secretly stealing U-235, although no one had figured out how they could have extracted it from uranium ore.

French scientists finally discovered that the U-235 had been destroyed by fission, about 1.7 billion years ago. Back then, the fraction of U-235 had been much larger than it is now. (That's because it decays faster than U-238.) Instead of 0.7% of natural uranium, the current value, it would have been over 3%.

A 3% ratio is large enough to use in a nuclear reactor, provided that there is water around to serve as a moderator. That is what we now believe happened in Gabon. Water seeping into the ground moderated the neutrons and turned the uranium deposit into a natural nuclear reactor. When the reactor overheated, the water was vaporized, and the moderation stopped. So the reactor was self-regulating, and it didn't blow up. The power output has been estimated to be several kilowatts. Fifteen regions in three uranium ore deposits have been found in Gabon that were once nuclear reactors.

But the U-235 was burned, and it dropped below the (then) natural level of 3%. It produced plutonium and fission fragments. Eventually, the uranium dropped to a lower level and the reactor turned off. Remarkably, despite abundant groundwater, the plutonium and fission fragments drifted through the rock less than 10 meters over the next 1.7 billion years.[32]

Nuclear Reactor Fuel Requirements

To get a gigawatt of electric power from a nuclear reactor, for a year, you must consume some uranium. The amount is surprisingly small: about 1 ton of U-235, which (if pure) takes a volume of about a cubic foot. This has to be extracted from ordinary uranium that would fill up a cube 2 meters on a side. If you are interested in how I got this number, you can read the following optional calculation.

OPTIONAL: URANIUM FUEL CALCULATION

We want to calculate the amount of U-235 needed to run a 1-GW power plant for a year. We'll do a simplified calculation that will give us an approximate

---

[32] For more information about the Gabon reactor, do a Web search, or see http://www.ocrwm .doe.gov/factsheets/doeymp0010.shtml.

answer. As I said earlier, each fission of U-235 produces about 200 MeV of energy. Let's convert that to joules (J). 1 eV = 1.6×10^{-19} J, so 200 MeV = $200 \times 10^6 \times 1.6 \times 10^{-19} \approx 3 \times 10^{-11}$ J.

How many do we need for a gigawatt-year of energy? A year[33] is 3×10^7 sec. A gigawatt is 10^9 J/s. So the number of joules in one year is E = $10^9 \times 3 \times 10^7 = 3 \times 10^{16}$ J.

So the number of fissions needed N is the energy needed divided by the energy per fission: $N = (3 \times 10^{16}$ J$)/(3 \times 10^{-11}$ J per fission$) = 10^{27}$ fissions. So we need 10^{27} atoms of U-235 to produce a gigawatt for a year.

We assumed that all of the energy goes into electric power. But that isn't true—only about a third does. So we really need 3×10^{27} U-235 atoms.

One mole contains 6×10^{23} atoms. So we need $(3 \times 10^{27})/(6 \times 10^{23}) = 5000$ moles. Each mole weighs 235 g (since there are 235 protons and neutrons in each atom). So the weight of U-235 that we need is $5000 \times 235 \approx 10^6$ g = 1 ton of U-235. Uranium has a density of 19 g/cm^3. So the amount of U-235 needed, 10^6 g, is $10^6/19 \approx 50,000$ cm^3, which is a cube with sides of 37 cm, a little more than a foot. So remember it this way: the amount of U-235 required is about a cubic foot.

This U-235 is found in natural uranium, but it is only 0.7%—i.e., it is 0.007 of the natural uranium. So the amount of natural uranium it takes to run a nuclear reactor for a year is about 1 ton/0.007 = 140 tons = 140×10^6 g. With a density of 19 g/cm^3, this works out to $(140 \times 10^6)/(19) = 7.4 \times 10^6$ cm^3, which is a cube with sides of about 2 m.

Nuclear Waste

The fission fragments from uranium all come from the uranium, so their weight is comparable. Thus, a year of operation of a nuclear power plant will produce about one ton of fission fragments. There may be a comparable amount of plutonium produced. It is potentially valuable for use as a fuel for other reactors, but it is presently considered (by the United States) to be part of the waste. That was done to avoid the "plutonium economy," mentioned earlier and discussed further in the following. Plutonium is much less radioactive than the fission fragments, since its half-life (24,000 years) is so long. But it lasts for a long time.

If they were concentrated, the fission fragments would take up a few cubic feet of volume. But it is expensive to concentrate such highly radioactive material, and so they are normally mixed in with larger amounts of unspent fuel, primarily U-238. This fuel with its fission fragments makes up the high-level radioactive waste of nuclear energy.

Most of the fission fragments are radioactive. They are the same particles that caused radioactive fallout. Some of them have half-lives of a few seconds. Some have half-lives of years. We already discussed Strontium-90, which makes up 5% of the fission fragments and has a half-life of 28 years.

If the reactor is turned off (by removing the moderator, or by putting in special control rods that absorb neutrons), then the chain reaction stops, but the reactor will still produce heat from the radioactive decay of the remaining fission

---

[33] That's the number you get if you take 60 seconds per minute, 60 minutes per hour, 24 hours per day, and 365 days per year: $60 \times 60 \times 24 \times 365 = 3.16 \times 10^7 \approx 3 \times 10^7$ seconds.

fragments. So the reactor continues to produce power, although the power level continues to decrease. It is this heat that leads to the China syndrome, discussed shortly.

A plot of the radioactivity from fission fragments versus time is shown in figure 5.13. Study this plot. It contains very important information for anyone concerned about nuclear power or nuclear waste. The left side shows the level of radioactivity, compared to the radioactivity of the uranium that was removed from the ground when it was mined. Notice that when the reactor is operating, the radioactivity is over a million times greater than that of the original uranium. As soon as the reactor is turned off, the radioactivity of the chain reaction stops, but there are so many fission fragments that the radioactivity only drops to 7.3% of the preceding level, nearly 100,000 times above that of the original uranium. But much of this comes from fission fragments with short half-lives, and they quickly decay away. After one year, the radioactivity has dropped to 8000 times that of the mined uranium. After 100 years, it is only 100 times greater, and after 10,000 years, it is actually less.

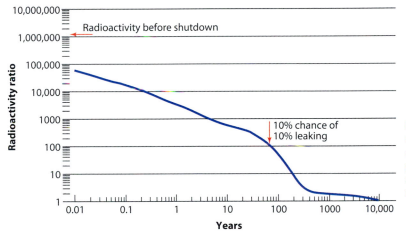

Figure 5.13 Radioactivity of nuclear waste, compared to the uranium removed from the ground that was used for the reactor. The plutonium waste is not included in this figure.

The plot is a little misleading, because it assumes that the plutonium in the waste is not included. In the United States, this is not done (it is in France), and the presence of plutonium leads many people to argue the waste must be stored for a very long time—since the half-life of plutonium is 24,000 years. But many scientists say that the plutonium should not be included in with the list of dangerous radioactive elements because it is very insoluble in water and will barely contaminate groundwater. Moreover, the LD50 dose for consuming plutonium is very large, about a half gram. Plutonium is dangerous if turned into small particles and breathed in, but not from contact with groundwater.

Yucca Mountain

What should we do with the radioactive waste of a nuclear power plant?

Some people say bury it. Put it back into the ground. But what if it gets into the groundwater? Most people assume that that would be horribly bad.

Therefore, they argue, it must be put in a very stable geologic mine, some place where it will undergo no disturbance for 10,000 years. Such a location has been prepared in Nevada. Tunnels have been bored into Yucca Mountain, as a test location for the storage of nuclear waste. But opponents say that even this site can't be certified for 10,000 years. Who knows what kind of government we will have then! That's a long time—10,000 years ago, agriculture had just been invented

You will hear people say that Yucca Mountain is inadequate, that the nuclear waste problem is unsolved because we can't guarantee that we can store it for 10,000 years. They are assuming that the waste must not offer any risk. When plotted versus the radioactivity of the mined uranium, as earlier, the relative risk is put in perspective. Don't forget that the original uranium, before it was mined, was in the ground unprotected. Look again at the plot: after 100 years, the radioactivity is only 100 times worse than before the uranium was mined. If it only had a 10% chance that 10% would leak, that's the same as saying 1% gets out, on average. (That point is marked on the plot.) Then the danger would be no greater than that of the original uranium! After 300 years, the radioactivity has dropped to the point that it is only 3 times greater than that of the original uranium (which, of course, was in the watershed of Colorado—the source of drinking water for Los Angeles and San Diego).

What do you think? Keep in mind that there is a real public fear of radioactivity that makes it very difficult to make a rational decision. Any governor who accepts radioactive waste storage in his state is very likely to be challenged by people who feel that any level of radioactivity is too much. As a potential future president, how would you handle this? How can you balance the real risks versus the perceived risks, and still be reelected?

Some people say avoid all this complication—just put the waste into rockets, and send it into the Sun! But people who say this are ignoring the possibility of an accident. What is the probability that the rocket will fail, and fall back to Earth, releasing all of the radioactivity. That probability may be much higher than the probability that nuclear waste will leak from the Yucca Mountain site.

Also remember that it is not a solution to say do nothing. We have lots of waste from nuclear plants. Right now, most of it is stored in a building near the reactor. Something must be done with it.

Is waste storage a technical problem? Many scientists think it is and are trying to find a clever technical solution. But it is possible that the issue is dominated by the public perception problem. The politician must find a solution that seems safe even to those who are unaware of the natural radioactivity in the environment. It is a very tough political problem.[34]

Coal burning plants bury their waste in the ground. They are not very radioactive, but the ashes are very high in carcinogens. What if these get into the groundwater? How safe is coal, as an alternative to nuclear?

When the decision was made to define plutonium as waste, and not "reprocess" it into nuclear fuel for reactors, one reason was the desire to avoid the plutonium economy, which we already discussed briefly when we were talking about breeder reactors. The fear was that plutonium would become a common

[34] See www.muller.lbl.gov/TRessays/26-Witch-of-Yucca-Mountain.htm.

material, widely used in power plants all around the country, and it might be easily be diverted to terrorist nuclear weapons.

Another reason to avoid reprocessing is that, according to some calculations, it isn't worth the cost. It is cheaper to find new uranium than it is to reprocess the plutonium. At the time that calculation was done, the enormous public opposition to waste storage didn't yet exist, so the costs of building facilities such as Yucca Mountain were not taken into account.

Many people still fear that the proliferation of nuclear material, and the possible diversion for use into nuclear weapons, is a strong reason to avoid expansion of nuclear power. Others argue that the decision not to reprocess was made at a time when oil was cheap and we could dismiss nuclear power as dangerous and unnecessary. This is an excellent topic for discussion.

The China Syndrome

The term *China syndrome*, originally invented by someone with a strange sense of humor, describes the worst conceivable nuclear reactor accident. (Most people seem to think that there is something worse: a reactor becoming a nuclear bomb. But, as I described earlier, that is not possible because the uranium is not sufficiently enriched.)

In the China syndrome, the water that is usually being boiled by the chain reaction suddenly leaks away. There is no water to boil. What would happen in this "loss of coolant" accident? Can you guess?

The first thing is surprising to most people: the chain reaction stops. The reason is that the cooling water is also a moderator; it slows neutrons. So when the water is gone, the neutrons are not moderated. That means that most neutrons are absorbed on U-238, which does not give a chain reaction. So the chain reaction stopped.

> **Interesting flub by a senator:** When the Chernobyl nuclear reactor underwent a similar accident, the Russians announced that the chain reaction had stopped. The chairman of the Senate Intelligence Committee announced on television that this was a "blatant lie." I cringed. He was confusing the chain reaction with the decay of the remaining fission fragments. He knew that the radioactivity hadn't stopped but didn't realize that the Soviets were being completely honest. The fact that the chain reaction had stopped was important; it meant that the level of power being produced had dropped enormously. (Remember this, if you become a senator!)

The chain reaction stops, but there is still the "waste heat" from the fission fragments. Without the cooling water, the reactor gets hotter and hotter. The fuel finally melts. It melts through its containers and forms a puddle at the bottom of the steel reactor vessel. The fuel puddle keeps on getting hotter and hotter. The steel reactor vessel melts. The fuel falls into the ground. It keeps on getting hotter. The soil and rock melt. The fuel just keeps on going—all the way "to China."

No, obviously it won't reach China. (Besides, China isn't on the other side of the Earth from the United States.) It won't get too far, because it spreads out, and that allows it to cool. But in doing this, it has broken through the steel vessel that is supposed to keep it from the environment. Any gases that are in the fuel pellets will escape into the atmosphere. It is these gases (and some volatile elements, such as iodine) that caused the most damage at Chernobyl.

There is a huge amount of radioactivity in the reactor—enough to kill 50 million people (if they ate it). Even a small amount leaked into the atmosphere can do enormous damage. As I stated in chapter 4, the number of expected deaths from Chernobyl, assuming the linear hypothesis, is about 24,000. It is difficult to imagine a worse accident than Chernobyl, so the 24,000 is a much more reasonable estimate than the 50 million. Of course, if the accident occurred in the midst of a very populated area, the consequences could be worse than at Chernobyl.

But 24,000 deaths is a pretty frightening number. Is nuclear power worth it? Why not just use something else, such as solar? Well, people may not want to use solar until it is as cheap as oil. (That will happen sometime in your lifetime—a Muller prediction.) So in the meantime, let's just use something safe like oil.

Is oil really safe? It pours lots of carbon dioxide into the atmosphere. The consequences of that are debated, but most people think the result will be serious global warming. How bad is that? How would you compare it to 24,000 deaths? Some people might argue that the Iraq war is one of the consequences of our use of oil. Why would we have bases in Saudi Arabia if oil weren't so important to us?

Incidentally, the Chernobyl power plant had a terrible design. It didn't even have a containment building, a structure that would trap the emitted radioactivity if the plant exploded or burned; plants in the United States all have these. If it did, there may very well have been virtually no deaths. So is it fair to think of U.S. nuclear power plants in terms of Chernobyl?

Other things are dangerous too. If you are unfamiliar with the tragedy of Bhopal, look it up on the Internet. In 1984, a gas leak from a chemical plant killed 5,000 people in the town of Bhopal in India. Some people have estimated that the total number of deaths from this accident will eventually reach 20,000.

THREE MILE ISLAND—FUEL MELTDOWN

The worst nuclear accident to occur in the United States happened in the nuclear power plant on Three Mile Island, near Harrisburg, Pennsylvania, in 1979. It began when pumps that fed external cooling water to the reactor failed. A backup pump had accidentally been left with a critical valve closed. Control rods were immediately plunged into the reactor core, so the chain reaction stopped. But energy from decaying fission fragments continued to heat the core. Other safety systems failed because of poor design or human error. (A technician turned off an emergency core cooling system because he mistakenly thought the reactor was full of water.) As a result, about 1/3 of the core melted. It did not melt through the steel containment vessel that held the fuel, so the China syndrome did not occur. However, some of the water that cooled the fuel had leaked into the concrete containment building, and radioactive gases dissolved in this water made the interior of the building very radioactive. In order to prevent pressure from building up, some of this gas was purposefully leaked to the outside environment. Calculations show that the number of expected cancers from this leakage (assuming the linear hypothesis) was about one.

This accident happened right after the movie *The China Syndrome* was released. Many people mistakenly thought the accident was as terrible as the one in the movie.

After the accident, many people measured the radioactivity in the region, and discovered it was extremely high—typically 30% above the national average. This caused enormous concern and was puzzling since the amount of radioactivity released from the plant was far too small to account for this high level. It was finally determined that the high level was characteristic of the region and had been that way long before the accident. The radioactivity came from uranium in the local soil, which decayed to radioactive radon gas.

The most famous symbol of nuclear power reactors is the building shown in figure 5.14. But this building is really just a cooling tower, designed to cool the hot water after it was used to run the turbine. The two reactors are in the small containment buildings with rounded tops, in the lower center and lower right.

Figure 5.14 The Beaver Valley Nuclear Power Station in Pennsylvania. The two nuclear reactors are in the two containment buildings with rounded roofs in the lower center and lower right. The large towers emitting steam provide cooling for the reactors. The rectangular buildings contain the electric generators. The steam comes from water taken from a nearby river; it does not come in contact with the nuclear fuel, and it is not made any more radioactive by the power plant. (Courtesy of U.S. Nuclear Regulatory Commission.)

CHERNOBYL—A REACTIVITY ACCIDENT

The worst nuclear reactor accident of all time occurred in Chernobyl (in Ukraine) in 1986. We discussed this in chapter 4, but here I'd like to describe what actually happened. (I'll be brief; for more details, I recommend the Wikipedia article.[35])

The Chernobyl reactor used carbon for a moderator. During an experiment designed to test the safety of the reactor, the chain reaction began to grow out of control. This was in part due to operator error, and in part due to bad design. The reactor heated to such a point that water turned to steam, and that led to a steam explosion. The carbon was set on fire, and much of the radioactivity in the core (the fission fragments) left the reactor carried by smoke. It is estimated that 5 to 30% of the fission fragments in the core spread to the surrounding countryside.

This was not a meltdown from loss of coolant. It was a "reactivity" accident from a runaway nuclear chain reaction. Since the chain reaction depends on slow neutrons, it shut off as soon as the temperature got to a point to cause a small explosion. The subsequent fire spread most of the radioactivity that we described in chapter 4. The Chernobyl plant, unlike Three Mile Island, had no large concrete containment building to hold radioactivity released from the uranium fuel rods.

THE PARADOX OF EVACUATION

Most of the remaining radioactivity in the Chernobyl region comes from the decay of the fission fragment cesium-137. In the "closed zone" where nobody is allowed to live, the dose per year is a factor of about 10 to 15 above the natural background: about 3 rem per year. Someone living there for a decade would get a dose of about 30 rem. Recall that a cancer dose is about 2500 rem, so living in this exclusion zone will give them an additional chance of cancer of about $30/2500 = 1.2\%$. So their probability of dying from cancer will increase from about 20% to 21%.

Should the area continue to be closed to people? The increased risk sounds small. But suppose that a million people moved into the area. Then there would be an additional 1% of cancers, and that would kill 10,000 people!

As an individual, I might choose to take the increased risk of cancer rather then move elsewhere. But isn't it reasonable for the government to deny me this right? After all, they want to save an unnecessary 10,000 deaths.

I have no resolution for this paradox. I give it to illustrate how two reasonable people, in evaluating the risks, could reach dramatically different conclusions.

Controlled Fusion for Power

Uncontrolled fusion is used in the hydrogen bomb. Can we control fusion and use it to produce electricity? That has been a dream since the 1950s. In principle, instead of using rare uranium or plutonium as the fuel, a fusion power plant

---

[35] en.wikipedia.org/wiki/Chernobyl_disaster.

could use the hydrogen that is abundant in the water of the oceans. True, the hydrogen would have to be extracted, but that would take very little energy compared to the enormous energy that would be released in the fusion.

The fuel could be ordinary hydrogen, but for practical reasons, the first fuel will probably use the heavy hydrogen isotopes. That's because a reaction between deuterium (heavy hydrogen) and tritium (doubly heavy hydrogen, with two extra neutrons per nucleus) doesn't require as high a temperature. Deuterium is found naturally in water, at about one part in 6000, and it is not hard to extract. Tritium is rare, but it can be made by hitting lithium with neutrons. Remarkably, the neutrons can be generated in the fusion power process, so it will be possible for the plant to generate its own tritium. So the first power plants will really use deuterium and lithium as their original fuel.

The amount of fuel needed is impressively small. For a gigawatt power plant, the total weight of the deuterium + tritium would have to be only 100 kg per year. Notice that this is about 10% of the weight of U-235 needed for a year of operation of a nuclear fission power plant.

There are several technologies that are being worked on to make controlled fusion. Remember, the main problem in fusion is the fact that the hydrogen nuclei repel each other from their electric charges. In the hydrogen bomb, they overcome this repulsion by getting very high kinetic energy from the primary fission bomb. That's why the hydrogen bomb is called *thermonuclear*. A similar approach is possible for controlled fusion: make the hydrogen very very hot, with a temperature of millions of degrees C. The problem, of course, is that any material that is that hot has high pressure and tends to explode. Moreover, the hot hydrogen will heat any physical container that you use to hold it.

There are three ways to address this problem. The first is to make the hydrogen gas work at a very low density, so the pressure will not get high. This is the *Tokamak approach*, named after the Russian device that first achieved some success. The second method is to let the hydrogen explode, but to keep the explosions small. That's the *laser method*. Last, there is a third speculative way to do it: keep the hydrogen cold, but do some other trick to make the fusion work. That's called *cold fusion*. We'll discuss each of these in turn.

TOKAMAK

In the Tokamak approach, the hot hydrogen is not solid or liquid, but a gas. The gas is so hot that the hydrogen atoms lose their electrons, so technically the gas is a *plasma*—that is, it is made up of electrons and nuclei that are not bound into atoms. The gas is too hot to be held in an ordinary container, so magnets are used. As long as the nuclei are in motion, the magnetic field puts a force on them to hold them in. The nuclei would just be protons if this were ordinary hydrogen, but since heavy hydrogen is used, the nuclei are deuterons (one proton and one neutron attached) and tritons (one proton with two neutrons attached). Some people say the hot hydrogen plasma is held in a *magnetic bottle*.

Tokamaks are big and expensive and still very experimental. A very large Tokamak called ITER (for International Thermonuclear Experimental Reactor) is under construction in France and is due to be operational around 2016. A drawing of it is shown in figure 5.15. Notice the person standing at the lower right.

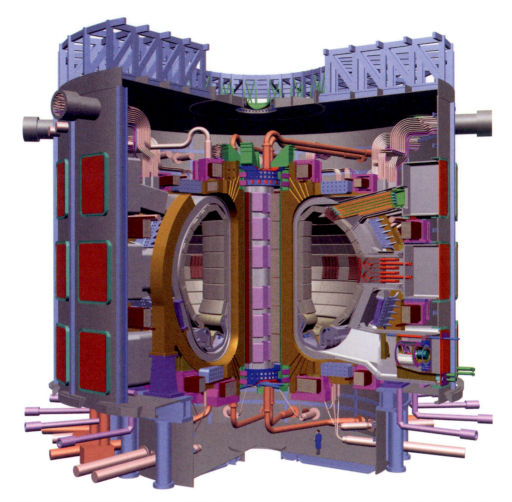

Figure 5.15 ITER Tokamak design. Notice the person standing near the bottom (just to the right of center). ITER was originally an acronym for International Thermonuclear Experimental Reactor. It is under construction in France and is scheduled to be completed by 2018. (Illustration courtesy of U.S. Department of Energy.)

The inside part that you see here is shaped like a doughnut and is kept (when running) at very high vacuum; the vacuum space has a volume of 840 cubic meters. The hydrogen plasma circles around in the doughnut and is heated by changing the magnetic field. When it gets hot enough, the deuterium and the tritium begin to fuse to make helium and neutrons. Most of the energy is carried off by the neutron and is absorbed by a "blanket" of lithium. The heat from this lithium is used to generate electricity. The neutron also breaks the lithium nucleus, creating new tritium nuclei to be used as fuel. You can get more than one neutron from each reaction if you let the neutron hit a material such as beryllium, which breaks up and "multiplies" the number of neutrons.

ITER has the goal of producing a half gigawatt of thermal power for 8 minutes, using a half gram of DT (deuterium + tritium) fuel. That is a good step toward the eventual design of a power reactor, if it works as hoped. But a true power Tokamak is probably at least 20 years in the future, and some people think that it will take much longer. Some skeptics say, "Controlled fusion is the energy source of the future, and it will always be the energy source of the future."

LASER FUSION

Lasers have the ability to deliver large amounts of power to small objects. Because of this, the U.S. Department of Energy has conducted a major program to see whether large lasers (filling a big building) could heat a small pellet of deuterium and tritium hot enough to ignite thermonuclear fusion. This would be a safe way to have fusion, since the amount of fuel can be quite small. (In a thermonuclear bomb, the fission bomb has to be big enough to have a chain reaction, so it is impossible to make one with only a small amount of fuel.)

This method has not yet proven practical. The program at the Lawrence Livermore National Laboratory is now called NIF, for National Ignition Facility.[36] The facility fills a building the size of a football stadium with 192 large lasers. A photo of the building is shown in figure 5.16.

Figure 5.16 The National Ignition Facility (NIF) laser building at Livermore. The goal is to ignite thermonuclear fusion of tritium and deuterium using lasers to power the implosion. (Photo courtesy of U.S. Department of Energy.)

The lasers deliver a power of 500 trillion (5×10^{14}) watts—that is 1000 times the electric generating power of the United States. But it delivers that power only for 4 nanoseconds. (Remember, a nanosecond is a billionth of a second, the time it takes light to travel one foot.) The energy released in this time is 1.8 megajoules, concentrated into a region about a cubic millimeter in size. The purpose is to make the core of a tiny capsule so hot that nuclear fusion will take place, without having to use a fission bomb to generate the high temperature. This kind of controlled thermonuclear fusion (CTF) may be used someday for

[36] You can read more about it at lasers.llnl.gov.

electric power generation, but the NIF facility can't be cycled fast enough for that purpose. Fusion ignition is scheduled for 2010.

The energy sounds less impressive if you calculate that 1.8 megajoules is the same as the energy in 1.5 ounces of gasoline. So it isn't a lot of energy. The purpose of the NIF facility is to deliver this modest amount of energy very, very quickly, so the fusion will take place before the capsule has a chance to cool off by radiation.

COLD FUSION

For fusion to take place, two nuclei must touch. That is hard to achieve, since the electric repulsion (both nuclei have a positive charge) is very strong. One solution is to give the particles extremely high velocity, enough to overcome the repulsion. In thermonuclear fusion, this velocity comes from heating the material.

As an alternative, you could use high voltage (we'll talk about this in chapter 6) to accelerate nuclei one at a time. In fact, accelerators that do this are used to create small amounts of fusion for many purposes, including the preparation of radioactive isotopes for medicine, and in order to create a source of neutrons for use in measuring rock in oil wells ("oil well logging"). Such devices usually use deuterium (D) and tritium (T), two isotopes of hydrogen, since these fuse at relatively low energy. D and T fuse to make helium and a neutron, and that neutron can be very useful; for that reason, these machines often called *DT neutron sources*. In addition, could they be used to generate useful power? As of today, all existing DT neutron sources take more power than they give back, so they are not practical; however, inventors keep on trying. Someday, a version of such a device might prove practical for power.

Another way to get fusion without requiring high temperatures was discovered by Luis Alvarez and his colleagues in 1957. There is an elementary particle called a *muon* that is created in the atmosphere by cosmic rays. It has negative charge, and when it slows down, it sometimes sticks to a nucleus. When it sticks to a hydrogen (or heavy hydrogen) nucleus, it cancels the proton charge. This electrically neutral nucleus can then wander around through the fluid (just from its thermal energy) until it gets close, very close, to another hydrogen nucleus. Then the nuclear force brings the two nuclei together in fusion. Most of the time, the muon would be ejected, so it was free to "catalyze" another fusion.

This kind of cold fusion was a big surprise at the time. Rather than heat the nucleus, just cancel its electric charge! Nobody had predicted that this would happen, but once it was seen, it was soon explained. (Alvarez said its explanation confounded them until they discussed it with Edward Teller, famous for his invention of the hydrogen bomb.)

It turned out that there was no practical way to make use of this muon catalyzed fusion. The problem was that the muon would sometimes stick to the fused pieces, and then it would not catalyze any further reactions. Scientists are still experimenting with different pressures and temperatures in the hope that muon catalyzed fusion will someday work, but I am not optimistic.

The fact that this approach almost worked has given people hope that other approaches could. In 1989, two chemists Stanley Pons and Martin Fleishman thought that they had achieved cold fusion using a palladium catalyst, but their discovery was based on a mistaken interpretation of their data.

Other cold fusion has been reported from time to time. Although there is no proof against it (after all, Alvarez did see a kind of cold fusion), most people are very pessimistic. The reason is that there are no other suitable replacements for the muons used by Alvarez, and any other chemical process typically has energy per atom that is a million times too small to allow the nuclei to approach each other. The whole field is distorted by the fact that anyone who discovers cold fusion will soon (1) win a Nobel Prize, (2) become a multibillionaire, and (3) be known in history as the person who solved the world's energy needs. As a result, when someone sees something that looks like cold fusion (but isn't), it is so exciting that there is a strong tendency to want to believe that a real discovery has really been made, and to keep all the details secret—but that means that they can't be checked by other scientists.

Classified Facts and Atomic Secrets

Most of the material about nuclear weapons was once classified, but it is now accessible to the public.[37]

In writing this chapter, I drew heavily from a paper by Richard Garwin titled "Maintaining Nuclear Weapons Safe and Reliable under a CTBT," dated 31 May 2000. *CTBT* stands for Comprehensive Test Ban Treaty.[38]

A fascinating historical book on nuclear weapons is *The Making of the Atomic Bomb* by Richard Rhodes (Simon & Schuster, 1995). Another is *The Los Alamos Primer* by Robert Serber (University of California Press, 1992). Robert Serber was one of the principal designers of the nuclear weapons used in World War II, and his primer is based on the once highly classified lectures he gave in Los Alamos to introduce physicists to nuclear weapons design.[39]

Chapter Review

The doubling law takes you from small numbers to extremely high numbers in a relatively small number of generations (e.g., 64, as in the squares on a chessboard). Chain reactions can involve doubling, tripling, or any other factor greater than one (such as 1.4). The classic chain reaction is the one that takes place in an atomic bomb. This is made possible by the fact that in a neutron-induced fission, two or three additional neutrons are released, and these can trigger further fissions. In 64 to 84 generations, a few hundred grams of material (6×10^{23} atoms) can be split.

---

[37] The Department of Energy has a Web site that lists previously classified material: www.osti.gov/opennet/. In particular, see the document RDD-7 posted at www.fas.org/sgp/othergov/doe/rdd-7.html.

[38] www.fas.org/rlg/010216-aaas.htm.

[39] For accurate detailed information, it is hard to beat the Web site for the Federation of American Scientists at www.fas.org/nuke/intro/nuke/index.html.

Other chain reactions include the growth of the fetus, the growth of a cancer, and the spread of a virus (both biological and computer). The population bomb was thought by Malthus to be similar, but it is presently slowing down. (Population explosions do occur after mass extinctions, and when foreign animals are introduced into a land in which they have no natural enemies, such as rabbits in Australia.) The concept of a chain reaction was developed into a practical tool called the *polymerase chain reaction*, or PCR. PCR is very valuable in biology and can be used to identify people from their DNA. Other examples of chain reactions include avalanches (rock, snow, and electrons—as in lightning and sparks). The math of the chain reaction is identical to that of compound interest and to Moore's law of computer technology growth.

Atomic bomb is the common name for a nuclear weapon that is based on the chain reaction of U-235 or Pu-239. U-235 is a rare (0.7%) isotope of uranium that is difficult to separate. In World War II, Lawrence did this with a Calutron, and Saddam Hussein chose this same approach in his weapons program. Plutonium is manufactured in nuclear reactors, and it is easy to separate from other chemicals, but it requires a difficult design (using implosion) to use in a nuclear weapon. The bomb explodes when the uranium or plutonium is collected into a *critical mass*, a blob of material big enough that most neutrons produced hit a nucleus and trigger fission instead of leaking out.

A thermonuclear bomb, also called a *hydrogen bomb*, is a three-stage weapon in which the fission primary ignites a secondary reaction that contains two isotopes of hydrogen (deuterium and tritium). The fission fragments from the primary (and from a uranium container) are the most dangerous part of the residual radioactivity. If the bomb was exploded at low altitude, then dirt mixed with the fission fragments causes these radioactive pieces to fall out rapidly, and that could cause greater death than the explosion itself. The worst fallout is strontium-90.

At the peak of the Cold War, the United States and the Soviet Union had over 10,000 nuclear warheads that could be launched at the other country. Nuclear weapons are no longer tested (in part, to reduce proliferation), and *stockpile stewardship* refers to the problem of making sure that these weapons still work without testing them.

Nuclear reactors are based on the chain reaction, but they normally work with a neutron multiplication of 1, so the reaction doesn't grow. Nuclear reactors use moderators to slow the neutrons. This increases the probability that a neutron will be attracted to a nucleus. If the moderator is lost (e.g., the water leaks out), then the chain reaction stops. If the reactor runs away (because of operator error, setting the neutron multiplication number greater than 1), then the reactor will burn or explode, releasing energy roughly equivalent to that of a few pounds of TNT. A nuclear reactor that depends on a moderator cannot explode like an atomic bomb. If the fuel collects at the bottom of the reactor, it will continue to heat from the remaining radioactivity of the fission fragments, and this can lead to the *China syndrome*.

Nuclear waste consists of the long-lived fission fragments. It takes about 10,000 years for the radioactivity to drop below the level of the original uranium that was removed from the ground, but except for the plutonium, most of the radioactivity is gone after a few hundred years. Proponents of nuclear power argue that the nuclear waste is far safer, since it is placed in special locations isolated from groundwater, unlike the original uranium.

Controlled thermonuclear fusion for power generation is based on deuterium extracted from seawater and tritium that is manufactured from lithium. Experimental programs are under way to study this. The main programs involved magnetic confinement (Tokamak, with ITER being one of the biggest), and laser heating of pellets (with NIF, the National Ignition Facility at Livermore, being the largest). Cold fusion is theoretically possible, but so far has been achieved successfully only using muons.

Discussion Topics

1. Yucca Mountain: What are the pro and con arguments for nuclear waste storage at this site? What are the alternatives? Do you conclude that we should stop the production of nuclear waste? What do we do with the present waste?
2. Suppose that we were to build a large number of nuclear reactors. Discuss the potential dangers of a "plutonium economy." How would those dangers compare to the dangers of "being addicted to oil"?
3. Some people sarcastically say, "Controlled fusion is the power source of the future—and it will remain the power source of the future for the indefinite future." Why are some people so pessimistic about the future of fusion power? Can you find people who are optimistic? Can you reconcile the two?

Internet Research Topics

1. Oklo prehistoric nuclear reactor in Gabon: What more you can learn about the ancient nuclear reactor located there?
2. Moore's law: What are the current predictions?
3. Nuclear proliferation: What countries are suspected of developing nuclear weapons? Which are suspected of plutonium weapons, and which of uranium weapons? What methods are they suspected of using for uranium enrichment?
4. Look up cold fusion. What are people claiming? If their claims are correct, why do we not yet have operating cold fusion power plants?
5. Will we run out of uranium to operate nuclear reactors? See what you can find out. Some people say yes; others say that uranium is not the major cost of nuclear power, so we can extract it from low-grade ores; others say that we can even afford to extract it from seawater, and that could provide a supply that could truly last for thousands of years.

Essay Questions

1. Discuss the similarities between the nuclear chain reaction and the spread of an epidemic. Put in the relevant numbers, and describe what ultimately limits the growth for each case.

2. There is a great deal of misinformation regarding nuclear power plants. Describe what the public thinks is true, but isn't, and what the public thinks is true, and is.

3. Fission fragments are important to understand, both for discussions of nuclear weapons and for nuclear reactors. Describe fission fragments and the key roles they play in these systems. What are their features that make them important?

4. What are the impediments to a terrorist building his own nuclear weapon? Can't a high school student do it? Discuss both the problems in obtaining materials and the difficulty of designing and manufacturing the weapons.

5. What are the prospects for controlled thermonuclear fusion? What approaches are being tried? What are the potential advantages over fusion? What are the difficulties?

6. What is meant by the "plutonium economy"? Why do people want to avoid it? What policy decisions have been triggered by the fear of it?

7. What is the purpose of a nuclear reactor? How does it work? What are the potential dangers? Give historical examples. Are there things that many people think are true about reactors that, in Muller's analysis, are not?

8. A country such as Iran may be developing a nuclear weapon. What kind of fission weapon might they attempt? Describe the steps and methods that a country such as Iran would have to use to make a fission weapon.

9. A nuclear chain reaction and a polymerase chain reaction seem to be very different, yet they both use the term *chain reaction*. Explain what they both are, what they have in common, and yet how they are different. Give examples of how both chain reactions are used.

Multiple-Choice Questions

1. Recent reports discussing the population bomb
   A. verify that the population will expand above 20 billion
   B. indicate that the explosion is slowing
   C. show that it doesn't matter since food supply grows equally fast
   D. show that the population of the world is now decreasing

2. A one-kiloton nuclear weapon, exploded at ground level, would destroy
   A. about 1 square kilometer of a city
   B. most of a small city (e.g., San Francisco)
   C. most of a large city (e.g., New York City)
   D. many cities, if they were within 100 miles of each other

3. For an atomic bomb, the number of doublings required is closest to
   A. 10^{23}
   B. 235
   C. 80
   D. 16

4. An implosion type bomb is required for
   A. a U-235 bomb
   B. a Pu-239 bomb
   C. a thermonuclear bomb
   D. a boosted fission weapon

5. Moore's law relates to
   A. urban legends
   B. computer viruses
   C. biological viruses
   D. computer chip capability

6. Which of the following does not represent exponential growth?
   A. a computer virus
   B. spread of smallpox
   C. epidemic of flu
   D. illness from anthrax

7. The statement that the Chernobyl accident will kill 24,000 people is based on
   A. measurements of leukemia and thyroid cancer near Chernobyl
   B. the linear hypothesis
   C. the concept of a chain reaction
   D. the fact that radioactivity is contagious

8. A moderator is something that
   A. slows neutrons
   B. slows fission fragments
   C. fissions more easily than U-235
   D. speeds up the chain reaction

9. The dangerous radioactivity in fallout comes from
   A. neutrons
   B. neutrinos
   C. gamma rays
   D. fission fragments

10. Fallout is much worse if the bomb
   A. is exploded near the ground
   B. is exploded at high altitude
   C. undergoes fewer generations
   D. contains no fissile material

11. A nuclear reactor cannot explode like a nuclear bomb because
   A. it contains too much uranium
   B. it contains no uranium
   C. the nuclear reactor depends on slow neutrons
   D. it is carefully designed to shut down quickly

12. According to the text, the least bad place to put nuclear waste is
   A. in the Sun
   B. all the way into outer space
   C. in our food
   D. underground

13. A breeder reactor is designed to produce
   A. U-235
   B. tritium
   C. U-238
   D. plutonium

14. The number of nuclear weapons that the United States had was closest to
   A. one million
   B. ten thousand
   C. one thousand
   D. several hundred

15. PCR has been used to (choose all that are appropriate):
   A. identify children of Thomas Jefferson
   B. check the guilt of convicted murders
   C. identify victims of 9/11
   D. identify the fathers of children

16. The Hiroshima bomb used
   A. a gun design
   B. thermonuclear fusion
   C. implosion
   D. boosted fission

17. If the water moderator is lost from a nuclear reactor,
   A. the reactor will explode like an atomic bomb
   B. radiation from fission fragments will continue to produce heat
   C. all of the power of the reactor will immediately go to zero
   D. the chain reaction will increase, causing a reactivity accident

18. The material that might be "reprocessed" from nuclear waste is
   A. tritium
   B. U-238
   C. U-235
   D. Pu-239

19. Which of the following is not a good example of the doubling law?
   A. lightning
   B. avalanche
   C. nuclear chain reaction
   D. future population growth

20. The Canadian Candu reactors use
   A. light water
   B. deuterium
   C. thermonuclear fusion
   D. tritium

21. Plutonium can explode with fewer generations than can uranium because
   A. plutonium fission releases more neutrons
   B. plutonium fission releases more energy
   C. plutonium does not require a moderator
   D. plutonium turns into uranium

22. In a nuclear power plant, the material that runs through the turbine is
    A. fission fragments
    B. electrons
    C. neutrons
    D. steam

23. The number of deaths from the Nagasaki bomb is estimated to be
    A. 100 to 500
    B. 50,000 to 150,000
    C. 2 to 3 million
    D. 12 million

24. The smallest container that could contain one critical mass of plutonium is
    A. a tablespoon
    B. a coffee mug
    C. a large suitcase
    D. the trunk of an automobile

25. The bomb dropped on Hiroshima used as its fuel
    A. uranium
    B. plutonium
    C. hydrogen
    D. deuterium and lithium-6

26. Saddam Hussein planned to enrich uranium using
    A. Calutrons
    B. centrifuges
    C. lasers
    D. gaseous diffusion

27. A centrifuge enrichment plant takes an area of about
    A. one living room
    B. one large classroom
    C. one large building
    D. about one square mile

28. PCR involves
    A. tritium
    B. carbon-14 (radiocarbon)
    C. heavy water
    D. DNA

29. One of the most dangerous radioactive materials from fallout is
    A. U-235
    B. Pu-239
    C. Sr-90
    D. deuterium

30. The term *China syndrome* refers to
    A. the accident at Chernobyl
    B. meltdown of the nuclear fuel
    C. production of plutonium in China
    D. nuclear terrorism

31. Depleted uranium is a useful substance because
    A. it is radioactive
    B. an artillery shell made from it is very penetrating
    C. it is poisonous (but not very radioactive)
    D. it can be converted to U-235

32. PCR was used to learn about
    A. the critical mass of uranium
    B. descendants of Sally Hemmings
    C. the security of Yucca Mountain
    D. locations of oil underground

33. Another name for the H-bomb is
    A. dirty bomb
    B. stealth bomb
    C. fission bomb
    D. thermonuclear bomb

34. A nuclear explosion similar to the one produced by North Korea would destroy an area of about
    A. the size of a college campus
    B. the size of a medium-size town
    C. the size of a major city
    D. the size of a small state (e.g. Massachusetts)

35. Cold fusion
    A. despite claims, has never been observed
    B. has been seen in laboratory experiments
    C. is the key mechanism used in the Tokamak
    D. is used by Livermore in the NIF (National Ignition Facility) project

36. At Three Mile Island (choose all that are correct)
    A. there was a reactivity accident (runaway chain reaction)
    B. over 200 people are expected to die from the released fission fragments
    C. some of the uranium fuel melted
    D. despite newspaper reports, no radioactivity was released

37. What distinguishes a breeder reactor from other reactors is that
    A. it uses slow neutrons
    B. it cannot melt down
    C. it uses hydrogen as its fuel
    D. it makes more fuel than it uses

38. Jupiter is not a star because
    A. it isn't made of hydrogen
    B. it isn't massive enough
    C. it is too far from the Sun
    D. it isn't made of helium

Electricity and Magnetism

Electricity Is . . .

- the cause of lightning, which delivers power much greater than that of a nuclear power plant
- used for all the computations done in a laptop computer
- used for radio communication, and to send telephone signals through wires
- the most convenient (and often the cheapest) way to transport energy, at least for short distances
- able to enter our homes when needed by the flick of a switch, through nationwide circuitry so complex that it can collapse in a few seconds
- so safe that we have outlets all over our homes, and yet it is still used as a gruesome method of execution for humans and was once used to kill a "bad" elephant
- used by the nerve cells in our bodies to send signals
- responsible for nuclear fission energy, since the fission fragments get their energy from electric repulsion

The twentieth century could rightly be called the century of electricity. (Of course, it might also be called the century of autos or of airplanes or of quantum physics or of antibiotics.) Most of what we call "high-tech" consists of the enslavement of electricity to do our purposes.

Equally mysterious is magnetism. Magnets also play a central role in our high-tech world.

Magnetism Is . . .

- something that was once a military secret
- the force that pushes things around in "electric" motors

- used to store information on computer hard drives
- the main way used to generate electricity
- what Saddam Hussein planned to use in his Calutrons to get U-235
- used to determine the ages of sedimentary rocks
- used to run loudspeakers and earphones

Moreover, radio waves, light waves, x-rays, and gamma rays carry half of their energy in electricity and the other half in magnetism.

It was once thought that magnetism was totally unrelated to electricity. We now know that magnetism is a subtle aspect of electricity.

But what is electricity?

Electricity

Electricity usually means the movement of electrons. These tiny particles, about 1/2000 of the mass of atoms, can exert a huge force—the *electric force*—on other electrons and on atoms.

Take two electrons and place them a centimeter from each other. Make sure that nothing else is around. Since each has mass, the force of gravity will attract the two electrons to each other. But the electric force between them is repulsive; it pushes them apart. Moreover, this repulsive force is stronger than the attractive gravity force by a factor of approximately

$$4170000000000000000000000000000000000000000$$

I wrote the number that way to be dramatic; the same number can be written as 4.17×10^{42}. So this electric force *completely* overwhelms gravity.

Now consider an electron placed a centimeter from a proton. They will attract each other, not repel. Yet this force will be *exactly* the same as the repulsive force between two electrons. (Incidentally—we still don't know why the proton has exactly the opposite charge of the electron.)

Electric Charge

The property of the electron that gives its force has a name: the *electric charge*. By convention, the charge of the proton is

$$q_p = 1.6 \times 10^{-19} \text{ Coulombs}$$

You won't need to know that number. The charge on the electron is exactly opposite that on the proton. In equation form we say $q_e = -q_p$. By putting the minus sign in front, we keep track of the fact that the force it exerts is opposite to the force exerted by a proton. We say that the electron charge is negative. (It is -1.6×10^{-19} Coulombs, but all you have to know is the sign.)

If we combine an electron with a proton to make a hydrogen atom, the total charge is 0. So the hydrogen atom does not "feel" an electric force from other particles, since the force on the proton and the force on the electron will be opposite and cancel each other. We say that the hydrogen atom is *neutral*. Neutral means that the total charge is zero, even if it is made up of pieces that have charge.

Neutrons have mass similar to that of the proton, but neutrons have charge 0. Why? Is it possible that the neutron is similar to a hydrogen atom? A hydrogen atom consists of a proton with charge +1, and an electron with charge –1, and the two cancel. Could a neutron also have interior charges that cancel?

We now know that the answer is yes. The neutron consists of three quarks. One of them is the u quark (also called the "up" quark), and the other two are d quarks (for "down"). We write this as udd. The u quark has a charge of +2/3 (in terms of the proton charge), and each of the d quarks has charge –1/3. So the total neutron charge is 2/3 – 1/3 – 1/3 = 0 proton charges. That's why the neutron is neutral.

The proton consists of uud, with a total charge of 2/3 + 2/3 – 1/3 = +1 proton charges.

Charge Is "Quantized"

As far as we know, all charges in nature are exact multiples of the quark charge. We don't know why. This is stated in physics by saying "charge is quantized." Particles can have charge –1/3, +1/3, 1, 2, etc., but cannot have charge 1/2, 4/5, or 1.22. We don't know why this is true.

You might guess that the reason is that all particles are made of quarks. But that isn't true. Electrons are not made of quarks.

A new and, as of yet, unproven theory is that all particles are made of objects called *strings*. If this theory is true, then the reason behind quantization is simply that all particles are really made of the same kind of thing.

Electric Current—Amps

When charged particles move, we call it *electric current*, in analogy to water current. For water, we measure current in gallons per second, or in cubic meters per second. For electric current, we measure current in electrons per second. A more practical unit is the ampere, or amp. One amp is 6×10^{18} electrons per second. Don't memorize this number, but you should know that the current is a measurement of electrons per second.

The current that flows through a lightbulb is typically about one amp. Wires in your house carry up to about 15 amps. The current is divided among all the systems that use electricity, such as your refrigerator, lights, TV, and computer. One bolt of lightning has thousands of amps.

The current from a flashlight battery is also about 1 amp. The main reason that the flashlight bulb is not as bright as a typical lightbulb is that the filament is shorter, so there is less to glow.

OPTIONAL: AN AMP FOR A DAY

Here is an interesting coincidence. Suppose that you let one ampere flow for a day. How many electrons total were there? One amp is 6×10^{18} electrons per second, and there are 86400 seconds per day.[1] The total number is the product of these two numbers: $6 \times 10^{18} \times 86400 \approx 5 \times 10^{23}$. That's almost one mole,[2] the number of electrons in one gram of hydrogen. Think of it in the following way: if you were to take a gram of hydrogen, and remove the electrons, you would have enough to make a flow of one ampere for one day. (I can think of no reason why this knowledge would be useful for a future president. That's why this section is optional.)

Wires: Electron Pipes

Metals have a wondrous property: electrons can flow easily right through the solid inside of a piece of metal. (Glass has a similarly wondrous property: light can pass right through it.)

Recall from chapter 4 that the nucleus takes up very little space in an atom, no more than a mosquito takes in a football stadium. The rest of the space is taken up by electrons. For metals, one of the electrons in each atom is not permanently attached, so it can move from one metal atom to another.

Electrons can move easily inside a piece of metal, but they can't easily leave the surface of the metal. They are held back by the attraction of the positively charged nuclei. Free movement of electrons can take place only if the moving electrons are replaced by other electrons. For this reason, electric current usually flows in circles or closed paths.

Have you noticed that most electric cords (e.g., those for a lamp) have two wires in them? The second one is for the electrons to return. Some computer wires are called *coax cables*. They also consist of two conductors, but instead of two wires, they have one wire surrounded by a cylindrical metal tube. (*Coax* derives from "coaxial"; it means that the axis of the wire is the same as the axis of the tube.) The tube serves as the electron "return path."

When a bird lands on an electric power line, some electrons will immediately flow into the bird. But with nowhere to go, the electrons soon repel other electrons from coming, so the flow will stop. Very few electrons are needed to stop the flow.

Likewise, if a person hanging on an electric power line were to touch nothing else, he would be safe. If he touches another wire (which could be the return path for the electric power), then a large current could flow through him.

Watch a mechanic attach a wire to an automobile battery. He'll be careful not to touch anything else, particularly not the metal of the car, with his other hand. That's because one side of the battery is usually attached to the metal of the car, and the mechanic does not want his body to serve as a return path. A car battery can deliver 100 amps of electric current, and that can be dangerous.

Even though the electrons in electric current move in circular paths, they can be used to carry energy and information. As electrons move through wires, you can remove some of the energy from them, much as a mill wheel can take energy from a stream of water. To send information, you vary the amount of current flowing in the circle. In a similar way, you can signal someone with a hose by

---

[1] 60 seconds per minute, 60 minutes per hour, 24 hours per day, gives $60 \times 60 \times 24 = 86,400$ seconds per day. Multiply this by 365 days per year to get 3.15×10^7 seconds per year.

[2] A mole is 6.02×10^{23} atoms.

turning the water on and off. Telephone wires carry sound signals by varying the current to match the vibrations of sound.

Resistance to Electric Current Flow

The easiest way to remove electron energy from current is simply from the friction caused by the electron flow. Such friction is called *electric resistance*. Some metals, such as tungsten, have lots of resistance. The filament of an ordinary incandescent lightbulb is made of tungsten. When current flows through it, the resistance (friction) heats the filament enough to make it glow. Thus, electric current is first turned into heat and then into light. (We'll discuss this more in the next chapter.)

Of course, you don't want to heat the wires that go from the wall outlet to the bulb, so those are usually made out of copper or another metal with low resistance.

Materials that conduct electricity well (such as most metals) are called *conductors*. Materials that don't conduct electricity very well (such as plastics, rocks, or wood) are called *insulators*. But between metals and insulators is a group of materials called *semiconductors*. These are materials that can be made to turn from conductors to insulators and back, by applying electricity in a special way. Their ability to control electric flow is what makes them so useful in electronics from stereo systems to computers. We'll discuss these in more detail in chapter 11.

Fuses and Circuit Breakers

Wires in your house are typically made out of copper, a metal with low resistance, so that they will not waste the energy of electric current. If the current is high, however, then the wires can get hot enough to start a fire in the walls. For this reason, most house wiring has a device that prevents the current from exceeding a safe value, typically 15 amps (enough for about 15 lightbulbs.) The two kinds of devices used are called *fuses* and *circuit breakers*.

A fuse is a short length of high-resistance material that melts when too much current flows through it. When it melts, it breaks the connection of the wires, and the current stops flowing. To get the current flowing again, the fuse must be replaced. In common usage, to "blow" a fuse means to send enough current through it that the metal inside melts or vaporizes.

A typical circuit breaker has a wire formed into a bimetallic strip (see chapter 2). When the bimetallic strip heats beyond the allowed limit, it bends away and breaks the connection to another wire. Unlike the fuse, the circuit breaker can be reset (the bimetallic strip placed back in contact with the wire) after it has cooled.

Superconductors

Superconductors are materials that have zero resistance—they don't impede electricity at all! Rings of superconductors have had currents flowing in them for decades, with no energy source. The phenomenon is similar to the Earth going around the Sun; if there is no friction, it will just go on forever.

Unfortunately, all known superconductors have the zero resistance property only at low temperatures. If we could find or manufacture a "room temperature" superconductor, it would revolutionize the way we use electricity. Right now, much energy is wasted by conducting electricity through resistive wires, and a real room-temperature superconductor would revolutionize the way energy is transported.

How can electrons flow inside a metal with zero friction? The answer was not known for many decades, but we now understand that the secret lies in quantum mechanics. We'll discuss this further in chapter 11.

The easiest way to cool a wire is to put it in a cold liquid. The original superconductors were kept cold by immersing them in liquid helium. The liquid is made in special refrigerators, and then transported to the customer in dewars (glass containers that are similar to Thermos bottles). Liquid helium boils at a temperature of 4 K—i.e., only 4 degrees above absolute zero. So as long as there is liquid helium, the temperature is low. Recall from chapter 4 that helium comes from alpha particles in the Earth's crust, and we collect it from oil and natural gas wells. When these wells run out, we will have no further source of helium. (The Sun is 10% helium, but that's not easy to get.)

Thirty years ago, most of the helium from wells was discarded, because the need for it wasn't great enough to justify the expense of trapping it. United States law now requires the oil and gas companies to recover and store the helium, because of anticipated needs for future superconductors.

"HIGH-TEMPERATURE" SUPERCONDUCTORS

In 1987, the Nobel Prize in physics was awarded to Georg Bednorz and Karl Muller for their discovery of certain compounds that become superconducting at relatively high temperatures. Right now, the highest temperature superconductor works at a temperature of about 150 K, equal to −123 C or −189 F. That's pretty cold for something called *high temp*, but it is the best anyone has done.

Part of the reason scientists use the word *high* for this temperature is that it's higher than the boiling temperature of liquid nitrogen, which is 77 K. Recall that nitrogen is about 80% of air; it is extremely abundant, especially when compared to helium. Nitrogen can be liquefied for about a dollar a quart, making its cost comparable to that of milk (and some bottled water brands). Superconductors that can be kept sufficiently cold with liquid nitrogen are, in principle, much more practical.

So why aren't we using such superconducting wires for all of our power transmission? The answer is that the high-temperature superconductors are all pretty brittle, and it has been difficult to manufacture useful wires from them. Nevertheless, it is being done for some special applications. An experiment to see if such wires can be used for commercial electric power transmission is currently underway by the Detroit Edison power company.

Of course, if liquid nitrogen is used for cooling, then some power is lost— the power needed to produce replacement liquid nitrogen when it boils off. So such transmission lines do use energy.

There is a limit to the amount of current that superconducting wires can carry. That's because high current creates very strong magnetic fields (to be discussed shortly), and strong magnetic fields can destroy superconductivity just as much as can high temperatures. The current they carry depends on the cross-sectional area; some materials have been reported that can carry several million amperes per square centimeter of area.

> **Amusing fact:** According to theory, highly compressed hydrogen should become a metal. It is even possible that the core of the planet Jupiter consists of superconducting hydrogen.

Volt: A Measure of Electron Energy

Amps tell you how many electrons are flowing past a point each second. *Volts* tell you the energy of the electrons. The energy unit called the electron volt, abbreviated[3] as eV, is defined as

$$1 \text{ eV} = 1.6 \times 10^{-19} \text{ Joules}$$
(don't memorize)

Whereas a Calorie is a typical amount of chemical energy for a gram of material, an eV is a typical amount of energy for a single atom or molecule. That fact is useful and worth memorizing!

$$1 \text{ eV} = \text{typical energy for a single atom or molecule}$$

> **Jargon:** If a piece of metal has a large number of electrons, each with energy of 1 eV, you'll hear people say that the metal *is at* one volt, or sometimes they'll say it has a *voltage* of one volt, or maybe that it has a *potential* of one volt. It is OK to refer to the energy of an electron in volts, rather than in eV. Physicists will use these two terms slightly differently, but it is not important here. Remember that when a piece of metal is at one volt, it means that every electron in that metal has that energy.

Here are some key numbers about volts:

Typical energy of an electron in an atom	1 volt
TNT energy per molecule	1 volt
Flashlight battery	1.5 volts
U.S. house voltage	110 volts
European house voltage	220 volts
Voltage inside CRT TV tube	50,000 volts
Alpha particle from nucleus	1,000,000 volts

Low-volt electrons are not very dangerous. A flashlight battery has a typical voltage of 1.5 V. You can read that on the label. It produces electrons with an energy of 1.5 V. You can hold such a battery in your hand with no danger; if

[3] The *V* in eV is usually capitalized. The justification is that it was named after a person, Alessandro Volta. Yet the word *volt* usually is not capitalized. The traditions are not consistent.

you touch the metal leads to your tongue, you'll feel a tingle. Don't do that with a higher voltage battery or the higher energy electrons might burn your tongue.

Finger Sparks and Static Electricity

The sparks that sometimes fly from your finger to a doorknob are often called *static electricity*. This occurs because your feet rub on the ground in such a way that electrons come off and stick to your body. These electrons are static in the sense that they stay there, on your body, until you walk up to a good conductor like a metal doorknob. You'll pick up even more electrons if you rub your shoe on a thick carpet. You can also rub electrons onto a comb by running the comb through your hair. Try doing that—run the comb through several times quickly, and then put the comb near some very small (mm size) pieces of paper. The electrons on the comb will attract the bits of paper.

If the air is moist, the static electricity leaks off your body into the air. But on a very low humidity day (which means there is very little moisture in the air), the air is a poor conductor, and the electrons stay on your body. They can move around inside your body, since your salty blood is a pretty good conductor of electricity. But when you have these excess electrons and you put your finger near a piece of metal, they will jump off, creating the flow of current we call a spark.

For that spark, the voltage was probably between 40,000 and 100,000 volts! Yet it doesn't kill you because the current is low, limited by the small number of electrons you picked up. Yet a similar voltage in the back of an old CRT TV set is very dangerous. That's because the amount of current that can flow to you is much greater.

To know the power, you must know the energy per particle AND the number of particles per second. The same is true with flowing water: you need to know the velocity of the water *and* the number of gallons flowing every second.

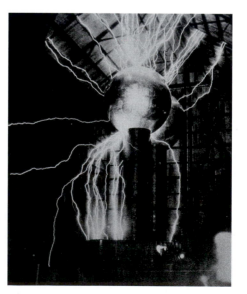

Figure 6.1 Large Van de Graaff generator at MIT. (Photo courtesy of U.S. Department of Energy.)

Finger sparks can be automated for a physics demonstration by using a device called a Van de Graaff generator. In this device, a band of rubber rubs against a piece of wool continuously and the charge is taken off by a wire attached to a metal sphere. In a few seconds the sphere can reach 100,000 V. Yet the sparks are not dangerous because the amount of charge is so small. Large Van de Graaff generators were the first method used to reach a million volts (figure 6.1).

Electric Power

The power delivered by electrons depends on the energy of the electrons, and the number per second that arrive. The first is the voltage, and the second is the current. Multiply these together and you get the power.

Let's do this calculation for a small, 1-volt battery delivering 1 amp. The energy of a 1-volt electron is 1.6×10^{-19} J. The number of electrons per second is 1 amp $= 6 \times 10^{18}$ electrons per second. Multiply these together to get $1.6 \times 10^{-19} \times 6 \times 10^{18} \approx 1$ J/s $= 1$ watt (W). That's not a coincidence. The numbers were chosen to make this work out exactly.[4] So here is the important conclusion:

$$\text{power (in watts)} = \text{volts} \times \text{amps}$$

Here is another practical example: suppose that you have a lightbulb that uses 110 V and carries a current of 1 amp. Then the power is $110 \times 1 = 110$ W. If you run that bulb for an hour, you use a total energy of 110 watt-hours $= 0.11$ kWh. Remember: 1 kWh costs about 10¢. So 0.11 kWh costs about 1¢.

Another example: A flashlight battery works with 3 volts (two batteries in a row) and uses about 1 amp. That means it uses a power of 3 volts $\times$ 1 amp $=$ 3 watts. If the batteries last for an hour, then the energy they delivered was 3 Wh.

Note that high voltage does not always mean high power. If the amps are tiny, then high voltage can be safe. That's why I can let large sparks from a Van de Graaff generator jump to my hand without it hurting (at least, without it hurting enough to make me admit it hurts).

The Energy in Finger Sparks and Lightning

I mentioned earlier that the energy of electrons in a finger spark can be 40,000 volts or more. But there aren't usually very many of these excess electrons on your body, typically not much more than about 10^{12} of them.[5] That may seem

---

[4] The energy in 1 eV is not exactly 1.6×10^{-19} J. A more accurate number is that 1 eV $\approx$ $1.60217733 \times 10^{-19}$ J. An ampere is not exactly 6×10^{18} electrons per second. A more accurate number is 1 amp $\approx 6.2415064 \times 10^{18}$ electrons per second.

[5] For those of you who have studied electrical engineering, here is the way I did the calculation. I assumed the electrons had an energy of $V = 40,000$ eV. I assume that the capacitance of your hand was about $C = 10$ picofarads. Then the charge in Coulombs is $Q = CV$. Divide by 1.6×10^{-19} to get the number of electrons. The energy in joules is $E = 1/2 \ CV^2$.

big, but it is much less than the number of atoms in a gram of material. The current is low enough to keep the power low.

In fact, if those electrons flowed out of at the rate of 1 milliamp (i.e., one thousandth of an amp, one thousandth of the current you get in a lightbulb), you would run out of electrons in only 1/1000 of a second. The total energy of the electrons is 0.01 J, less than 2 micro Calories (2 millionths of a Calorie). It is not important that you know these numbers. It is important for you to know that high voltage is not dangerous if there isn't much current and if it doesn't last for very long.

In contrast to the little finger spark, lightning has both high voltage and high current. For typical lightning values, 10 million volts at 100,000 amps, we get a power of a terawatt, 10^{12} watts, equal to 1000 gigawatts. (A large commercial power plant is 1 gigawatt.) But the power only lasts for about lasts for 30 millionths of a second; that means that the energy is $10^{12} \times 30 \times 10^{-6} = 30 \times 10^6$ joules. Divide that by 4200 (the joules in one Calorie) to get that it is about 7000 Calories. That's the energy in 7000 gm = 7 kg of high explosive. That means that lightning can kill, or knock down a tree, but it seems too low to really be a useful source of energy.

Frog Legs and Frankenstein

In 1786, Luigi Galvani, one of the pioneers of electricity, discovered that when he applied small sparks from static electricity to the legs of dead frogs, the legs twitched. Later, he hung frog legs on metal hooks outside his house during thunderstorms. (Electricity was not easy to get in those days; Galvani had not yet invented the battery. But Benjamin Franklin had already discovered that lightning was electricity.)

Galvani thought that he had made the frog leg come alive. He hadn't. He had just delivered a signal to the muscle that made it contract. But he believed that he had discovered a secret of life, and he called it "animal electricity." For some fascinating drawings of his experiments, look up "Galvani frog" on the Web.

In 1817, Mary Shelley, inspired by Galvani's experiment, created one of the first science fiction classics, *Frankenstein*. Just as Galvani thought electricity could bring a dead frog leg to life, Shelley's fictional character Dr. Frankenstein thought he could bring a dead person to life by using lightning.

The story of Frankenstein became a symbol of what could happen when scientists develop new technology without anticipating its applications. In honor of Frankenstein, today some people use the derisive term *frankenfood* for food that has been genetically altered.

House Electric Power

The electricity that comes to your home is usually kept by the power company at a average[6] of 110 volts. If you have no lights turned on, no refrigerator, no

---

[6] By the *average voltage*, I mean the RMS value. If you are interested, *RMS* stands for "root mean square" value. It is calculated by squaring the voltage, averaging it (since house current oscillates

heaters, no TV, no anything, the voltage is still 110 volts—although the current is zero. The power company works very hard to keep the voltage at 110 V even when you start using more appliances. The voltage doesn't change, only the current. The power you use is equal to $P =$ volts $\times$ current, with volts $= 110$, so in the United States, your power is

$$\text{watts} = 110 \times \text{amps}$$

Appliances are usually marked with their power requirements in watts. If you want to figure out how many amps they take, just divide the power by 110 V.

$$\text{U.S. household amps} = \text{watts}/110$$

A bright lightbulb that uses 110 W takes 1 amp. A heater that uses 550 W takes 5 amps. Amps add (they represent the number of electrons per second), so if you have both the lightbulb and the heater, then you will have a total of $1 + 5 = 6$ amps coming into your home. If you use more than 15, your fuse might blow. (By "blow," I mean that the wire inside the fuse melts, stopping all further current from flowing.)

In Europe, the typical house voltage is 220 V rather than 110 V. That means that for typical power, the voltage is higher and the current is less. Higher voltage makes the electricity more dangerous than in the United States, but lower current means that there is less energy lost in the wires that deliver electricity to the outlet. (Or, alternatively, it means that they can use cheaper wires without getting too much heating.)

If you want to keep a 15-amp fuse from blowing, then you should limit the power of your electric appliances to 15 amps $\times$ 110 V $= 1650$ W. One electric heater can use this much. Appliances such as toasters tend to use high current for short periods, but that is enough (when used together with a heater) to blow a fuse.

High-Tension Power Lines

Most long-distance transmission of electricity is done at extremely high voltage, several tens of thousands of volts, although some get as high as 500,000 volts. At these high voltages, you can sometimes hear the crackle of small sparks coming from the wires. Sometimes people refer to these lines as *high-tension lines*. That's not because people who live near them get tense, but because *tension* is an old synonym for voltage. It is still used in the UK.

There is an important reason that we use high voltage for such lines. Recall that power = voltage $\times$ current. So high-voltage lines have less current (for the same power delivered) than do low-voltage lines. But heating from resistance depends only on the current, not on the voltage. So if we use high-voltage lines, then we can reduce the amps, and that reduces the loss of power from resistive heating.

---

60 times per second), and then taking the square root of this average value. In statistics, the average of the squares is called the *variance*.

Since high voltage can make electricity dangerous, there are special devices that raise the voltage V and lower the current I, while keeping the power P unchanged (i.e., V times I remains unchanged). Such a transformer is called, aptly, a *transformer*. We'll talk about how they do their work after we have discussed magnetism. Some transformers are near homes, so they don't lower the voltage until they are as close as possible. Many of these transformers are filled with an insulator known as PCBs.[7] When it was discovered that PCBs can cause cancer, an ongoing campaign began to eliminate these liquids and replace them with something that was less carcinogenic.

Electric Forces at Different Distances

Near the beginning of this chapter, we considered the force between two electrons a centimeter apart. I said that the electric force was stronger than the gravity force by a factor of 4.17×10^{42}.

Now suppose that we put the electrons 2 cm apart. The force of gravity is then 4 times weaker (because gravity is an inverse square law). And so is the electric force!

If, instead of 1 cm separation, you put the electrons 1000 centimeters apart, then both forces are weaker by a factor of $1000 \times 1000 = 1$ million.

This similarity in the distance behavior has intrigued a lot of people. It means, for example, that an electron orbiting a proton bears a lot of similarity to the Earth orbiting the Sun. That's why you'll often hear people describing atoms as little solar systems. The analogy is not perfect, however, because when you get to the small dimension of an atom, quantum physics becomes important. We'll talk more about that in chapter 11.

Magnets

You're probably familiar with magnets, such as those used to stick messages on refrigerators. Magnets are truly strange, and I strongly recommend that you play with several. A magnet will attract a piece of iron, but it can either attract or repel another magnet, depending on the orientation of the two.

According to the ancient author Pliny (who lived from AD 23 to AD 79), the word *magnet* in Latin comes from Magnes, the name of a shepherd who noticed that his iron staff and nails from his boots were attracted to certain rocks.

The simplest magnets have two ends, one of which is called N or the *north pole* (because if you hang it from a string, it will orient itself to face the North Pole of the Earth), and the other called S or the *south pole*. Play, and you'll discover that two north poles repel each other, that two south poles repel each other, but that a north pole will attract a south pole. The repulsion seems par-

---

[7] *PCB* stands for "polychlorinated biphenyl." *Polychlorinated* means that the molecule contains multiple molecules of chlorine. *Biphenyl* means that the organic molecule has two phenylgroups attached, each consisting of a benzene ring with one hydrogen removed. Even most physicists don't know this kind of chemistry. (I had to look it up too.)

ticularly mysterious, because it is so unlike gravity. But it is very similar to electricity because the like charges repel and opposite charges attract.

Permanent magnets are materials that keep their magnetism. You can also create a *temporary magnet* by using electricity. Every time that electric current flows it creates magnetism. A magnet made with electric current is called an *electromagnet*. You can turn its magnetism on and off by changing the current.

Lodestones, Kissing Stones, and Compasses

The first known magnets were natural rocks containing iron ore, known in English as *lodestones*. A magical feature of these stones is that if you suspend them (by a string, or by floating them on a piece of wood), they tend to rotate until one end is pointing north. This became an enormously important discovery, since it could be used to tell direction. It was called a *compass* and was so valuable that it was originally a deeply held military secret. Even on a completely cloudy day, far out at sea, you could tell which direction was north. The word *lodestone* derives from the Old English word *lode*, which means "way or path"; a lodestone helps you find your way. The impact that the magnetic compass had on history is difficult to know. In 1620, Francis Bacon ranked it with gunpowder and the printing press as the three inventions that had revolutionized the world. (He must have meant the "recent" world, since he didn't include earlier inventions such as the wheel or controlled fire. Freeman Dyson once argued that an even more important invention was hay.)

For hundreds of years, nobody understood why one end of the lodestone points north. Some people assumed that the lodestone felt some attraction toward the North Star. The secret turned out to be that the Earth itself is a large magnet, and the north pole of the lodestone was being rotated by the magnetism of the Earth.[8] The "north-pointing pole" of the lodestone was referred to as simply the north pole of the magnet. The other end was called, naturally, the south pole of the magnet.

Another major discovery was that you could make new magnets from iron. You can do this yourself by rubbing a needle on a magnet; be careful to rub only one direction, and not back and forth. The needles made into magnets this way could then be used for compasses. Another way was to apply a strong magnetic field (perhaps from an electromagnet) to a piece of iron. After the electromagnet was removed (or turned off) the iron retained some of this "remnant" magnetism.

A second magical feature of lodestones is their force of attraction to each other. Because of this property, the Chinese called them *tzhu shih*, which means "loving stone." The French word is similar: *aimants*, literally, "stones that like each other." Of course, the attraction depends on the orientation. Magnets placed N to N, or S to S, dislike each other.

---

[8] There are records of magnets being used in China in the first century. The first records in Europe date from a manuscript written in 1187 by Alexander Neckam. In 1600, William Gilbert (the physician to Queen Elizabeth I) figured out that the Earth was a giant magnet. He wrote, in Latin, "Magnus magnes ipse est globus terrestris." That can be poetically translated as, "A magnificent magnet is the terrestrial globe."

Magnetism from Moving Charge

We now know that the force of magnets—what we call magnetism—is really another aspect of electricity. It is a force between electric charges that occurs only if the electric charges are moving. For that reason, you can think of magnetism as a force that occurs between electric *currents* rather than between stationary charges.

The magnetic force law[9] is similar to the electric force law, and to the gravity force law. It states that if you have two short lengths of wire, each carrying current, then the force between them is inverse-square. That means that if you double the distance, the force will weaken by a factor of four. It is more complicated, however, because the force is not a simple one of attraction or repulsion, but can be in a different direction that depends on the orientation of both segments.

To calculate the force between long wires, you have to add together all the forces between each pair of wire segments. For long wires, there are a huge number of such pairs, and that makes the problems complicated. For simple cases (e.g., two long straight wires), the total force can be worked out mathematically; the result is that two parallel wires carrying current in the same direction will attract each other. For more complicated cases, such as wires wrapped in large loops, the calculation is usually done on a computer.

Magnetism turns out to be the most useful tool for generating electric current. If you move a wire through a magnetic field, then the magnetism puts a force on the electrons in the wire, and they will move along the wire. This is how an electric generator works.

Permanent Magnets

Modern permanent magnets are used for refrigerator magnets, for magnetic compasses, and for door latches. They certainly don't *seem* to have any electric current. Moreover, they don't *seem* to have anything to do with electricity.

But now we know that permanent magnets do get their magnetism from electric currents. But the electric currents are extremely well hidden. The electric currents for permanent magnets are inside the electrons!

It was discovered in the twentieth century that all electrons spin. That means that the charge within the electron is also spinning, and that is an electric current. This makes every electron into a tiny magnet.

It is hard to detect this magnetism if there are many electrons and they are all spinning in random directions, because then the magnetism tends to cancel. This is the case for most materials. But in a few materials, known as *ferromagnets* (iron is the most prominent example), the electrons from different atoms tend to line up and have the same spin direction. Then the magnetism adds. These are the materials from which we make permanent magnets. They are permanent because they retain their magnetism without any need for additional power.

---

[9] It is usually called the Biot-Savart law.

You can imagine why it was hard to discover this. Who could have suspected that electrons—all electrons—spin? In fact, we now believe that it is impossible to stop this spin. Electrons always spin. We can change the direction of their spin, but we cannot stop it.

Magnetic Monopoles?

As I mentioned earlier, in some ways magnets behave like electric charges. North poles repel, just as like electric charges repel, and opposites attract. This has led many people to speculate that there must be magnetic charges, similar to electric charges. These hypothetical objects are called *magnetic monopoles*. Permanent magnets behave as if they have a concentration of such charges at their ends.

Yet we know this is not really true. All present permanent magnets actually work because of currents flowing within their electrons.

If you take a magnetic needle, one end will be the north pole, and the other end will be the south pole. You might think that you can break off the north pole by cutting the needle, but if you do that, new poles form at the broken ends—so each piece continues to have one north and one south pole each. Magnets appear to always have both north and south poles, no matter how they are made. That's because a broken magnet still consists of rotating electric currents, and those always produce north and south poles.

Some physicists have speculated that even though all known magnetism comes from currents, that doesn't mean that magnetic monopoles are impossible. Many projects have been made to search for them, or to try to make them. Some theories (e.g., superstring theories) predict that they should exist or, at least, that it should be possible to make them. Searches have been made in materials that have been exposed to extremely energetic collisions, since those may have created monopoles. Materials studied have included lunar rock (exposed to energetic cosmic rays for billions of years) and metals placed at the end of large particle accelerators ("atom smashers").

If magnetic monopoles could be made, they would be valuable. They could be accelerated to very high energy by ordinary magnets, and this could be a convenient way to create radiation (which would have applications in medicine and elsewhere).

The Short Range of Magnetism

Because magnets have both north and south poles (until monopoles are discovered), once you get a reasonable distance away from one, the forces from the two poles tend to cancel. This cancellation makes the force between two magnets fall, not with an inverse square law but with an inverse fourth law. If you go twice as far away, the force is reduced by a factor of $2^4 = 16$. So when the separation is doubled, the force is $2 \times 2 \times 2 \times 2 = 16$ times less. When tripled, the force is $3 \times 3 \times 3 \times 3 = 81$ times weaker.

The result is that magnets are very useful for short distances but don't work very well for larger distances. You may have noticed this if you tried to pick up an object with a magnet. Unless the magnet is close to the object, it has very little net force on it. Contrast that with what you see in cartoons, where magnets are depicted as being able to lift things at great distances.

Electric and Magnetic Fields

It was once thought that one electric charge put a force directly on other electric charges. Now we know that there is something intermediate that happens. The electric charge creates something that we call an *electric field* that fills up space. It is this field that puts the force on the second charge.

Gravity works the same way. Mass creates a gravitational field. When a second object is in that gravity field, it feels a force from the field. In other words, there is no direct force between the two masses. Rather, one mass creates a field, and the other mass feels that.

The situation is similar to two people pulling on ends of a rope. One person pulls on the rope, and the rope pulls on the other person. The two people don't directly touch each other.

The way we know electric fields really exist is from the behavior when you suddenly remove one of the charges. The force on the other charge is still there, if only for a short time.

We also know that the field can be made to vibrate, a phenomenon that gives rise to something known as an *electromagnetic wave*. (This is analogous to shaking the rope.) It turns out that light, radio signals, and x-rays are all examples of electromagnetic waves.

The key idea here is that charge produces an electric field, and this electric field can produce a force on other charges. Likewise, moving charges (currents) produce a magnetic field, and this field can exert a force on other moving charges.

Magnetic fields can be visualized by sprinkling iron filings near a powerful permanent magnet. In figure 6.2, a piece of glass was placed above the magnet, and the filings were sprinkled on the glass.

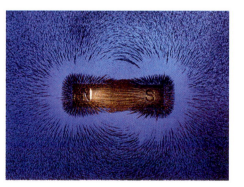

Figure 6.2 Magnetic fields visualized by their effect on iron filings. (Courtesy of NASA.)

If there is a magnetic field in a vacuum, is it still a vacuum? That is partly a matter of definition. There are no particles there, but the magnetic field does contain energy. Is space really empty if it contains energy? Normally, we define a vacuum to be a region of space with no (or few) particles, and we don't worry about whether a field is present.

The magnetic field is easy to visualize because of the way it lines up iron filings. It is possible but much harder to "see" strong electric fields since they tend to produce sparks, and when that happens the charges move in such a way to reduce the field.

Electromagnets

If you put a wire into the right geometry, you can arrange it to exert a very strong electrical force on other currents, or on a permanent magnet. A common geometry to do this is called a *solenoid*. It is just wires wrapped around a cylinder. Turn on the electricity, and you have a strong magnet. Turn it off, and the magnet is turned off. Reverse the current, and the magnetism is reversed (i.e., the north pole becomes a south pole).

Electromagnets have lots of uses. In automobiles, they are used to lock and unlock doors. (If you click the door switch, a solenoid electromagnet pulls a permanent magnet.)

Small electromagnets are used in speakers and earphones to create sound. Typically, such devices have a small permanent magnet, and an electromagnet. Electric current goes through the electromagnet and that causes an attraction between it and the permanent magnet. Then the current is reversed, the magnetism of the electromagnet is reversed, and now the two magnets repel. Usually, the electromagnet is made very lightweight and it can move back and forth in response to these reversing forces. The electromagnet vibrates in a way that follows the oscillations of the current. In an earphone or speaker, a piece of paper or metal foil attached to the electromagnet oscillates along with it, and that pushes against the air, making the air vibrate. Air vibrations reach the human ear, and we hear them as music. (We'll discuss this further in chapter 7.)

Superconducting Electromagnets

Large, strong electromagnets require high currents, and that means that a lot of power is wasted in resistive heating. For that reason, many such devices are now made using superconducting wires. Although some energy must be used for the refrigerators that keep the wires cool, that turns out to be much less than you would lose from the resistance of ordinary wires. Figure 6.3 shows a large superconducting magnet used at a particle accelerator (commonly known as an *atom smasher*) at Fermilab in Illinois.

Superconducting magnets are also widely used in medicine to provide the strong magnetic fields needed for magnetic resonance imaging (MRI). (We'll discuss MRI further in chapter 9.) The magnetic field of such magnets provide a force on the nuclei of hydrogen atoms that makes them wobble around the direction of the field. This wobble can be detected, to create an image of the distribution of hydrogen.

Figure 6.3 Superconducting magnet at Fermilab. (Photo courtesy of U.S. Department of Energy.)

The Electromagnet in the Earth

We believe that the magnetism of the Earth comes from large currents flowing in the liquid iron core. (We know the core is liquid from seismic data, discussed in chapter 7.) The flow is complicated, so the Earth's field is complicated. A recent computation of the Earth's magnetic field lines is shown in figure 6.4.

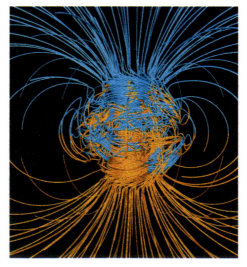

Figure 6.4 The magnetic field lines of the Earth, as computed by physicists Gary Glatzmaier and Paul Roberts. Deep inside the Earth, the field is very tangled and complex, but on the surface, it is relatively simple. (Reproduced by permission from Gary Glatzmaier.)

Magnetic Materials—The Special Role of Iron

I said earlier that permanent magnets are made from materials in which a large number of the electrons are spinning in the same direction. Ordinary iron is not normally a permanent magnet because its electrons, even though they are spinning, are all spinning in different directions.

But if you apply an external magnetic field, e.g., by an electromagnet, then that puts a force on these spinning electrons. For iron atoms, it tends to make the electrons all spin in the same direction, and that makes the iron into a magnet, as long as there is current flowing in the external electromagnet. We say that magnetism is *induced* in the iron.

That's why a permanent magnet can pick up a paperclip. When you bring the permanent magnet near the paperclip, magnetism is induced in it, and then the permanent magnet and the paperclip attract each other.

Here is how an electromagnet can lift a piece of iron, as in a junked car. The electromagnet is turned on, and it makes a strong magnetic field. This magnetism aligns the electron spins in the iron of the car, turning it into a magnet. It is an induced magnet. For iron, the two magnets (the electromagnet and the induced iron magnet) attract each other.

Induced magnetism can also be used to make magnetic fields that are much stronger. If you place iron inside the cylinder of an electromagnet, then the weak magnetism of the current is strongly enhanced by the induced magnetism of the electron spins. And it doesn't stop there. The induced magnetism of some of the atoms induces even more electrons to spin in the same way. The strength of the magnetism grows dramatically, until the magnetism is hundreds of times stronger than it would have been without the iron. This kind of magnetic amplification is so useful that most electromagnets use iron cores.

Remnant Magnetism

Imagine that you have an electromagnet that is applying its field to a piece of iron. When the electric current is turned off, and there is no externally applied magnetic field, then most of the induced magnetism goes away. But usually some of the electrons remain lined up with each other, so there is a small remnant (that means *remaining*) magnetism.

Remnant magnetism can be very useful—e.g., to make permanent magnets—or it can be a real nuisance. If you bring an iron screwdriver close to a strong magnet, it becomes magnetized; when you take it away, there may be some remnant magnetism left. If that is true, the screwdriver may attract screws or little bits of iron, and that can be useful or annoying. Old watches (in pre-electronic days) would become magnetized if brought close to a magnet, and then the pieces within the watch would attract each other, and that was usually enough for the watch to stop working. Watch repair experts would fix the watch by putting it back in a changing magnetic field that would slowly reduce the magnetization to zero.

Magnetic Recording

Induced magnetism is also the basis behind magnetic recording, and that includes videotape, computer hard drives, and MP3 players. In these devices, a very small electromagnet induces magnetism in a small region of a magnetic material. In the adjacent region, it can induce similar magnetism or a reversed magnetism. The signal is stored in the magnetic material by these small regions. For example, if adjacent regions have the upward direction induced in a series of north and south magnetic poles N, N, S, S, N, then this could be a way of recording the digital signals 1, 1, 0, 0, 1. This is the basic principle for all magnetic recording.

A computer hard drive has magnetic material distributed on the surface of a rotating disk. As the disk moves under the electromagnet, different places have different induced magnetism. These days, these regions are typically a micron or smaller in size. That's the kind of drive used in many iPods.

The magnetic recording can be "read" by another wire. When a moving magnet passes a wire, it makes a small amount of electric current flow, and that current can be detected. In modern hard drives the wire is a special material in which the resistance of the wire depends on the magnetic field. By measuring that resistance, the wire gives information about the magnetic field.

Heat Destroys Magnetism: Curie Temperature

If you heat a permanent magnet, the atoms and electrons bounce around faster and faster. This can cause the atoms to change their orientation, and the electrons within them to change the direction of their spin. Pierre Curie, the husband of the more famous Madame Marie Curie, discovered that at a certain temperature, all permanent magnetism disappears (since the electron spins get mixed up). Every material has its own *Curie temperature* at which this happens.

> **Remember:** If you heat up a permanent magnet to its Curie temperature, then its magnetism goes away.

Rare-Earth Magnets

In the last few decades, a particularly strong type of permanent magnet was invented. The first was made out of a compound called samarium cobalt. Samarium is an element known as a *rare earth*. Since its discovery, other similar compounds have been found, and these magnets are often called *rare-earth magnets*.

These magnets are so strong that they can be dangerous. If you break one (maybe by dropping it) and it breaks in such a way that the two pieces repel each other, then pieces can go flying apart at such high velocities that they can hurt someone. When used in earphones, they are packaged in such a way as to prevent the magnet from being struck with a shattering blow.

The earphones once used by your grandparents were big and bulky. Now, thanks to rare-earth magnets, high-quality earphones can be small and light; similarly for loudspeakers and motors.

Finding Submarines

A submarine is made of steel, and when it sits in the Earth's field, it becomes a big magnet. During World War II, scientists realized that you might be able to find submarines deep under water by detecting this magnetism. Because magnetic fields get weak at large distances (by a factor of $1/r^3$), this method doesn't work for very deep submarines, but it is still used when submarines are within a few hundred meters of the surface.

Because this method works so well, submarines are specially treated every time they come to port to remove any remnant magnetism they may have picked up.

Electric Motors

Electric motors are really based on magnetism. In an electric motor, the wires are wound in such a way as to create a strong magnetic field. In the simplest version of a motor, this magnetism is used to pull or push on a permanent magnet. If the current is periodically reversed, then the alternating pushes and pulls can be made to move the magnet in a circle. That is how an "electric" motor works—by magnetic forces.

It is not necessary to use a permanent magnet. Many electric motors use two electromagnets, one which is stationary and one which rotates. The electric current is switched in such a way that the force of one magnet on the other pushes the rotating magnet in circles.

As long as thick wires are used, the electric resistance can be small, and electric motors can be very efficient—i.e., they can turn the electric power into mechanical motion with very little loss to heat. Hybrid automobiles use the electricity stored in batteries to drive the wheels with electric motors.

Electric Generators

The most effective way to make electricity for commercial use is by moving a wire through a magnetic field. When this is done, it is called an *electric generator*. Essentially, all the electricity that you use is made this way. You also use some electricity from batteries (in flashlights and in your auto), but that is only a very little bit compared to the rest.

A wire made of metal has electrons in it that can move. When you move this wire through a magnetic field, then the electrons move with the wire. Moving electrons, just as with any current, feel a force from the magnetism. If you move the wire perpendicular to the length of the wire, then the force of the magnetism will be along the wire, so the electrons will be pushed along the wire—that is, current will flow along the wire.

At nuclear power plants, the nuclear chain reaction is used to produce heat, and that turns water into steam. The steam drives propellers (technically called a *turbine*), and those are used to drive wires through a magnetic field, producing electricity.

In a coal-burning power plant, the coal is burned to produce heat—and from there on, the power plant works the same way, ultimately producing electricity by pushing wires through a magnetic field.

In a gasoline burning power plant, or one that uses natural gas, the fuel is burned to produce heat, and from there on the process is the same.

In a hydroelectric plant, water coming from the reservoir is used to turn wheels, and these push wires through magnetic fields, and so on.

Once your automobile has started, it no longer needs a battery. From then on, all the electricity it needs (for spark plugs, and to light the headlights) is made from the gasoline engine, which turns an axle called a crankshaft, which turns a wheel that moves wires through a magnetic field.

Dynamos

To work well, a generator needs a strong magnetic field. For small generators, the field can be made out of permanent magnets. But for big generators, the magnets must be electromagnets. Guess where they get the electricity to run the electromagnets.

That's right. They get the electricity from the generator! When this is done, the generator is called a *dynamo*.

This sounds paradoxical, but it really works. Most large generators are dynamos. That sounds like you are getting something for nothing, but that isn't true. It takes energy to push the wire through the magnetic field,[10] and all the electric energy that emerges (in the current of the wire, and in the magnetic field) comes from the energy that you put in.

The North Pole Is a South Pole

As we discussed earlier, the Earth is a great magnet. That's why compasses point toward the poles. But the Earth's magnetism is not perfectly aligned with the axis of the Earth's spin, so the direction that the compass points is not true north but a different location. The magnetic pole is located at a latitude of about 75 degrees, in northern Canada near Baffin Island. Maps often have a little symbol on them that shows the difference between magnetic north and true north.

The situation is much worse on some of the other planets. On Uranus and Neptune, the magnetic poles are 60 degrees away from the poles of the rotation axis.

You should be aware of a semantic problem in our terminology. The north pole of a compass needle points toward the Earth's magnetic pole. But the

[10] The current that flows in the wire interacts with the magnetism and produces a force that resists the motion. That's why you have to do work to move the wire.

north pole of a magnet is attracted to a south pole of another magnet. Thus, magnetically speaking, the magnetic pole that is up in Canada is really a south magnetic pole!

Einstein's Mystery

When William Gilbert deduced that the Earth was a magnet, he naturally assumed that it was a permanent magnet, perhaps from large deposits of lodestones. But we now know that rocks below the Earth are hot, from the Earth's radioactivity. At a depth of about 30 kilometers, the temperature is higher than the Curie temperature, so all magnetism must disappear. These paradoxes led Albert Einstein to list the origin of the Earth's magnetism to be one of the greatest unsolved problems of physics.

We now believe that we know the answer: the Earth is a dynamo. We don't know in detail how this works, but we understand the general picture. The early Earth (4.5 billion years ago) was very hot, and most of the iron melted and sank to the center. It is still there; if you go about halfway to the center of the Earth, the material changes from rock to molten iron. Moreover, this iron is in constant flow from heat that is being released from a small solid iron core deep within. This flowing iron behaves like a dynamo. When liquid iron moves in a magnetic field, electric currents flow (just as in a moving wire). The arrangement of flow in the core is such that these electric currents circle around to create the magnetic field, just as they do in a commercial dynamo generator.

This picture is verified by computer and mathematical models, but it is hard to be sure, since the center of the Earth is far harder to reach than the surface of the Moon.

The Earth Flips . . . Its Magnet

As ocean animals die and drift to the bottom of the sea, they eventually form new layers of rock. These rocks become slightly magnetized by the Earth's magnetic field, and then they hold that magnetism for millions of years. If we study the layers of rock, and measure their ages (from potassium-argon dating, as discussed in chapter 4), we can read the history of the Earth's magnetism.

From these records, we have learned that the strength of the magnetic field changes slowly with time. But much more startling is the discovery that from time to time, the magnetism of the Earth flips! That means that if you took a present-day magnetic compass back into the past, the north-pointing needle would point south instead of north.

The last flip was almost a million years ago, and such flips (at least in recent times) seem to occur, on average, once or twice every million years. The flip takes several thousand years to happen, but in the geologic record that seems very fast.

Now you can understand the following paradoxical thought: "The Earth's North Pole is a south pole. However, about a million years ago, it was a north pole." Try it on your friends.

We don't know why the magnetism flips, but several theories have been proposed. It turns out that the actual flow of liquid iron that drives the dynamo doesn't have to change. Instead, only the electric current has to reverse. When that happens, the magnetism will flip too. There are some theories that attribute the change to the chaotic behavior seen in some dynamo models.

My favorite theory (also unproven) is that the flip magnetism consists of two steps: a destruction of the dynamo flow (perhaps triggered by avalanches of rock at the liquid/rock boundary), followed by a rebuilding of the dynamo in the opposite direction.[11] Of course, it won't always rebuild in the opposite direction; sometimes it will happen in the same direction. When that happens, it's not called a flip, but an *excursion*. Many excursions have been observed in old rock records; there appear to be more of them than flips.

Flipping Magnetism and Geology

The fact that the Earth's magnetism flips every million years or so has been enormously useful in geology and related fields such as climate study. It is valuable because we often cannot measure the age of a rock from its radioactivity. For example, rocks often don't contain enough potassium for the potassium–argon method to be used. However, most rocks formed under the sea preserve a record of the Earth's magnetism. We can see a pattern in the layers, almost like a fingerprint, with some flips coming close to each other in time, and others with wide spacing. Once this pattern is known, then we can correlate the patterns at different locations around the Earth. We don't know how old a layer is, but at least we know it is the same age as another rock somewhere else on Earth.

But we can do even better. If we search long enough, we'll probably find a rock that was formed near a volcano. Volcanic ash contains lots of potassium. If we can use potassium-argon dating to obtain the age of this one rock, then we immediately know the age of rocks all around the world that are at the same position in the geomagnetic pattern.

This is also important because these other rocks often contain unique records of their own. Some of them record the patterns of previous climate. If you put all this together, you can figure out when the last ice age occurred on Earth, how long it lasted, and how quickly it ended. In this way, much of our knowledge of the past has used the Earth's magnetic field flips.

Earth's Magnetism and Cosmic Radiation

Just as electrons flowing in a wire feel a force from a magnetic field, cosmic rays coming from space feel a force and are deflected by the Earth's magnetic field. This prevents a large number of these particles from hitting the top of the Earth's atmosphere. Some people have speculated that when the Earth's field

---

[11] This theory is my favorite, in part, because it is my theory. It was published in the journal *Geophysics Research Letters* and is available online at www.muller.lbl.gov/papers/Avalanches_at _the_CMB.pdf.

collapses (as during a magnetic reversal), life on Earth will be exposed to this deadly radiation. This idea has been widely spread by science fiction movies such as *The Core* (2003).

If the field collapses, then it is true that more of the cosmic rays will hit the Earth's upper atmosphere. But the atmosphere is the true shield, and even without the field, the radiation that reaches the Earth's surface will increase by only a few percent.[12] Thus, the field collapse will not significantly affect life.

In fact, the north magnetic pole of the Earth, in northern Canada, currently has no magnetic field production whatsoever, because all the field lines point inward (and therefore they don't deflect cosmic rays). Yet the cosmic radiation at that location is only a tiny bit higher than at the equator. That's because the atmosphere stops most of the radiation.

Beware—even many good scientists have fallen into the trap of thinking that it is the Earth's magnetic field that protects us from cosmic rays. There was a recent *NOVA* program that assumed this and talked about the possible impending disaster if the Earth's magnetic field is in the process of reversing. But it ain't so.

Transformers

An electric generator works by moving a wire past a magnetic field. It would work equally well if the magnet were moved past the wire.[13] In fact, the magnets don't actually have to move; this works equally well if their magnetic field is just changing, and that can be done by changing the current in an electromagnet.

If all the ideas in the previous paragraph are put together, we get one of the great inventions of all time: the *electric transformer*. In a transformer, there is a coil of wire called the *primary*. Changing electric currents in this primary create a changing magnetic field. The changing magnetic field passes through a second coil of wire called the *secondary*, and it causes current to flow in the secondary.

One remarkable fact about a transformer is that it can pass energy from the primary coil to the secondary coil very efficiently, with almost none being lost. The primary and the secondary don't touch each other. The energy is all passed through in the form of magnetism!

What makes the transformer so valuable is the fact that the number of loops of wire in the primary and secondary can be different, and the result is that the voltage and current in the two coils will be different. A transformer transforms high-voltage electricity to low-voltage electricity, or the other way around. It is transformers that take the high voltage from power lines and reduce the voltage to make it safe for our homes. And they all work using magnetism.

---

[12] At the geomagnetic pole, the Earth's field presently gives no shielding whatsoever. This results in a much stronger cosmic radiation at the top of the atmosphere, and yet the radiation at the bottom of the atmosphere is only slightly greater than elsewhere on Earth.

[13] This is true, but it isn't really obvious. The discovery that it was true was led to Einstein's postulate that the laws of physics are identical regardless of the way you are moving, and that lead him to the theory of relativity.

If there is any iron near the transformer, then that iron may vibrate as the magnetic field changes. You can often hear a "hum" from a transformer that is doing this. Of course, that hum means that some energy is being lost from electricity to sound, so high-quality transformers are built so that this doesn't happen.

The Tesla Coil

Nikola Tesla, a scientist who worked with Thomas Edison, invented a very high voltage transformer we now call a *Tesla coil*. One of his tricks was to make the current change very rapidly, and that generated very high voltages in the secondary. A Tesla coil can be used for a dramatic demonstration in the classroom, with continuous sparks over a foot long. At the same time, the sparks are not particularly dangerous. When the transformer raises the voltage of electricity, it must also lower the current—since the power is current times voltage, and the power doesn't change. So a Tesla coil can create extremely high voltage sparks, but they release relatively low power.[14]

Magnetic Levitation

Ordinary iron, when exposed to a magnet, becomes a magnet itself and is attracted to the original magnet. But some materials behave differently. When exposed to magnetism, they become magnets themselves, but in the opposite sense. The part that is exposed to the north pole of the magnet becomes a north pole itself, and instead of being attracted, it is repelled.

Such materials are not common, and that is why our experience is that magnets "attract" things. Liquid oxygen is one of the materials that is repelled by ordinary magnets. But superconductors are also repelled. When exposed to magnets, currents start flowing inside superconductors in just such a way as to create a repulsive force. If you place a small superconductor on top of a magnet, the force can make the superconductor *levitate* above the magnet, with the repulsive force countering gravity.

If you have a changing magnetic field, created by an electromagnet with alternating current, then levitation can be done with ordinary metals. The changing magnetism will cause currents to flow in the metal, and these currents will create magnetism that repels the original magnet. This approach can be used to levitate large objects.[15]

Levitation can also be done with moving magnets. If a strong magnet (samarium cobalt, or a strong electromagnet) is moved over a conductor, then electrical currents will be induced in the conductor. Those create a magnetic field that repels the original magnet. This approach is used commercially in magneti-

---

[14] A description of the demonstrations used in the Berkeley Physics Department can be seen at these Web pages: www.mip.berkeley.edu/physics/D+75+04.html and www.mip.berkeley.edu/physics/D+75+08.html.

[15] A description of the demonstration we do in class can be found at www.mip.berkeley.edu/physics/D+15+24.html.

cally levitated trains in Japan and elsewhere (figure 6.5). At slow velocities, there is no levitation (since the induced magnetism requires a rapidly changing, or moving electrons). As the train moves faster, the magnet moving over the rails induces stronger and stronger currents, until finally the magnetic repulsion lifts the wheels off the tracks. The advantage of magnetic levitation is that it avoids all the friction of contact. However, the currents flowing in the rails do lose some energy to electrical resistance, and that can be a serious limitation. Superconducting rails would avoid this problem, but they have to be kept cold. With all these problems, magnetic levitation has not proven to be as successful as some futurists have predicted. That could change if we ever develop room-temperature superconductors.

Figure 6.5 A futuristic shot of the Sydney Monorail train flying over Darling Harbor at night. (Courtesy of Shutterstock.)

Rail Guns

In chapter 3, we discussed the limitations of launching objects into space using chemical fuels. The problem was the exhaust velocity of such fuels was only 1 or 2 km/s, so it was hard to use them to push objects that had to go 11 km/s. But using magnetism, we can overcome that limit. The device that does this is called a *rail gun*.

The simplest version of a rail gun consists of two long parallel metal rails, just like those used for railroads. A high voltage is placed across the ends of the rails, and a piece of metal (called a *sabot*) is placed across or between the two rails. High current flows from the end of one rail, down the rail, across the metal, to the second rail, and back. The high current in the rails creates a strong magnetic field, and this puts a force on the current flowing through the metal sabot. As a result, the sabot is pushed down the rails. Theoretically, rail guns can launch a sabot at extremely high velocities.

Rail guns are under development by the U.S. Navy as a way of shooting down missiles attacking a ship, and they may one day be used to launch materials from the Moon.

AC versus DC

Most of our homes use *alternating current* electricity, abbreviated AC. In AC, the current is constantly changing, cycling its flow from positive to negative and then back again, 60 times every second. That's what we mean when we say that house current is 60 cycles—that is short for 60 cycles per second. There are 60 minutes in an hour, 60 seconds in a minute, and 60 cycles in a second.

A new terminology is to use the name *Hertz* to mean "cycles per second." Hertz is abbreviated Hz. In the United States, we use 60-Hz electricity. In Europe, they use 50 Hz.

Batteries give DC, or *direct current*. So why do we use AC in our homes? The answer is because AC works naturally with transformers. High voltage (and low current) is used in high-tension power lines to bring electricity to our homes. But before it enters the home, a transformer changes it to a relatively low voltage of 110 V and relatively high current of up to about 15 amps.

It was not always obvious that our electrical system would be based on AC. In the late 1800s, Thomas Edison believed the future would be DC. His rival, Nikola Tesla, was a believer in AC (figure 6.6). I'll give the gruesome details (including the execution of an elephant) in the next sections.

Figure 6.6 Thomas Edison (left) and Nikola Tesla (right).

In the end, Tesla won. We use AC, not DC, and our power plants are located far away, not on every street corner. Our wall plugs deliver 110 volts at 60 Hz. Many of our homes have a separate set of wires for 220 V, used on devices that take much more power, such as air conditioners.

The Edison–Tesla Conflict

Here's the story of how we adopted AC power instead of DC.

In the late 1800s, Thomas A. Edison had invented the lightbulb. This had such a great impact on the world that even today, cartoonists use an image of a lightbulb suddenly appearing above someone's head as an indication that the person had a great idea.

The man who most disliked Edison's invention was a John D. Rockefeller, who had made a fortune selling oil. At that time, oil was used almost exclusively for heating and lighting. Electricity (which could be made by burning coal—which boiled water, which ran a turbine, which ran a generator) could conceivably make his oil virtually worthless. Fortunately for him, right about that time improvements in oil-driven engine technology (in particular, the internal combustion engine) made possible a new invention: the auto-carriage, also known as the automobile. So Rockefeller's fortune was preserved.

Edison wanted to "electrify" New York City. His vision was to put metal wires on poles above the city streets to carry current to every house. Because some energy is lost in those wires (from resistance), the energy could not be transported very far. But he saw that as creating no real problem: he would place an electric power generator in every neighborhood, so the wires would never be more than a few blocks long.

Edison had hired a very talented engineer named Nikola Tesla. But Tesla quit in a huff. Tesla claimed that Edison had patented all of Tesla's ideas in the name Edison, and had not given Tesla the monetary rewards that he had promised.

Tesla had become enamored with the idea of *alternating current*, or AC for short. In AC, the voltage and the current oscillated, positive and then negative and then positive again, 60 times every second. If one used AC instead of Edison's DC (short for "direct current"), then you could make use of a wonderful invention called the *transformer*. (The transformer was invented in 1860 by Antonio Pacinotti. Recall that transformers used to generate extremely high voltages are often called *Tesla coils*.) A transformer used the fact that a wire with current in it creates a magnetic field. If the current varies, then the magnetic field varies. A changing magnetic field will create a current in a second wire. The amazing part of all this is that the voltage in the second wire could be very different from the voltage in the first wire. What the transformer transforms is the voltage.

Start with low-voltage AC, put it through a transformer, and what comes out is high-voltage AC. The advantage of high-voltage AC is that it carries power with very little electric current. That means that there is very little power loss in the wires, so the power can be sent for long distances using long wires. There would be no need to have electric generating plants in every neighborhood. When the electricity got close to a home, it could be transformed again, to convert the electricity to low voltage, which is less dangerous to use. A small transformer could be placed on the top of the pole that supported the wires. (Most neighborhoods today have just these transformers on the pole tops. When they burn out or otherwise fail, the neighborhood is left without electricity, and the transformer must be replaced or repaired. The local electric company usually does this within a few hours.)

AC turned out to have such an advantage (no neighborhood power plants) that it completely won out over Edison's DC. Tesla got the support of George Westinghouse, and their system turned into the one we use today. The voltage in our homes is only 110 volts AC. The voltage changes from positive to negative and then back to positive 60 times per second—i.e., 60 Hertz, abbreviated 60 Hz. In Europe, they use the slower frequency of 50 Hz, which is why their lights and their televisions flicker.[16]

But Edison did not give up without a fight. He tried to convince the public that high voltage was too dangerous to use in cities. He did this with a series of demonstrations of the danger, in which he invited the public to watch as he used the Westinghouse/Tesla high-voltage system to electrocute puppies and other small animals. Eventually, he put on a demonstration using high voltage to kill a horse. Edison had also invented a motion picture camera, and so he was able to make a movie of the electrocution of an elephant. I find the movie horrifying. The name of the elephant executed was Topsy, and she was a "bad" elephant who had been condemned to die for having killed three men (including one who fed her a lit cigarette). Apparently, the Society for Prevention of Cruelty to Animals approved of the execution, since they thought it would be inhumane to hang Topsy. Look up Topsy on Wikipedia for details. In an unrelated quote, Edison said, "Nonviolence leads to the highest ethics, which is the goal of all evolution. Until we stop harming all other living beings, we are still savages."[17]

The ultimate horror, of course, was to show that high-voltage electricity could kill humans. To do this, Edison convinced the State of New York to switch from hanging its condemned inmates to electrocuting them. He also argued that this method of execution was more humane—a conclusion that most modern observers think is exactly backward. But New York adopted the method, and then so did several other states. Despite the publicity created by all these things, the advantages of AC won the day, and that is what we use now.

---

[16] Our eyes don't notice flickering if it is faster than about 55 Hz. I think the Europeans made a dumb mistake, all for the purpose of trying to be a little more metric than the United States. For a while, they also tried 50 seconds to the minute, and 50 minutes to the hour, but they gave up—people couldn't get used to it. But the 50 cycles per second remained. Our peripheral vision is significantly faster than our central (foveal) vision, so some people see flicker, even for 60 Hz, out of the corner of their eye. That can really annoy you if you are in a house with old flickering fluorescent lights.

[17] I found a copy of the movie, but I don't recommend that you view it. If you can't resist, it is at available http://muller.lbl.gov/movies/Topsy.html. The Topsy story was part of the inspiration that led to the Walt Disney movie *Dumbo*.

Chapter Review

Electricity is the flow of electrons, or other similar particles that carry *electric charge*. By convention, the electric charge on the electron is -1.6×10^{-19} Coulombs (a number that you don't have to know). The proton has an equal and opposite charge. Quarks, hidden inside the nucleus, have charge 1/3 or 2/3 of this value. Atoms usually have zero net charge, since the electrons and protons balance. (If they don't, the object is called an *ion*.) The flowing of charges (usually electrons) is called *electric current* and is measured in amperes. One ampere is a Coulomb of charge every second. Current usually flows in loops; otherwise, charge builds up and the resulting force slows the flow.

Current can flow in gases, in vacuum, and in metal. When electrons do this, they usually lose some energy, and that is called *electric resistance*. The power lost is determined by the current flow. Insulators are materials that are poor conductors (high resistance). Superconductors, which require very low temperatures, have resistance = 0. High-temperature superconductors require temperatures of 150K, equal to −189 F.

Voltage measures the energy of the electrons. Power is voltage × current. High voltage is not particularly dangerous unless the current is large enough to give high power.

Batteries are rated by amp hours. That is actually the total charge they can deliver. Multiply the amp hours by voltage, and you get the total energy available in watt-hours.

In our homes, we use AC (rather than DC) because the voltage can be changed easily using transformers. High voltage (low current) is used to bring the electricity to our homes, but the voltage is lowered to make it safer before it comes in.

The equations for electric force look similar to those of gravity. There are two laws, one for charge and one for current. The force drops with the square of the distance, so things 10 × farther away have 100× less force. But there are differences. Two charges with the same sign repel and with opposite signs attract. For electrons, the electric force is much greater than gravity. When the force is between currents, we call it *magnetism*. Permanent magnets arise when the flow of electric charge within a large number of atoms is all in the same direction. Permanent magnets are used in magnetic compasses. No magnetic monopoles have ever been found, but the search continues. Electromagnets are made by making currents flow, typically in loops. They are used in automobile door locks, speakers, and earphones. When electromagnets are used to make a shaft spin, it is called an *electric motor*. Strong permanent magnets like samarium cobalt have made small earphones and motors possible. Iron, when placed in a magnetic field, strengthens the field, unless the iron is warmer than its Curie temperature. Some materials remain magnetized after being exposed to magnetic fields, and these are used for magnetic recording.

When a wire passes through a magnetic field, currents flow in the wire, and this is used for electric generators. If the current is used to make the magnetic field stronger, the generator is called a *dynamo*. Dynamos are used for the generation of commercial electric power. The core of the Earth has a natural dynamo, and that makes the Earth into a magnet. The magnetism of the Earth flips, on average, several times every million years. That discovery is very useful in geology for determining the age of rocks.

Transformers change voltage and current, while wasting very little power. A Tesla coil is a transformer that produces very high voltages.

Magnetic levitation uses repelling magnetic fields. These fields are sometimes generated by moving metal or by AC current. Rail guns can accelerate metal to high velocities more efficiently (with less wasted energy) than can rockets.

Discussion Topics

1. Read the following passage, taken from *Popular Science* in 1892, and see what sense you can make of it, given the modern understanding of electricity.

   We know little as yet concerning the mighty agency of electricity. Substantialists tell us it is a kind of matter. Others view it not as matter, but as a form of energy. Others, again, reject both these views. One professor considers it "a form, or rather a mode of manifestation, of ether." Another professor demurs to the view of his colleague, but thinks that "nothing stands in the way of our calling electricity ether associated with matter, or bound ether." Higher authorities cannot even yet agree whether we have one electricity or two opposite electricities. The only way to tackle the difficulty is to persevere in experiment and observation. If we never learn what electricity is, if, like life or like matter, it should remain an unknown quantity, we shall assuredly discover more about its attributes and its functions.

   The light which the study of electricity throws on a variety of chemical phenomenon cannot be overlooked. The old electrochemical theory of Berzelius is superseded by a new and wider theory. The facts of electrolysis are by no means either completely detected or coordinated. They point to the great probability that electricity is atomic, that an electrical atom is as definite a quantity as a chemical atom. The electrical attraction between two chemical atoms being a trillion times greater than gravitational attraction is probably the force with which chemistry is most deeply concerned.

   —*Popular Science*, February 1892 (quoted in the February 1992 edition)

2. Will we someday look back on the twentieth century as a time when we used an "ancient form of energy transmission called electricity"? Are there better ways to carry energy? What about pipelines carrying fuels? What about mechanical transport? What about lasers? What else can you think of?

3. Batteries were once the best way to produce reliable electricity. What would the world be like today if the electric generator had not been invented, and we still used batteries?

4. How many electric motors (including solenoids that open doors) are there in a typical modern automobile? How many can you or your classmates think of?

Internet Research Topics

1. Where is magnetic levitation being used around the world? What other places are considering adopting it? Can you find Web pages that say it is not as good an idea as others think? What kind of magnets do they use?
2. Some people regard Tesla as one of the greatest geniuses of all time. In his later years, he claimed to have invented ways of transmitting electric power without wires. See what you can find out about this. Why don't we use these ideas today?
3. How long are the longest power transmission lines? Are people considering making them even longer? Why? Why not just put the source of power near where it is most needed, for example, near cities?
4. What kind of batteries are least expensive? Are there new developments in batteries that will bring down the cost? Will batteries be a useful way to store intermittent electricity, such as electricity generated by solar or wind plants?

Essay Questions

1. Electricity and magnetism seem different, but electricity can create magnetism, and magnetism can create electricity. Describe how this is done. Give examples illustrating how this is used in practical applications.
2. Magnets make nice toys, but magnetism is also indispensable for modern life. There are numerous uses of magnetism in applications where many people in the general public do not even know that it is involved. Describe several of these, and say a few words about each one to show what the magnetism does in each case.
3. Discuss superconductivity. What is its value, and what are its limitations? How might it prove important in the future?
4. Why is the electricity from the wall AC rather than DC? Describe what important physical principles go into making the choice. Describe how this happened historically, and what you think might happen in the future.
5. Describe how energy contained in coal is transformed into electric energy, how it gets to your home, and then how it is converted into energy in the forms that you need. Be as detailed in each stage as you can.

Multiple-Choice Questions

1. Current is measured by:
   A. volts
   B. calories
   C. amps
   D. watts
   E. ohms

2. A volt is a measure of
A. energy per electron
B. number of electrons per second
C. force on the electron
D. density of electrons

3. Magnetism comes from
A. magnetic monopoles
B. moving quanta of light
C. quantization of charge
D. moving electric charge

4. In a permanent magnet, the magnetism comes from:
A. the spin of the electrons
B. the motion of the protons
C. the high voltage inside the atom
D. the electron avalanche

5. Magnetic monopoles
A. are found at the ends of magnets
B. are produced by cosmic rays
C. are found in the Earth's core
D. have never been found

6. Rail guns push the projectile using
A. high-energy explosives
B. electric fields
C. magnetic fields
D. protons

7. Telsa wanted to use a transformer to distribute electricity because
A. it allowed there to be a power plant in every neighborhood.
B. it made high voltage less dangerous
C. it decreased the current going through long wires to reduce the loss of electricity.
D. it was given strong support by Thomas Edison, who popularized it

8. You can demagnetize a natural magnet
A. by placing it on a TV or computer screen
B. by placing it in a very strong electric field
C. by heating it above the Curie point
D. none of the above

9. What discovery now allows for headphones to be small and light?
A. dynamos
B. rare-earth magnets, such as samarium cobalt
C. transformers
D. maraging steel

10. Which of the following use or consist of permanent magnets? (Choose all that apply.)
A. lodestones
B. transformers
C. compasses
D. earphones

11. Theoretically, the Earth should not have a permanent magnet in its core because
    A. its interior is too hot
    B. the Earth spins
    C. there is not enough iron
    D. magnets do not occur naturally

12. The Earth's magnetism comes from
    A. a dynamo in the core
    B. permanent magnets in the core
    C. iron in the crust
    D. monopoles near (but not at) the north and south geographic poles

13. The Earth flips its magnetism, on average, approximately
    A. once every 11 years
    B. twice every million years
    C. once every billion years
    D. never (at least not yet)

14. The north geographic pole of the Earth
    A. is a south magnetic pole
    B. has always had the same polarity
    C. is exactly at the magnetic pole
    D. none of the above

15. Compasses point north because
    A. the North Star attracts them
    B. the Earth has an electric charge
    C. there are electric currents in the iron core of the Earth
    D. there are magnetic monopoles near the North Pole

16. Electricity will heat a wire if it has
    A. high voltage
    B. high current
    C. high frequency
    D. DC rather than AC

17. Static electricity occurs when two surfaces rub against each other and
    A. protons flow from one to another
    B. electrons flow from one to another
    C. positrons flow from one to another
    D. neutrons flow from one to another

18. Europe uses higher voltage in homes because it
    A. creates less heat in home wiring
    B. carries greater power
    C. is less dangerous
    D. works better for DC (used in Europe instead of AC)

19. Lights and TVs in Europe tend to flicker because Europe uses
    A. 220 volts—the United States uses 110 volts
    B. 110 volts—the United States uses 220 volts
    C. 50 Hertz—the United States uses 60 Hertz
    D. 60 Hertz—the United States uses 50 Hertz

20. When an object loses some electrons, it
   A. has a negative charge
   B. has a positive charge
   C. has a low resistance
   D. glows (e.g., lightbulb wire)

21. To get a spark, you normally need:
   A. high voltage
   B. high current
   C. low resistance
   D. alternating current

22. A fuse is used in a house to prevent
   A. large power surges
   B. house wires from overheating
   C. illegal use of electricity
   D. too much voltage entering the house
   E. too much energy usage

23. The Earth's magnetic flips are used for
   A. creating new permanent magnets
   B. proving that the Earth has a solid iron core
   C. generating useful power
   D. geologic dating

24. In a metal, electrons
   A. are confined to a single atom
   B. always point in the same direction
   C. can move freely
   D. do not exist

25. The aspect of electricity that would make it most dangerous is high
   A. voltage
   B. current
   C. frequency
   D. power

26. A certain kind of generator uses its own electricity to strengthen its magnetic field. Such a generator is called a
   A. superconductor
   B. transformer
   C. samarium
   D. dynamo

27. The original superconductors were cooled with
   A. liquid nitrogen
   B. liquid helium
   C. freon refrigerators
   D. hydrogen gas

28. Room-temperature superconductors
   A. are used in advanced computers
   B. are used to carry electric power

C. are used for strong magnets

D. have no practical applications

29. High-temperature superconductors operate at approximately:

A. room temperature

B. 4 K (liquid helium temperature)

C. –123 C (liquid nitrogen temperature)

D. 2000 C

30. We use AC instead of DC because

A. 110 volts AC is less dangerous than 110 volts DC

B. AC is cheaper than DC

C. AC can use transformers

D. DC can use transformers

31. Topsy was executed with AC

A. to show the dangers of high voltage

B. because DC would not have worked

C. because it was the only way to get high current

D. AC delivers more power than DC

32. Edison wanted

A. a power plant on every city block

B. high-voltage DC

C. alternating current

D. to abolish the electric chair

33. The current through a flashlight bulb is

A. about 110 amps

B. about 1.6×10^{-19} amp

C. about one amp

D. about 15 amps

34. A stereo speaker produces sound by the force

A. of a magnetic field on electric current

B. of an electric field on an electric charge

C. of a magnetic field on an electric charge

D. of an electric field on a magnetic charge

35. The light from an ordinary lightbulb comes from the fact that

A. the wire contains phosphor

B. the wire is heated by electricity

C. the high voltage causes small sparks

D. electricity is a wave

36. A Tesla coil is a kind of

A. transformer

B. dynamo

C. radio transmitter

D. sensor for the Curie point

37. Submarines can be detected by their

A. magnetism

B. electric charge

       C. backscattered x-rays

       D. MRI signal

38. At the Curie temperature

       A. molecular movement stops

       B. fusion takes place in the sun

       C. magnetism disappears

       D. fission takes place in a bomb

39. Most electric power is generated by

       A. static electricity

       B. a wire moving through a magnetic field

       C. moving a wire in a strong electric field

       D. chemical means (batteries or fuel cells)

40. What invention made it possible to change low voltage AC to high voltage AC?

       A. Van de Graaff generator

       B. samarium-cobalt magnets

       C. transformer

       D. inverter

41. To reduce losses from resistance, electric power lines use

       A. very high current

       B. very high voltage

       C. very high power

       D. very low voltage

42. An example of a transformer is

       A. a rail gun

       B. a dynamo

       C. a Tesla coil

       D. a Van de Graaff generator

43. Better motors and earphones have recently been made possible by

       A. stronger magnets

       B. wires with less resistance

       C. higher voltage batteries

       D. using light rather than electricit

Waves

Including UFOs, Earthquakes, and Music

Two Strange but True Stories

The following two anecdotes, "Flying Saucers" and "Rescuing Pilots," are actually closely related, as you will see later in this chapter. They both will lead us into the physics of waves.

Flying Saucers Crash near Roswell, New Mexico

In 1947, devices that the U.S. government called "flying disks" crashed in the desert of New Mexico. The debris was collected by a team from the nearby Roswell Army Air Base, which was one of the most highly classified facilities in the United States. The government put out a press release announcing that flying disks had crashed, and the story made headlines in the respected local newspaper, the *Roswell Daily Record*. Take a moment to look at the headlines for 8 July 1947 (figure 7.1).

The next day, the U.S. government retracted the press release and said that their original announcement was mistaken. There were no flying disks, they claimed. It was only a weather balloon that had crashed. Anybody who had seen the debris knew it wasn't a weather balloon. It was far too large, and it appeared to be made from some exotic materials. In fact, the object that crashed was *not* a weather balloon. The government was lying, in order to protect a highly classified program. And most people could tell that the government was lying.

The story I have just related sounds like a fantasy story from a supermarket tabloid—or maybe like the ravings of an antigovernment nut. But I assure you, everything I said is true. The story of the events of Roswell, New Mexico, is fascinating, and not widely known, since many of the facts were classified until

Figure 7.1 Serious newspaper headlines from the respected *Roswell Daily Record*. This was not a joke. RAAF stands for "Roswell Army Air Force." (U.S. Air Force, *Roswell Report*, 1995.)

recently. In this chapter, I'll fill in the details so that the Roswell story makes sense.

Incidentally, if you are unfamiliar with the name *Roswell*, that means that you have not watched TV programs or read any of the other voluminous literature about flying saucers and UFOs. Try doing an Internet search on "Roswell 1947" and see what you find. Be prepared to be astonished.

Now for the second anecdote.

Rescuing Pilots in World War II

The true story of the flying disks began with an ingenious invention made by the physicist Maurice Ewing near the end of World War II. His invention involved small objects called *SOFAR spheres* that could be placed in the emergency kits of pilots flying over the Pacific Ocean. If a pilot was shot down, but he managed to inflate and get on to a life raft, then he was instructed to take one of these spheres and drop it into the water. If he wasn't rescued within 24 hours, then he should drop another.

What was in these miraculous spheres? If the enemy had captured one and opened it up, they would have found that the spheres were hollow with nothing inside. How could hollow spheres lead to rescue? How did they work?

Here's the answer to the SOFAR question: Ewing had been studying the ocean, and he was particularly interested in the way that sound travels in water. He knew that the temperature of the water got colder as it got deeper—and that should make sound travel slower. But as you go deeper, the pressure gets stronger, and that should make the sound travel faster. The two effects don't cancel. When he studied it in detail, he concluded that the sound velocity would vary with depth. His most interesting conclusion was that at a depth of about 1 km, the sound travels slower than at any other depth. As we will discuss later, this implies the existence of a *sound channel* at this depth, a layer that tends to concentrate and focus sound and keep it from escaping to other depths. Ewing did some experiments off the coast of New Jersey and verified that this sound channel existed, just as he had predicted.

The SOFAR spheres were hollow and heavier than an equal volume of water. They sank but were strong enough to hold off water pressure until they reached the depth of the sound channel. At that depth, the sphere suddenly collapsed with a bang. That sent out a pulse of sound that could be heard thousands of kilometers away. From these sounds, the Navy could figure out the approximate location of the downed pilot and send out a rescue team.

It turns out (this wasn't known back then) that Ewing's little spheres used the same phenomena that whales use to communicate with other whales: the focusing of sound in the sound channel. We'll discuss this shortly.

At the end of World War II, the same Maurice Ewing proposed a second project based on the same idea. This project was eventually given the name *Project Mogul*. It used "flying disks" for a highly classified purpose: to detect nuclear explosions. It made use of a sound channel in the atmosphere. But the flying disks crashed in Roswell, New Mexico, in 1947, made headlines, and became part of a modern legend.

To explain these stories, we have to get into the physics of sound. And to understand sound, we have to talk about waves.

Waves

All waves are named after water waves. Think for a moment about how strange water waves are. Wind pushes up a pile of water, and the pile creates a wave. The wave moves and keeps on moving, carrying energy far from the place where the wave was created. Waves at the coast are frequently an indicator of a distant storm. But the water from that distant storm didn't move very far, just the wave. The wind pushed the water, and the water pushed other water, and the energy traveled for thousands of miles, even though the water only moved a few feet.

You can make waves on a rope or with a toy called a *Slinky*. (If you've never played with a Slinky, you should go to a toy store as soon as possible and buy one.) Take a long rope or a Slinky, stretch it across a room, shake one end, and watch the wave move all the way to the other end and then bounce back. (Water waves, when they hit a cliff, also bounce.) The rope jiggles, but no part of it moves very far. Yet the wave does travel, and with remarkable speed.

Sound is also a wave. When your vocal cords vibrate, they shake the air. The air doesn't move very far, but the shaking does. The shaking moves as far as the ear can hear and farther. The initial shaking air around your vocal cords makes the air nearby shake also, and so on. If the shaking reaches someone else, then it causes his eardrums to shake, which sends signals to his brain and causes him to hear you.

For a nice animation of a sound wave, showing how the molecules bounce back and forth but create a wave that moves forward only, see http://www .kettering.edu/~drussell/Demos/waves/wavemotion.html.

When you look at this, make a point of watching one black dot. If you watch carefully, you'll see it actually oscillates back and forth.

If the sound wave hits a wall, it bounces. That's what gives rise to echoes. Sound waves bounce just like water waves and rope waves.

A remarkable thing about all these kinds of waves is that the shaking leaves the location where it started. Shake some air and you create a sound, but the sound doesn't stay around. A wave is a way of transporting energy long distances without actually transporting matter. It is also a good way to send a signal.

It turns out that light, radio, and TV signals also consist of waves. We'll get to that in the next chapter. What is waving for these? The traditional answer is "nothing" but that is really misleading. A much better answer is that there is a "field" that is shaking —the electric and magnetic fields. Another correct answer is that "the vacuum" is what is shaking. We'll discuss this further in the chapter on quantum physics (chapter 11).[1]

Wave Packets

Waves can be long with many vibrations, as when you hum, or they can be short, as in a shout. We call such short waves *wave packets*. You may have noticed that water waves often travel in packets. Splash a rock into a pool, and you'll see a bunch of waves moving out, forming a ring that contains several up and down oscillations. That's a packet. A shout contains many oscillations of the air, but these oscillations are confined to a relatively small region. So that too is a wave packet.

Now think about this: short waves act in a way very similar to particles. They move and they bounce. They carry energy. If the packet were extremely short, maybe you wouldn't notice that it was really a wave. Maybe you would think that it was a small particle.

In fact, the theory of quantum mechanics is really a fancy name for the theory that all particles are really little packets of waves. The packets for an electron and a proton are so small that we don't normally see them. What is waving in an electron? We think it is the same thing that is waving for light: the vacuum.

---

[1] Here is a brief summary of the answer: When it was discovered that light is a wave, physicists didn't know what was waving, but they gave it a name: the "Aether." (I spell it this way to distinguish it from the chemical *ether*, which is totally different.) Most modern physicists believe that the Aether was shown not to exist, but that isn't true. The distinguished theoretical physicist Eyvind Wichmann pointed out (in a class I took from him at Berkeley) that the Aether was only shown to be invariant under the laws of special relativity, and therefore was unnecessary. But then quantum mechanics started giving it properties: it can be polarized, and it carries dark energy. Wichmann says that the Aether never went away from physics—it was made more complex, and simply was reborn with a new name: the vacuum.

So when you are studying sound, water, and earthquakes, you are really learning the properties of waves. That will be most of what you need in order to understand quantum mechanics.

Sound

Sound in air results when air is suddenly compressed—for example, by a moving surface (such as a vibrating vocal cord or bell). The compression pushes against adjacent air, and that pushes against the air in front of it, and so on. The amazing thing about sound is that the disturbance travels, and the shaking of the original air stops. The energy is carried away very effectively.

Sound is generated in air when something compresses it in a local region. This could be the vibrating of vocal cords, a violin string, or a bell. The compressed air expands, and compresses the air next to it. The air never moves very far, but the compression is passed on from one region to the next. This is depicted in figure 7.2. Each little circle represents a molecule. The wave consists of compression and expansion of regions of the gas.

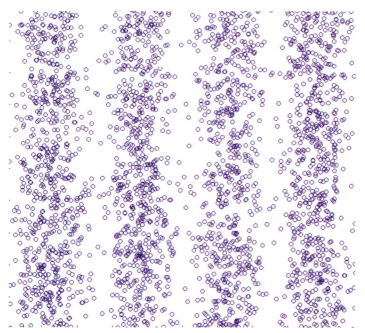

Figure 7.2 Air molecules in a sound wave.

Each molecule shakes back and forth and doesn't travel very far. But the waves travel forward. Look at the diagram, and imagine that you are looking at a series of water waves from an airplane. But the waves in sound come not from up and down motion, but from compression and dilation. When these compressions reach your eardrum, they make it vibrate. Those vibrations are then passed on through the rest of your ear to nerves and then to the brain, where the vibrations are interpreted as sound.

To understand this, it is easiest to watch a movie, such as the one I referred to earlier at www.kettering.edu/~drussell/Demos/waves/wavemotion.html.

A wave is moving from the left to the right. But if you watch one molecule, you'll see that it is shaking back and forth, and never travels very far. It bangs into a nearby molecule, and transfers its energy.

That is the key aspect of waves. No individual molecule travels very far, but the energy is transferred. The molecules pass on the energy, from one to the next. It is the energy that travels long distances, not the particles. Waves are means for sending energy without sending matter.

Sound waves can travel in rock, water, or metal. All those materials compress slightly, and this compression travels and carries away the energy. If you hit a hammer on a railroad rail, then the metal rail is momentarily distorted and the distortion travels down the rail. If someone puts his ear to the rail a mile away, he will hear the sound. The best way to hear the sound is to put your head against the rail. The vibration in the rail will make your skull vibrate, and this will make the nerves in your ear respond—even if none of the sound is actually in air.

Because steel is so stiff, it turns out that sound travels 18 times faster in steel than in air. In air, sound takes 5 seconds to go 1 mile; in steel, sound will go that same distance in less than 1/3 second. In the olden days, when people lived near railroad tracks, they could listen to the track to hear if a train was coming, and they could even estimate the distance to the train by the loudness of the sound.

For sound to travel, the molecules of air have to hit other molecules of air. That's why the speed of sound is approximately equal to the speed of molecules. We discussed this fact in chapter 2. But in steel, the molecules are already touching each other. That's why sound in steel can move much faster than the thermal velocity of the atoms in the steel.

Sound travels in any material that is springy—i.e., that returns to its original shape when suddenly compressed and then released. The faster it springs back, the faster the wave moves. The speed of sound in water is about 1 mile per second, but it varies slightly depending on the temperature and depth of the water.

Note that a sound wave in water is a different kind of wave than the water wave that moves on the surface. In water, sound travels *under* the surface, in the bulk of the water. It consists of a compression of the water. Water waves on the surface are not from compression, but from movement of the water up and down, changing the shape of the surface. So although they are both in water, they are really very different kinds of waves. You can see surface waves easily. You usually cannot see sound waves. Surface waves are slow and big. Sound waves are microscopic and fast.

The speed of sound in air doesn't depend on how hard you push—that is, on how intense the sound is! No matter how loud you shout, the sound doesn't get there any faster. That's surprising, isn't it?

Why is that true? Remember, at least for air, the speed of sound is approximately the speed of molecules. The signal has to go from one molecule to the next, and it can't do that until the air molecule moves from one location to another.

(The added motion from the sound vibration is actually very small compared to the thermal motion of the molecules.) When you push on the air, you don't speed up the molecules very much; you just push them closer to each other.

But the speed of sound does depend on the temperature of the air. That's because the speed depends on the velocity of the air molecules, and when air is warmer, the velocity is greater.

Table 7.1 gives the speed of sound in several materials.

Table 7.1   Speed of Sound in Various Materials

Material and temperature	Speed of sound
Air at 0°C = 32°F	331 m/s = 1 mile for every 5 seconds
Air at 20°C = 68°F	343 m/s
Water at 0°C	1402 m/s = 1.4 km/s
Water at 20°C	1482 m/s = almost 1 mile per second
Steel	5790 m/s = 3.6 miles per second
Granite	5800 m/s

There is no need to memorize this table. But you should remember that sound moves faster in solids and liquids than in air. And you should know that the speed of sound in air is about one mile every five seconds.

Sound traveling in rock gives us very interesting information about distant earthquakes. We'll come back to that later in this chapter. Observations of the surface of the Sun show that sound waves arriving from the other side, traveling right through the middle of the Sun. Much of our knowledge of the interior of the Sun comes from the study of these waves. (We detect them by sensitive measurements of the surface of the Sun.) Sound has been detected traveling through the Moon, created by meteorites hitting the opposite side. On the Moon, we use instruments that were left behind by the *Apollo* astronauts.

There is no sound in space because there is nothing to shake. A famous tagline from the science fiction movie *Alien* (1979) is, "In space, nobody can hear you scream." Astronauts on the Moon had to talk to each other using radios. Science fiction movies that show rockets roaring by are not giving the sound that you would hear if you were watching from a distance—since there would be no sound.[2]

Transverse and Longitudinal Waves

When you shake the end of a rope, the wave travels down its length, from one end to the other. However, the shaking is sideways—i.e., the rope vibrates

---

[2] To enjoy the movie, I always assume that the microphone is located on the spacecraft, so although we are watching the rocket pass, we are hearing sound as if we were on the rocket. A similar conceit was used in the movie *Downhill Racer*. We watched the racer (Robert Redford) from a distance, yet we heard the chattering on his skis on the ice as if we were skiing. Similarly in movies, we often see people from afar but hear their conversations as if we were right next to them.

sideways even though the direction that the wave is moving is along the rope. This kind of wave is called a *transverse* wave. In a transverse wave, the motion of the particles is along a line that is perpendicular to the direction the wave is moving.

For an illustration of a transverse wave, go back to http://www.kettering .edu/~drussell/Demos/waves/wavemotion.html, and look at the second illustration on that page.

A sound wave is different. The vibration of the air molecules is back and forth, in the same direction that the wave is moving. This kind of compressional wave is called a *longitudinal* wave. In such a wave, the motion and direction of the wave are both along the same line.

This may seem peculiar, but water waves are even stranger.

WATER-SURFACE WAVES

Water waves (the term we will use when we mean the ordinary surface water waves—as opposed to water sound waves) gave all waves their name. If you are swimming or floating and a water wave passes by, you move slightly back and forth as well as up and down. It is worthwhile to go swimming in the ocean just to sense this. In fact, for most water waves, the sideways motion is just as big as the up and down, and you wind up moving in a circle! But when the wave has passed, you and the water around you are left in the same place. The wave, and the energy it carries, passes by you.

For a nice illustration of the motion of particles in a water wave, take a look again at www.kettering.edu/~drussell/Demos/waves/wavemotion.html, but this time scroll down to the third animation. Look at one particle, maybe one of the blue ones, and watch how that particle moves. Does it move in a circle?

When there is a series of waves following each other, we call that a *wave packet*. The distance between the *crests* (the high points of the waves) is called the *wavelength*. Waves with different wavelengths travel at very different speeds. Those with a short wavelength go slower, and those with a long wavelength go faster. In deep water (when the depth is greater than the wavelength), the equation is as follows:[3]

$$v \approx \sqrt{L}$$

In this equation, v is the velocity in meters per second (m/s), and L is the wavelength in meters (m), and the squiggly equals sign $\approx$ means "approximately equal to." So for example, if the wavelength (distance between crests) is $L = 1$ m, then the velocity is about $v = 1$ m/s. If the wavelength is 9 m, the velocity is 3 m/s. Does that agree with your image of ocean waves? Next time you swim in the ocean, check to verify that long waves move faster.

That equation is remarkably simple, but it is correct only for deep water— that is, for water that is much deeper than a wavelength.

---

[3] For the physics major: The standard physics equation for deep water waves is $v = \text{sqrt}[gL/(2\pi)]$, where $g = 9.8$ m/s^2 is the acceleration of gravity (from chapter 3). Putting in $g = 9.8$ gives $v \approx 1.2 \, \text{sqrt}(L) \approx \text{sqrt}(L)$.

SHALLOW WATER WAVES

When the water is "shallow" (the depth D is much less than the wavelength L) then the equation changes to

$$v = 3.13\sqrt{D}$$
$$\approx \pi\sqrt{D}$$

where D is the depth in meters.[4] Note that all shallow water waves travel at the same velocity, determined only by the depth of the water, regardless of the wave's wavelength. The speed of shallow water waves depends only on the depth of the water. This might match your experience when you surf on relatively long waves in shallow water.

If the wavelength is very long, then we have to regard even the deep ocean as shallow. This is often the case for tsunamis.

TSUNAMIS (TIDAL WAVES)

A *tsunami* is a giant wave that hits the coast and washes far up on the shore, often destroying buildings that are within a few hundred meters of the beach. Tsunamis were traditionally called *tidal waves*, but a few decades ago scientists (and newspapers) decided to adopt the Japanese word, and now it is more commonly used.

Underwater earthquakes and landslides often generate tsunamis. These waves usually have a very high velocity and a very long wavelength. In the deep ocean, they may have a very low amplitude, so they can travel right under a ship without anyone on board even noticing. But as they approach land, they are slowed down, and the energy is spread out over a smaller depth of water. As a result, the height of the wave rises. The rise can be enormous, and that is what causes the damage near the coast.

In Pacific islands (such as Hawaii), you'll see sirens mounted on poles near the beaches. If an earthquake is generated within a few thousand miles, these sirens will be sounded to warn the residents to evacuate. A tsunami could arrive within a few hours.

If a very large earthquake fault moves underneath deep water, the wave it creates can be very long. For a large tsunami, a typical wavelength is 10 km, although some have been seen with wavelengths of 100 km and more. That means even in water with a depth of 1 km = 1000 meters, a tsunami is a *shallow* water wave! (Recall that a shallow water wave is one in which the wavelength is greater than the depth.)

The velocity of the tsunami can be calculated from the shallow water equation. In water 3 km deep, $D = 3000$ meters, so the velocity is $v = 3.13$ sqrt(3000) ≈ 171 meters per second. That's 386 miles per hour, about half as fast as the speed of sound in air. A tsunami that is generated by an earthquake

---

[4] The second equation is only approximate. I wrote it using the symbol π to make it easier to remember, even though you don't have to remember it.

1000 mi away will take 2.6 hours to arrive. That's enough time to give warning to coastal areas that a tsunami is on its way.

YOU CAN OUTRUN A TSUNAMI

Imagine a tsunami moving at 171 meters per second, with a wavelength of 30 kilometers. Imagine that one crest of the wave passes you. The next one is approaching you from 30 km away. Even with its speed of 171 m/s, it will take $t = d/v = 30{,}000/171 = 175$ seconds to reach you, nearly three minutes. The water will fall for the first 87 of these seconds, and then rise for the next 87. Thus, although these waves travel fast, they are slow to rise and fall. That's why tsunamis were called tidal waves. If you are in a harbor, and there is a small tsunami, it might take several minutes for the water to rise and fall, and it gives the appearance of a rapid tide. The image of a huge breaking wave hitting the shore is largely fictional; most tsunamis are just very high tides (with ordinary waves on top) that come and wash away everything close to the shore.[5] That's how they do their damage. If the ocean rises 10 meters, it destroys everything, even if it takes 50 seconds to reach its peak. If you are young and healthy, you can usually outrun the rising water as it comes in. If you are not fast enough, then you get swept up in a very large volume of water, and dragged out to sea when the wave recedes. Small tidal waves are frequently observed as slow (175 second) rises and falls in harbors. Boats tied to docks are often damaged by these slow waves, as they rise above the dock and get thrown into other boats. Many captains take their boats out into the harbor or out to sea when they are alerted that a tsunami is coming. In Japanese, the word *tsunami* means "harbor wave."

The Equation for Waves

Recall from the previous section that if the wavelength is L, and the velocity is v, then the time it takes between crests hitting you is $T = L/v$. The time T is called the *period* of the wave. This calculation is true for all waves—sound, tsunamis, deep water waves, even light waves. It is the fundamental relationship between velocity, period, and wavelength.

$$T = L/v$$

If the period is less than one second, it is usually more convenient to refer to the number of crests that pass by every second. That is called the *frequency* of the wave, f, and it is given by $f = 1/T$. Putting these into the preceding equation, we get $1/f = L/v$, or

$$v = fL$$

You don't have to memorize this equation, but we will use it a lot, especially when we discuss light. Light in vacuum has a speed $v = 3 \times 10^8$ m/s, a

---

[5] The tsunami in the movie *Deep Impact* (1998) is particularly inaccurate. It shows a giant wave breaking over Manhattan Island. But the harbor of New York City is relatively shallow; there is no place for that much water to come from, unless a giant wave broke far out to sea.

number we usually call *c*. Since we know *c*, the equation allows us to calculate the frequency whenever we know the wavelength, or the other way around.

Sound Doesn't Always Travel Straight

Sound waves, whether in air or in ocean, often do not travel in straight lines. They will bend upward or downward, to the left or to the right, depending on the relative sound speed in the nearby material. Here is the key rule:

Waves tend to change their direction by bending their motion toward the side that has a slower wave velocity.

To understand why this is so, imagine that you are walking arm-in-arm with a friend. If your friend is on your left side and slows down, that pulls your left side backward and turns you toward the left. If your friend speeds up, that pulls your left arm forward and turns you to the right (and also turns your friend to the right). The same phenomenon happens with waves. A more complete description of this is given in the optional section at the end of the chapter about Huygens's principle.

This principle can be demonstrated in a large classroom by having students raise their hands as soon as their neighbor raises a hand. The location of students' hand-raising moves throughout the classroom like a wave. This procedure is popular among fans in sporting events, where it is also called a "wave." If the students in one part of the room are told to be a little slower, then the wave will bend toward them as it spreads across the room.

The direction changing rule is true for all kinds of waves, including sound, water surface waves, and even earthquakes and light.

EXAMPLE: "NORMAL ATMOSPHERE"

Here is an example from the atmosphere. At high altitude, the air is usually colder. That means that the velocity of sound at high altitude is slower than it is at low altitude.

Now imagine a sound wave that is initially traveling horizontally, near the surface of the Earth. Above it, the velocity is lower, so it will tend to bend upward. This is shown in figure 7.3.

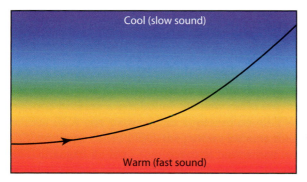

Cool (slow sound)

Warm (fast sound)

Figure 7.3 Cool atmosphere near the warm ground. Sound bends upward.

Notice that the sound bends away from the ground toward higher altitude. It bends upward. That's because the air above it has a slower sound velocity.

SOUND IN THE EVENING

When the Sun sets, the ground cools off rapidly. (It does this by emitting infrared radiation; we'll discuss this further in chapter 9.) The air does not cool so quickly, so in the evening, the air near the ground is often cooler than is the air that is up higher. This phenomenon is called a *temperature inversion* because it is opposite to the normal pattern of the daytime. When there is a temperature inversion, sound tends to bend down toward the ground, as shown in figure 7.4.

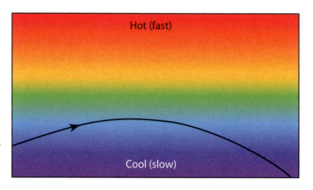

Figure 7.4 Sound path near the ground when there is an inversion.

SOUND DURING THE DAY, AGAIN

Now let's look at the morning situation again, with warm air near the ground and cold air up high. But let's draw many sound paths, all coming from the same point. This is done in figure 7.5.

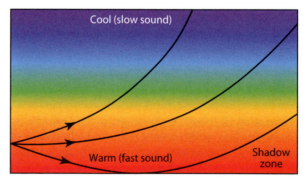

Figure 7.5 The shadow zone. Sound coming from the point on the left cannot reach the region labeled "shadow zone" because it is blocked by the ground.

The solid line at the bottom edge of the figure represents the ground. Note that it blocks certain paths—the ones that drop too steeply. In the lower right corner is a small region that none of the paths can reach, since to reach this region the sound waves would have to go through the ground. (We'll assume for now that the ground absorbs or reflects sound, and does not transmit it—at least, not very well.) If the sound were coming from the point on the left, and you were standing in this shadow zone, then you wouldn't hear any sound at all. You are in the sound shadow of the ground.

This diagram shows why mornings tend to be quiet. Sounds bend up toward the sky, and if you are near the ground, there is no way that most of them can reach you. You won't hear distant automobiles, birds, waves, lion roars . . .

SOUND IN THE EVENING, AGAIN

In figure 7.6, I've redrawn the evening situation, with the inverted temperature profile (cold at the bottom, warm at the top).

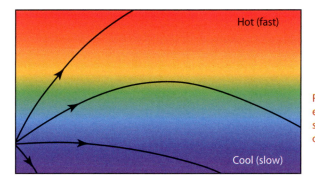

Figure 7.6 Sound paths in the evening. Notice that there is no shadow zone, so distant sounds can be heard at long distances.

Notice that there is no shadow zone. No matter where you stand, there are paths by which the sound can reach you.

Have you ever noticed that you can hear more distant sounds in the evening than in the morning? I've noticed that in the evening I can often hear the sound of distant traffic, or of a train; I rarely hear such sounds in the morning. (This phenomenon first mystified me when I was a teenager living 1/4 mile from the beach. I noticed that I could hear the waves breaking in the evening, but almost never in the morning.)

The explanation is in the preceding diagrams: In the evening, sound that is emitted upward bends back down, and you can hear sound from distant places. There is no shadow zone.

If you happen to be a wild beast, then the evening would be a good time to search for prey, since you could hear it even when it was far away. Of course, it can hear you too.

FORECASTING A HOT DAY

There are times when I wake up in the morning and hear distant traffic. Then I know that it will probably be a hot (and maybe smoggy) day. I learned this from experience long before I figured out the reason why.

The reason is that hearing distant sounds means that there is an inversion— i.e., the high air is warmer than the low air. The sound diagram for an inversion in the morning is identical to the sound diagram shown earlier for the evening.

Inversions are unusual in the morning, but they do happen. The presence of an inversion in the morning leads to a special weather condition. On normal (no inversion) days, hot air is near the ground, and cold air is above it. Hot air is less dense than cold air, so it tends to float upward. (In the same way, wood floats on water if it is less dense than water.) So the hot air tends to leave the ground, replaced by cooler air from above.

But if there is an inversion—i.e., there is hot air above and cold below—then the air above is less dense than the air at the ground. So *convection*, the floating of the ground air upward, doesn't take place. With no place to rise to, the hot air accumulates near the ground, making for a hot day. Smog and other pollutants also accumulate. The weather forecast on the radio or TV will often announce that there is an *inversion*. Now you know what that means: the normal temperature profile is inverted—i.e., it is upside down, with the cool air near the ground and the warm air above.

Inversions frequently happen at the end of a hot day. The ground cools more rapidly than does the air. (That's because it emits more infrared, or IR, radiation; we'll discuss that in the next chapter.) The air near the ground is cooled by contact with the cool ground, and the air above remains warm (unless there is turbulence from wind). For people who are sensitive to smog, the announcement of an inversion is bad news. For people who love hot weather, it is good news.

DISTANT GUNSHOTS IN BAGHDAD

A general stationed in Baghdad was filmed by a TV news organization, and he commented on the fact that gunfire in Baghdad increased significantly just as the Sun was setting. He didn't understand why this was the case, but he adjusted his patrols because of it.

One of our graduate student instructors, Joel Mefford, heard the interview and wondered, was the increase real, or was it simply due to the fact that in the evening, when sounds bend downward, the general was able to hear distant gunfire that would not be heard during the morning and afternoon? We don't know, but the example shows how a little understanding of physics can have immediate and practical applications.

The Sound Channel Explained: Focused Sound

Now let's get back to the mysterious sound channel in the ocean that Maurice Ewing exploited for his SOFAR system in World War II.

In the ocean, the temperature of the water gets cooler as we go farther down. This would make the speed of sound less. But, as mentioned earlier, the water is also getting more and more compressed (i.e., denser) because of the increasing pressure. This tends to make sound go faster. When these two effects are combined, we get a gradual decrease in sound velocity as we go from the surface to about 1 km of depth, and then the sound velocity increases again. This is illustrated in the figure 7.7. Darker means slower sound (just as it did in the atmosphere diagrams).

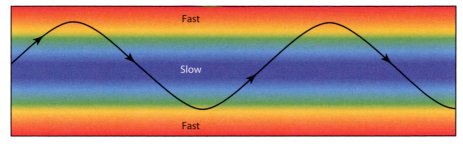

Figure 7.7 A wave channel.

I've also drawn the path of a ray of sound. Notice that it always bends toward the slow region. The path I drew starts with an upward tilt, bends downward, passes through the slow region, and then bends upward. The path oscillates up and down, but never gets very far from the slow region, the 1-kilometer-deep sound channel.

> **Exercise:** Draw some other paths, starting at different angles. What happens if the ray starts out horizontally? Vertically?

HOW SOFAR SAVED DOWNED PILOTS

Let's return now to the magic of Ewing's SOFAR spheres. As I stated earlier, they were hollow, and yet they were made of heavy material. Since they weighed more than an equal volume of water, they didn't float, but sank. Ewing designed the spheres to be strong enough to withstand the pressure of water down to a depth of 1000 meters. At this depth, the spheres were suddenly crushed. (Like an egg, the round surface provides lots of strength, but when it breaks, it breaks suddenly.) The water and metal collapsed and banged against the material coming in from the other side. It's like a hammer hitting a hammer, it generates a loud sound. The energy released from a sphere with radius of 1 inch at a depth of 1 kilometer is approximately the same as in 60 mg of TNT. That doesn't sound like a lot—but it is about the same amount you might find in a very large firecracker.

In the air, the sound of a firecracker doesn't go far, perhaps a few kilometers. But at a depth of 1000 meters, the ocean sound channel focuses the sound. Moreover, the sound channel is quiet. Sound doesn't get trapped unless it originates within the sound channel itself. (Can you see why?) Any sounds created in the sound channel by whales or submarines stay in there, so the sound doesn't spread out as much as it would otherwise. Microphones placed within the sound channel can hear sounds that come from thousands of kilometers away.

During World War II, the Navy had arranged for several such microphones placed at important locations, where they could pick up the ping of the imploding Ewing spheres. They could locate where the implosion had taken place by the time of arrival of the sound. If the sound arrived simultaneously at two microphones (for example), then they knew that the sound had been generated somewhere on a line that is equally distant from the two microphones. With another set of microphones, they could draw another line, and the intersection of the two lines gave the location of the downed pilot.

> **Historical note:** *SOFAR* supposedly stands for "SOund Fixing And Ranging." Fixing and ranging was Navy terminology for determining the direction to a source (that's the fixing part) and its distance (ranging). Despite all this, I suspect that the acronym was forced, and the real name came about because the channel enabled you to hear things that were *so far* away. Some people still refer to the sound channel as the SOFAR channel. I learned about the SOFAR spheres from Luis Alvarez, who knew about them from his scientific work during World War II. I have spoken to several other people who remember them, including Walter Munk and Robert C. Spindel. Spindel believes that the spheres contained a small explosive charge to enhance the sound. We have not yet found any historical documentation that verifies this.

WHALE SONGS

What does the sound channel look like? The word *channel* can be misleading, since it brings up a vision of a narrow corridor. It is not like a tube. It is a flat layer, existing about a kilometer deep, spreading over most of the ocean. Sound that is emitted in the sound channel tends to stay in the sound channel. It still spreads out, but not nearly as much as it would if it also spread vertically. That's why the sound can be heard so far away from its source. It tends to get focused and trapped in that sheet.

In fact, the sound channel is like one story in a very large building, with ceiling and floor but without walls. Sound travels horizontally, but not vertically. If sound is emitted at the surface of the ocean, then it does not get trapped. So the sound of waves and ships does not pollute the sound channel. The sound channel is a quiet place for listening to SOFAR spheres and other sounds that are generated in the sound channel.

Whales discovered this, probably millions of years ago. We now know that whales like to sing when they are at the sound channel depth. These songs are hauntingly beautiful.[6]

You can find other recordings on the Internet, and you can buy recordings on CDs. Nobody knows what the whales are singing about. Some unromantic people think that they are saying nothing more than "I am here."

The Cold War and SOSUS

During World War II, the part of the military that used submarines was called the "Silent Service." This reflected the fact that any sound emitted by a submarine could put it in danger of detection, so submariners trained themselves to be very quiet. Someone in a sub who drops a wrench makes a sound that is

---

[6] If your computer has the right software, you can listen here to the recorded song of the humpback whale at www.muller.lbl.gov/teaching/physics10/whale_songs/humpback.wav and of the gray whale at www.muller.lbl.gov/teaching/physics10/whale_songs/gray.wav.

unlike any other in the ocean. (Fish don't drop wrenches.) The wrench clatters against the hull, and the hull carries the sound to the water, and the vibrations of the hull send the sound into the ocean. Ships on the surface, and other submarines, had sensitive microphones to listen to possible sounds emitted from submarines.

The presence of the sound channel did not remain secret for long, but its properties did. In the period from the 1950s to 1990s, the United States spent billions of dollars to put hundreds of microphones into the channel at locations all around the world. These microphones carried the signals back to an analysis center, and then the world's best computers analyzed them. The system was called *SOSUS*, an acronym for "SOund SUrveillance System." The magnitude of the SOSUS effort was one of the best-kept secrets of the Cold War. Effective use of SOSUS required the Navy to make extensive measurements of the ocean and its properties, and to update the temperature profile of the ocean all around the world. (The ocean has weather fronts analogous to those in the atmosphere.)

FURTHER READING

If you really want to know more about this subject, one of the best introductions is the novel *The Hunt for Red October* by Tom Clancy. (Not the movie. The movie skips all the interesting technology.) When this novel came out in 1984, much of the material in it was still classified. Clancy had a talent for reading documents, talking to people, and figuring out from what they said and what was really true. The book was so detailed and so accurate (although it does have some fiction in it and some errors) that new people joining the submarine service were told to read the book in order to get a good picture of how operations worked! Many of the details of the SOSUS system were finally declassified in 1991, seven years after Clancy's book was published. The SOSUS system was one of the largest and most expensive secret systems any nation ever built.

Back to UFOs: A Sound Channel in the Atmosphere!

Soon after he did his work in the ocean, Maurice Ewing realized that there should be a sound channel in the atmosphere too! His reasoning was simple: as you go higher, everyone knows the air gets colder. Mountain air is colder than sea-level air. The temperature of the air drops about 4 F for every 1000 feet of altitude gain.

That means that the velocity of sound decreases with altitude. So sound waves should bend upward. But he also knew that when you get to very high altitudes, the temperature begins to rise again. Starting at about 40,000 to 50,000 feet, the air starts getting warmer, and sound will bend downward. The temperature variation with altitude is shown in figure 7.8.

Remember that the speed of sound depends on the temperature of the air. When the temperature is low, so is the speed of sound. That means that the speed of sound is fast at both high and low altitudes, and slower at about 50,000 feet.

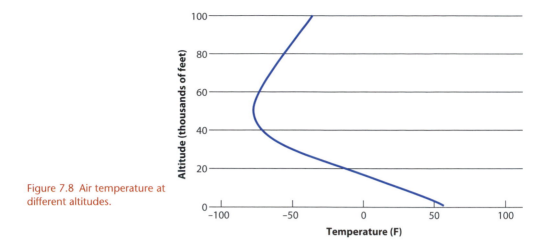

Figure 7.8 Air temperature at different altitudes.

Look at figure 7.7 again, the diagram that showed sound moving in a wiggly line through the ocean. Exactly the same diagram can be used for sound in the atmosphere. That means that there is a sound channel in the atmosphere, centered at about 50,000 feet. (The exact altitude depends on latitude, as well as on the season of the year.) This is what Ewing figured out. He had an important U.S. National Security application in mind to take advantage of this realization.

But first, we need a little more physics. Why does the atmosphere get warmer above 50,000 feet?

OZONE: THE CAUSE OF THE HIGH-ALTITUDE HEATING

Why is high altitude air hot? The reason is the famous ozone layer. At about 40,000 to 50,000 feet, there is an excess of ozone, and this ozone absorbs much of the ultraviolet radiation from the sun. Ultraviolet light, also called UV, is that part of sunlight that is more violet than violet. This light is there, but invisible to the human eye. The ozone layer protects us, since ultraviolet light can induce cancer if absorbed on the skin. We'll talk more about the ultraviolet radiation in chapter 9.

At the end of the twentieth century, scientists began to fear that the ozone layer could be destroyed by human activity, and that would let the cancer-causing, ultraviolet radiation reach the ground with greater intensity. In particular, the scientists worried about the release of certain chemicals into the atmosphere called CFCs (chlorofluorocarbons, used in refrigerators and air conditioners). CFCs release chlorine and fluorine, and these catalyze the conversion of ozone O_3 into ordinary O_2. (To balance the equation, two molecules of O_3 turn into three molecules of O_2.)

The use of CFCs was outlawed internationally, and that was expected to solve the problem. For this reason, the human destruction of the ozone layer is no longer considered an urgent problem. For more, see chapter 9.

Ewing's Project Mogul and His Flying Disks

Maurice Ewing had an urgent application for his predicted atmospheric sound channel: the detection of nuclear tests in Russia. In the late 1940s, the Cold War had begun, and there was growing fear in many countries of the totalitarian communism represented by Russia. The Russians had great scientists, and there was widespread belief that they would be building an atomic bomb soon. At that time, Russia was a very secret and closed society. In fact, Stalin was starving to death 30 million *kulak* farmers, and he could get away with it because he controlled information going in and out of the nation. In 1948, George Orwell wrote *1984*, expressing his fears of such a government.

Ewing realized that as the fireball from a nuclear explosion rose through the atmospheric sound channel, it would generate a great deal of noise that would travel around the world in the channel. (Not all of the sound is generated when the bomb detonates. The roiling fireball continues to generate sound as it reaches the atmospheric sound channel.) Ewing argued that we should send microphones up into the sound channel to detect and measure any such sound. That way we could detect Soviet nuclear tests, even with microphones in the United States!

The microphones that he used were called *disk microphones*. The actual microphone was strung by springs or cords in the center of a disk; this careful mounting was needed to keep vibrations from coming from the support into the sensitive microphone. (The sound generated by such vibrations is called *microphonics*.) Figure 7.9 shows Mohandas Gandhi using a disk microphone in the 1940s (the same era when Project Mogul had its flying disk mikes).

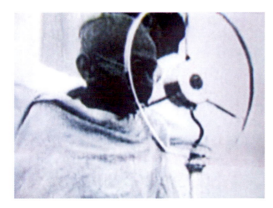

Figure 7.9 Mohandas Gandhi with a disk microphone.

You'll also see lots of disk microphones in the movie *The Aviator* (2004). The disk microphone became a symbol of the new radio era, and formed the basis for the famous RCA test pattern that appeared on all TV screens; when RCA created a TV network called NBC, the pattern became the NBC logo (see figure 7.10). NBC changed its logo to a peacock only when it started broadcasting in color.

Figure 7.10 Early NBC logo based on the disk microphone.

Ewing's idea was to string together a long line of microphones, fly them under a high-altitude balloon, have them pick up the sounds in the sound channel, and then radio the sounds back to the ground. Their shorthand for the flying disk microphones became "flying disks." (The word *flying* was not confined to airplanes; it was equally used by ballooners when they went up.) The Mogul balloons were huge, and the string of microphones was 657 ft long, longer than the Washington Monument is high.

The project was a success. The system detected American nuclear explosions, and on 29 August 1949, it detected the first Russian test.

THE ROSWELL CRASH OF 1947

One of the Project Mogul balloon flights crashed near the Roswell Army Air Force base on 7 July 1947. It was recovered by the U.S. Army, who issued a press release stating that "flying disks had been recovered." The *Roswell Daily Record* had headlines the next day. We referred to these at the beginning of this chapter: "RAAF Captures Flying Saucer."

The fallen object was not a flying saucer, it was a complex balloon project that carried flying disk microphones to pick up Russian nuclear explosions. The program was highly classified, and the press release said more than the security people considered acceptable, so the next day the press release was "retracted." A new press release stated that what had crashed was a "weather balloon." It wasn't a weather balloon. The U.S. government was lying.

THE GOVERNMENT FINALLY TELLS THE TRUTH

In 1994, at the request of a congressman, the U.S. government declassified the information they had on the Roswell incident and prepared a report. Popular articles appeared in the *New York Times* and in *Popular Science* (June 1997).[7]

Should the U.S. government ever lie? This is just the sort of issue that you should confront *before* you become president! It is a good discussion question.

---

[7] The official U.S. government report (synopsis only) on Project Mogul is also available at http://muller.lbl.gov/teaching/physics10/Roswell/USMogulReport.html. See also the official U.S. government report on the Roswell Incident at http://muller.lbl.gov/teaching/physics10/Roswell/RoswellIncident.html.

HOW DO WE KNOW THE GOVERNMENT ISN'T LYING NOW?

Many people believe that the official government report on Project Mogul is just an elaborate cover-up. They believe that a flying saucer really did crash, and the government doesn't want the public to know. Maybe I am part of this conspiracy, and part of my job is to mislead you into believing that flying saucers don't exist! (According to the 1997 movie *Men in Black*, the job of the Men in Black is to make sure that the public never finds out.)

I suggest the following answer: The people who continue to believe that Project Mogul never happened probably don't understand the remarkable science of the ocean and atmosphere sound channels. I could not have invented such a wonderful story. It has too many amazing details. In contrast, it is relatively easy to make up stories about flying saucers. Those don't require much imagination. So here is my hypothesis: It is possible to distinguish the truth by the fact that it is more imaginative and more fascinating!

Of course, I might be lying. The photograph in figure 7.11 shows me at the UFO Museum in Roswell, New Mexico, in 2007.

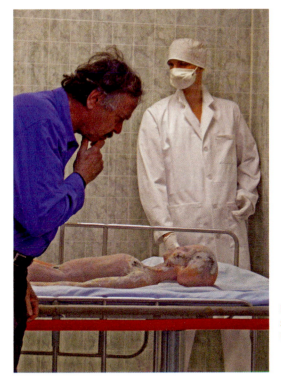

Figure 7.11 The author examines something interesting at the UFO Museum, Roswell, New Mexico.

Earthquakes

When a fault in the Earth suddenly releases energy, it creates a wave in the ground. The location where an earthquake starts is called the *epicenter*. Most people who experience an earthquake are far from the epicenter and are shaken by the wave that starts at the epicenter and shakes them as it passes by.

The epicenter of an earthquake can be located by noting when the earthquake wave arrived at several different locations—just as the SOFAR spheres were used to locate downed pilots in World War II. Moreover, the epicenter is often deep underground, so even someone who is standing at the latitude and longitude of the epicenter can be standing over 15 miles away from it (i.e., above it).[8]

Huge amounts of energy are released in earthquakes, often greater than in our largest atomic weapons. That shouldn't surprise you. If you are making mountains shake over distances of tens or hundreds of miles, it takes a lot of energy. In 1935, Charles Richter found a way to estimate the energy from the measured shaking. His scale, originally called the *Magnitude Local*, became known as the *Richter scale*. An earthquake with magnitude 6 is believed to release the energy equivalent of about 1 million tons of TNT. That is the energy of a large nuclear weapon. Go up to magnitude 7 (roughly that of the Loma Prieta earthquake that shook San Francisco and the World Series in 1989), and the earthquake releases energy 10 to 30 times greater.

Why do I say a factor of 10 to 30? Which is it? The answer is that we don't really know. Magnitude is not exactly equivalent to energy. For some earthquakes, a magnitude difference of 1 unit will be a factor of 10, and for others it will be a factor of 30. It is easier to determine magnitude than it is to determine energy, and that's why magnitude is so widely used.

In table 7.2, I give the approximate magnitudes of some historical earthquakes in the United States. I rounded them off to the nearest integer.

Table 7.2 Earthquake Energy Release

Earthquake	Approximate magnitude	Megatons of TNT
	6	1
San Francisco area 1989	7	10 to 30
San Francisco 1906	8	100 to 1000
Alaska 1999	8	100 to 1000
Alaska 1964	9	1000 to 30,000
New Madrid, Missouri, 1811	9	1000 to 30,000

Waves transport energy from one location to another. The velocity of an earthquake wave depends on many things, including the nature of the rock or soil in which it is traveling (granite? limestone?) and its temperature (particularly for earthquakes traveling in deep rock).

An especially deadly effect occurs when a wave moves from high-velocity material into low-velocity material, such as from rock to soil. When a wave slows down, its wavelength (the spacing between adjacent crests) decreases.

---

[8] *Shallow* earthquakes are defined to be those less than 70 kilometers deep.

The energy is still there, but now squeezed into a shorter distance. That increases the amplitude of the shaking. Even though the energy carried by the wave is unchanged, the effect on buildings becomes much stronger. This is what happened in downtown Oakland in the 1989 Loma Prieta quake. The earthquake wave passed right through much of Oakland without causing great damage, until it reached the area near the freeway. This region had once been part of the bay and had been filled in. Such soft ground called *landfill* has a slow wave velocity, so the amplitude of the earthquake increased when it reached this ground. The most dangerous areas in an earthquake are regions of landfill. The Marina District in San Francisco is also landfill, and that is why it was so extensively damaged.

> **Personal story from the author:** My daughters were at the Berkeley WMCA when the 1989 Loma Prieta earthquake hit. One of them told me that she was thrown up against the wall by the earthquake. I said to her, "No, Betsy, that was an illusion. You weren't thrown against the wall. The wall came over and hit you."

LOCATING THE EPICENTER OF AN EARTHQUAKE

You already know that you can measure the distance to a lightning flash by counting the seconds and dividing by 5. The result is the distance to the lightning in miles. But here is another trick: As soon as you feel the ground shaking, and as you are ducking for cover, start counting seconds. When the bigger shaking finally arrives, take the number of seconds and *multiply* by 5. That will give you the distance to the epicenter (the place where the earthquake started) in miles.

Why does that work? To understand it, you should know that in rock, there are three important kinds of seismic waves. These are the *P wave*, the *S wave*, and the *L wave*.

THE P WAVE (PRIMARY, PRESSURE, PUSH)

P stands for "primary" because this wave arrives first. This is a longitudinal (compressional) wave, as is ordinary sound. That means that the shaking is back and forth in the same direction as the direction of propagation. So for example, if you see that the lamppost is shaking in the east–west direction, that means that the P wave is coming from either the east or the west. Some people like to use the memory trick that the P wave is a pressure wave—i.e., it is like sound because it is a compression and rarefaction, rather than a transverse motion. The P wave travels at about 6 km/s = 3.7 mi/s. That is a lot faster than the speed of sound in air (which is 300 m/s = 0.3 km/s).

THE S WAVE (SECONDARY, SHEAR)

S stands for "secondary" because this wave arrives second. This is a transverse wave. That means that the shaking is perpendicular to the direction of propagation. If the wave is traveling from the east, then this implies that the shaking is either north–south, up–down, or some angle in between. Some people like to use the memory trick that the S wave is a shear wave—i.e., it can only propagate in a stiff material that does not allow easy shear motion (sideways slipping). Liquids do not carry shear waves. We know there is a liquid core near the center of the Earth because shear waves do not go through it. The S wave travels at about 3.5 km/s = 2.2 mi/s.

THE L WAVE (LONG, LAST)

L stands for "long." These are waves that travel only on the surface of the Earth. Like water waves, they are a combination of compression and shear. They are created near the epicenter when the P and S waves reach the surface. They are called *long* because they tend to have the longest wavelength of the three kinds of seismic waves. It is the L wave that usually does the most damage, because the wave traveling on the surface often retains the biggest amplitude since it is not spreading out into three dimensions. The L wave travels at about 3.1 km/s = 2 mi/s. Some people like to use the memory trick that the L wave is the last to arrive. (Careful with memory tricks. The L wave is *not* a pure "longitudinal" wave!)

Figure 7.12 shows the shaking of the ground caused by a distant earthquake. Look at the wiggly line that crosses the image near the top. That is the first line, and it shows the shaking measured by the seismograph. Later lines are below this one. The little circle shows when the earthquake actually took place, at 9:27:23 UT. (*UT* stands for "universal time," and it is the time at Greenwich, UK.) The first shaking, due to the P wave, actually reaches the seismograph about 11 minutes later (shown as point 2). The S wave arrives about 10 minutes after that. There is no evidence for an L wave.

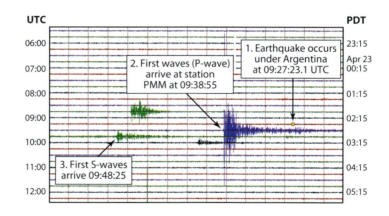

Figure 7.12 A seismogram recording an earthquake. Each horizontal line spans 15 minutes.

For a nice animation of the L wave (also known as the *Rayleigh wave*) go to http://www.kettering.edu/~drussell/Demos/waves/wavemotion.html, and look at the fourth animation on the page. If you look at the blue dots, you'll see that they move in circular-like patterns (they are actually ellipses). At the top of the wave, the circle moves backward—that is, opposite to the direction that the wave is moving! That's the opposite of what you saw in a water wave. And, if you go deep enough, the motion is forward. Very strange.

DISTANCE TO THE EPICENTER, AGAIN

Let's return now to the method of estimating the distance to the quake. As you are ducking under a table, start counting seconds from when you felt the first tremor—i.e., the P wave. (You can get very good at doing this if you live in California long enough.) When the S wave arrives, you know:

For every second, the epicenter is about 8.4 km = 5 mi away

This is the rule I mentioned earlier. Thus, if there is a 5-second gap between the waves, the epicenter was $5 \times 5 = 25$ mi away. You may even be able to estimate the direction from the P wave shaking—the back and forth motion is in the same direction as the source. If you are lucky enough to have an earthquake during a lecture, then maybe you can watch the professor doing this. (This equation is not true for travel through the deep Earth, where the velocities are faster.)

For those of you who like math, can you see how I got the value of 8.4? It is based on the P and S velocities. Hint: The distance a wave travels is equal to the velocity multiplied by the time. This calculation is optional (not required) and relegated to a footnote.[9]

There are small earthquake waves passing by all the time, just as there are small waves everywhere you look on the ocean surface. To see the waves recorded for the last few hours, look at the University of California Berkeley seismograph record at http://quake.geo.berkeley.edu/ftp/outgoing/userdata/quicklook/BKS.LHZ.current.gif. This is an extremely interesting link to keep on your computer if you live in California; it is something you can check any time you think you might have felt a quake. (Wait a little while before checking; the online plot is updated only every few minutes.) You'll see it there, even when it is not reported on the news. The quakes that occur every day in this region are also available on a map, at http://quake.wr.usgs.gov/recenteqs/.

The Liquid Core of the Earth

Halfway to the center of the Earth, about 2900 kilometers deep (1800 miles) is a very thick layer of liquid. (The distance to the center of the Earth is 6378 kilometers.) You could say that the entire Earth is "floating" on this liquid layer. The layer is mostly liquid iron, and the flow of this liquid creates the Earth's magnetic field (as discussed in chapter 6). The liquid is so hot, 1000 C, that if we didn't have the rock blanket between it and us, the heat radiation from the core would quickly burn us to a crisp.

---

[9] Suppose that an earthquake is at a distance d from where you are standing. The P wave moves with a velocity v_p. The time it takes the wave to reach you is $T_p = d/v_p$. The S wave moves with a velocity v_s. The time it takes to reach you is $T_s = d/v_s$. First you feel the P wave, and you start counting seconds. Then the S wave arrives. The time difference that you measured is $T = T_s - T_p$. According to our equations, this is $T = T_s - T_p = d/v_s - d/v_p = d(1/v_s - 1/v_p) = d(1/2.2 - 1/3.7) = d(0.184)$. Solving for d gives $d = T/0.184 = 5.4\ T$. We approximate this as $d = 5\ T$.

So much for curious facts—the real question for now is: How could we possibly know all this? The deepest we can drill is only a few miles. Nobody has ever gone to the core. Volcanoes don't come from regions that deep. How could we possibly know?

The interesting answer is that we know from watching signals from earthquakes. Thousands of these happen every year, and they are studied by earthquake detectors all around the Earth. The largest earthquakes send strong waves that travel down through the bulk of the Earth, and are detected on the opposite side.

An interesting aspect of the earthquakes is that *only the P waves pass through the core*. The S waves are all reflected! That is the wonderful clue. P waves are longitudinal pressure waves, and they travel through rock, air, or liquids. But S waves are transverse shear waves. Shear waves travel through solids, but they don't go through liquids or gases. That's because liquids and gases moving in the transverse direction can just slip past the rest of the liquid or gas; it doesn't exert much shear. So the fact that the P waves pass but the S waves don't gave one of the clues that there is a liquid core. Scientists also measured the speed at which the waves travel, and from this they can rule out gas and many kinds of liquids. They measure the density of the core from its contribution to the mass of the Earth, and they also see the magnetic field that the core creates. From all this, they were able to rule out every possible liquid except iron, although there could be liquid nickel mixed in with it.

We believe that the iron melted on the Earth when the Earth first formed. Most of the iron sank to the core, since it was denser than the other rock. The liquid iron is still in the core, and it hasn't yet completely cooled off. The very center of the core, called the *inner core*, is under great pressure. Even though it too is hot, the inner core has been compressed into a solid. If the pressure of all the weight of the Earth were removed, it would turn into a liquid, or possibly into a gas.

> **Discussion question:** How do we know that the liquid core has a solid center? (Or rather, how did scientists figure that out?) For the answer, see the footnote.[10]

Bullwhips

In a bullwhip,[11] the thickness of the whip is tapered toward the end. When the whip is snapped, a wave begins to travel down the whip to the end. Because the end is thin, the velocity of the wave increases near the end. The loud "crack" that you hear from the bullwhip is a sonic boom that occurs when the velocity of the wave exceeds the speed of sound.

---

[10] When a compressional wave hits the depth of the inner solid core, it breaks up into two waves. From the behavior of these waves, we know that one of them is a shear wave. So although the shear wave didn't travel through the outer core, shear waves are generated in the inner core. That means that the inner core must be made of a solid.

[11] If you don't know what a bullwhip is, then you might watch the opening scene in the movie *Indiana Jones and the Raiders of the Lost Ark* (1981) in which Indiana Jones uses a bullwhip to "whip" a gun out of the hand of a bad guy.

Note this difference: In earthquakes and tsunamis, the added danger comes because the wave enters a region in which it slows. In the bullwhip, the crack comes because the wave speeds up.

Waves Can Cancel (or Reinforce)

Suppose that you are very unlucky and are standing right in the middle of two earthquakes. One is to the north, and it takes you up, down, up, down, up, down, etc. The other earthquake arrives from the south, and it shakes you down, up, down, up, down, up, etc.—exactly the opposite of the shaking of the first wave. What will happen? Will the up from one be canceled by the down of the other?

The answer is yes! If you are unlucky enough to be between two such waves, then try to be lucky enough to be at just the right place for them to cancel. You are depending on the fact that the two waves arrive with exactly opposite shakings.

Of course, if you were standing at a different location, the waves would arrive at different times, and they might not cancel. Suppose the first wave gave you up, down, up, down . . . and so did the second wave. Then the ups would arrive together, as well as the downs, and you would be shaken twice as much.

This circumstance is not as unlikely as you might think. Even if there is only one earthquake, parts of the wave can bend, and so you can be hit by the same earthquake but from two different directions. If you are lucky, the two waves will cancel, but a short distance away, they can add. This phenomenon was see in the 1989 Loma Prieta earthquake that shook Berkeley, Oakland, and San Francisco. There were buildings where one side was shaken badly (causing that side to fall down) and the other side of the building was undamaged. This was probably due to the arrival of the wave from two directions at once, and the cancellation of the wave at the lucky end of the building.

If two waves are traveling together in the same direction but have different wavelengths (or frequencies), then the same kind of cancellation can happen. Take a look at the two different waves shown in figure 7.13, one red in color and one blue. The curves show the amount that the ground moves up and down (in centimeters) at different times due to the red earthquake and the blue

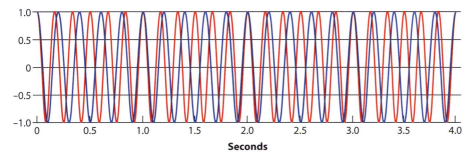

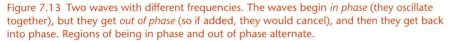

Figure 7.13 Two waves with different frequencies. The waves begin *in phase* (they oscillate together), but they get *out of phase* (so if added, they would cancel), and then they get back into phase. Regions of being in phase and out of phase alternate.

earthquake. Zero represents the original level. The blue earthquake shakes the ground upward (to 1 cm), and downward (to −1 cm). So does the red earthquake. So far, we have not considered the effects when added together.

First look at the blue wave. At zero seconds, it starts at the maximum value of 1. It oscillates down and up, and by the time it reaches 1 second, it has gone through 5 cycles. (Verify this. Try not to be distracted by the red wave.) We say that the frequency of the blue wave is 5 cycles per second = 5 Hz.

Now look at the red curve. In 1 second, it oscillates up and down 6 times. The frequency of the red wave is 6 Hz.

Suppose that you are shaken by both waves at the same time. At time zero, you are shaken in the upward direction by both the blue and red waves; their effects add, and you will move up by 1 + 1 = 2 cm. Look at what happens at 0.5 second. The red wave is pushing you up by 1 cm, and the blue wave is pushing you down by 1 cm, so the two effects cancel, and at that instant you will be at level ground.

Note that there are also times when both waves are pushing you down. There's no place when they are both exactly at their minima, but they come pretty close at about 0.1 s. At this time, both red and blue waves are down near −1 cm, so the sum effect will be to lower the ground by a total of 2 cm.

BEATS

If we add the red and the blue waves, point by point, we get the oscillation shown in figure 7.14.

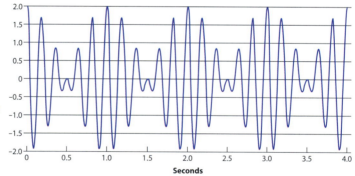

Figure 7.14 Beats from two waves, when they are added together, alternatively reinforce and then cancel.

The curve is taller because it ranges between 2 and −2. The shaking is not as regular, because of the alternating reinforcement and cancellation. Try counting cycles, and see what you get.

You probably got a frequency of 6 Hz (that's what I got). But some of the cycles are much bigger than others. Mathematically, we would not try to characterize this oscillation by a single frequency; it is a superposition (sum) of two frequencies.

If you felt this combination of waves under you, would say that the shaking was modulated, with the biggest shaking taking place every 1 second (at 0, 1, 2, 3, 4, . . .). These are called the *beats*. The beat frequency is given by this elegant equation:

$$f_{\text{BEATS}} = f_1 - f_2$$

where f_1 and f_2 are the two frequencies that make up the signal (i.e., they are the frequencies of the light and dark waves). If the number comes out negative, ignore the sign; that's because beats look the same if they are upside down.

To demonstrate beats, you can listen to two tuning forks with slightly different frequencies. The demonstration that we use at Berkeley is described at www.mip.berkeley.edu/physics/B+35+20.html. For a very nice computer demonstration of how water waves can interfere, look at the UCLA site at http://ephysics.physics.ucla.edu/physlets/einterference.htm. This site needs a fairly up-to-date browser with Java installed. You may already have that without knowing it, so it is worth trying. Move the red dots around, and then click on "calculate." Waves will come out of the two spots. These waves will add ("reinforce") at some locations, and cancel at others.

Music: Notes and Intervals

For sound, we refer to their frequency by their *pitch*. A high note, it turns out, has high frequency; a low note has low frequency (so no memory tricks are needed). A musical note usually consists of sound waves that have one dominant frequency. We name these notes by letters of the alphabet. The white keys of a piano are designated A, B, C, D, E, F, G, A, B . . . with the eight letters that repeat in cycles. The middle white key on a piano is known as *middle* C; the A above it has a frequency of 440 Hz (at least when the piano is tuned to the *Just scale*). The reason that the letters for the keys repeat (so, for example, there are 8 keys on a piano keyboard that correspond to the letter A) is because, to most people, two consecutive A keys sound similar. They are said to be an *octave* apart. In fact, when you go one octave higher (8 notes), the frequency is exactly doubled. The A above middle A has a frequency of 880 Hz. The next A has a frequency of 1760 Hz. Normal human hearing is quite good up to 10,000 Hz, and some people can hear tones as high as 15,000 to 20,000 Hz.

If two notes are played at the same time, and their frequency differs by just a little bit, then you will hear beats. Suppose that you have a tuning fork that you know has a frequency of 440 Hz. You play the A string on a guitar, and listen to it and the tuning fork together. If you hear 1 beat per second, then you know that the guitar is mistuned by 1 Hz; it is either 441 Hz or 439 Hz. You adjust the tension on the string until the frequency of the beats gets lower and lower. When there no longer are beats, the string is "in tune."

The interval between the A note and the higher E note is called a *fifth* because it consists of five notes: A, B, C, D, E. Likewise, middle C and the higher G make a fifth: C, D, E, F, G.

A violin is tuned so that the fifth has two frequencies with a ratio of exactly 1.5. So with the A tuned to 440 Hz, the E above it has a frequency of 660 Hz. This combination is also considered particularly pleasant, so many chords (combinations of notes played simultaneously, or in rapid sequence) contain this interval, as well as octaves.

Another pleasant interval is called the *third*. A and C make a third. The ratio of notes for a perfect third is 1.25 = 5/4. The pleasant reaction of the sound is believed to be related to the fact that these frequencies have ratios equal to those of small, whole numbers.

But that rule isn't exact. A particularly unpleasant interval is the *tritone*, in which the frequencies have the ratio of 7/5. The tritone is used in music to make the listener temporarily uncomfortable, and so it is considered dissonant. It is also used in ambulance sirens to make a sound that you can't easily ignore.

Vibrations and the Sense of Sound

As I said, the A on a piano (the one above middle C) vibrates 440 times per second. The A below that is 220 Hz (approximately, since piano tuning is "stretched" a bit). The next lower C is 110 Hz, and the one below that is 55 Hz. That's pretty slow. Find a piano, play that note, and try singing it. Can you sense that your vocal cords are vibrating only 55 times per second? You almost feel that you can count the vibrations, but you perceive the tone as a tone, not as a collection of vibrations.

Ordinary house electricity oscillates 60 times per second, from positive to negative and back. Sometimes this causes a buzz in electronics, or in a faulty lightbulb. The buzz is actually 120 times per second, since both the positive and the negative excursions of the current make sound. Do you remember hearing such a buzz? Can you hum the buzz, approximately? That is 120 Hz.

Remember the sound of the "light saber" in the *Star Wars* movies? That sound is 120 Hz. It sounds like a faulty fluorescent lightbulb. In fact, it was made by picking up the buzz from electrical wires.

Understanding sound is made much easier by the fact that waves with different frequencies all travel at the same velocity. That means that low tones and high tones will arrive together, no matter how far you are from the source. That doesn't happen with water waves; for them, different frequency waves travel at different velocities.

NOISE-CANCELING EARPHONES

Because sound is a wave, it can be canceled just like the shaking of an earthquake. So some smart people have made earphones that have a built-in microphone on the outside. This microphone picks up noise, reverses it, and then puts it into the earphone speakers. If done correctly, the reversed sound exactly cancels the noise, and the wearer hears "the sound of silence." On top of this quiet, the electronics can put music into the earphones. Since the music does not reach the outside microphone, it is not canceled.

I have a set of Bose noise-canceling earphones, and I use them mostly on airplane flights. The result is that I can listen to high-quality classical music, or to a typical airplane movie, and hear it as clearly as I would in a movie theater, without distracting noise.

There are even more expensive versions of noise-canceling earphones that are used by professional pilots and others who work in very noisy environments. It would be very nice to be able to cancel noise over a much larger region—e.g., in an entire room. However, that is probably not possible, at least not from a single small speaker. The reason is that the wavelength of sound (see next section) is typically 1 m. If the noise is not coming from the same location as the speaker, then although the sound could be cancelled in one location, it would probably be reinforced in a different location. That is not a problem for earphones, since the entire earphone is so small. Noise cancellation for an entire room might be possible if the walls of the room were made out of loudspeakers, or if they otherwise could be caused to vibrate to cancel any noise that might otherwise pass through.

Wavelength of Sound

Let's apply our wave equation to sound. Recall that the equation is

$$v = fL$$

Let's use this equation to figure out the wavelength of sound for middle C on a piano. That has $f = 256$ Hz. The velocity of sound in air is about 330 meters per second. So the wavelength is $L = v/f = 330/256 = 1.3$ meters.

Does that seem long to you? It is large compared to the typical size of a head, so the wave is moving your two eardrums together.

Suppose that we go up by three octaves. That means the frequency is doubled 3 times—i.e., increased by a factor of 8. Since the sound velocity v is the same in the wave equation, that means that the wavelength will be reduced by a factor of 8, from 1.3 meters, to 0.16 meters = 16 cm. That's smaller than the distance between your ears. So for this frequency, the eardrums on the opposite sides of your head may be vibrating opposite to each other.

Doppler Shift

When an object is approaching you, you'll hear a higher frequency than the one they emit. That's because each time a crest is emitted, or a trough, the object is closer to you than it was for the previous cycle. So you hear them closer together. Likewise, if the object is moving away, you'll hear a lower frequency. This effect is called the *Doppler shift*, and it is extremely important in radar and in cosmology since it allows us to detect the velocity of very distant objects.

When a car or truck goes by, listen to the sound. I'm not sure how to describe it in words—something like "shhhhh–ooooo." (Sorry. That's the best I could come up with. Suggestions for better ways to write this would be appreciated.) But the important thing to note is that the pitch of the sound drops just as the car goes by (that's the change from the shhhhh to the ooooo). That's the Doppler shift.

The Doppler shift is seen in all waves, not just sound. The Doppler shift in light means that an object moving away from you has a lower frequency. In astronomy, this is referred to as the *red shift*. It was from the red shift that Edwin Powell Hubble 1889–1953) discovered that the Universe is expanding away from us.

The Huygens Principle —Why Waves Bend Toward the Slow Side

Imagine that you are in an airplane and that you are watching waves on the ocean. Draw lines on the crests of the waves—i.e., on the highest points. Suppose the waves are moving to the right. The image will look like figure 7.15.

Figure 7.15 Crests of waves. Imagine, for example, that you are in an airplane looking down at water waves. Each line represents the high part (crest) of a wave. Halfway between the crests are the troughs, where the water level is lowest.

Look carefully at this image. The lines are the crests—i.e., they are the high points of the waves. The waves are all moving toward the right. That means that if we had a movie, each crest (each line) would move toward the right. Between the lines are the low points of the water waves, called the *troughs*. They move too.

Recall that the distance between the crests is called the *wavelength*. In figure 7.15, the wavelength is the separation of the lines.

Now imagine the waves moving to the right, but the ones near the top of the picture moving slower than the ones at the bottom. The lines would have to distort for this to be true. Figure 7.16 shows what this would look like.

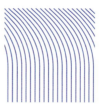

Figure 7.16 Waves bend when a region (the upper part in this figure) has a slower wave velocity. That could happen, for example, if the water is shallower in that part. Notice that the direction of the waves in the upper region is toward the top of the page; waves bend toward the slow wave side.

The waves near the top are moving to the right, but slower than the ones near the bottom. They will arrive at the right edge later. Notice how the slowing tends to bend their direction. But also notice that the waves near the top are becoming diagonal. The crests are no longer straight up and down. The direction of the wave is perpendicular to the crest. So the wave is no longer moving from left to right, but is also moving slightly upward. The direction has changed toward the side that has slower velocity.

The same thing would happen with a marching band (assuming that adjacent band members held hands), seen from above, if the field near the top was muddy and the members marched slower than the ones near the bottom. This is illustrated in figure 7.17.

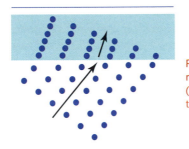

Figure 7.17 A marching band enters a slow region of the field. If they try to stay together (as waves do), then the direction bends toward the slower side.

Notice how the direction of the marching band (the arrows) changes once the band enters the muddy field—presuming that they try to stay lined up with each other. For waves, they do stay lined up, because each wave is generating the next one. That is a big abstraction, but many people find the preceding diagram helpful anyway. This method of explaining wave direction change is called the *Huygens principle*.

Spreading of Waves

Any wave, when passing through an opening, spreads. If not for such spreading, we usually would not hear people shout when they talk from behind a corner. Figure 7.18 on the left shows a wave (coming from the left) going through a hole and spreading out. On the right is a photo of waves passing through a gap in a floating log.

Figure 7.18 Waves passing through an opening will spread. Initially, all the waves are moving in the same direction (left to right, in the left image). But after the waves pass through the opening, their directions change, and some of them move upward and some downward. (Photo by Michael Leitch.)

There is a simple formula for this spreading of waves. The only thing you need to know is the wavelength of the wave L and the diameter of the opening

D. Then the size S of the wave when it goes a distance R will be given by the approximate formula

$$S = \frac{L}{D} R$$

You don't need to learn this formula, but it will be useful for us to calculate wave spreading. It becomes very important for light, because it limits the ability of telescopes to *resolve* objects. As you will see in the next chapter, this spreading is what prevents spy satellites from being able to read license plates.

This spreading equation is true for all kinds of waves, including sound waves and earthquake waves. The same equation works for them all. Let's take sound as an example. We showed earlier that for the tone of middle C, the wavelength is $L = 1.3$ meters. Suppose that this sound passes through a doorway that is 1.3 meters in diameter. If not for the spreading, the wave would be only 1.3 meters in diameter even after it went 10 meters past the door. But from the equation, there will be spreading. The amount of spreading will be large:

$$S = \frac{L}{D} R$$
$$= \frac{1.3}{1.3} 10$$
$$= 10 \text{ meters} = 33 \text{ feet}$$

The spreading is so great that you can hear a person on the other side of the door even if you can't see him. The spreading of light will be much less, because L will be much smaller. We'll talk about the spreading of light in the next chapter.

Chapter Review

Waves travel in many materials, such as water, air, rock, and steel. Even though the material only shakes and none of the molecules move very far, the wave moves and carries energy over long distances. Waves are longitudinal when the direction of vibration is along the direction of the wave. Longitudinal waves include sound waves and the P earthquake wave. Waves can also be transverse. This means that the shaking is perpendicular to the direction of motion. An example is a wave on a rope. Water waves are both transverse and longitudinal. Light waves travel in a vacuum or in a material such as glass. They consist of a shaking electric and magnetic field. Light waves are transverse. Electrons and other particles are actually waves too, but they are so short (*wave packets*) that this was not discovered until the twentieth century. The fact that particles are waves is called the *theory of quantum mechanics*.

If a wave repeats, then the number of repeats per second is called the *frequency*. For sound, frequency is the tone—i.e., high pitch or low pitch means high frequency or low frequency. For light, frequency is color. Blue is high frequency, and red is low frequency. The wavelength is the distance between crests of the wave.

The velocity of the wave depends on the material it is passing through. Sound travels about 1 mile in 5 seconds through air, but 1 mile per second in water, and even faster in rock and steel. Light travels at 1 foot every billionth of a second—i.e., 1 foot per typical computer cycle. That is 186,000 miles per second.

The speed of sound depends on the temperature of the air. In hot air, sound travels faster. If sound is traveling horizontally, but the air above or below has a different temperature, then the direction of the sound will bend toward the side that is slower. This phenomenon causes sound to get trapped beneath the ocean and is exploited by whales to send sounds thousands of miles. It also was used by the military for SOFAR (locating downed pilots) and for SOSUS (to locate submarines). If four different microphones can pick up the same sound, then the source of the sound can be located. The same principle is used using radio waves for GPS.

A sound channel in the atmosphere is created because of the high-altitude heating caused by the ozone layer. Project Mogul took advantage of the sound channel in the atmosphere. It was designed to detect Soviet nuclear tests. When the flying microphone disks crashed near Roswell, New Mexico, in 1947, stories began to spread about flying saucers.

When the ground is cool, sound bends downward, and that lets us hear distant sounds. When the ground is warm, sound bends upward, and we do not hear distant sounds.

The velocity of sound waves does not depend on their frequency or wavelength. If it did, it would be hard to understand speech from someone standing far away. But the velocity of water waves does depend on the frequency and wavelength. Long-wavelength water waves travel faster than short wavelength ones. Very long wavelength water waves, usually triggered by earthquakes, are called *tsunamis* or *tidal waves*.

Earthquakes begin when a fault ruptures at the epicenter, but they then travel as waves to distant places. The Richter scale gives a rough idea of the energy released. One point in the Richter scale is about a factor of 10 to 30 in energy released. The P wave is a compressional wave that travels fastest. Next comes the S wave (transverse), and last the L wave. The time between the P wave and the S wave can be used to tell the distance to the epicenter. The fact that S waves do not travel through the center of the Earth enables us to deduce that there is liquid there, probably (from the velocity we measure) liquid iron.

Waves can cancel, and that gives rise to beats (in music) and to strange effects, such as buildings that feel no shaking because of the fact that two canceling earthquake waves approached the building from different directions.

Discussion Topics

1. Does GPS require that the GPS receiver transmit any signals? How is that considered important for the soldier? But if GPS doesn't transmit signals, how can it be used as an emergency system on a cell phone to have someone locate you?

2. Should the government ever lie to its own population? Should the rules on this be different during peacetime and war? What about during a "cold war"?

3. Suppose that it is true (as the author believes) that the Rosswell incident had nothing to do with extraterrestrials. Will it be possible to convince the public that this is true? How would you do that? Is it important to do so? Are there other things that people believe in that are equally difficult to change their minds about?

Internet Research Topics

1. Look up "Roswell" and see what you find. (Be prepared to find lots of links.) What can you find out about the Roswell museum (where the photo of Muller with the alien was taken)? Can you find sites that explicitly dispute the account of the Roswell events that is discussed here? Which do you think is right?

2. What can you find out about SOSUS? Is it still active? What is it used for? Do Russian submarines still patrol the deep oceans? What fraction of the world's oceans did SOSUS cover?

3. The Doppler effect has numerous applications, in science, in engineering, and in many practical problems. Look up "Doppler" and see what you can find. What application surprises you the most?

Essay Questions

1. Everybody knows that water waves are waves, but in fact, there are many different kinds of waves in the world that are not obviously waves. Give examples of as many such phenomena as possible. For each, cite evidence that you might present to a skeptic to show that these phenomena really are waves.

2. Sound doesn't always travel in straight lines. The bending of its direction as it travels gives rise to many interesting phenomena. Give examples, and include details that would help a new student understand them.

3. Waves have a peculiar property: they can cancel. Give examples of cancellation for sound, light, water, and earthquakes. Explain why cancellation is so difficult to observe for light.

4. Most people have never heard of a *sound channel*. Give two examples of sound channels. Describe how the velocity of sound varies with location and how this affects the direction of the waves. Give examples of the practical use of these channels.

5. Discuss the noticeable and/or important effects that arise from the different velocities of waves as they travel though different parts of the same material.

6. Discuss the noticeable and/or important effects that arise from (1) the cancellation and reinforcing of waves or (2) the different

velocities of waves as they travel through different parts of the same material

7. Water waves and sound waves are both waves, despite the fact that they appear to be dissimilar to most people. Describe the way that they are both waves, properties they both share. What properties of sound make it clear that sound is really a wave?

8. Everyone knows that an earthquake is the shaking of the ground. Describe the ways in which it acts like a wave. How can the S and P waves be used to determine the location of the epicenter and the nature of the interior of the Earth?

9. Describe the properties of sound underneath the ocean surface. Describe how sound moves, and the implications of this for life (wild and human) under the water.

10. Describe how sound travels in air near the surface of the Earth. How does it depend on time of day and weather conditions? What interesting phenomena can an observant person notice?

11. According to the text, what were the "flying disks" that crashed near Roswell, New Mexico? Describe the formerly classified program that they were to be used for. Be sure to include all the relevant physics.

12. The SOFAR spheres used by the Navy had remarkable properties. Describe how they work, and how they were intended to be used. Be sure to include all the relevant physics.

Multiple-Choice Questions

1. Beats measure
   A. frequency
   B. the difference between two frequencies
   C. loudness
   D. the presence of noise

2. Waves tend to bend in to the side
   A. with slower wave velocity
   B. with higher wave velocity
   C. that is upward
   D. that is downward

3. The fastest earthquake wave is the
   A. L wave
   B. S wave
   C. P wave
   D. They all travel at the same speed.

4. The slowest earthquake wave is the
   A. L wave
   B. P wave
   C. S wave
   D. They all travel at the same speed.

5. An *octave* refers to two frequencies that differ by a factor of
A. 1.5
B. 2
C. 8
D. sqrt(2)

6. When two different waves pass through an opening of the same size, which one will spread more?
A. smaller wavelength
B. larger wavelength
C. higher frequency
D. lower frequency

7. The fastest sound wave is:
A. low frequency
B. middle frequency (voice)
C. high frequency
D. They all travel at the same speed.

8. You can measure the distance to the epicenter by measuring
A. the amplitude of the P wave
B. the amplitude of the S wave
C. the frequency of the L wave
D. the time between the P and S waves

9. Sound travels fastest in
A. air
B. water
C. rock
D. vacuum

10. When an opening gets smaller, a wave that passes through it
A. spreads more
B. spreads less
C. stays the same
D. changes its wavelength

11. The very center of the Earth is
A. pure rock
B. liquid rock
C. liquid iron
D. solid iron

12. Which kind of earthquake wave is purely longitudinal?
A. L wave
B. P wave
C. S wave
D. They are all longitudinal.

13. Which of the following statements about earthquakes is true?
A. S waves are fastest and cause the most destruction.
B. P waves are fastest, and L waves cause the most destruction.

C. L waves are slowest, and P waves cause the most destruction.

D. P waves are fastest and cause the most destruction.

14. The L wave is often the most damaging because
   A. it stays on the surface, so it doesn't spread out very much
   B. it moves slowest, so it has the greatest energy per mile
   C. the S and P waves carry too little total energy
   D. it arrives first, before people have a chance to take cover

15. An earthquake wave does its worst damage when it reaches an area that
   A. slows it down
   B. increases its frequency
   C. decreases its frequency
   D. adds additional energy

16. Landfill is dangerous because
   A. the frequency of an earthquake increases
   B. the wavelength of an earthquake increases
   C. they tend to focus earthquake energy
   D. the amplitude of the earthquake increases

17. When there is an atmospheric *inversion*, sound tends to
   A. become focused
   B. bend upward
   C. bend downward
   D. become absorbed

18. The ocean sound channel
   A. is very quiet
   B. is very noisy
   C. is radioactive
   D. focuses earthquakes

19. Sound tends to bend toward the side with
   A. colder air
   B. warmer air
   C. denser air
   D. less dense air

20. The same note is played on two pianos. Beats are heard once per second. From this, we deduce that
   A. at least one of the pianos is out of tune (the notes are at the wrong frequency)
   B. both pianos are out of tune
   C. the pianos have been accurately tuned
   D. the pianos will sound especially pleasant if played together

21. We know that the inner part of the Earth is liquid because
   A. no S waves move across it
   B. we can detect the flow of material from the emitted sound

C. at such great pressures, everything becomes liquid
D. neutrinos pass through it and show the pattern

22. A magnitude 9 earthquake, compared to a magnitude 8 earthquake
    A. has twice the energy
    B. has 10 to 30 times the energy
    C. has velocity 2× faster
    D. has velocity 10 to 30× faster

23. You feel the tremors of an earthquake. Ten seconds later, you feel another shaking. The distance to the epicenter is about
    A. 2 miles
    B. 5 miles
    C. 10 miles
    D. 50 miles

24. If we double the frequency of sound, the wavelength is
    A. doubled
    B. halved
    C. unchanged
    D. quadrupled

25. Beats demonstrate that
    A. sound is a wave
    B. sounds bends
    C. sound bounces
    D. sound spreads

26. The velocity of sound is approximately
    A. 1000 feet per second
    B. 1 mile per second
    C. 5 miles per second
    D. 186,282 miles per second

27. The speed of sound in air
    A. is always the same
    B. increases if you shout louder
    C. depends on frequency
    D. increases as air temperature increases

28. Sound waves are
    A. transverse
    B. compressional (longitudinal)
    C. a combination of transverse and compressional
    D. rotational

29. On a typical day, sound emitted near the ground tends to bend
    A. upward, toward the sky
    B. downward, toward the ground
    C. not at all; it goes straight

30. You are more likely to hear distant sounds when
    A. the air near the ground is warm and the air above it is cool
    B. the air near the ground is cool and the air above it is also cool

C. the air near the ground is warm and the air above it is cool

D. the air near the ground is cool and the air above it is warm

31. Because of evaporation, the air above the surface of a lake becomes cool. Sound in the air above the lake will tend to

A. bend upward away from the surface

B. bend downward toward the surface

C. go in a straight line parallel to the surface

D. go alternatively up and down

32. To have a sound channel, there must be

A. a minimum in the velocity of sound

B. a maximum in the velocity of sound

C. a decrease of the velocity with depth

D. an increase of the velocity with depth

33. SOFAR took advantage of

A. the sound channel in the ocean

B. the sound channel in the atmosphere

C. the magnetic field of the Earth

D. the uncertainty principle

34. Water waves are

A. pure transverse waves

B. pure longitudinal waves

C. both transverse and longitudinal

D. compressional

35. The sound channel in the ocean carries sound a long distance because

A. the ocean doesn't absorb sound at that level

B. whales listen to the sound and sing it again, increasing its volume

C. the pressure of the ocean at that depth makes sound louder

D. the sound doesn't spread out in the up or down directions

36. As you travel deeper into the ocean, the water temperature

A. decreases with depth

B. increases with depth

C. does not change with depth

D. first gets colder, and then gets warmer

37. A pianist plays two keys: middle C, and the C above middle C (i.e., an octave higher). The speed of sound for the higher frequency, compared to that for the lower frequency, is (careful, this may be a trick question)

A. the same

B. 2× faster

C. 2× slower

D. sqrt(2) faster

38. As you move to a higher altitude, the temperature of the air

A. first gets cooler, then warmer

B. stays constant, then gets cooler

C. first gets warmer, then cooler

39. The atmospheric sound channel would not exist, if not for
   A. thunderstorms
   B. carbon dioxide
   C. ultraviolet light
   D. infrared light

40. The ozone layer is created by
   A. carbon dioxide
   B. lightning
   C. sunlight
   D. chlorofluorocarbons

41. SOSUS refers to
   A. a method of rescuing pilots designed during World War II
   B. a project to detect nuclear explosions
   C. a system for detecting submarines
   D. a system using many artificial Earth satellites

42. Which of the following statements was true about Project Mogul?
   A. It was concerned with the atmosphere.
   B. It resulted in the first nuclear bomb.
   C. It led to the discovery of nuclear fission.
   D. It involved the invention of integrated circuits.

43. According to this text, the flying disks that crashed near Roswell were
   A. advanced U.S. space vehicles
   B. alien flying saucers
   C. microphones
   D. U-2 airplanes

44. Very long wavelength water waves
   A. travel slower than short ones
   B. travel faster than short ones
   C. travel the same speed as short ones
   D. travel faster if they have high amplitude and slower if they have low amplitude

45. Whales and fiber optics both make use of what principle?
   A. Huygens
   B. Heisenberg
   C. Moore
   D. Curie

46. When an earthquake at sea starts a tsunami or tidal wave, the initial height is relatively small. What accounts for the towering wave that breaks near the shore?
   A. The wave builds up energy as it moves.
   B. The wavelength increases.
   C. The depth increases.
   D. The wave moves faster.
   E. The wave moves slower.
   F. Its appetite for destruction is whetted.

47. A water wave has a wavelength of 10 meters and a frequency of 2 cycles/sec. Its velocity is
A. 5 meters per second
B. 10 meters per second
C. 20 meters per second
D. 50 meters per second

48. The sound from a passing car sounds like a high pitch, but as it passes, it gets lower. That's an example of
A. the Huygens principle
B. wave cancellation
C. the Ewing principle
D. the Doppler shift

49. Thunderclouds tend to rise until
A. they rain out all their water
B. they reach air that is colder than they are
C. they hit the carbon-dioxide layer
D. they reach air that is warmer than they are

50. The atmospheric sound channel is created because of
A. carbon dioxide
B. conduction
C. convection
D. ultraviolet light

51. In the daytime, sound tends to
A. bend upward
B. bend downward
C. go straight
D. create a mirage

52. Most damage from an earthquake usually comes from the
A. S wave
B. P wave
C. L wave
D. M wave

53. For sound waves, low-frequency waves travel
A. faster than high frequency
B. slower than high frequency
C. the same speed as high frequency
D. The speed depends on the amplitude of the wave, not on the frequency.

54. The piano note A above middle C is 440 Hz. The next higher A has a frequency
A. 660
B. 880
C. 550
D. 1320

CHAPTER

Light

High-Tech Light

Light is full of surprises. Even in the mid- to late twentieth century, many of the following applications were not anticipated:

- **Fiber optics.** We once thought that we would send all our information by satellite, relaying microwave signals to carry millions of telephone conversations at the same time. But we found a better way: send the signals with light, using fiber optics buried under the sea floor.

    Why is light so much better than microwaves? How do fiber optics work? Why are they sometimes called *light pipes*?

- **Multispectra.** With our eyes, we see what appears to be an infinite number of different colors. But multispectral cameras can see many more. They see colors that can indicate the health of the vineyard or the status of moisture in the soil of China.

    What do they see that we don't?

    (**Hint:** They see the same light we see. They just turn it into a larger number of colors. But what does that mean? Don't we already see an infinite number of colors?)

- **Spy satellites.** Suspicious nuclear facilities can be photographed in other nations using spy satellites. The satellites must fly low, in order to get good photos. But that means that they are over the object of interest for less than a minute (since they are moving 5 miles per second).

    Why can't they fly high, get a view that lasts, and just use a powerful telescope? What feature of light makes low-Earth orbit the preferable one?

    (**Hint:** It is related to the fact that light is a wave.)

- **Laser-induced nuclear fusion.** The most powerful way of delivering energy to a small spot is with light. Laser beams directed at small pellets of tritium and deuterium can cause the isotopes to combine, a process called *nuclear fusion*, releasing energy.

What properties of *light* makes it the best way to heat pellets to the extremely high temperatures required?

- **Computer screens.** If you look at a white computer (or TV) screen up close, with a magnifying glass, you won't see any white at all. You'll see red, green, and blue spots. Try it.

  Why does it look white from a distance? What is color, really? Where do the colorful sparkles of diamonds come from? Or the color of rainbows (which comes from water droplets)? Why do music CDs and movie DVDs show all the colors of the rainbows in sunlight?

What Is Light?

As the preceding examples suggest, light is a puzzling phenomenon, with properties that seem to make no sense.[1] Yet if you understand them, they have important uses and applications that you otherwise would never guess.

The key to understanding the behavior and properties of light is to recognize that light is actually a wave. But light doesn't *seem* to be a wave. And if light is a wave, what is waving? For water waves, it is the water that is waving. For earthquakes, it is the Earth that is quaking. For sound, the air is shaking. But what is moving when a light wave waves?

Here's the answer, and don't worry if it sounds abstract: Light is a wave consisting of vibrating electric and magnetic fields. Since both shake, we often call light an *electromagnetic wave*. Previously in this book, I've mentioned that "the vacuum" can vibrate. The electric and magnetic fields are part of this vacuum. They are the aspects of the vacuum that wave when light is present.

If you shake the air, a wave of sound is emitted. If you shake the ground (by the sudden release of a fault), an earthquake is emitted. If you shake some water, a water wave is emitted. If you shake an electron, an electromagnetic wave is emitted. A vibrating electric and magnetic field moves together away from the electron, carrying energy. When this electromagnetic field hits an electron. it puts a force on it, in the same way that sound exerts a force on your eardrum, or an earthquake exerts a force on a building.

If light is a wave, why doesn't it look and feel like a wave? The answer: Because its frequency is extremely high, and its wavelength is very short. The average wavelength for visible light is about 0.5 microns $= 0.5 \times 10^{-6}$ m. Recall that the diameter of a human hair is 25 to 100 microns. So the crests of light waves are so close that you can't easily detect the individual crests. From this wavelength, we can calculate the frequency of light using the equation for waves:

$$v = fL$$

---

[1] Even Isaac Newton (1642–1727), the inventor of physics, came up with an incorrect theory for light, although he got almost everything else in physics right.

We now set v = speed of light = 3×10^8 m/s, equivalent to 1 feet per typical computer cycle.[2] Then $f = v/L = (3 \times 10^8)/(0.5 \times 10^{-6}) = 6 \times 10^{14}$ cycles per second = 6×10^{14} Hertz. That's very fast. Every nanosecond (ns), or every computer cycle, such light vibrates nearly a million times. No wonder we don't usually notice that it is a wave.

The high frequency of light is the key feature that has made light the main system for carrying information. Most of the Internet and most of our telephone systems send their signals in the form of light, guided by fiber optic light pipes. To understand why, we must delve a little into the theory of information.

Information Theory

Computers keep all their information stored in terms of the numbers 0 and 1. Each stored number is called a *bit* of information. If you want to send the letter *A*, then you combine 8 of these bits into a code that represents the letter *A*. The most widely used code is called ASCII.[3] In this code, the binary bits for the letter *A* are 00001010. The letter *B* is 00001011. Note that only one bit is different. *C* is 00001100. (No, you don't have to learn these.) Everything that the computer does is translated into strings of 0 and 1 for computational purposes. (Yes, you should know that.)

Modern communication works the same way. If you want to send a telephone conversation across the United States, the electronics will first encode it into a long string of 0's and 1's, and then send these. The more such signals you can send per second, the more information you can send. To send a signal using light, one way is to turn it on and off, with "on" representing 1, and "off" representing 0. The number of bits that you can send every second is called the *baud rate*, R.

The *theory of information* was developed by Claude Shannon in the middle 1940s, and he discovered the most important results. Perhaps the most significant of these is that you cannot really send signals faster than the frequency of the wave that you are using. This important equation says that the rate at which you can send information is limited by the frequency of the signal[4]:

$$R = f$$

This is Shannon's basic discovery. In this equation, f is the frequency of the signal (either light, radio waves, or microwaves), and R is in bits per second (baud). His equation says that the bits per second that can be sent is approximately equal to the frequency of the wave being used.

---

[2] Recall that this also works out to be equal to 30 cm/ns (1 ns = 10^{-9} s). Since 30 cm is about 1 foot, and a nanosecond is about 1 computer cycle (for a 1-GHz computer), this means that the speed of light is about 1 foot per computer cycle. If you have a faster computer—e.g., 2 GHz—then light travels only 15 cm—i.e., 6 inches per 1 cycle.

[3] ASCII stands for American Standard Code for Information Interchange.

[4] This footnote is for those who are interested in some of the details. What Shannon really proved was that the information rate was given by $R = B \log_2(S/N)$. In this equation, $B = f_H - f_L$ is called the *bandwidth*. (f_H is the highest frequency you can use, and f_L is the lowest; $\log_2$ means logarithm to base 2.) This means that B is less than the higher frequency f_H. S/N is the signal-to-noise level. The $\log_2$ factor is a bit greater than 1, so that makes R a little bigger. If $f_H \gg f_L$, then $R \approx f_H$.

Remember: The frequency tells you the number of bits per second you can send.

Shannon's equation has a simple interpretation: You can't make a wave vary faster than its own frequency f. You can turn it on and off, but not faster than f times per second.[5] So the maximum number of bits you can send every second is the frequency. (In principle, you could change the signal faster, but then you would be giving it a higher frequency, and that might be too high to travel on the wire efficiently.)

If you want to send a largest number of bits per second, you need the highest frequency signal you can conveniently use. Phone lines typically use a frequency of 1 MHz or lower. Television and radio use typically use GHz frequencies, capable of sending billions of bits per second. Light has a frequency up to 6×10^{14} Hz, a factor of 600,000 higher. You could send a beam of light from one person to another, and turn it on and off 10^{14} times per second. Each pulse would consist of six oscillations of light and would carry 1 bit. You would be sending 10^{14} bits per second. That's why light has become the primary way to send large amounts of information quickly. The Internet is based on light.

You can't send light through metal wires, so scientists had to invent a special wire for light, called a *light pipe*.

Light Pipes and Fiber Optics

The original light pipes were just a long thin cylinders of glass. As these became thinner and longer, they were called *fibers*. The light would travel through the glass; if it hit the edges, it bounced back in. If there were any scratches on the surface, some light would be lost. That led to the invention of the *graded index fiber*, based on the same principle as the sound channel.

In a graded index fiber, different glasses are used, with the highest index (slowest light) near the central axis, and the lowest index (fastest light) near the outer surface. Any light that deviated from going along the axis would be focused back, just as in the sound channel. The fibers could be very small in diameter, typically 1 millimeter, but very long in length, many kilometers (figure 8.1).

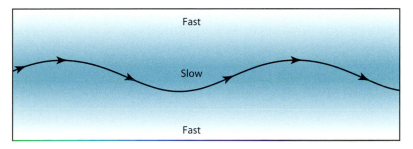

Figure 8.1 Fiber optic light pipe, with graded index that creates a light channel. An actual fiber may have a diameter of 1 mm and a length of 10 km.

[5] In Shannon's theorem, the bandwidth is actually the *channel capacity*—i.e., the maximum rate that you can vary the signal. (See the previous footnote.) If you make a signal vary faster than f, you are actually changing f.

Color

Even though light vibrates very fast, our eyes can still distinguish light of different frequencies. The common word for light frequency is *color*. Red light has a wavelength of about 0.65 micron = 650 nanometers (nm), and that works out to a frequency of 4.6×10^{14} Hz. Blue light has a wavelength of about 0.45 micron = 450 nm, and that means that its frequency is about 7×10^{14} Hz. The color of light of different wavelength is illustrated in figure 8.2.

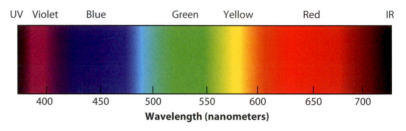

Figure 8.2 Color and wavelength. Notice that red has a longer wavelength than blue. One nanometer is 10^{-9} m = 10^{-3} micron, so 500 nm is 0.5 micron.

The colors indicated in figure 8.2 resemble the colors of the rainbow, and that's because they are. *White light* consists of a mixture of all these colors. When they all arrive into our eye, our brain calls the color "white." White light does not consist of a single frequency, but of a mixture of light of different frequencies. Some people would put it this way: "White is not a pure color." That statement is true if by "pure" you mean that it contains only a single frequency of vibration.

The rainbow is created when sunlight passes through raindrops. This process of bending is called *refraction*, and we'll discuss it further in a moment. The bending is different for the different colors, so they emerge from the raindrop in different directions, and that's what makes the rainbow.

Notice in figure 8.2 that beyond red, at long wavelengths, is a region marked *IR* That stands for "infrared." It is light, but just not visible to the human eye. Off to the left (very short wavelengths) is *UV*, which stands for "ultraviolet." That is also invisible to our eyes. We'll talk more about these invisible colors in the next chapter. UV is, in fact, the color that is most responsible for the ozone layer in the atmosphere. It is the color that produces the worst sunburns.

In chapter 4, I said that x-rays and gamma rays are also light. They have very short wavelength, way off to the left of the chart in figure 8.2. The wavelength of a 100 keV x-ray is about 0.01 nm and of a 1-MeV gamma ray is about 10^{-3} nm (10^{-12} m). Their frequencies are therefore much higher than that of visible light.

Radio waves and microwaves are also light. They are far off to the right in the figure 8.2 color chart, on the long-wavelength side. A typical wavelength for a TV broadcast signal is 3 meters. That means that its frequency is $c/f = 3 \times 10^8/3 = 10^8$ Hz = 100 MHz. Figure 8.2 has its lower axis labeled in nanometers, and 3 meters is 3 billion nm, far off the scale shown.

Color Sensors in the Human Eye

Look at figure 8.2 again. It contains all the colors of the rainbow. Do you notice that some "colors" are missing? Where is magenta, or cyan? And, of course, white is missing. It turns out that none of these are pure colors, but are mixtures of colors, in the same way that you get a mixture of notes if you hit several keys of the piano at the same time.

Many animals do not sense color. They can see only that something is brighter or dimmer. This is sometimes described by saying that they see "in black and white"—but they see gray too. Humans can sense color, but even our ability is very limited. Our eyes have four kinds of sensors. The ones we call *rods* are also found in most animals. Rods sense brightness, but not color. The ones we call *cones* come in three varieties: red, green, and blue.

The range of sensitivity for each of the cones is shown in figure 8.3.

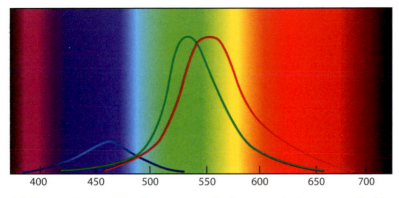

Figure 8.3 The sensitivities of the blue, green, and red cones in the human eye. The blue peaks near 450 nm, the green near 525, and the red near 550.

Notice that the red cone has its maximum sensitivity to green light, not to red! In fact, the red cone sensitivity is remarkably similar to that of the green cone. So how does the eye distinguish red from green? Think about it for a moment. Can you figure it out?

The answer is that the eye gets signals from both the green and the red sensors. If the signal from the green cone is stronger, the brain calls the signal green. Look again at the diagram. To call a signal red, the signal from the red sensor has to be twice as strong as the signal from the green. Only in the red region is the red line twice as high as the green line.

Notice also that green light is detected by all three cones—most strongly by the green cones, a bit weaker by the red cones, and weakest of all by the blue cones. When the brain gets this combination of signals, it sees the color as green. If it receives a strong red, a weaker green, and no blue at all, then it tells you the color is yellow. (Can you see that in the diagram?)

Suppose that all three sensors receive strong signals. Then our eye interprets that as white. Figure 8.4 shows the intensity of different colors emitted by the Sun. When the three sensors of the eye see this, it appears white.

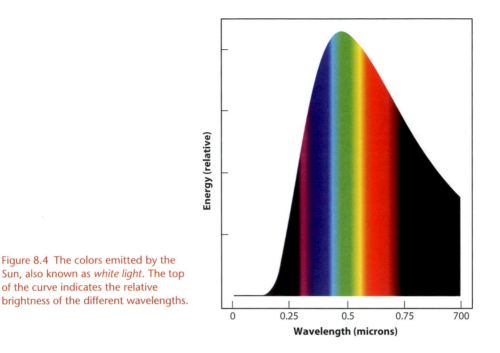

Figure 8.4 The colors emitted by the Sun, also known as *white light*. The top of the curve indicates the relative brightness of the different wavelengths.

Look at the solar colors in the plot. You'll see that there is quite of bit of light in the invisible IR portion. In fact, about half the energy reaching the surface of the Earth is in this IR. That will turn out to be important for issues involved in global warming, so you should remember it.

The combination of colors in sunlight is not the only combination of color that appears white to the human eye. Any combination of colors that stimulates the red, green, and blue cones in the same way will give the sensation of white. Since the eye has only three color sensors, it is easily fooled.

False White

The easiest way to fool the eye is by using color dots. Find a white part of a computer screen, and look at it closely. Use a magnifying glass, or if you are nearsighted take off your glasses and look very close. You'll see that, up close, the white isn't white at all, but consists of little red, green, and blue spots. Unlike natural sunlight, the computer screen has no pure yellow, no orange, no blue-green, and yet your eye, with its limited ability to discern the different components, cannot distinguish it from true white. Your eye was fooled by a computer system that adjusts the three colors to stimulate the blue, green, and red cones in just the right amount to make your brain think it is seeing white light. If this is done well, your eye can't distinguish between this "false" white and pure white sunlight. Of course, a scientific instrument that measures the intensity at many frequencies can easily tell the difference.

Color Blindness

Are you color blind? About 5% of males and 0.5% of females do not have both red and green sensors, or have less sensitivity in one of these sensors. (In a class of 500 students, that means 25 males and 2.5 females, on average.) They are called *color blind* even though they can see many colors. But they often cannot distinguish red from green. That's because with just two sensors (say, for example, the blue and green sensors), light in the green–red region is not detected by the blue sensor. With only one signal, not a ratio, the brain can't guess what frequency the light is.

Have there been any great painters who are known to be color blind? Would you want to have a color-blind person design the colors for your house? Or choose a shirt for you to wear?

We Are All Color Blind—Multispectral Cameras

Suppose that you had four different color cones in your eyes, instead of just three. Then an amazing thing would happen: things that used to look like they were the same color would now be different—just as red and green are different to the non-color-blind person but are indistinguishable to someone who is color blind. The white of a piece of paper illuminated by sunlight would appear different from the white of a computer screen. That's because the three colors on the computer screen can fool only three cones. To fool four cones, you would need four colors of dots on the screen. In fact, some people have been found with four different color cone types.

Cameras are built these days that do exactly this. They can have ten, a hundred, or even a thousand different color sensitivities. Two fields that look like they are identical shades of green to us, look different to the cameras. They can use this multiple-color ability to detect things that we miss, such as disease in crops, or to identify different kinds of rocks. These systems are called *multispectral cameras*. Multispectral cameras flown in satellites are available, for a price, to photograph and analyze your farm. They are being used by wineries in California to detect the effects of sharpshooter beetles on the vineyards. They will be an even more important technology in the future as we learn to notice patterns and identify the meaning of many multispectral colors. Right now, we are not very good at this because so much of our color experience is based on just three colors.

So, in a sense we are all color blind. People called color blind are really just a little more color blind than the rest of us. But if we had four cones—blue, green, yellow, and red—what would the world look like?

Perception: A Discussion Topic with No Answer

Does a color-blind person perceive red as red, and green as red, or does he perceive red as green, and green as green? Think about it.

Does the question make any sense? Can it be answered? Is it a question in the realm of physics, or isn't it?

Many scientists would say that the question is meaningless. We can measure which regions of the brain are stimulated by different colors. But that doesn't answer the question about the color-blind person. Does he see red or green?

Many scientists are fond of saying that if there is no way to answer a question, even in principle, then the question is meaningless. Do you agree?

A student recently drew my attention to a special case: a woman who is color blind in only one eye! This is described in the book *Psychology*, by Gleitman, Fridlund, and Reisberg (W.W. Norton, 2007). In one eye, she could see both red and green. To her other eye, the color in the color-blind eye matched the green color in her non-color-blind eye.

Does that answer the question?

She saw red in her color-blind eye as looking the same as green in her normal eye. But maybe her normal eye saw red as green and green as red, and so she actually saw both as red. . . .

I consider this to be a question that is not answerable in the realm of science.

Printed Color

Printed colors work in a similar way to computer screens, but because the inks are usually laid on top of each other, the colors used have to be different. That's because the colors in the ink absorb rather than emit, so they are taking the light away from the reflected light. The colors often used for printing are cyan, magenta, and yellow. They are not pure frequencies, but mixtures. Cyan dye absorbs all light except for the mixture of frequencies we call cyan; that's why cyan is reflected off the paper. For black, you would have to have all three colors laid down on white paper, and it is cheaper and easier to just use a fourth ink colored black. That is the fourth color in four-color printing. Most magazines use four-color processes. Look at some color magazine photos with a magnifying glass and you'll see little dots with these colors.

But people often don't look at magazines in white light. Fluorescent lights typically have more blue in them than sunlight, and that affects the colors perceived by the eye. Some printing processes use more than the standard four colors in an attempt to achieve the right effect under different lighting conditions.

Colors of an Oil Slick

The fact that light is a wave is demonstrated by the fact that two light waves can cancel and reinforce. Such cancellation is the cause of the colors seen on a thin film of a soap bubble. It is also the cause of colors seen on an oil slick. Mix a little bit of water and motor oil on the top of a black plate, and you'll see subtle colors. I did this, and enhanced the colors a little for the image in figure 8.5.

We now understand the oil slick colors in the following way: Since oil is less dense than water, a few drops of oil will float on the surface. As the oil spreads out (motor oil does this better than cooking oil), it can make a very thin layer. In fact, it is common for that layer to be only a few microns thick, comparable to the wavelength of light. That fact enables us to notice the cancellation of

Figure 8.5 Colors seen on an oil slick. Different colors come from the varying thickness of the oil layer.

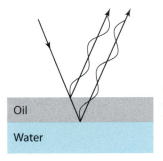

Oil

Water

Figure 8.6 Light reflecting off water that has a layer of oil on top. (Light also reflects from the bottom of the water level; that is not shown.)

light waves. Look at figure 8.6, which represents the cross section (viewed from the side) of a greatly magnified oil slick.

The light enters on the left side. I've shown its direction by the line; I haven't shown the individual oscillations. Some of it reflects off the top of the oil, and an equal amount reflects off the top of the water. (Some light continues to penetrate the water, but we don't show that.) There are two reflected waves. The two waves will overlap. For the reflected waves, I've drawn the oscillations. Note that I've drawn them so that they approximately cancel. When this happens, the total wave reflected is zero.

If the wavelength of the incoming light was different, but the thickness of the oil layer was the same, then the two waves might reinforce rather than cancel. In fact, white light is full of waves of different colors. Some cancel, some reinforce. The ones that reinforce are the ones we see in the reflected light. Since the thickness of the oil slick is different at different locations, the colors that reinforce are different at different locations, and that is what gives rise to the large variation in colors from an oil slick.

Most people think an oil slick is ugly. That's because they associate oil slicks with pollution, such as spilled oil, or decaying vegetable matter in a lake. But when I see an oil slick, I think that Newton, if he had been clever enough, would have recognized from the colors that light must be a wave.

Similar layers are coated on camera lenses to minimize reflected light. Light that bounces off the outer surface is canceled by light that is bouncing off the bottom of the thin coating. Such lenses are more expensive, but they produce better photographs.

Images

Of course, the most remarkable feature of light is that we use it to see. The eye manages to give us amazing detail about objects, even ones that are far away. This is enabled by the fact that we can create an "image" of an object in our eye. The concept of *image* is also critical for understanding holograms, mirrors, cameras, microscopes, and telescopes. We'll begin with the simplest device that can create an image, the pinhole camera.

Pinhole Camera

A pinhole camera is the simplest kind of camera anyone has ever used.[6] It makes use of the fact that light travels in straight lines, more or less. It does that only because its wavelength is very short, so that a packet of light behaves very similar to the way a particle would behave.

A pinhole camera consists of a box with a small hole on one side and film on the back, as shown in figure 8.7. Light from the object passes through that hole and lands on the back. In the diagram, you see that light from the head lands low on the film, and light from the feet lands high. If you were to look at the back you would see an "image" of the object.

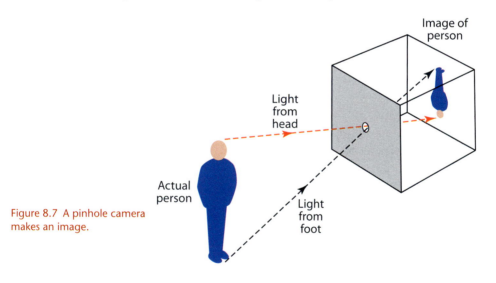

Figure 8.7 A pinhole camera makes an image.

Notice that the image is inverted, compared to the object. Of course, when the film is developed, nobody notices that the image is upside down; they just turn the picture over!

You can make a filmless pinhole camera to play with, using any box, a small hole, and a piece of paper for the back. It works best with waxed paper (available in supermarkets—it was a predecessor to plastic wrap) rather than ordinary paper, since waxed paper transmits the light better, so you can see the

---

[6] Some people claim that the pinhole camera was invented in Pinole, California, and that's where the name came from, but I don't believe it.

image through it. (Paper soaked in vegetable oil also works well.) If the pinhole is too large, the image gets blurry. That's because light from several places on the object can reach the same place on the paper. If the pinhole is too small, then the image is very dim and hard to see.

A camera works in exactly the same way, except that the pinhole is replaced by a lens, and the paper replaced by something that records the light—film. We'll discuss lenses in a moment; their main advantage is that they let more light in than does the pinhole.

Brief History of Photography

If the image makes a permanent change on the material placed on the back of the camera, then we have created a photograph. This was originally done with chemicals on a plate placed in the back of the camera. The first known photograph was taken by Joseph Niepce in 1827.

Niepce used a chemical called *bitumen* that hardened when light hit it. If he then washed away all the soft material, he was left with a thin sculpture that resembled the object (e.g., a person's face) that was preserved. At the time, even this crude image was hailed as a miracle of technology, since the only other way to capture someone's face was by hiring an artist, or by tracing a shadow. Most people thought that the captured image was remarkably realistic, even though it was very crude by today's standards.

Louis Jacques Daguerre improved on this with metal plates. His artistic ability was also extraordinary, and many of his daguerreotypes have become famous. The plates were improved by William Talbot and George Eastman, who used silver halide. The silver halide breaks up when exposed to light, and the silver particles are released. The process of "developing" the plate consists of removing the unexposed silver halide and then reversing the exposure. (The silver particles give a black appearance, and yet they were exposed to light. So a second image must be made to reverse the process and give a realistic black-gray-white image.) Their cameras made a sound that they described as "ko-dak," and the Kodak company was named after that.

In the twentieth century, the photographic plates were replaced by flexible photographic film that was coated with the same chemicals. Photographic film (or just film for short) became one of the major uses for silver! I wouldn't invest in silver today, since digital cameras are making silver-based films obsolete.

More on Pinhole Cameras

You can actually build a pinhole camera and take photographs using ordinary photographic film. There are clubs that do this. Not surprisingly, there is even a Web site that sells pinhole cameras.[7]

Of course, you could make one yourself very easily, since nothing special is needed. I described the basic method earlier. For a pinhole camera that uses film, the hardest part is keeping stray light away from the film so that the only light that reaches it is through the pinhole. The pinhole doesn't let in much light, so long exposures are needed.

[7] www.pinholecamera.com.

If you use a larger pinhole, then the exposure can be shortened. But a larger pinhole means that the light from any point on the object gets spread over a region of the film that is as large as the pinhole (or even slightly larger). This makes the image blurry. So when taking a pinhole picture, you have to decide how long you can hold the camera (a shaky camera also blurs the picture) and how big a hole you can use.

In figure 8.8, I show a class pinhole camera demonstration with several different pinhole sizes. You will notice that when the pinhole size is made larger, the image stays the same size, and gets brighter, but more blurred.

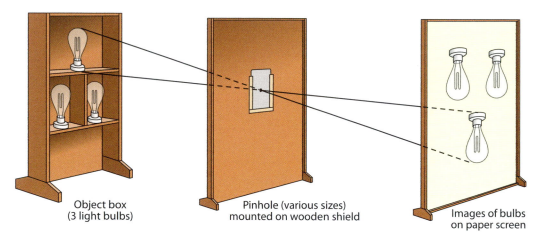

Object box
(3 light bulbs)

Pinhole (various sizes)
mounted on wooden shield

Images of bulbs
on paper screen

Figure 8.8 Another example of a pinhole camera.

Human Sight

Your eye acts very much like a pinhole camera. The "pupil" acts like the pinhole, and the retina (with its rods and cones) acts like the film. Each sensor sends a signal to the brain, which then interprets the image.

In fact, the eye uses a lens just behind the pupil to focus the light better than the pupil would by itself. We'll discuss lenses soon.

But here is the key thing I want you to recognize: the eye is a device for measuring the brightness and color of light for different directions. That is a simple statement, but ponder it. You eye tells you, for every direction in its field of view, the color and brightness of the light coming from that direction. This is all that an eye measures.

When looking at a three-color computer screen, the eye is fooled into thinking that equal mixtures of red, green, and blue make white. But there is an even more marvelous way to fool the eye. It is called a *mirror*. It fools the eye into thinking an object exists, where it doesn't.

We'll talk a lot more about the human eye later in this chapter, after discussing lenses.

Mirrors

Mirrors are wonderful miracles that we pay little attention to since they are so common. A *mirror* is a surface that is very good at reflecting light. When our eye sees light coming into it, it has no way of knowing if the light is coming directly from an object or is being bounced off a mirror. The spot behind the mirror from which the light appears to be coming is also called an *image*, even though it is quite different from the pinhole camera image. Here is the key difference: there is actually no light present at the image location. You can see that in figure 8.9.

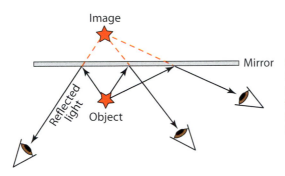

Figure 8.9 Reflections from a mirror. Light from the object bounces off the mirror. No matter where the observer is, the light appears to be coming from a location behind the mirror, at the location of the "image."

Study figure 8.9. Light from the object bounces off the mirror and into the eyes of the three people shown. Notice that they all see the light exactly the same way that they would see it if no mirror were present, but instead the light appears to be coming from the location marked "image." That's what makes the image from a mirror so compelling. Even if you move your head around to different locations, the position of the image doesn't change its location. The image behaves just like a real object.

If you are mathematically inclined, you might like to try to prove that all bouncing light from an emitting point will appear to be coming from a single point on the other side. You will have to assume that the angle of bounce is equal to the angle of incidence on the mirror.

MIRRORS AND MAGIC

We take mirrors for granted because high-quality mirrors are so abundant in modern society. And yet we can still be fooled by them. They are favorites of magicians and of those who create illusions at amusement parks. Suppose that you want to make it look like a ghost is sitting next to a person. (This is done in the Haunted Mansion at Disneyland.) The trick is to use a *half-silvered* mirror, a mirror that reflects half of the incident light and transmits half.

If you look in such a mirror, you'll see your own reflection, but you'll also see transmitted light. If the object behind the mirror is made to look like a ghost, you'll see you own reflection, and what appears to be the reflection of a ghost sitting right next to you. The ghost will disappear as soon as the light shining on the ghost model (behind the mirror) is extinguished.

So this illusion is taking advantage of the fact that we are so accustomed to mirrors, when we see someone in the mirror that looks like us, we assume that all the light is being reflected. The illusion would not work on someone who had never seen a mirror.

Corner Reflectors

When light bounces into a corner, with each wall a mirror, then it is reflected in a path parallel to the path it was projected in. This is easiest to see if it hits only two mirrors, as shown in figure 8.10. For those of you who enjoyed geometry, I leave it as an exercise to show that if the mirrors form a right angle with each other, the returning light will be parallel to the incoming light, regardless of the angle that it makes when hitting the first mirror.

Figure 8.10 Mirrors at right angles make a retroreflector—that is, a combination that sends light back in the same direction that it came from.

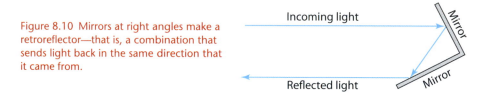

This only works if the two mirrors make a 90-degree right angle with each other. Remember, the angle that the incoming light makes with the surface of each mirror is the same as the angle the reflected light makes with the same surface.

I show this in class a demonstration of corner reflectors. I shine a laser beam on a mirrored corner, and the beam comes right back at me, displaced sideways by a little bit. The corner reflector that I use has three mirrors. A diagram illustrating this appears in figure 8.11.

The light actually bounces off all three flat surfaces. There is an analogous situation if you throw a frictionless ball against the corner of a room—the ball will come right back at you. This would be useful knowledge for certain indoor sports, including racquetball and squash, except for the fact that the balls in

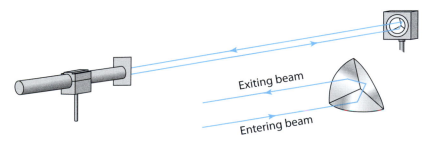

Figure 8.11 Corner cube retroreflector. The laser beam bounces off two or three mirrors and then travels backward along a path parallel to the incoming path.

such sports have significant friction. When the balls spin, they bounce at different angles, and so the corner reflection rule (that the direction of rebound is parallel to the incident direction) fails.

There are optical devices other than corners that do the same thing—i.e., reflect light back in the direction of the source. The general term for such an object is a *retroreflector*. I'll show later that cameras and eyes are retroreflectors, and that accounts for the annoying artifact in photographs known as red-eye.

CORNER REFLECTORS FOR RADAR

Radar is a form of light (electromagnetic wave) but with a lower frequency and much longer wavelength than visible light. A typical radar wavelength is 1 centimeter to 1 meter (versus 5×10^{-5} cm for visible light).

You can make corner reflectors for radar too. Ordinary metal works well; there is no need to polish the metal to mirror-like shine since the wavelength is so much longer.

Corner reflectors are extensively used in radar. For example, if you are flying an airplane using radar for navigation, a corner reflector placed near an airport runway can help you locate it. As you point your radar emitter in many different directions, there will be only one direction in which the radar comes right back—when it is aimed at the corner reflector. Radar corner reflectors on boats or hung from balloons can make it easy for radar to locate them.

The most notorious use of corner reflectors for radar came when they were made part of the immense microphone string used for Project Mogul. Such objects were so unusual that the nearby residents thought that they must be extraterrestrial in origin. An image of the Project Mogul string, with the corner reflectors visible, is shown in figure 8.12. Each module has eight corners, in case the direction shifts with the wind.

Figure 8.12 Project Mogul corner reflectors hanging from the balloon. When the reflectors were found on the ground after the crash, some people concluded that they were alien devices (from the Air Force report, 1995).

CORNER REFLECTORS ON THE MOON

In the 1970s, an experiment was done to bounce light from the Earth off the Moon. A powerful laser was used to shine light on a small spot, and the spot was observed with a telescope. To make sure that as much light as possible came back toward the telescope, the astronauts placed corner reflectors at their landing site. By recording the time it took the light to travel both ways, scientists were able to measure the distance to the Moon within an accuracy of a few centimeters. That precise measurement may sound silly, but by using this extreme accuracy as the Moon went around the Earth, we were able to detect the very small changes in the orbit that are predicted from Einstein's general theory of relativity.

A photo of the array of corner reflectors on the lunar surface is shown in figure 8.13.

Figure 8.13 Array of corner reflectors placed on the Moon by astronauts. In the rectangle, there are 10 rows of small corner reflectors oriented to reflect light to the upper right. You can also see the footprints of the astronauts in the lunar dust. (Photo courtesy of NASA.)

STEALTH

In radar, the radio receiver emits a strong signal and then picks up the reflection. If there is a corner on the target, then the signal returned will be very large. If you don't want an object to be detected by radar, then you should not have any right angles on it. If you look at modern military stealth aircraft, you'll notice that they don't have right angles. Even the tail is tilted with respect to the wings, so a right angle doesn't form. This is all part of the "stealth" technology: don't have any inadvertent corner reflectors!

The other trick to stealth is to cover the airplane with material that absorbs the radar, rather than reflects it. Such materials are said to be "black to radar." The word *black* is used as a metaphor for something that does not reflect. An object is black if it doesn't reflect visible light. But no material can be made completely black. If it could, then the absence of corner reflectors would not be necessary.

Slow Light

Light does not always travel at the speed of light. That paradoxical statement was purposely worded to be confusing so that you will remember it. Let me explain to you what I really mean.

When scientists use the term *the speed of light*, we usually mean the speed of light in *space*, denoted by the letter c, which is approximately 186,282 miles per second. That's fast enough to get to the moon in about 1.3 seconds. This speed is also approximately equal to 300,000 km/s, and to 1 ft/ns. (Recall that a nanosecond is one billionth of a second, about the time it takes your computer to make a computation.)

But that is only the speed of light when it is traveling in a vacuum. When light enters materials, it travels at a slower speed. In the air, it travels at about 99.97% of c. In water, light travels at only 75% of c! In glass, it travels even slower, at about 2/3 c. That's still pretty fast, of course. But it is not as fast as the speed of light in vacuum. In some exotic materials, physicists have managed to have the speed of light come down close to zero.

The quantity c is not only the speed of light in vacuum, but it is also the speed of gravity waves, and the speed of anything that has zero rest mass. We'll talk more about that when we discuss relativity theory in chapter 12. Perhaps a better name for c would be "the velocity constant from relativity theory." Or it might be called "the speed in vacuum of massless particles." Or, for reasons that I'll discuss in chapter 12, "the Einstein constant." But, for purely historical reasons, it is usually referred to as the *speed of light*. Just remember, the actual speed of light when it goes through materials is not always c.

If neutrinos had been discovered before light, we might have used the term *speed of neutrinos* instead of speed of light. For many decades, we believed that neutrinos had zero mass. But evidence published recently suggests that some neutrinos actually have a tiny but nonzero mass. We know that there are three kinds of neutrinos, which we call the *electron neutrino*, the *muon neutrino*, and the *tau neutrino*. It is still possible that the electron neutrino actually is massless, but we can't really be sure of that.

Now that we know that some neutrinos have mass, it is reasonable to ask whether particles of light have mass. (As you will see, quantum mechanics says that every wave is also a particle, and the particle associated with light is called a *photon*.) We think . . . that light particles probably do *not* have mass. We do know that if they do have mass, it is much smaller than the mass of any other known particle.

The Index of Refraction n

The speed of light in different materials, such as glass or water, can be described by its value compared to the speed of light in a vacuum. The index of refraction *n* is defined as

$$n = c/v$$

where *v* is the velocity of the light in the material, and *c* is the speed of light in vacuum.

From this, you see that the speed of light in a material that has index *n* is given by $v = c/n$. The index of refraction of sea-level air is about 1.0003, so the speed of light in air is $c/1.0003 = 0.9997\ c$. The index of refraction for water is about 1.33, and for glass is about 1.5. So in glass, $v = c/1.5 = (2/3)\ c$. In glass, light travels at $v = 2/3$ its speed in a vacuum.

Mirages

On a hot day, when you look along a road, you sometimes see what appears to be a puddle of water on the road. In the desert, you see what looks like a lake close to the horizon. But in both cases, nothing is actually there; these are examples of an optical illusion called a *mirage*.

Remember how sound is bent by air? When the ground is warm, sound tends to bend away from the ground; when the ground is cool, sound tends to bend toward it. Exactly the same phenomenon occurs with light. When the air is hot (e.g., above a hot road), then the speed of light is faster than in cold air. The result is that light will tend to bend upward. Blue light from the sky can be bent upward in this way, and give the illusion that the blue light is coming from the ground—i.e., from a pool of water. A diagram illustrating this is shown in figure 8.14.

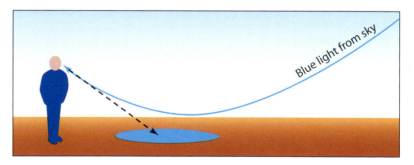

Figure 8.14 Mirage. When you look down toward the ground (dashed line), you see blue light and think that you are seeing water. In fact, you are seeing blue light from the sky that has been bent upward by the hot air near the ground. (For real mirages, the bending is much less, and the pool of water appears to be near the horizon.)

The light ray bends, but your eye doesn't know that. So you assume that the blue is coming from the ground; you interpret this as a puddle of water.

SPLITTING THE RED SEA

According to the Judeo-Christian Bible, Moses invoked the power of God to push aside the waters of the Red Sea so that the Jews could escape from Egypt. Another possible interpretation is that the ancient Jews assumed that Moses had split the sea but had been fooled by a mirage.[8]

DIAMONDS, DISPERSION, AND FIRE

When light enters a diamond, unless it hits exactly perpendicular to the surface, it bends. This happens because the part of the light wave that hits the diamond first is slowed, and the rest of the light is bent in that direction.

The bending is exactly the same effect that we had in the mirage, and it is completely analogous to the bending of sound that we had in the sound channel, and in hearing distant sounds in the evening. Light bends toward the direction in which it is traveling more slowly. The only difference here is that the diamond has a surface, so the light enters it suddenly. But the bending is exactly the same as when light enters a region of air that has a different index of refraction.

This is illustrated in figure 8.15. The index of refraction of a diamond is 2.4, so the speed of light in a diamond is $c/2.4$. Green light enters the triangular piece of diamond and is bent. Note that on entering, it is bent toward the diamond, and again on leaving. Both times, it bends toward the right, since that is the side in which light travels more slowly. It is the right side of the narrow beam that enters the diamond first.

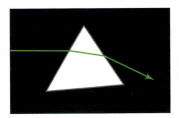

Figure 8.15 Light passing through a prism of glass.

Do all colors of light bend the same amount? The answer is approximately, but not exactly. The speed of light depends on frequency—i.e., on color—so different colors bend different amounts. This effect is called *dispersion* since it means that white light is separated (dispersed) into different colors. This is what gives rise to the many different colors, and that's what makes the diamond so beautiful.

---

[8] This reading is definitely *not* required! I warn you: it is written in the first person by the character Jesus, who has been enslaved. If that offends you, don't read it. In the excerpt, his good friend and mentor Simon gives him a lesson about how people fool themselves. If you are interested in reading part of a chapter of a novel that describes this possibility, then go to the selection of the novel at www.richardmuller.com/pages/RedSea.txt.

Dispersion is illustrated in figure 8.16. Now I have a white beam coming in from the left side and hitting a triangular piece of glass called a prism. I've put everything over a black background. The white beam of light breaks up in the prism (since different colors move at different speeds) into red, green, and blue.

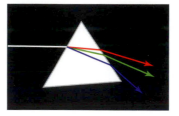

Figure 8.16 Different colors (red, green, blue) bend differently.

The fact that the three colors are separated is simply a result of the fact that in glass (or in a diamond), the three colors all move at slightly different velocities. So red, green, and blue light do not have exactly the same index of refraction—they are a tiny bit different. The index of refraction for a material such as diamond, when listed in a table, is usually given for a middle color (yellow), and the dispersion number is the difference between the indices of refraction for blue and red. In the gem business, dispersion is called *fire*. A gem with high dispersion has strong fire.

Originally, when people looked through a prism at other people, they saw the other people, but they appeared to be surrounded by an aura or halo of colored light. There are glasses you can buy today at novelty stores that do the same thing. The ghost-like colors that surrounded people were called *spectra*—meaning "ghosts." Even today, when a scientist makes careful measurements of the different frequencies present in a beam of light, the terminology persists: he says he is measuring *spectra*.

Table 8.1 lists, for example (you are not required to know these numbers!), the indices of refraction for ordinary glass, for water, and for diamond.

Table 8.1   Index of Refraction for Different Colors

	Glass	Water	Diamond	Cubic zirconia
Red	1.514	1.331	2.410	2.22
Yellow	1.517	1.333	2.417	2.23
Blue	1.523	1.340	2.450	2.28

From the table, you can see that the dispersion for glass (the difference in index for red versus blue) is $1.340 - 1.331 = 0.009$. But the dispersion for diamond is $2.450 - 2.410 = 0.040$. That is more than four times greater than for water or glass! It is this high value of the dispersion that gives diamond its *brilliance*—i.e., the fact that it appears to sparkle with different colors when you move it slightly.

COUNTERFEIT DIAMONDS AND ADVICE FOR THE ENGAGED

Cubic zirconia, called CZ for short, is a human-made crystal that is much cheaper than diamond, but has even more fire. It is often sold under the name of "counterfeit diamond." Look it up on the Internet. Whereas the dispersion (fire) for diamond is 0.040, the dispersion for CZ is in the range of 0.060 to 0.066. Because of this high dispersion, in sunlight CZ glitters with significantly more color than does diamond. Since it is primarily such fire that made diamonds so desirable, then CZ is more beautiful than diamond, at least according to the traditional evaluation.

It is also much, much cheaper. Gem-quality diamonds cost about $30,000 per gram, or about $6,000 per carat. One carat of CZ costs about $20.[9] Wow! Greater beauty at 1/300 the cost! Different processes produce CZ with different dispersions. Which value of dispersion would you guess is the most valued? The one with the most fire, the most beauty—i.e., 0.066?

Nope. Most people prefer the lower value. Why? Can you guess?

Here is the fascinating answer: people prefer the smaller dispersion, and its lesser fire, because it looks more like a "genuine" diamond! Many people don't like CZ that sparkles too much, because then everyone knows that it isn't diamond. The irony is that the diamond was initially admired because it had more dispersion than any other gemstone. It was also very expensive. Now we have something that is prettier, but inexpensive, so many people don't want it. How can you show people that you love them by giving them something beautiful but cheap? Diamonds are admired because they are expensive, and they are expensive because they are admired. Put another way, diamond is expensive because it costs so much. Some day I predict the price of diamonds will plummet because their value has no real basis.[10] It is just another bubble. Tulips once sold for fantastic prices, because they were so expensive, and then the tulip market crashed.[11]

Tulips were not forever, and I predict that neither are diamonds. So when you get engaged, save money by getting the prettier stone and save a fortune. Laugh at all those who waste money enriching the De Beers diamond cartel, and send me an e-mail message telling me that you took my advice![12]

Swimming Pools, Spearing Fish, and Milk Glasses

Because water has an index of refraction of 1.33, light leaving the surface is bent, and that gives rise to many illusions. One of the strangest of these is the swimming pool illusion: when observed from an angle, the pool appears to be much less deep than it really is. We can see why in figure 8.17. Light from an object on the bottom of the pool is bent as it emerges from the surface (note: it is bent *toward* the water), and this makes it look as if the object on the bottom is not as deep as it truly is.

---

[9] See, for example, http://e14k.com/czinfo.htm.

[10] Unless you value a diamond because it is the hardest known material. But that is not a very romantic feature. And new counterfeit diamonds are also very hard.

[11] For a description of the tulip mania, see http://en.wikipedia.org/wiki/Tulip_bubble.

[12] My e-mail address is ramuller@berkeley.edu.

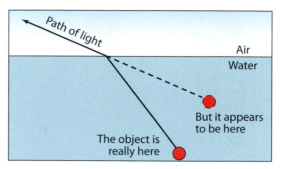

Figure 8.17 Light bends when leaving water.

When you look at your friend who is standing in the water just a few feet from you, the effect can be comical, since his feet might appear to be only a short distance below his head.

The effect also causes potential problems for people who are, for example, trying to catch fish. Suppose that you have been trapped on a remote island and are trying to survive by throwing a spear at a fish, as happened in the movie *Cast Away* (2000). Look at the diagram in figure 8.17. Let the red object be the fish. Where should you aim? Above or below the apparent location of the fish?

The answer is aim below. The fish will appear to be shallow, but it is in fact deeper. Don't aim at the location where it appears to be.

The same phenomenon occurs with milk glasses, but sideways. Take a round glass and fill it with milk. Now look at the side of the glass. Do you notice how the milk seems to come right out to the edge of the glass, as if the glass had zero thickness? This can also be explained by the diagram in figure 8.17. The "object" is now the milk. The "depth" of the object is now the thickness of the glass. But the light, when it hits the surface, bends toward the glass. So someone looking at the white light from the milk sees it at a grazing angle, as if the milk is really much closer to the outside of the glass than it really is.

Rainbows

Dispersion (fire) in water droplets is responsible for one of the most beautiful of natural phenomena, the rainbow. Rainbows are actually much harder to understand than you might guess. Any time you are looking at a rainbow, you are actually looking at small round droplets of water, and the Sun is behind you. The water could be the spray from a waterfall, from a sprinkler, droplets of fog, or rain from a cloud that you don't even notice. Whenever it has been raining, and the storm moves away and the Sun comes out, look for a rainbow on the side opposite you from the Sun. If rain is still falling over there, then you should see a rainbow.

Small raindrops are spherical, and when light enters the side of the front surface, it bends, bounces off the back, and then comes out with separated colors. The paths are shown in figure 8.18.

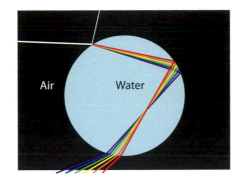

Figure 8.18 Light passing through a raindrop breaks up into the colors of the rainbow (violet, blue, green, yellow, and red).

Actually, some of the light leaves the back side, but that isn't shown in the diagram since it doesn't contribute to the rainbow. You can see how the different colors bend and separate whenever they pass through a surface. (I exaggerated the separation.) But how does this pattern of separation lead to the rainbow?

Notice the emerging light. It is going in a particular direction. Unless there is a person standing in that direction, there will be nobody to see the color. Someone standing directly to the left of the droplet will get no reflected light (except a little off the surface.)

Every drop spreads its colors out, but only certain drops send light in the right way to reach any particular person. Those are the drops that appear colored and make up the rainbow. For each drop, only one color bends at the right angle to reach the person who is looking. If the person is standing in the path of the blue ray, then the drop will appear blue. The red light will miss his eye; it will pass below his eye. (Can you see that in the diagram? The red light comes out lower than the blue light.)

There will be some drops that are high enough above the blue drops that the red light coming from them will hit the observer's eye. These drops will look red to the observer. They won't look blue, because for these drops, the blue light passes above his head.

Lenses

A *lens* is a wonderful invention that takes spread-out light coming from one source and focuses it all to one spot. It does this by having a curved surface, so the light at the edges bends more than the light in the center. This is illustrated in figure 8.19. Light is coming in from the left, from a great distance, so all the rays are parallel to each other. When they enter the curved glass surface, they are bent as shown. They are further bent when they leave the glass, and they all come together at the focus. (They don't stop when they get to the focus unless there is a piece of film there to stop them.)

The marvelous thing about a lens is that it takes light that is spread out and brings it to a point. That's why it can be used to concentrate sunlight and start a fire. But it is also the reason that the pupil of our eyes can be larger than a

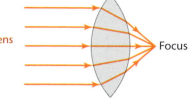

Figure 8.19 Light passing through a lens focuses on a spot.

Focus

pinhole. If we widen the pinhole opening in the pinhole camera, the image gets blurred. But if I put a lens in the wide opening, the broad beam of light is brought to a small focus. That way I can let in a lot of light, yet still get a non-blurred image. Otherwise, the camera—or eye—works the same way as a pinhole camera.

Eyes

The human eye actually has two lenses. One is called, appropriately, the *lens*. The other is called the *cornea*. These are shown in figure 8.20. The iris is the pretty part of the eye, the colored part that opens or closes to let more or less light in.

Figure 8.20 The structure of the human eye.

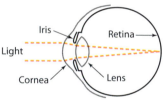

The cornea does most of the focusing. But the lens is variable. If squeezed by the muscles of the eye, it can change the amount it focuses.

Why should we need a lens of variable strength? The reason is that nearby objects need more focusing than distant objects. If you are looking at a flower, and it is close to your eye, then the light from each petal is spreading out slightly by the time it reaches your eye. If it is to be brought to a good focus on the back of your eye, such light has to be bent more than, say, the nearly parallel light from a distant star. This process of squeezing the lens to make it accommodate (able to focus on) nearby objects is called accommodation.

Nearsightedness

If your cornea is too curved (or if the retina is too far back), then it is easier to focus on nearby objects, but hard or impossible to focus on distant objects. When that happens we say the person is *nearsighted*. That can be fixed by re-shaping the cornea so that it doesn't focus as much. (That's what Lasik surgery does.) Or if you don't want someone to carve away at your eye, you can wear contact lenses or eyeglasses.

There is speculation that some people become nearsighted by doing a lot of reading when young. This could conceivably come about by the continual squeezing of the lens, until it begins to keep the squeezed shape. But most experts dispute this process and say it doesn't happen. It is just as plausible that children who are naturally nearsighted find reading to be less stressful on their eyes than children whose eyesight is "normal."

Farsightedness—and Aging

As you grow older, the flexibility of your lens is lost. Eventually, you cannot squeeze your lens enough to be able to focus on nearby objects. (There is an exception. Some people are nearsighted before they age, and when they lose their accommodation, they focus only on nearby objects.)

Everyone loses their accommodation as they age (the process begins when we are about age 15), and so all of us (except those who are already nearsighted) eventually become farsighted.

People who are over age 35 cannot squeeze their lens enough to focus over the whole range of desired distances. Typically, they might wear bifocal lenses in their eyeglasses. Bifocals are really just two lenses of different strength placed one above the other. The person wearing them uses the lower, stronger lens for reading and the upper, weaker lens for more distant objects.

When you get to age 35, you might consider buying "reading glasses." These are inexpensive ($10) and are available at most drugstores. Sometimes they are half lenses so that the reader can look over them when he or she wants to see something that is farther away than the book. People who wear such half lenses are readily identified as age 35 or older.

Since only older people wear bifocals, those who want to "look young" don't want to wear them. Eyeglass makers go to great extremes to conceal the fact that eyeglasses are bifocal by hiding the line that divides the two lenses.

On the other hand, actors can look older by wearing half lenses, or by picking up a newspaper and holding it away from them—indicating that they are farsighted, and therefore older. It is interesting that even young people recognize this behavior, even if they don't realize why the person looks older. If you see only older people acting that way, then you begin to associate the behavior with age.

Another way to avoid the "old" look of bifocals is to have two different lenses for your two eyes, one that focuses near and the other far. Then you can pretend to be younger than you are. This is what *bifocal contacts* usually means. No matter what distance you see, it will be with only one eye at a time.

You may find some older people who are proud of the fact that they don't need glasses, even for reading. In every case that has been studied, that has meant that their two eyes are different. One focuses in close (useful for books) and one far away (useful for reading signs).

Red-Eye and Stop Signs

The eye also works as a retroreflector, sending a beam of light that hits it back to the source of the light (just as a corner reflector does). This happens because of the focusing property of the eye. Look at the diagram in figure 8.21. The incoming light (arrow coming from left, toward the lens) is focused on the surface at the right. Some of it bounces, but that bounced light scatters in all directions. But all of the bounced light that hits the lens is deflected to return to the source of the light. You can see that in figure 8.21—the light coming out is parallel to the beam of incoming light.

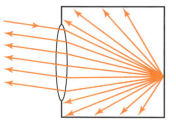

Figure 8.21 A retroreflector based on a lens. Even if the light scatters in different directions when it hits the back, all of the light that passes back through the lens winds up going in the same direction.

When you take a photograph of a person, much of the light from the flash returns directly to the flash. If the lens of the camera is near the flash, then much of this light goes right back into the lens, creating the ugly phenomena we call *red-eye*. Figure 8.22 shows a photo of my daughter Elizabeth showing the red-eye phenomena: the light that enters the pupils of the eyes tends to come right back toward the camera, making the center of the eyes appear bright when they should be dark.

Figure 8.22 Red-eye, an annoying effect that shows the center of the eye to be very bright, is a consequence of the fact that an eye is a lens retroreflector.

A small retroreflector can be made from a spherical bead. If the glass has the right index of refraction ($n = 2$), then light hitting the front surface is focused on the back surface; it reflects and it then refracts back to the source, as shown in figure 8.23.

These beads can be as small as grains of sand and are now produced very cheaply. If these grains are glued to the surface of a road sign, then they will reflect the light from headlights right back at the car. As a result, they make road signs look very bright to the person in the automobile. Beads such as these are

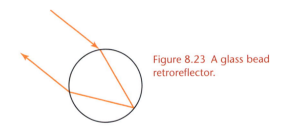

Figure 8.23 A glass bead retroreflector.

also used to coat surfaces (such as night clothing) that you want to look bright in headlights.

In Yosemite, if you shine a flashlight on a tree that contains a bear, all you will see clearly reflected are the two bright eyes. I vividly remember seeing those two points of light way up in a tree, deep in the wilderness, and hearing him munch the chocolate he had stolen from us.

Telescopes and Microscopes

We've learned everything we need to know to understand telescopes and microscopes. In both of these systems, a lens is used to form an image of the object. If the object is far away, we call the system a *telescope*; if it is small and nearby, the system is a *microscope*.

Look at the diagram for the pinhole camera again (figure 8.7). If the object is very close to the pinhole, then the image will be larger than the object. That's what a microscope does. To be able to look at this image, up close, a second lens called the *ocular* is placed in front of the eye.

For a telescope, an image is made of a distant object. That image is usually placed just in front of the eye. It could be put on a translucent screen. Then an *eyepiece* is put in front of the eye so that it can focus up close, and see details of the object. It turns out that the translucent screen is unnecessary. The light forms a focus at the place where the screen would be, and continues on, into the eyepiece.

Keck and Hubble Telescopes

In astronomy, the problem is not only magnification, but also collecting enough light to see a dim distant object. That is why most astronomers talk about their telescopes in terms of their diameter. The "most powerful" telescope in the world is the 10-meter telescope at the Keck Observatory. It uses a curved mirror to focus light instead of a curved lens. The 10 meters refers to the diameter of the mirror, and it is the ability of such a large mirror to gather so much light that makes it "powerful" for astronomy.

Even though the Keck telescopes are on a mountaintop, their images are blurred from turbulence in the air above them. Any ground-based telescope suffers from the fact that air pockets form little lenses, and these make continually changing distortions in the focus. That is one of the primary reasons that we put some astronomical telescopes in space. The most important of these is the Hubble telescope. Its diameter is only 2.4 meters, but because it is above the atmospheric distortions, it can focus much better, and see things that are much smaller in their apparent size (e.g., the diameters of distant stars and galaxies) than can ground-based telescopes. The distortion from the air pockets does lead to one pleasant feature—that's what makes stars twinkle.

You can tell a planet from a star by the fact that it doesn't twinkle. Why? Actually, a planet is wide enough that the top of the planet twinkles out of step with the sides and the bottoms. So the twinkling tends to cancel.

Spreading Light—Diffraction

When making powerful telescopes and microscopes, the wave nature becomes important again. As we discussed in chapter 7, any wave, when passing through an opening, tends to spreads out when it emerges on the other side. In physics, this spreading is called *diffraction*. The equation for S, the amount of spreading, is

$$S = \frac{L}{D} R$$

where L is the wavelength, D is the diameter of the opening, and R is the distance beyond the opening. Notice that small wavelength waves spread less. (If the wavelength L is small, then the spreading S is small too.) That is the reason humans hardly notice spreading of light. For visible light, L is only a half micron, so most of the spreading is usually negligible. It becomes important only for really small details, since the spreading causes blurring. That happens with telescopes, and for fine detail with cameras.

A similar equation can be used to determine the closest objects that an optical system (camera, eye, or telescope) can distinguish. That distance B, sometimes called the *resolution*, is given by the same equation we have for S:

$$B = \frac{L}{D} R$$

where B is the separation of the objects (the blur distance or resolution), L is the wavelength, D is the diameter of the lens or opening, and R is the distance to the objects.

Now we will apply this equation to spy satellites.

Spy Satellites

Suppose that we want to put the Hubble space telescope in geosynchronous orbit, to observe whatever is happening in the mountains of Iran. We want it to be geosynchronous so that it will always be above the same place. What will we be able to see? Will we be able to recognize people? Can we read license plates? The answer turns out to be no. Diffraction will cause so much blurring that we will not be able to see details that are smaller than 7 meters, or about 23 feet!

Here is the calculation. For geosynchronous orbit, we set $R = 22,000$ mi $= 35,000$ km $= 3.5 \times 10^7$ m. Let the wavelength of light correspond to the value for visible light: 0.5 micron $= 5 \times 10^{-7}$ m. The diameter of the Hubble telescope is 2.4 meters. So the resolution on the ground will be

$$B = (L/D)\ R = (5 \times 10^{-7}/2.4)\ 3.5 \times 10^7$$
$$= 7\ m$$
$$= 23\ ft$$

So two objects on the ground separated by 23 feet would be blurred together in the image!

Suppose that we put the Keck telescope into geosynchronous orbit. It is about 4 times bigger than the Hubble, so it could do 4 times better, or about 6 feet. Better, but still not very good.

Now let's do the calculation for the Hubble 2.4-meter telescope in LEO (low-Earth orbit), a height $H = 150$ mi $= 240$ km $= 2.4 \times 10^5$ m. The only quantity that we change in the previous calculation is H. So for a low-Earth orbit, we get

$$B = (2 \times 10^{-7})\ H$$
$$= (2 \times 10^{-7})(2.4 \times 10^5) = 0.05\ m$$
$$= 5\ cm = 2\ in$$

So a low-Earth satellite could almost read a license plate (at least if it is facing upward). But a geosynchronous satellite misses by a big factor, due to wave spreading as the light passes through the aperture of the telescope.

Resolution of the Human Eye

We can use the same equation to estimate the resolution of the human eye. In daytime, the aperture of the pupil is 5 mm $= 0.005$ m. Again, we use the wavelength of light of 0.5 micron $= 5 \times 10^{-7}$ m. From this, we get that two objects can be distinguished if

$$B = (L/D)\ R = (5 \times 10^{-7}/.005)\ R = 10^{-4}\ R$$

This is satisfied if $B/R = 10^{-4}$. From trigonometry, this is equivalent to an angle of 0.007 degrees $= 0.25$ minutes of arc. Excellent vision for humans (20/20 vision) means that the person can resolve things at 1 minute of arc. This is approximately the spacing of rods in the retina of the eye. So we can't quite see at the diffraction limit.

If we had closer rods in our eyes, we would have better eyesight. This is probably what gives certain animals, such as eagles, better vision than humans.

Holograms

Holograms are considered very mysterious by many people, but in fact they are no more mysterious than are mirrors. When a light wave hits a mirror, it makes the electrons in the metal surface shake. Under such shaking, they emit waves, and these emitted waves are what give us the image.

Suppose that we could get the electrons to shake in just the right way, without having an incident light wave? Then they would still emit the light that gives the virtual image. That is exactly what a hologram does. It "records" the wave that, on a mirror, would give a virtual image. Then, when hit with light, it reproduces the wave that gives the image.

A hologram is made by shining laser light on the thing being photographed, for example, on you. Then the light reflected from you falls on the film. In addition, a second beam of light shines on the film directly from the laser. The laser light and the light reflected from you interfere with each other, leaving a microscopic pattern of dark and light spots (after the film is developed).

Later, when you want to view the hologram, you shine light directly on hologram. The light hits the bright and dim spots, and when it comes off it has just the right pattern to reproduce the original light. Since the light coming off is the same as the light coming off the original object, an observer can't tell the difference. Of course, the image doesn't look completely real, since it reproduces the light coming off an object illuminated with a laser.

A hologram could not work except for the fact that light is a wave. A hologram is, in essence, a frozen mirror. It preserves the kind of wave that is reflected and then re-creates it when hit with light.

Polarization

Light is a transverse electromagnetic wave. That means that the direction of the electric field is perpendicular to the direction that the wave is moving, just as the shaking of a rope wave is perpendicular to the direction of the rope.

Electromagnetic waves, like rope waves, are polarized. If the electric field points in the vertical direction, we say the wave is *vertically polarized*. (Remember, the wave is transverse. That means that the direction of the electric field is perpendicular to the direction the wave is moving, just as with a rope wave.)

The wave can also be *horizontally polarized*. By convention, any other direction of polarization is seen as a combination of vertical and horizontal. For example, a wave polarized at 45 degrees from the horizontal can simply be said to be oscillating in the vertical and horizontal directions at the same time.

Polarization has become extremely important in modern technology. Liquid crystal displays (LCDs; used on computer and TV screens) are based on switchable polarizers (we'll discuss this soon). Polarization gives us a fascinating insight into materials, rocks, and microscopic creatures. We'll discuss these applications in a moment.

Ordinary light (from a lightbulb, or from the Sun) usually consists of many different waves all coming at the same time. As light comes from different atoms in the source, each little part of the wave can have a different polarization. Such light is said to be *unpolarized*. The light can be made to have a single polarization by passing it through a material called a *polarizer*. To understand this, it is useful to think about rope waves passing through a picket fence. In the diagrams in figure 8.24, the fence will pass only waves that are polarized in the matching orientation.

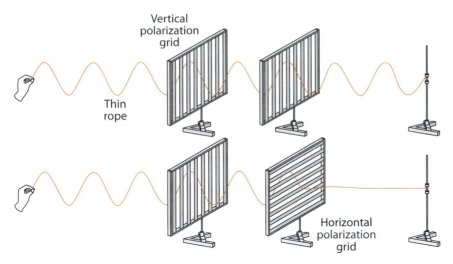

Figure 8.24 Rope waves can be polarized by passing them through grids. (Based on a drawing by Rusty Orr.)

Thin films that do the same thing were invented by Edwin Land, and trade-marked under the name *polaroid*. (The Polaroid company later made cameras, but their initial work was solely polaroids.) Polaroid films are also used today in sunglasses. As with the ropes, the film passes only one kind of polarization at a time.

Unpolarized light, when it passes through a polaroid, emerges polarized. If it then hits a second polaroid, it will pass through—provided the polaroid is oriented in the same direction. (A little is always absorbed, since the polaroids aren't perfect.) If the second polaroid is oriented perpendicular to the first, then the light is stopped. This is illustrated in the diagram in figure 8.25.

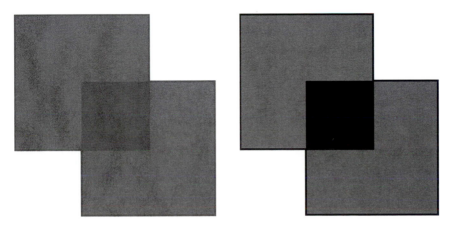

Figure 8.25 Crossed polarizers. Two sheets have light passing through them. In the left figure, both polarizers pass vertical light. In the right figure, the top polarizer passes vertical light, and the bottom one passes horizontal light; no light gets through the overlap region.

Polaroid Sunglasses

There are other ways to polarize light. When light bounces off the surface of glass or water or even asphalt (and other nonmetals), then the light tends to become polarized in the horizontal direction. If you are fishing and want to see into the water, and you don't want to be distracted by light reflecting off the surface of the water, then you can take advantage of the fact that the bouncing light is polarized. Wear sunglasses made out of polaroid film, with the film oriented such that it will pass only vertically polarized light. Then it will stop the reflected light, but half of the light coming from the fish will still be visible.

Polaroid sunglasses advertise that they "cut glare." What they really cut is light that has been reflected off nonmetal surfaces. When light bounces off a metal it does not become polarized, so such sunglasses don't help for that kind of glare.

Crossed Polarizers

When light passes through a transparent material such as plastic, then internal (and normally invisible) stresses in the material can rotate the angle of polarization. Moreover, different colors will be rotated by different amounts. If a horizontal polarizer is placed below the object, and a vertical one is above it, then no light will be transmitted unless there are stresses inside the object that rotate the light. This effect is seen in the image of a plastic CD case in figure 8.26.

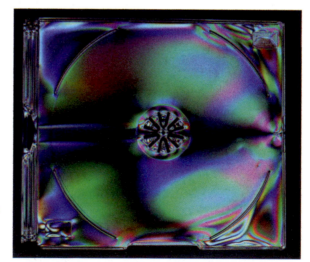

Figure 8.26 Plastic CD case, seen between two polarizers. Different colors show different amounts of stress left in the plastic when it was manufactured; black regions have little or no stress. You can see this in real life by crossing two pairs of polarizing sunglasses until no light gets through the combination (as in figure 8.25) and then putting a piece of stressed plastic between.

This effect can be useful in the design of engineering structures. You build a model out of plastic, and view it with crossed polarizers. Then you put a force on the model. The regions of the model that are stressed the most will show up in color. This way you can determine which parts of the structure are most likely to break, and the design can be changed (if necessary) to relieve some of that stress.

Liquid Crystals and LCDs

Liquid crystals are materials that act like polaroid film, except that their ability to polarize can be turned on and off with an electric voltage. If you have crossed polarizers, and one is a liquid crystal, then the amount of light coming through can be changed with an electronic signal.

Many thin-screen computer displays and flat TV displays take advantage of this. The term *LCD* refers to "liquid crystal display." Laptop computer displays are usually liquid crystal displays with fluorescent lights or light emitting diodes (LEDs) behind them. If the polarizers are completely crossed, then no light gets through. If they are parallel, then the maximum light comes through. If they are oriented (electronically, remember) at 45 degrees, then half of the light comes through.

Each pixel is colored red, green, or blue, to match the sensors in the retina of the eye. By adjusting the amount of light for each pixel, the screen can duplicate the sensation of color seen by the human eye.

If you haven't already looked at a computer screen with a magnifying glass (I suggested you do it earlier in this chapter), do it now. Pick a region that is white. When you look up close, you'll see no white whatsoever—only red, green, and blue spots. As you back away, so that you can't resolve these individual colors, they all blend into white. It is a most marvelous optical effect.

3-D movies

When you look at a nearby object, your two eyes are seeing it from a different angle. Your brain notices this, and interprets it to mean that the object is close. If the object is far away, the light enters the two eyes from nearly the same direction. Your brain interprets this as meaning that the object is far.

To watch a 3-D movie, you wear special glasses. The most common of these are polaroid glasses, in which the two polaroids are crossed—so they see different light. One could be vertical and the other horizontal.[13] The projected movie actually consists of two separate movies, one for each eye. One movie shows what would be seen by your left eye, and the other shows what would be seen by your right eye. It is the fact that these two images are different that gives the 3-D image.

There are other types of 3-D imaging that don't use polarizers. Some computer screens designed for 3-D blink back and forth between two images. Special blinking goggles then allow you to see with each eye only the image meant for that eye. 3-D postcards are actually double images, with each image divided into strips, and the two images alternating. Over the double image is a series of plastic ridges that bends one image toward your left eye, and the other toward your right eye, so each eye gets a different image.

---

[13] In practice, one is usually oriented at 45 degrees to the right of the vertical, and the other at 45 degrees to the left. That way, they are still 90 degrees different from each other.

Chapter Review

Visible light is an electromagnetic wave with wavelength of about 0.5 micron and frequency of about 6×10^{14} Hertz (cycles per second). According to Shannon's theorem, this high frequency allows light to carry a high number of bits per second, which is valuable for communications.

Color is the frequency of the light. We detect this by having three sensors in our eyes, sensitive to red, green, and blue. We can fool the eye into thinking that we have full spectrum of white light with a computer screen by exciting all three of these sensors. Multispectral analyzers do a more complete job, and see colors that we can't distinguish. Printed colors use magenta, cyan, and yellow, since they absorb light rather than emit it.

The fact that light is a wave can be seen in the colors of a soap bubble or an oil slick. Light reflecting from the front and back surfaces reinforces and cancels (i.e., interferes) and gives rise to the colors.

The simplest device to make an image is a pinhole camera. The image is upside down. If the pinhole is large, the image is brighter but blurred. Mirrors make "virtual" images, which appear real, but have no light passing through them. Three mirrors arranged as a corner will reflect light back to the source. That configuration is useful when you want to send light somewhere and have it bounce back toward you. This has been done on the Moon with visible light and is useful for radar (a form of low-frequency light).

Stealth is a system used by the military to avoid detection by bounced radar. It is based on two ideas: no corner reflectors, and high absorption (i.e., it is "black").

When traveling through air, water, or glass, light travels slower than the speed c in the vacuum. The slowing factor is called the *index of refraction* (n). Light bends just the same way that sound bends, toward the slower side. That explains mirages and lenses. A lens, by focusing light, allows a camera to have a "big pinhole" without too much blurring. The blurring of light of wavelength L for an aperture of diameter D at a distance R is given by $B = (L/D) R$. Such blurring means that a spy satellite at geosynchronous orbit would not be able to distinguish objects on the ground that are closer than 28 feet. But from low-Earth orbit, they can see objects as close as 2 inches.

Different colors have different values of n, and that explains prisms, diamonds, and rainbows. This property is called *dispersion*, and among jewelers is called *fire*. Fire makes diamonds and cubic zirconia crystals colorful and beautiful.

Eyes are like cameras. They have two lenses, the cornea and the flexible "lens." As we get older, the lens loses its flexibility, and we use reading glasses and bifocals to compensate.

Large astronomical telescopes such as Keck need large apertures to collect the light from dim objects. Space telescopes such as Hubble can see smaller objects because they don't have distortions from the Earth's atmosphere.

Holograms work by reproducing the wave that would be present if the object were reflecting its light off a mirror.

Light is transverse, and if all the light in a wave is vibrating in the same direction we say it is *polarized*. Polarization can be described as horizontal, vertical, or some combination. Polaroid film can be used to turn unpolarized light into polarized. Light reflected from a nonmetallic surface becomes polarized, so such glare can be reduced by polaroid sunglasses. Plastic materials placed between crossed polaroids show their internal stresses, and that is useful for analyzing such stress.

Discussion Topics

1. To what extent do most people depend on sight more than they do on sound? Why? What are the physics characteristics of light that makes it able to give more detailed information than does sound?
2. In the chapter, I gave examples in which bending of light gives illusions, such as the "shallow" illusion of a swimming pool, and the mirage on a road. Can you think of other examples in which the bending of light gives illusions? What about the reflection of light?
3. Discuss the "beauty" of jewelry. Is it truly pretty—or does its value derive from its expense? Think of examples other than diamonds. Why do some people consider "cultured" pearls less desirable than natural pearls? What about rubies and sapphires? What about other things of beauty? Would we find rainbows more beautiful if they were rarer?
4. Think about 3-D vision. Close one eye. Do you still see things in 3-D? Maybe a little bit? To what extent is 3-D perception simply a way that the brain describes its conclusions to you?
5. Does the description of a hologram as a frozen mirror help you understand holograms? Or does it just make you more confused about mirrors? If holograms were abundant and mirrors were rare, which would be considered the more amazing device?

Internet Research Topics

1. Old 3-D movies (and old 3-D comic books) were based on a red–green system, rather than on a polarizing system. In fact, some images of other planets were published by NASA using a similar system, since they can be transmitted on the Web. See if you can find such pictures. Do you understand how they work? You may be able to find red/blue plastic glasses to use. Can you find a Web site that sells such images and such glasses? If you have the glasses, then you can draw 3-D images yourself using colored pens or crayons.
2. Rather than wear eyeglasses or contact lenses, some people have operations on their eyes. How do these operations work? What do they do the eye? If you know people who have had this done, talk to them and find out what was done. Do they change the cornea or the lens?
3. Look up ads for diamonds. See if you can spot the subtle ways that they try to suggest that diamonds are better than the competitors, such as cubic zirconium. Do they explicitly mention cubic zirconium (CZ) by name? What can you find out about the De Beers company, and the diamond market?

Essay Questions

1. Newton thought that light was a particle, but we now know it consists of waves. What behavior makes light appear to be a particle? How do we know it is a wave?
2. Some people are surprised by the statement that "light does not always travel at the speed of light." What does that mean? What are the implications of "slow light"? What are the phenomena that result, and what practical applications come from the fact that light sometimes travel slower than the value $c = 3 \times 10^8$ m/sec.
3. If you were designing a spy satellite, to take photos of the ground, what considerations would you take? Discuss the height of the satellite and how you would pick it, depending on needed resolution and time over target.
4. We are all color blind. Discuss the meaning of that statement. How can instruments do better than human eyes? What is the value of such instruments?
5. A scientist argues that a fly has better vision than a human, in the sense that the fly can distinguish things that are closer together than can a human. Is that possible? Discuss the relevant physics.
6. In some science classes, they teach that light travels in straight lines. But this is not always true. Why it is widely believed that light behaves this way? Give example that show the situations in which it is not true. Use numbers whenever possible.

Multiple-Choice Questions

1. An indication that light is a wave is (choose all that are appropriate)
   A. that it has dark bands in the diffraction pattern
   B. that it passes through glass
   C. that it reflects off surfaces
   D. its high speed

2. The colors of an oil slick indicate that
   A. light is a wave
   B. light bends when entering material
   C. light changes its wavelength when passing through oil
   D. oil is made of many different chemicals with different colors

3. The *index of refraction* measures
   A. the frequency of light
   B. the speed of light
   C. the period of light
   D. density of the glass

4. Light waves are
   A. longitudinal
   B. transverse
   C. circular

5. The resolution of a human eye is about
   A. 4 mm
   B. 2 cm
   C. 1 micron
   D. 1/60 degree

6. Which of the following has the lowest index of refraction?
   A. water
   B. glass
   C. air
   D. crystal

7. As the wavelength of light decreases, the frequency
   A. decreases
   B. increases
   C. stays the same

8. In one millionth of a second, light will travel (caution: possibly a trick question)
   A. about 1 foot
   B. about 1000 feet
   C. from one side of a computer chip to the other
   D. from the Earth to the Moon

9. A piece of glass is shaped like a pyramid. The side of the pyramid is tilted at 45 degrees with respect to the horizontal. A beam of light is inside the glass, moving horizontally. When it emerges from the tilted surface of the glass, it will be moving
   A. perfectly horizontally
   B. in a direction that is tilted upward (so it will eventually go into space)
   C. in a direction that is tilted downward (so it will eventually hit the ground)

10. A stop sign is bright when you shine the headlamps of your car on it. It most likely is covered with
   A. fluorescent paint
   B. phosphorescent paint
   C. tritium
   D. small glass spheres

11. Diamonds sparkle in many colors because
   A. light travels very slowly
   B. light travels very fast
   C. there is color-dependent absorption
   D. light velocity depends on color

12. Which of the following are retroreflectors? Choose all that apply.
   A. bicycle reflectors
   B. human eyes
   C. stop signs
   D. animal eyes

13. Two soap bubbles from the same soap appear to be different colors. They probably have
   A. different sizes
   B. different absorption

C. different temperature

D. different thickness

14. If *c* is the speed of light in vacuum, then the speed of light in glass is approximately
    A. 1.5 *c*
    B. (2/3) *c*
    C. *c*
    D. 0.999 *c*

15. Stealth bombers are undetected by radar, in part because
    A. of their low heat emission
    B. of their high speed
    C. they are made of translucent materials
    D. of the absence of corner reflectors

16. Rainbows show different colors because
    A. droplets of water have different sizes
    B. in water, different frequencies have different velocities
    C. the direction of light "spreads" because the wavelength is short
    D. droplets of water change the color of air molecules

17. Which of the following is polarized?
    A. blue sky light
    B. direct sunlight (yellow)
    C. light from a candle
    D. light from a TV screen

18. A fiber can send much more information per second than a wire because
    A. light has a very high frequency
    B. electricity travels better in glass than in a wire
    C. sound travels very rapidly in glass
    D. light travels faster than electricity

19. Which color of light would give the highest number of bits per second in fiber optics?
    A. red
    B. white
    C. blue
    D. infrared

20. Shannon is famous for having invented or discovered (among other things):
    A. GPS
    B. *Global Hawk*
    C. the sound channel
    D. the bit

21. The human eye has cones that detect
    A. red, yellow, blue
    B. cyan, magenta, yellow
    C. yellow, green, red
    D. green, blue, red

22. People squint to see better because squinting
   A. reduces the light
   B. bends the lens to make it stronger
   C. reduces the blur size
   D. They don't see better. They only think that they do.

23. Old people need reading glasses because their
   A. pupils can't contract as well
   B. eyes become less sensitive to visible light
   C. lenses become less flexible
   D. they forget how to read

24. Red-eye in photographs comes from
   A. the film detecting IR
   B. poor focus on the eye
   C. light of the flash reflecting off the retina (the back of the eye)
   D. light of the flash reflecting off the cornea (the surface of the eye)

25. Red-eye demonstrates that
   A. light is a wave
   B. air absorbs blue more than red
   C. camera film is sensitive to red
   D. the eye is a retroreflector

26. Your eye has
   A. one lens, called *the lens*
   B. two lenses—the lens and the cornea
   C. two lenses—the lens and the retina
   D. no lenses, but behaves like a pinhole camera

27. Light is polarized when it bounces off (choose all that are correct)
   A. water
   B. glass
   C. air

28. Sunglasses can help you see a fish underwater because
   A. reflected light is polarized
   B. light from the fish is polarized
   C. they dim the light, making your pupils dilate
   D. they cut the blue light of the water surface

29. Polaroids are used for 3-D movies because
   A. they reduce glare from the surface
   B. they give a different image to each eye
   C. they polarize the light
   D. they reduce the dispersion

30. Stress in plastic can be detected by looking at
   A. the transmission of different colors
   B. the reflection of different colors
   C. interference
   D. polarized light (crossed polaroids)

31. In the pinhole camera, more blurring occurs if
A. the hole is made very large (but not if it is very small)
B. the hole is made very small (but not if it is made very large)
C. the hole is either very large or very small
D. Never, since there is no lens in a pinhole camera.

32. Holograms depend on the fact that light
A. contains red, green, and blue
B. is quantized
C. is a wave
D. can be focused

33. A successful fisherman will throw the spear
A. above the fish image
B. below the fish image
C. right at the fish image
D. It depends on how close the fish is.

34. To make a spy satellite that can read a license plate, you would have to
A. put it in a higher orbit
B. cover it with a polarizing filter
C. use a larger mirror
D. Do nothing; they already can read license plates.

35. The wavelength of visible light is closest to the diameter of
A. a human hair
B. a red blood cell
C. an atom
D. a nucleus

36. The Keck telescope is "powerful" because
A. it gives larger magnification
B. it can use UV light
C. it has a bigger mirror to collect light
D. it has a bigger focal length than others

37. When we say a man is color blind, we usually mean that
A. everything looks black and white (or gray)
B. unlike others, he can't see ultraviolet or infrared
C. he can sense only three colors
D. he can't distinguish red from green

38. The Keck uses what to focus light?
A. computers
B. a lens
C. a mirror
D. It doesn't focus light.

39. Light in a vacuum goes 1 foot every nanosecond (billionth of a second). In water, in one nanosecond it will go about
A. 1 foot
B. 1.5 feet
C. 0.66 foot
D. 0 feet (light doesn't travel through water)

40. Present-day 3-D movies make use of which technology?

    A. polarizers

    B. holograms

    C. dispersion

    D. mirages

41. A computer screen does not have spots that are colored

    A. red

    B. green

    C. blue

    D. white

42. For a good telescope (like Hubble) in geosynchronous orbit (HEO), a typical resolution (for objects on the surface) is about

    A. 4 inches

    B. 3 feet (about 1 meter)

    C. 20 feet

    D. 300 feet

43. A major use of fiber optics is for

    A. communication

    B. making flashlights brighter

    C. focusing light for lasers

    D. traffic lights

CHAPTER

Invisible Light

An Opening Anecdote: Watching Illegal Immigrants Cross the Border in Darkness

In 1989, I had an opportunity to spend a night with the U.S. Border Patrol that guards the U.S./Mexican border near San Ysidro. After touring their facilities and having dinner, we went to a hillside overlooking the border just as the Sun was setting. Many people were gathering on the Mexican side. There were stands selling them tacos and hot dogs, primarily (I was told) to people who had been caught the night before and had to spend a day waiting until nightfall to try again.

It began to get dark, and the other side of the border was getting crowded. I could still see everyone clearly. Suddenly, one boy ran to the fence, climbed over, and ran to hide on the U.S. side. That seemed to trigger an avalanche. Hundreds of people swarmed toward the fence, young and old, some with ladders, and in a few minutes they were all across and disappearing into the gullies of the desert on the U.S. side.

My host, the border patrol, did nothing for a while, and then drove me along a dirt road up to a hilltop. By the time we got there, it was dark. The lights of Tijuana twinkled in the distance, but the desert between us and the border was dark. On the back of a jeep, the border patrol mounted a special pair of binoculars, and they were surveying the darkness. The binoculars were cooled with liquid nitrogen, and hooked to a battery. These were *night-vision* binoculars. I was allowed to use them myself and scan the countryside. Through the binoculars, I could see mostly blackness, the outline of the hills, and in the valleys (from our high vantage point), clusters of people glowing in the dark. Their faces and hands were bright, and the rest of their bodies somewhat dimmer. They were waiting. At one location, they had lit a small fire (visible even with the unaided eye as a small reddish spot), and that was extremely intense white in the binoculars.

"What are they waiting for?" I asked.

"Their guides," the border patrol officer replied. The immigrants had been given simple maps to show them how to get away from the border, to a location that they could find less than a mile in from the border. This is where they would be met by the guide they had hired.

They waited a long time, and so did we. Finally, after over an hour, the groups started moving through the gullies. I wondered if they knew how easily we could see them in the absolute darkness. As one group approached a road, we drove over to them. They heard the car coming and waited.

I asked, "Why don't they run?"

The border patrol officer answered, "Because it's too dangerous. They might get lost, and besides, if they do get caught, we will only send them back to Mexico, and then they can try again tomorrow."

—Richard A. Muller

For a nonphysics observation on these events, see the footnote.[1]

What were the mysterious binoculars? How could they see in the dark? Where was the illumination coming from? Why did they require liquid nitrogen?

In figure 9.1, I show an image similar to what I saw that night.

Figure 9.1 Infrared image of two people cutting a fence and one person climbing over. Notice that the warm parts of the image (face and hands) are brighter than the cooler clothing. (Copyright Indigio Systems, used with permission.)

[1] This is not the place to judge the wisdom of the border policy in 1989. A few years later, the border was tightened. But in 1989, I did ask one border patrol officer what he thought of those trying to sneak into the United States. "They are really decent people," he said. "They come here, work hard, and send most of their earnings back to their families in Mexico. I wish I could get my son to work that hard." The life of the border patrol officer struck me as being similar to that of Sisyphus. And, as Albert Camus argued in his essay, *The Myth of Sisyphus*, Sisyphus was happy. See http://muller.lbl.gov/teaching/Physics10/pages/sisyphus.html.

Infrared Radiation

The mysterious binoculars I used at the border were an optical system that could see infrared light. Infrared is light that has a wavelength longer than that of visible light. It is often called IR. The longest wavelength of visible light that we can see has a wavelength of about 0.65 micron (μ, or μm since it is a millionth of a meter). Infrared light has a wavelength from 0.65 micron up to about 20 microns. Since its wavelength is longer, its frequency is *lower* by the same factor.

Humans emit infrared radiation because we are warm. The electrons in our atoms shake, since they are not at absolute zero. A shaking electron has a shaking electric field, and a shaking electric field creates an electromagnetic wave. The effect is analogous to the emission of all other waves: shake the ground enough, and you get an earthquake wave; shake water, and you get a water wave; shake the air, and you get sound; shake the end of a Slinky, and a wave will travel along it.

The result is that everything emits light, except things that have temperature at absolute zero. Everything glows, but most things emit so little light that we don't notice. And, of course, we don't notice if most of the light is in a wavelength range that is invisible to the eye.

What are some of the things that emit a noticeable glow from heat? Here are some: a candle flame, anything being heated "red-hot" in a flame, the Sun, the tungsten filament of a 60-watt lightbulb, a kiln used for firing pottery. This glowing is often referred to as *thermal radiation* or as *heat radiation*.

Heat Radiation and Temperature

The amount of heat radiation can be calculated using the physics laws of thermodynamics. The results are shown in figure 9.2. The horizontal axis is the wavelength, and the vertical axis is the intensity of radiation emitted in a square meter. You'll notice that the units are kilowatts per square centimeter in the visible band.

Examine this plot, because it tells us a lot about heat radiation. The straight vertical lines show the wavelengths that correspond to blue, green, and red. Light that is within these lines is visible—that's called the *visible band*. The light at smaller wavelengths is invisible ultraviolet light; the light at longer wavelengths is invisible (except to those special binoculars) infrared light. This kind of radiation is also called *blackbody radiation*. It gets this name because it turns out that good absorbers (things that are black) are also good emitters. Black things don't reflect much light, but their shaking electrons are good at radiating it.

Each curve is marked with a temperature. The lowest temperature is 3000 K, the highest is 7000 K. So only very hot temperatures are present in this plot.

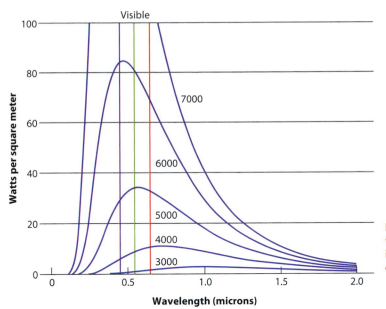

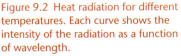

Figure 9.2 Heat radiation for different temperatures. Each curve shows the intensity of the radiation as a function of wavelength.

RED-HOT

Look at the curve for 3000 K. It has very little power in the visible band. Most of the power is in the infrared region. And the light in the visible band is stronger in the red than in the blue. An object that is heated to 3000 K glows red. We say it is *red-hot*.

WHITE-HOT

Now look at the curves for 5000 K and 6000 K. The temperature of the surface of the Sun is near the 6000 K curve. Notice that there is a lot of red, green, and blue. Although the blue is higher, the levels are all high. It is this almost-equal combination of red, green, and blue, that we call *white*. The Sun is *white-hot*. (I think it is fascinating that with all the nuclear processes going on within the Sun, in the end, it is only the high-temperature shaking electrons on the surface that are responsible for the light!)

BLUE-WHITE

The plot doesn't show the top of the 7000 K curve, but you can guess that the blue part is the most intense at this temperature.

The fact that the color changes with temperature is summarized by the following important law: we can calculate the wavelength L of the most intense light from the temperature from the color law[2]:

$$L = \frac{3000}{T}$$

[2] In physics texts, the color law is called the Wien displacement law.

In this equation, if T is in degrees Kelvin, then L is in microns. So, for example, if the temperature is $T = 6000$ K, then the maximum emission should occur at $3000/6000 = 0.5$ microns. You already knew this was true (the Sun is nearly 6000 K, and sunlight is about 0.5 microns). We'll find this law very useful when we get back to explaining the glowing images of the immigrants through the night-vision binoculars.

There is a surprise in this law. Most people think of the color red as hot and blue as cool. That is the way the terms are used in art, and they derive their sense from flames and water. But for flames, the hottest flame glows blue-hot, not just red-hot. Gas stoves usually give off a blue flame; candles, which are cooler than gas stoves, give off a red flame. Potters can estimate the temperature of their kilns by the color of the glowing stones that line it. Astronomers can use the color of stars to estimate their temperature.

Here is the physics behind the law. When things are hotter, the electrons in them shake faster. Remember that it is the shaking of electrons that creates the electromagnetic waves. Hotter electrons give higher frequency (bluer) radiation. Higher frequency is shorter wavelength.

COOL COLORS—WHEN BROWN PAINT IS WHITE

Buildings are warmed, in part, by the sunlight that is absorbed on their roofs. If the roofs are white, that means that much of the sunlight is reflected. White roofs can significantly lower the air-conditioning cost. But people don't like white roofs. Neighbors complain that they are too bright.

There is a remarkable way to reflect half of the light even with a dark (black or brown) roof. The trick is to use a paint that reflects the infrared radiation but absorbs the visible. Take a look at figure 9.2 again, and concentrate on the curve that is labeled 6000 K—because that is the temperature of the Sun and shows the radiation of sunlight. Compare the visible part of the spectrum to the long-wavelength infrared part. It turns out that over half the power is in the infrared! (It is not a high part of the plot, but it is much wider.) So if you make a paint that reflects in the infrared, but absorbs in the visible band, then the paint will look dark to the human eye, and yet reflect most of the incident power.

This is now being done, particularly in regions of the country that are very hot. Physicists sometimes like to describe such paint with a phrase that they understand but that is very confusing to people who don't understand invisible light. They say that the paint is *white in the infrared*. Can you see what they mean by this?

This is a good topic to research on the Web. A common term for the concept is *cool roofs*.[3]

Total Power Radiated

One thing that is pretty clear from these plots is that when objects get hot, they emit much, much more radiation than when they are cool or even medium warm. Look at the total radiation emitted at 3000 K and compare it to the

---

[3] See, for example, http://eetd.lbl.gov/HeatIsland/CoolRoofs/.

amount at 6000 K. The amount at 6000 K is much more. Actually, it turns out to be 16 times more! The rule is that the total amount of radiation is proportional to the fourth power of the temperature[4]:

$$P = A \, \sigma \, T^4$$

In this equation, A is the surface area in square meters (bigger areas emit more radiation), T is the absolute Kelvin temperature, and the Greek letter σ (sigma) is a constant $= 5.68 \times 10^{-8}$. But the important fact is the fourth power. This means that if something is twice as hot, the radiation it emits is increased by a factor of $2^4 = 16$. If it is 3 times hotter, the radiation is $3^4 = 81$ times stronger. So the amount of radiation increases a lot when the temperature goes up just a little.

The surface of the Sun has a temperature of about 6000 K. That's about 20 times hotter than room temperature. How much brighter is the Sun that the surface of the Earth? The total power emitted goes up by the factor $20^4 = 20 \times 20 \times 20 \times 20 = 160{,}000$. That's why the Sun is so bright.

TUNGSTEN LIGHTBULBS

Ordinary screw-in lightbulbs have a hollow glass sphere with a small tungsten wire inside. Electricity flowing through the wire makes it very hot, typically about 2500 K. The light from such a filament has much more red light than blue light. Since the plot in figure 9.2 doesn't show the line at 2500 K very well, I've re-plotted it in figure 9.3, showing *only* the power emitted for a 2500 K object.

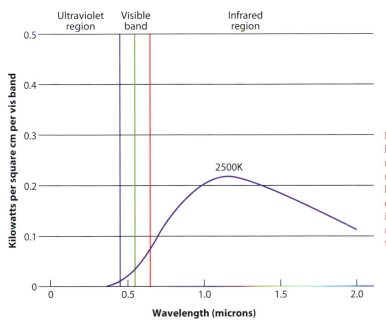

Figure 9.3 Radiation from a tungsten lightbulb. Notice that most of the radiation is in the invisible infrared region. That means that this kind of lightbulb is very inefficient in turning energy into light. (Infrared radiation is often called *heat radiation*, which means that we can say a bulb such as this produces more "heat" than light.)

[4] In physics texts, this law is sometimes called Stefan's Law, and σ is called the Stefan-Boltzmann constant.

Remember, we see light only if it is in the visible band. So it isn't really "white." Most people can see that, especially if they compare the light from a tungsten bulb to the light from a fluorescent light or to that from the Sun. Light from a tungsten bulb is reddish.

The fact that it is reddish makes many people think they look healthier in the light from an ordinary lightbulb. Bulbs used for photography usually have higher temperatures, typically 3300 K, in order to have more green and blue, but even so, they give reddish photographs unless additional corrections are made.

Lightbulbs use much more energy than they radiate. The "watts" label on the bulb tells you how much power you will use to light the bulb, not how much light comes out. Typically, only 16% is turned into visible light; most comes out in the invisible infrared region, is absorbed by the room, and is converted to heat.

OPTIONAL: CALCULATION

Here is an optional calculation that gives the 16% figure. If you break open a 100-watt bulb (as I did just before I wrote this), then you'll find that the filament has a diameter of about 1/10 mm, and a length of about 20 mm. Counting both sides of the filament, the area is about 4 mm^2 = 0.04 cm^2. If it operates at 2500 K, then we can calculate the power emitted in the visible band from the figure at the top of this page. It is 0.04 kW/cm^2 = 40 W/cm^2. Multiply this by the area to get the power is $40 \times 0.4 = 16$ W.

Some cameras can see this infrared radiation. They have special settings for photography in "total darkness." But what total darkness means that for us (no visible light) does not mean the same thing for the special cameras, since their sensors can "see" the infrared radiation. Such cameras often have small infrared lights on them to provide the light and illuminate the object that the camera is viewing.

When you use such a camera, the viewfinder will turn the infrared signal into a visible image. But, of course, there is no color, because all the camera is detecting is infrared. It does not detect how much red, green, or blue light is reflected off the object, only how much infrared is reflected.

HEAT LAMPS AND "HEAT RADIATION" (REALLY IR)

Suppose that we make a lamp that operates at about 1500 K to 2000 K. The amount of visible light will be almost zero, and there will be considerable infrared emission. Such lamps exist—they are called *heat lamps*. The watts emitted are mostly invisible, and yet they will be absorbed on your skin. A little bit of visible light is still emitted, and this is usually seen as a dull red glow. You can get the same warmth by sitting in front of a bright lamp. The advantage of the heat lamp is that you can warm part of your body without having a bright, distracting, visible light. Many people refer to infrared radiation as *heat radiation* because of the invisible warmth it can induce.

Return to Watching Illegal Immigrants

We are back to the example that I used at the beginning of this chapter. If everything that is hot also glows, let's think about humans. Our temperature is about 98 F = 37 C = 316 K. But our skin is cooler, about 85 F = 29 C = 302 K, which we'll approximate as 300 K. Using the color law, let's first figure out the wavelength that we radiate. Plugging in, we get that the wavelength is $L = 3000/300 = 10$ microns. That is pretty long wavelength, but it is still considered to be in the infrared region.

How much power is emitted? Let's use the total power law to compare the power emitted by our body-heat radiation to the power emitted at the surface of the Sun. Instead of 6000 K, the temperature is 300 K, a factor of 20 less. So the power emitted should be less by a factor of 20^4—i.e., by a factor of $20 \times 20 \times 20 \times 20 = 160,000$. So instead of emitting roughly $16\,kW = 16,000$ W/cm^2 (read off the chart), we should emit 16,000 divided by $160,000 = 0.1$ W/cm^2. What is our area? I estimate that I have a total surface area of about 500 cm^2. So my total radiation should be about 50 W.

That 50 W of radiation is quite a lot. The number shouldn't surprise you, however. You know that the presence of a person can significantly warm a small room. Moreover, you know that a person, even if inactive, eats about 2000 Calories per day. That is equivalent to a total power use of about 100 W. (For the calculation, see the footnote.[5]) What we have shown is as much as 50 W of that is just because you are warm. No wonder it was possible to devise an instrument that could detect this radiation and make a visible image for me to see through the binoculars. The infrared radiation was absorbed by a surface, and this was used to create a picture, like a TV picture, which is what I was really looking at through the binoculars.

Of course, we are usually surrounded by objects (e.g., clothing, a house, the ground) that are only a little cooler than we are, and we absorb radiation from them. So we don't have to supply all the energy from food. Yet this number illustrates why humans can barely survive on 1000 Calories per day. We burn them just to keep warm. In the United States, a teenager consumes about 3000 Calories per day, and an adult about 2000.

Sleeping under a Clear Sky (and Getting Wet from Morning Dew)

When I go backpacking, I often sleep out in the open without using a tent, so I can watch the stars and look for meteors. But I learned the problem with doing that: in the morning, the outside of my sleeping bag would be covered with dew. I'd have to wait until I had some direct sunlight to warm up the bag and let it dry out. However, I found that if I slept under the cover of a tree, I

---

[5] 2000 Calories per day is 2000×4200 joules per day = 8.4×10^6 joules per day. One day consists of 24 hours of 60 minutes, each with 60 seconds, so the number of seconds in a day is $24 \times 60 \times 60 = 86,400$ seconds. So the joules per second is $8.4 \times 10^6/86400 = 97$ watts ≈ 100 watts.

wouldn't be wet in the morning (although, I couldn't see as many meteors). What was happening?

The answer is that the tree, unlike the night sky, radiates infrared radiation just as I do (although not quite as much, since it is a few degrees cooler). So the outside of my sleeping bag was kept warm by the tree. When I have nothing but a black night sky above, then I radiate infrared radiation into the sky. Nothing comes back except a little starlight, but that isn't much. So the surface of my bag gets much colder when it is under a night sky than when it is under a tree. Water vapor tends to condense (form dew) on cool surfaces.

A cloudy sky can also keep the ground warm enough to keep dew from forming. Some people say that a cloud acts as a "blanket," but that doesn't really explain what is happening. In reality, a cloud thick enough to be opaque radiates infrared radiation at me, and that partially balances the infrared radiation that I emit, so I don't get as cold.

Remote Sensing of the Earth

Infrared satellites can measure the temperature of the surface of the Earth by the amount of infrared radiation emitted. Figure 9.4 shows satellite measurements of the temperature of the ocean surface, called sea-surface temperature (SST).

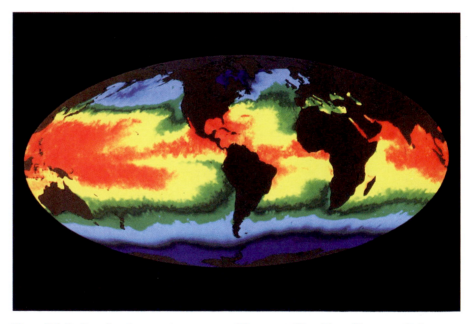

Figure 9.4 Earth surface temperature, measured from a satellite with an IR camera. Red indicates warmer; blue is cooler. (Courtesy of NASA.)

Weather Satellites

Modern weather satellites take images in visible light (wavelength 0.5 micron) and IR (near 10.7 microns). The visible measurements are useful for observing clouds, and the IR can be used to measure the temperature. Some modern satellites also include a kind of infrared multispectral that is sensitive to other wavelengths th at give additional information about the surface. One of the most important of these is a camera that is sensitive at 6.5 microns. This is a wavelength that can detect water vapor. Air that is high in water vapor will form clouds (and rain) if it is cooled, so a photo showing the flow of water vapor can be used to help predict where storms might form. Such an image is shown in figure 9.5 with a map superimposed. (No, the state outlines are not really visible from space.)

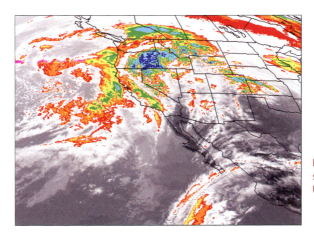

Figure 9.5 Weather satellite image showing water path over the western United States. (Courtesy of NOAA.)

Military Special Ops

The U.S. Special Operations Command has a motto: "We own the night."[6] The reason for this motto is that they plan to do almost all of their operations at night. They do all their training at night using night-vision viewers that are worn over their eyes. Some of these are simply *image intensifiers*—that is, they amplify the existing light. There are also two types of IR viewers. One type is sensitive to long-wavelength IR (about 10 microns) and picks up the infrared radiation emitted by warm people. The other type picks up short-wavelength IR between 0.65 and 2 microns. (That's still longer than visible light, but shorter than the long-wavelength IR.) What good is that? People don't emit much light in that range. Well, it wouldn't be very useful, if the Special Ops didn't carry infrared flashlights. These emit a light that can't be seen by the enemy, but can be seen through their infrared viewers. I found a set of such goggles, with built-in light, for sale on the Internet. They operate at a wavelength of 950 nm = 0.95 micron. That's almost twice the wavelength of visible light.

Infrared viewers are also valuable for detecting anything that is warmer than the surroundings. An infrared viewer, looking at an automobile, can tell if that automobile has been operated recently because it shows the warm region near the

[6] The Army has now adopted this motto too, since they now train in the same way for night combat.

engine. An infrared viewer can detect the heat from a campfire long after it has been extinguished. It can tell which caves have people inside them because those caves have warm air coming out of them. The U.S. military knows that the infrared viewers will be even more effective when the ground is covered with snow, because the background "clutter" from ground emissions will be reduced.

In addition to ground troops, the United States operates unmanned aircraft called *drones* or *unmanned air vehicles* (UAVs) that carry infrared imaging devices. The most impressive of these is the *Global Hawk*. If you want to see a very strange looking aircraft, do a Google search and look at an image of the *Global Hawk* aircraft. This drone can fly unmanned, directed by GPS, from Germany to Afghanistan, then circle over Afghanistan for 24 hours, and then fly back. In the meantime, its cameras (infrared and visible) and radar are constantly sending images back to the United States by radio. It flies at 65,000 feet, the same altitude as the classic U-2 spy airplane, and far above the ability of the Taliban to shoot it down. In fact, it is so high, that they probably will never know when it is watching them. From the angle of the tails, the drone appears to be incorporating stealth technology to make its radar signature weak. That is probably the reason the engine is mounted on the back of the plane, where it will not be easily visible to ground radar—after all, the many angles in the engine might reflect radar back to the source, just like a retroreflector.

Stinger Missiles

One of the real worries of the American troops in Afghanistan was the threat of Stinger missiles. These are missiles that can be launched from a device, held by a single soldier, and aimed at a low-flying airplane or helicopter. They weigh only 35 pounds, and can reach up to about 10,000 feet. These missiles were given to the Taliban by the United States back in the days when the United States was trying to undermine the Russian-imposed government.[7]

The reason that I am discussing the Stinger missiles in this section is because of the way they work. They have a device that steers them toward anything that is emitting strong infrared radiation. That means that anything hot, and up in the sky, such as the tail pipe of an airplane or a helicopter. In order to prevent a Stinger from hitting the airplane, the pilot will sometimes drop hot flares, and the missile will choose to go after those instead. But some of the Stinger missiles have special devices to help keep them from being fooled. Because the Stinger is so small, it can change direction more rapidly than the helicopter, so it is hard to outmaneuver.

Pit Vipers and Mosquitoes

Pit vipers are highly poisonous snakes that have regions on the sides of their faces called "pits." What makes them interesting to us is that the pits can detect

---

[7] You can read more about the Stinger at the Federation of American Scientists site, at www.fas .org. This is perhaps the best site on the Web on U.S. national security issues. To go directly to the page on Stinger missiles, go to www.fas.org/man/dod-101/sys/land/stinger.htm.

infrared radiation! So the pit viper can sense the closeness of prey even in "total darkness" (which refers, of course, to the absence of visible light). Figure 9.6 shows the head of a pit viper. The "pit" that is sensitive to infrared radiation is the small dark area at the intersection of the white and black arrows.

Figure 9.6 Pit viper head. The arrows point to the small "pit," which is sensitive to infrared light. The pit viper uses this to sense the proximity of warm prey. (Photo by J. V. Vindum.)

Mosquitoes are also attracted to infrared radiation.[8] Why don't other animals detect infrared radiation? Probably because their eyes work well. It is very hard for a warm-blooded animal to have an infrared sensor, since the animal is emitting so much IR itself. The pit viper, of course, is a cold-blooded snake.

Humans actually do have some ability to detect infrared light by the warming it causes on a surface. Put something hot near your lips, and they will sense the warmth even without touching. Do this and you are really detecting the heat generated by the absorption of IR. Put your face close to that of someone else (with permission), and you can sense the warmth. Most of this comes from the infrared radiation they are emitting. But be careful if you do this experiment—if you get too close, it could lead to unintended consequences.

Infrared radiation, even though it is invisible to our eyes, is just another form of light. It is similar enough (just a little longer wavelength) that it tends to bounce off glass and metal surfaces in the same way that visible light does. We can bounce IR off mirrors and focus it like visible light.

Infrared in the Home: Remote Controls

Suppose that you put a solar cell on your TV and made it turn on a switch whenever it senses a bright pulse of light. Then you could control your TV using a flashlight.

That's actually just what you do when you use the remote control for your television, except for the fact that the flashlight uses infrared light, so you don't see it. When you push a button on the controller, it flashes a series of pulses, with the number and relative spacing of the pulses depending on the function you pushed. A light cell (similar to a solar cell, but sensitive to IR) receives the signals and sends a pulse to the electronics that direct it to change a switch.

Look at a remote control device. You'll see a dark red plastic spot. That is designed to transmit the IR. Now look on your TV. You'll find a similar spot

[8] Here is a site I found on the Internet that discusses this www.howstuffworks.com/mosquito3.htm.

somewhere on it; that's the receiver. Cover it, and you'll find that the remote control no longer works. The IR "flashlight" is designed to come out in all directions. That way you don't have to carefully point the beam to make it hit the detector.

Some remote controllers use radio signals instead of IR, but they are generally more expensive. Most remotes use IR.

UV—"Black Light"

Black lights are lights that emit little or no visible light, and yet make other things glow. You can buy a black light at many hardware stores, and the salesperson will probably warn you that they are dangerous to look at. The reason is that black lights are actually glowing brightly at ultraviolet wavelengths, which are invisible to the eye, but still carry energy. The ultraviolet rays (UV) can burn the retina of your eye, and on the skin can cause sunburn and cancer.

Some chemicals, when they absorb UV, will then emit visible radiation. We call such chemicals *fluorescent* or *phosphorescent*. A fluorescent light is called that because of the interesting way it works. Electricity flowing in the gas of the light emits UV radiation. That radiation is absorbed by a coating on the inside of the lightbulb and is re-emitted as visible light.

The good-quality black lights that you can buy at a hardware store consist of an ordinary fluorescent bulb with the fluorescent coating removed (or, more likely, just never put there). A cheap (and not very satisfactory) black light can be bought at some stores, but it consists of an ordinary lightbulb painted black. Actually, it is black only to visible light, and the UV radiation emitted by the filament passes on through. Black lights made in this way emit only a tiny amount of UV and are not any fun to play with.

Black lights are popular during Halloween because of the way they can make certain things glow in an otherwise dark room. Geologists use black lights to detect certain minerals, which are known to fluoresce. If you've been to Disneyland, or some other exhibit that uses black lights, you've probably noticed that some clothing, particularly white clothing, glows very brightly, while colored clothing doesn't. Why is that? The answer is in the next section.

Whiter Than White

There was an advertisement a few decades ago for a laundry soap that made your clothes "whiter than white." That sounds absurd, doesn't it? Many physicists instinctively thought that the ad was nonsense and simply ignored the laws of physics. If all the light that is incident on a surface is reflected, then we say that that surface is white. How could anything be whiter than white?

It turns out that the physicists were wrong. Some clever laundry detergent people figured out a way to make clothing whiter than white! They added a fluorescent chemical to their soap and made it stick to the clothing. So after washing your white shirt, it would have this stuff stuck all over it. That's terrible, you say—the washing was making the shirt dirty! Yes, but it sold soap!

Why? How? Well, when people washed their white shirts, they noticed a difference. Hold the shirt out in sunlight, and it is brighter. In fact, you might even say, it looked "whiter." The shirt was reflecting all the red, green, and blue light, just as before. But in addition, it was absorbing the invisible UV radiation, and reemitting it as visible light. It was glowing, not in the dark, but in the sunlight. Anybody who looked at the shirt would notice it was brighter, and therefore (by the logic of that time) cleaner (even though it was coated with a chemical).

The shirt manufacturers caught on too, and they started adding fluorescent dyes to their white fabrics. Out in the Sun, which contains lots of UV (see the diagram in figure 9.2 for $T = 6000$ K), the brightness is very noticeable. It has less of an effect indoors, since tungsten filament bulbs operate at about 3000 K and don't have very much UV. The ads told you that if you really want to see if your clothing is clean, check it in sunlight.

Sunburn

UV light is responsible for sunburns, suntans, "windburn," and most cases of skin cancer. The form of UV that is most potent for burning and cancer has a wavelength of about 300 nm (0.3 μ). This kind of radiation is absorbed by ordinary window glass (window glass is "black to UV"). Sunglasses are much thinner than window glass, so UV gets through, and that can cause sunburn on your eyes; most sunglasses these days are made of a special glass that absorbs UV even better than window glass, and they are labeled with their UV absorption capabilities.

Germicidal UV Lamps for Disinfection

Black light lamps can be made to emit a large amount of UV with a wavelength of 254 nanometers; it turns out that this is a wavelength that is very effective at destroying DNA. That makes it very effective for destroying the ability of bacteria to reproduce. Such lamps are common at places that require absolute cleanliness, including operating rooms. They are called *germicidal lamps*.

UV Waterworks: Cheap, Clean Water for the Developing World

One of the most fascinating uses for UV lamps has been in rural India and other underdeveloped areas, where such lamps are being installed to shine on water that is pumped from contaminated rivers and wells. A few seconds under an intense UV lamp can be enough to cause mutations in the DNA of the bacteria, and as a result they cannot reproduce. The water must be passed under the UV lamp in a thin layer so that there isn't enough water to absorb the UV. The lamp is powered by a gasoline engine with an electric generator. It uses 60 watts. The cost for purification is about 4 cents per ton of water![9]

This program supplies water purification systems for poor villages all around the world. The microorganisms are killed with just a 12-second exposure to

---

[9] You can read more about this "UV Waterworks" project at http://eetd.lbl.gov/iep/archive/uv/.

the UV. Using the flow from a hand pump, the system can kill the bacteria in 4 gallons of water every minute.

Windburn—and Viewing Eclipses of the Sun

Your skin cannot tell when it is in bright sunlight except from the warmth. If it is a cold day, and particularly if the wind is blowing, your skin might feel cool, and you won't realize that you are getting a sunburn. Even more deceptive is a cloudy day. How can you get "sunburn" if you can't see the Sun? Well, you can. Just as the visible light isn't blocked but is spread out by the clouds (cloudy days can still be quite bright), the UV light is also not blocked. As a result, on a cool, breezy, cloudy day, you can get a bad sunburn without realizing it. Such a burn is often called a "windburn," which gives the misleading impression that you are being burned by the wind. You are being burned by UV, and it is the wind that keeps you from noticing.

During a partial eclipse of the Sun (when the Moon blocks the center of the Sun) people like to look at the Sun, and see the crescent shape. You can do this by looking through a piece of dark glass. But when you do this, you better make sure, first, that the glass is also dark in the UV. If it isn't, then you will be looking at a bright UV source, and that can sunburn your eye.

The Ozone Layer

When sunlight passes through air, the UV tends to break up some of the O_2 molecules into individual atoms, O and O. These atoms attach themselves to non-broken O_2 molecules to make O_3, also known as *ozone*. Ozone is a very strong absorber of UV radiation from the Sun.

If we didn't have any ozone, then we would expect the UV from the Sun to penetrate the atmosphere and reach the ground (and us). But when ozone forms at high altitude, it prevents the UV from reaching low altitude, and little UV makes it to the ground. As a result, little ozone forms at the ground. Most of the ozone is in the *ozone layer* at an altitude of 40,000 to 60,000 feet.

High-altitude ozone protects us from the UV. If the ozone layer were to vanish, then the UV would reach the ground, and problems with sunburn and skin cancer would be exacerbated.

Over the South Pole, the Sun rises only once every year, and when that happens, the ozone layer forms. This formation is studied with UV sensors in Antarctica. In the 1970s, scientists noticed that the amount of ozone was decreasing every year. This decrease became known as the ozone hole. A NASA plot of the ozone decrease is shown in the figure 9.7.

A plot of the actual numbers for the ozone depletion is shown in figure 9.8.

Was this decrease natural? Would it spread to the entire globe, or be restricted to Antarctica? Nobody knew, although some people thought that it might be to due to a pollutant introduced into the atmosphere by humans. In fact, that is what it turned out to be. I'll discuss that in the next section, on Freon and chlorofluorocarbons (CFCs).

The ozone layer absorbs enough UV to cause heating of the upper atmosphere. As a result, when clouds rise in the cool atmosphere, they stop rising

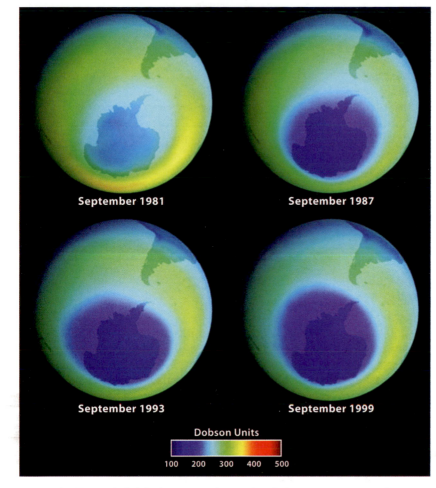

Figure 9.7 The "ozone hole" depletion over the South Pole from 1981 to 1999. The continent of Antarctica is prominent, and the southern tip of South America is in the upper right. Darker shading indicates ozone depletion. (Drawing based on DoE image.)

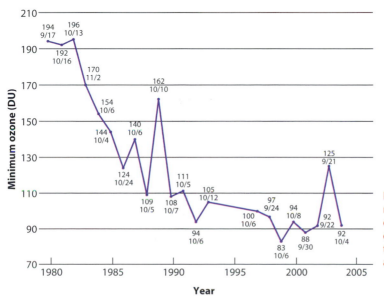

Figure 9.8 Plot of the ozone minimum reached each year over the South Pole. The decrease is believed to be due to CFC pollution in the atmosphere.

when they reach the ozone layer. (Warmer air up there is less dense than the warm air of rising cloud.) The cloud spreads out, creating the shape known as a *thunderhead*, or an anvil-shaped cloud. The image in figure 9.9 shows this spreading. To see where the ozone layer is, just look at the highest levels of the thundercloud, the altitude at which the cloud begins to spread.

Figure 9.9 Anvil-head cloud spreading at the ozone layer. (Photo by Cristina Ryan.)

Freon and CFCs

A chemical called Freon was in widespread use, in refrigerators and air conditioners, and as a solvent (cleaning agent). But Freon was a compound in a class called CFCs (chlorofluorocarbons) containing the elements chlorine and fluorine, and it was highly stable; it didn't decompose readily, so when it leaked into the atmosphere from defunct refrigerators and air conditioners, it stayed there for a long time. It eventually diffused into the high atmosphere, where it was hit by ultraviolet light and broken into its constituent chlorine and fluorine atoms. It turns out that chlorine is very effective at turning ozone back to ordinary oxygen, O_2; one chlorine atom can destroy 100,000 ozones. So Freon was, in effect, destroying the ozone layer.

The biggest effect was over Antarctica. Nobody knew why, until atmospheric scientists realized that certain crystals of nitric acid formed there in the early spring, and on the surface of those crystals, the chlorine was far more effective at destroying the ozone. So the biggest effect was seen first over Antarctica.

Nobody knew for sure whether the destruction of ozone would continue until it reached more populated areas, but the world was sufficiently worried that it outlawed the use of Freon. As a result, we expect that the ozone destruction will soon be halted, at least as soon as the existing Freon in the atmosphere is used up. The treaty to do this is called the Montreal Protocol.

Freon had also been used as a propellant for aerosol cans, for everything from shaving cream to insect repellant. It has been replaced for that purpose with other gases, including nitrous oxide. Some people still boycott aerosol products because they don't realize that the new ones are no longer dangerous to the ozone layer.

Some people say that the real lesson from this experience is that we can affect the atmosphere with human pollution, and that the effects are sometimes larger than we calculate. So we should be cautious. That brings us to another different, but similar problem: the potential atmospheric pollution from the burning of fossil fuels. This is the danger of *global warming*, and it is so important (and in the news) that we will devote the next chapter to the subject.

Seeing through Smoke and Dust—Firefighting and Infrared Astronomy

Waves with long wavelengths diffract easily past small obstacles. A long-wavelength water wave seems to pass right by a tower that sticks up out of the water. The same principle works for smoke particles and dust. Long-wavelength infrared light tends to diffract right around smoke particles and not be deflected.

This principle is becoming extremely important for firefighters. They can carry infrared cameras on their heads, with the screen in front of their eyes, when entering a smoky building. Much smoke consists of particles larger than the wavelength of visible light, and yet smaller than the wavelength of infrared. As a result, the infrared light passes right through the smoke, and the camera will show a scene that is not obscured by the smoke.

The same principle applies in astronomy. When astronomers want to be able to see through the dust that envelops the center of the Milky Way, they use infrared cameras. This technique was important in the discovery of the black hole that exists at the core of the Milky Way.

Electromagnetic Radiation—an Overview

Table 9.1 lists all of the forms of light including "invisible light." Some we have discussed in some detail, including visible, IR, and UV. Two of these, x-rays and gamma rays, are mentioned briefly in chapter 4.

The list is in increasing order of frequency. All the signals shown travel through a vacuum at the same speed of 3×10^8 meters per second = 186,282 miles per second = 1 foot per nanosecond (10^{-9} second = a typical computer cycle).

Table 9.1   Spectrum of Electromagnetic Waves

Name	Comments	Typical wavelength	Frequency
AM radio	Ordinary radio	300 m	1 MHz
TV, FM radio	Higher frequency for more bits/s	3 m	100 MHz
Microwaves	Radar, microwave oven, cell phones	10 cm	3 GHz
Heat infrared (IR)	Emitted by warm bodies (e.g., humans)	0.002 cm = 20 μm	15,000 GHz
Near infrared (NIR)	A strong color from the sun that we can't see	1 μm	3×10^{14} Hz
Visible light	Detectable by human eyes	0.5 μm	6×10^{14} Hz
Ultraviolet (UV)	Responsible for sunburn and can cause skin cancer	0.3 μm	10^{15} Hz
X-rays	Passes through water, not bone	10^{-9} m	3×0^{17} Hz
Gamma rays	Emitted by nucleus, can cause cancer, and are detectable from distant galaxies	10^{-11} m	3×10^{19} Hz

Radio Waves

Radio waves are electromagnetic waves at frequencies between several thousand cycles per second up to tens of millions of cycles per second. At these frequency ranges, they can carry only that many bits per second, and that is enough for ordinary radio. The highest frequencies are used for high-quality music (FM radio) and TV, since these signals require more information every second.

Because of their long wavelength (3 to 300 meters or more), they spread a lot when they pass through openings. (Recall that the spreading S depends on L/D. So large wavelength L means large spreading.) The longer waves can even spread around the curvature of the Earth, so they can be heard at long distances. That's why AM radio usually can be heard farther from the transmitting antenna than can FM or TV signals. FM and TV usually require "direct line of sight" since they don't diffract around mountains and buildings as much as the longer wavelength waves used for AM radio. The only reason that AM radio isn't used for everything is that the lower frequency can't carry as many bits per second, and so the sound on AM radio is not as good as on FM. (Signal compression methods have recently been used to improve this.)

Microwaves

Microwaves are very much like radio waves but with a shorter wavelength, typically only a few centimeters. Microwaves are used in radar, microwave ovens, and cell phones. Just like ordinary radio waves, they travel right through clouds and smoke, and that means that they can be used to send signals very reliably. Some of our satellites use microwaves to send telephone and Internet signals. Because their wavelengths are so small, microwaves can be bounced off dish antennas (and aimed at distant locations) without much spreading. A microwave dish acts like a lens. The microwaves are emitted near the focus; they bounce off the dish and come off parallel, aimed at some distant target. On mountain tops, you will see towers with small dish antennas that bounce microwaves from one dish antenna to another one that is perhaps 10 to 50 miles away. (The distance is really limited by the curvature of the Earth.) Because of their high frequency, they can carry many telephone conversations or many Internet signals simultaneously.

Radar

The word *radar* was originally an acronym for "radio detection and ranging." It is a method of emitting radio waves (primarily microwaves) into the sky and looking for reflections off metal objects such as airplanes. From the time it takes the signal to come back, the distance to the plane can be measured. Microwaves are used for radar because of their short wavelength. That means that they don't spread too much from a small antenna, so you can aim them accurately and determine the direction of the plane. Of course, it is also important for this application that they travel pretty easily through clouds and smoke.

Radar was first used to detect Nazi airplanes during World War II. It probably saved Britain from invasion in the following way: During the early parts of the Battle of Britain, the Nazis sent numerous bombers to attack London and other cities. Every time they reached the English shore, they were met by British fighter airplanes. The Nazis assumed, incorrectly, that Britain had thousands of such airplanes, since they were always patrolling the coast at the right place. In fact, there were not very many, but radar told the planes where to go to meet the incoming attackers. As a result of this deception, the Nazis overestimated the British strength and postponed an invasion that probably would have succeeded.

Radar can be used to locate airplanes, missiles, and rain clouds. Most large airplanes are equipped with radar to detect other airplanes and stormy regions. You can buy a relatively inexpensive radar system (for a few thousand dollars) for your pleasure boat, and use it to spot other boats and help you navigate through fog.

Radar Camera: SAR

Radar is now getting extensive use as an imaging tool and for mapping. It is possible to put radar in an airplane and look down at the ground. The received signal can be used to form an image of the ground. To make it even better, the image can be accumulated as the airplane flies, and the total signal (over five minutes of flight) can be put together to form the same kind of image that a huge radar receiver would have obtained—i.e., the image that you would have from a radar receiver that is several miles long (as long as the flight path). Such a system is called *synthetic aperture radar* (SAR) because it synthesizes the image. Figure 9.10 shows two SAR images: the Pentagon taken by a *Global Hawk* unmanned air vehicle, and a satellite SAR image of New York City.

Figure 9.10 Radar images. These were taken using synthetic aperture radar (SAR). The image on the left is the Pentagon building in Washington, DC, and the image on the right is New York City.

Microwave Ovens and "Radar Ranges"

The original microwave oven was invented by a man who stood in front of a radar emitter and found that it warmed him up. For this reason, the original microwave oven was called a *radar range*. We now understand that microwaves are absorbed by the water in your body, and their energy is converted to heat. In a microwave oven, the microwaves are confined by metal walls, so they bounce back and forth inside the oven, getting absorbed by whatever water-containing substance is inside.

You are always warned not to put metal inside a microwave oven. There are several reasons for this. The waves can create a high enough voltage across metal parts to create sparks that can do damage. But in addition, the fact that the metal will reflect the microwaves can lead to uneven heating of the food you are trying to cook.

Despite the fact that microwaves can warm you, it is now known to be very dangerous to your eyes. The reason is that the inside of your eyes have no good way to get rid of the heat that microwaves will create there. The man who was standing in front of the radar was lucky that he didn't damage his eyesight.

Cell Phone Dangers

Cell phones also use microwaves to carry their signals. Do you dare put a cell phone up to your ear? There are many urban legends about that. Low levels of microwaves cause only low levels of heating, and your eyes can handle that with no trouble. Microwaves do not have the ability to damage DNA in the same way as x-rays and gamma rays. Most of the fear that microwaves can cause cancer is simply a reaction to the fact that microwaves are sometimes called *microwave radiation*, and the use of the term *radiation* frightens some people.

X-Rays and Gamma Rays

X-rays and gamma rays are very high frequency, very short wavelength forms of light. We discussed these in chapter 4. X-rays are emitted by rapidly accelerating electrons. The x-ray machine that your dentist uses makes a beam of electrons and shoots them at a piece of heavy metal such as tungsten. When the electrons suddenly stop, they emit x-rays in the same direction that the electrons had been moving.

The same thing happens when a beam of electrons hits the front of an old-fashioned CRT TV screen. X-rays are emitted right toward the viewer. To stop this radiation, a special glass is used in the front of the TV, called *lead glass*. As its name suggests, this glass contains a lot of the heavy element lead, and that absorbs the x-rays.

Gamma rays are emitted by nuclei undergoing radioactive decay (i.e., the nuclei are exploding).

Because the wavelength is so short, and the frequency is so high, every wave packet of x-rays and gamma rays contain substantial energy. (We'll discuss this further when we get to chapter 11.) X-ray packets typically contain 50 to 100 thousand eV, and gamma ray packets contain 1 MeV or more. This high energy made them appear, to early scientists, to be energetic particles. Indeed, even energetic particles are waves, so the difference between them begins to blur. But unlike protons and electrons, gamma rays and x-rays are electromagnetic waves—i.e., they consist of electric and magnetic fields.

The most interesting applications of x-rays and gamma rays are for medical imaging. Indeed, when Wilhelm Conrad Roentgen first discovered x-rays (1895), he used them to make an image of the bones of his hand (figure 9.11). That astonished the world so much that many people assumed that "Roentgen rays" (as x-rays were originally called) were a fraud. People were unwilling to believe that such a miracle—seeing inside of the body—would be possible.

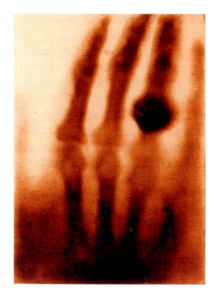

Figure 9.11 Roentgen's image of his wife's hand, one of the first x-ray images ever taken (1895).

Medical Imaging

Any radiation that can penetrate the body, and which moves in more-or-less straight lines, can be used to get information about what is going on inside the body. The value of that for medicine is obvious.

X-Ray Images

X-ray images are like shadows of semitransparent objects. They depend on the fact that heavy elements (such as the calcium in your bones) absorb x-rays more readily than do light elements (such as the carbon, hydrogen, and oxygen in your blood and muscles). X-rays are particularly useful for viewing bones.

To take an x-ray image, x-rays are emitted at a point, usually by hitting a heavy material such as tungsten with a finely focused beam of electrons. When the electrons suddenly stop, they emit x-rays. The object to be imaged, perhaps a broken bone, or (as in figure 9.12) a head, is placed between the emitter and a piece of film. The silver halide in photographic film is exposed by x-rays, and this is still the most popular way to detect them. Electronic recording, such as in a digital camera, is being put to use, but it doesn't offer the high resolution that photographic film does and most doctors still depend on that resolution for an accurate diagnoses.

One of the first x-ray images, taken by Roentgen, is shown in figure 9.11. An x-ray image of a human skull is shown in figure 9.12. Remember, it is similar to a shadow. X-rays tend to darken the film, so the background is black. The bright parts of the image are the regions in which fewer x-rays got through. So white regions represent bone; that makes the image easy to interpret.

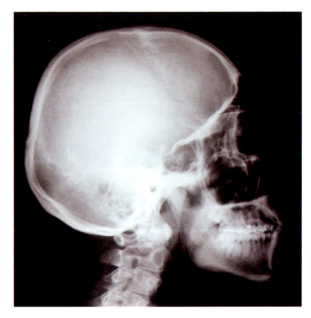

X-rays also have applications in industry and in national security. High-frequency x-rays and gamma rays can be sent through cars, trucks, and containers (from shipping) to look for heavy materials. They are particularly sensitive to the heaviest elements—such as uranium and plutonium. Thousands of containers enter the United States every day, and the government has begun a program to subject as many of these as possible to x-rays, in order to intercept such materials.

MRI (Formerly Called NMR)

Figure 9.13 is a *magnetic resonance image* (MRI) of a human skull. If you initially thought that it was a drawing, or a real skull that was physically sliced open, then you can probably understand the feeling of people who saw the first x-ray images a hundred years ago. In my mind, the image is a true miracle of modern physics. Such images have immense value to physicians and their patients.

Medical magnetic resonance imaging maps the distribution of hydrogen. The density of hydrogen is greatest in soft tissue (in carbohydrates and in water), and that's why you see the brain and tongue so well. It looks great, but notice that you can't see the teeth, as in the x-ray picture. Teeth do not contain much hydrogen.

Magnetic resonance imaging was formerly called *nuclear magnetic resonance imaging*, or NMR. However, the presence of the word *nuclear* frightened potential patients, and so it was dropped. (I'm not kidding. That's exactly what happened.)

The way it works is by detecting the wobble of the hydrogen nucleus—i.e., the proton. The patient is placed in a strong magnetic field, usually inside a

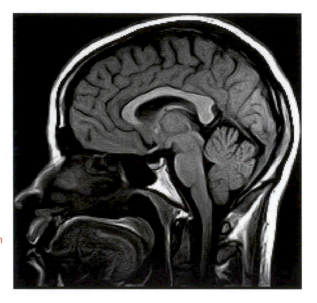

Figure 9.13 MRI image of human head. (Courtesy of NASA.)

cylindrical magnet. (Some patients don't like being placed in such a confined place.) The protons in the patient are also magnets (because they are spinning charges), and the external magnet tends to make them line up in the same direction. A radio transmitter is then used to create a rotating magnetic field, and that causes the protons to wobble. This wobble then can be detected with another radio receiver. The real trick is to identify the location from which the protons are creating this emission, and that is done by varying the frequency of the radio wave, varying the magnetic field, and using other tricks.

As far as we know, MRI does no damage whatsoever to the body. The magnetic field and radio waves do not affect DNA, and there are no known side effects. The only problem with MRI imaging is that it costs a lot, since the magnet is expensive. Some people think that the price will go down. Already some companies are offering relatively inexpensive MRIs, even as gifts that you can give to loved ones. It is not clear whether the MRI really is valuable as a way to scan healthy people for early signs of disease, or whether it will be used mostly to diagnose those with a high likelihood of problems.

CAT Scans (Computer-Aided Tomography)

Computer-aided tomography scans (CAT scans, or sometimes just CT scans) are made by taking x-rays from many different directions. That's the *tomography* part. As with other x-rays, it is most sensitive to heavy elements such as calcium. A computer can then combine all these images to calculate a 3-D map of the x-ray absorption. That's the *computer-aided* part of the name. Once all the data are in the computer, then the medical doctor can display any portion of the data. In the image in figure 9.14, the computer plotted the data as it would look if a thin slice were taken through the head.

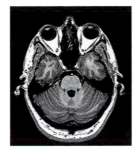

You'll notice that, even though x-rays are being used, the CAT scan contains lots of detail about the brain. The reason is that in the regions that have no bone, the contrast can be enhanced to show the very subtle differences in absorption that take place even in soft tissue.

The disadvantage of computer-aided tomography (versus MRI) is that it uses x-rays. Since many images must be taken, the x-ray dose can be large. CAT scans are often used in industry to look into structures to see if there are small cracks or other defects; in such cases, the dose is irrelevant.

PET Scans (Positron-Emission Tomography)

Positrons are a simple form of antimatter. (We'll discuss antimatter further in chapter 13.) Certain isotopes emit positrons (instead of alpha or beta particles), and these isotopes are extremely useful in medicine. A list of some positron emitters, along with their half-lives, is given here:

Carbon-11	20 minutes
Nitrogen-13	9 minutes
Oxygen-15	2 minutes
Fluorine-18	110 minutes
Iodine-124	4.2 days

To use PET, you prepare a material containing the positron emitter. For example, you could give the patient some iodine-124 (either in a pill, or injected into the blood). The iodine tends to concentrate at the thyroid gland of the patient. However, if part of the gland is not functioning right, that part may not get its share of iodine.

When iodine-124 emits a positron, the positron doesn't travel very far before it hits an electron. When that happens, the two particles disappear (that is called *annihilation*), and two gamma rays are emitted in exactly opposite directions from each other. Those gamma rays leave the thyroid and are detected with gamma detectors that measure exactly where they hit.

The instrument does not know where the gamma ray came from, but it knows that the origin was somewhere along the line connecting to two detection spots. To make an image requires millions of gamma rays. Each one gives the computer one line. The computer then calculates the distribution of I-124 in the thyroid that would give such a set of lines. That's the part called *tomography*. Any region of the thyroid that is not functioning properly will have a blank spot. A part that is hyperactive will be bright.

Positron emitters such as carbon-11 can be incorporated in biological compounds and then have their locations measured by PET scans. If you give the patient oxygen-15, you can detect the most active parts of the brain by seeing where the oxygen flows. The image in figure 9.15 shows such a PET scan, used to monitor the activity in the brain. Different shadings are generated by the computer to indicate the different intensities of signals coming from the various locations.

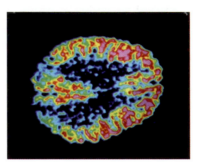

Figure 9.15 Imaging with antimatter: a PET image of the brain. (Courtesy of U.S. Department of Energy.)

The short half-life of the compounds used makes the method difficult to implement, since the radioisotopes decay so rapidly. But, on the other hand, they do not last long within the body, so the rem dose to the patient is usually small.

Thermography

We discussed the fact that objects of different temperatures emit different wavelengths of infrared radiation, as well as different intensities. A map of the IR emission therefore can give the medical doctor an image of the temperature of the skin.

This is a relatively new technique, and it is not completely clear what its value will be. However, it has already been found useful in diagnosing breast cancer, metabolic disorders, and certain kinds of injuries. Figure 9.16 shows the author of this book imaged with a thermographic camera. The colors were put in by the computer to indicate temperature.

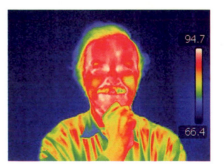

Figure 9.16 Thermograph of the author, taken with a camera sensitive to infrared light. The image is reproduced in "false color," with the scale on the right showing the relationship between temperature and color. Notice that hair and clothing are cooler than skin.

Ultrasound Imaging

We have talked about imaging with many different kinds of waves. Why not use sound waves? The basic problem is that most sound waves have very long wavelengths. A 1000-Hz sound wave, moving with $v = 330$ m/s, has a wavelength of $L = v/f = 33$ cm. For medical imaging, the problem is even worse. Recall (from chapter 7) that the velocity of sound in water is 1482 m/s, almost 1 mile per second. In body tissue, the velocity is about 1540 m/s. A 1-kilohertz sound wave would have a wavelength of 1.5 meters.

But if we go to very high frequency sound, say 100 kHz, then the wavelength is $1540/10^5 = 1.5$ cm. These high-frequency waves are called *ultrasound*, and their study and use is called *ultrasonics*. Because of the way that they can shake small objects, they are used for cleaning jewelry and other small objects and parts. Such devices are called *ultrasonic cleaners*.

Medical ultrasound generators use a frequency of 1 to 1.5 MHz, with a wavelength of 1 mm. They are completely inaudible, but can be detected with sensitive microphones. Such images are now commonly made to look at an unborn fetus. Ultrasound images can be used to detect possible problems long before birth. In figure 9.17, I show the image of the head of a baby, still in the womb, taken with ultrasound.

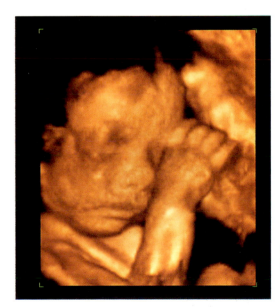

Figure 9.17 Ultrasound image of a fetus. (Courtesy of Shutterstock.)

Ultrasound—Sonar (Bats and Submarines)

Bats also use ultrasound, mainly to find prey and to avoid obstacles. Their ability to do this allows them to live in caves that are completely dark. Most bats can't see very well, and that leads to the expression "blind as a bat." But they don't need to. One flew into my living room a few years ago and zipped around the room like a trapped bird. But unlike birds, the bat crashed into nothing. In fact, I felt completely at ease watching him fly around, because I knew he wouldn't bump into me or anything else. Eventually, he found the open door to the outside and disappeared into the night.

Bats use frequencies just above the human audible range, about 20 kHz. In air, the wavelength is $L = v/f = 330$ m/20000 Hz $= 0.015$ m $= 1.5$ cm. That is good enough for them to locate flying insects,[10] to skim just above the surface of water to take a drink, and to avoid physics professors in their homes.

Submarines use a similar system for detecting objects and other submarines underwater. Sounds are emitted and their reflections mapped to determine the direction they are coming from. *Passive sonar* is more popular among submariners who don't want to be located by the sound they emit. Instead they listen to the sound that travels through the water (maybe in the sound channel) to detect noisy objects. The word *sonar* originated as an acronym for "sound navigation and ranging."

X-Ray Backscatter

X-rays are absorbed by heavy elements as they pass through matter. But some of them bounce backward, mostly off electrons. This bouncing is called *Compton scattering*, and the x-ray loses some energy (to the electron) when it does that.

The more electrons there are in the matter, the more backscatter there will be. Since there is one proton for every electron (and the number of neutrons is roughly equal to the number of protons), the result is that the density of electrons is approximately the same as the density of grams. So the amount of backscatter is roughly determined by the density of the material.

This backscatter has become important recently for some applications where you can't send the x-rays all the way through the object, or where you are more interested in seeing variations in density than identifying the heavy elements (such as calcium in bone). The spectacular image in figure 9.18 revealed illegal immigrants attempting to enter Southern Mexico from Guatemala in a truckload of bananas.

You can tell this is a backscatter image from the fact that you cannot see very far through the objects—the reflection is off the first few centimeters of material. Of course, some x-rays were also scattered backward off the truck wall, but as long as that scattering was uniform everywhere, it is just a uniform gray across the image, and that was removed by increasing the contrast. You can also see the columns that hold up the wall.

X-ray backscatter can also be used to search a person for concealed weapons. There are several objections to this, however. Even though the radiation dose is extremely low, many people fear it anyway, and will refuse to be x-rayed. But even if they overcome this fear, some people object because of a privacy concern.

---

[10] Bats could not "resolve" two close insects, but if they get a sonic reflection from a single bug, they can find the direction of the strongest signal; that gives them accuracy far better than a wavelength.

Figure 9.18 X-ray backscatter image revealing illegal immigrants attempting to enter southern Mexico from Guatemala in a truckload of bananas. (Image ©2006 American Science and Engineering, Inc., used with permission.)

The backscatter x-ray can easily penetrate light clothing and return an image of the surface of the skin, giving an image of a person who appears to be naked. I have been told that some countries who don't take individual rights seriously use x-ray backscatter on all airline passengers to detect weapons; that is plausible, but I have not verified this independently.

Picking Locks

X-ray backscatter can also be used to see what is inside the mechanism of a combination lock, in order to open it when you don't know the combination. I talked to a designer of advanced locks one time, and he showed me a special layer of heavy metal that they insert in front of the mechanism of their best safes put there specifically to defeat the known danger of x-ray backscatter attack.

Chapter Review

All objects that are above absolute zero in temperature emit light. The temperature law is that the maximum wavelength emitted is $L = 3000/T$. At room temperature, the peak is in the infrared ($L = 10$ microns). At 3000 K, the color is red, at 5000 it gets white, and at 7000 it is blue. The total power of radiation is proportional to the fourth power of the temperature, so doubling the absolute temperature results in 16 times as much power radiated. Tungsten lightbulbs emit reddish light because of their 2500 K temperature. Heat lamps emit primarily in the infrared. Humans emit very long wavelength infrared, but this can be imaged using special cameras. Infrared light emission from trees can warm the ground and help prevent the formation of dew. Military special ops use infrared viewers to be able to see at night. Stinger missiles home in on the infrared signal emitted by engine exhaust. Mosquitoes and pit vipers use IR to detect enemies and prey.

Ultraviolet (UV) radiation is light with a wavelength a bit shorter than that for visible light. "Black lights" emit UV. When UV hits a phosphor, the phosphor absorbs the UV and emits visible light. Such phosphors can make a material glow, and even look whiter than white. UV can cause suntan, sunburn, and skin cancer. It is also responsible for windburn. Most modern sunglasses block UV to prevent the eyes from damage. When UV from the Sun enters the atmosphere, it creates an ozone layer at an altitude of about 50,000 feet. The ozone is a strong UV absorber, and it keeps intense UV from reaching the ground. The ozone can be destroyed by chemicals known as chlorofluorocarbons, or CFCs, which were extensively used in refrigerators, air conditioners, aerosol sprays, and for cleaning, until they were outlawed.

The Earth is warmed by visible light and is cooled by IR emission. It is even warmer because of the greenhouse effect: the atmosphere absorbs IR and prevents the Earth from cooling as much as it otherwise would. Greenhouse absorbers include water vapor, carbon dioxide, and ozone. Humans have been adding carbon dioxide to the atmosphere, and this may be responsible for the observed global warming. A similar process makes the interior of automobiles get hot when they sit in the Sun.

Invisible light includes radio waves, TV and FM waves, microwaves, x-rays, and gamma rays. They are distinguished by their different frequencies and wavelengths. For all of them, however, $fL = c$, where c is the speed of light. Microwaves are used to heat water and for radar. They can damage eyes if they cause the interior to heat. X-rays and gamma rays have many applications for medical imaging, including computer-aided tomography (CAT) scans. Magnetic resonance imaging (MRI) was once called *nuclear magnetic resonance* (NMR). It gives good images of soft tissue containing hydrogen. Positron-emission tomography (PET) can show where chemicals (such as iodine) go in the body. Positrons emit gamma rays, in opposite directions, when they annihilate with electrons.

Infrared radiation can be used to image the temperature of the skin and is useful for detecting breast cancer and other physiological problems. Ultrasound can be used to image within the body—e.g., for images of a fetus. Bats use ultrasound to detect insects and obstacles. They bounce high-frequency sound off objects and then detect the bounce. When submarines use the same principle, we call it *sonar*.

X-ray backscatter is useful for looking into the surface of an object that is too thick to penetrate with x-rays.

Discussion Topics

1. Patients dying of cancer have refused to be diagnosed with nuclear magnetic resonance (NMR) imaging because of their fear of radioactivity. What are the real risks and benefits? What does this say about the way people feel about "nukes"?

2. On a cold morning, you'll sometimes see frost on cars. But cars that are parked under trees may not have any frost. Why?

3. Are some people afraid of ultrasound imaging? Why? Are there people on the Web who warn against it? Do you think they are right?

Is there a danger to the fetus? How great a danger? Are there good medical reasons to want an ultrasound image?

Internet Research Topics

1. Can some animals see in the UV light that is invisible to humans? Are there flowers and plants that have patterns of reflection in UV, meant for these animals? You can begin by looking at http://www.bio.bris .ac.uk/research/vision/4d.htm.
2. How do "tanning parlors" give people tans? Why do people doing this wear eye masks? Are there any dangers in getting a tan this way? What are the long-term consequences of getting a "healthy-looking" tan?
3. Find other images on the Web that were made using invisible light. X-ray images are common, but what about IR images? Can you find any taken in the UV? What are the most amazing medical images that you can find? Look for movies taken in the IR.

Essay Questions

1. Infrared radiation is also called IR. Describe how it is different from visible light. What emits IR? Can animals use it? Explain how emission of IR sometimes "wastes" energy. Give examples showing how IR is used in modern technology.
2. Modern medical imaging can do much more than take a simple x-ray. Define and explain four methods that are used for medical imaging. Explain the physics behind the techniques, describe what they measure, and say something about the applications.
3. Visible light is an electromagnetic wave. Name five other phenomena that are also electromagnetic waves but have different names. (Try to pick things that a nonphysics student would not know were "light.") How do these waves differ? Briefly describe applications for each of these waves.
4. Invisible light includes ultraviolet light, also called UV. What is this? How does it affect our everyday lives? What are its practical uses and applications?
5. Humans have been polluting the atmosphere in ways that could have lasting effects for future generations. What are the chemicals that have been put into the air that could do this? What effects could they have? Describe as much of the relevant physics as you can.
6. Describe the dangers and benefits of ultraviolet radiation. How can it cause harm, how can it be used for business, and how can it be used for beneficial purposes?
7. Discuss the military uses of invisible light.
8. How can invisible light be used to purify water?
9. Discuss what is meant by *cool colors*, especially with regard to paint. What problem does a cool color address? How does it work?

Multiple-Choice Questions

1. Which of the following has the highest frequency?
   A. gamma rays
   B. UV
   C. IR
   D. microwaves

2. Which of the following has the lowest frequency?
   A. TV
   B. AM radio
   C. visible light
   D. gamma rays

3. If you double the absolute temperature of an object, the wavelength of the emitted light
   A. gets longer by a factor of 2
   B. gets longer by a factor of 16
   C. gets shorter by a factor of 2
   D. gets shorter by a factor of 16

4. Which kind of light has the longest wavelength?
   A. red light
   B. blue light
   C. infrared light
   D. ultraviolet light

5. *Invisible light* includes
   A. UV, quarks, and gluons
   B. x-rays, UV, and IR
   C. blue, ultrablue, and ultrared
   D. quanta, antiwhite, and photons

6. As an object gets hotter and hotter, the color of its glow will change from
   A. white, to yellow, to red, to blue
   B. blue, to white, to yellow, to red
   C. red, to yellow, to white, to blue
   D. yellow, to white, to blue, to red
   E. none of the above

7. Radio waves and x-rays have the
   A. same frequency
   B. same speed
   C. same wave length
   D. same energy
   E. none of the above

8. If you double the absolute temperature, the total radiation
   A. is reduced to half
   B. doubles

C. increases by 4

D. increases by 16

9. Most of the energy of an ordinary tungsten-filament lightbulb is emitted in the color

   A. ultraviolet

   B. green

   C. red

   D. infrared

10. Heat radiation is also called

    A. IR

    B. UV

    C. white light

    D. gamma rays

11. A tungsten lightbulb has *100 watts* printed on it. About much power is actually emitted as visible light?

    A. 16 watts

    B. 40 watts

    C. 100 watts

    D. 220 watts

12. Which color star would be the hottest?

    A. blue

    B. red

    C. orange

    D. white

13. Sunlight is brightest in which color?

    A. UV

    B. red

    C. green

    D. blue

    E. IR

14. Which of the following exposes a person to the kind of radiation that can cause cancer? (Choose all that apply.)

    A. computer-aided tomography (CAT) scans

    B. magnetic resonance imaging (MRI)

    C. positron-emission tomography (PET)

    D. thermography

15. Because of their warmth, humans emit primarily

    A. sonic radiation

    B. infrared radiation

    C. ultraviolet radiation

    D. visible radiation

16. If you sleep under a tree, you won't get wet from morning dew. That's because the tree

    A. absorbs starlight

    B. keeps air from rising

C. emits infrared radiation

D. blocks the clouds

17. Dew forms when the ground cools
    A. by emitting UV
    B. by emitting IR
    C. by emitting visible light
    D. by conduction downward

18. A "heat-seeking" missile actually aims itself toward
    A. infrared light
    B. ultraviolet light
    C. visible light
    D. radio waves

19. The ozone layer is created by and absorbs:
    A. CFCs (chlorofluorocarbons)
    B. carbon dioxide from fossil fuels
    C. cosmic rays
    D. ultraviolet radiation

20. Ozone depletion is caused by (choose all that are correct)
    A. CFCs
    B. carbon dioxide
    C. fossil fuels
    D. cosmic rays

21. A few years ago, a certain kind of spray can was banned. These were the cans that used
    A. chlorofluorocarbons (CFCs)
    B. carbon dioxide
    C. ozone
    D. They were never banned.

22. *Multispectral* refers to
    A. UV, IR, and visible
    B. the many colors in visible light
    C. UV, IR, and x-rays
    D. "whiter than white"

23. The name *NMR* has been changed to
    A. CAT
    B. MRI
    C. ultrasound
    D. PET

24. Which of the following statements about the ozone layer is *not* true?
    A. The location of the ozone layer is the tropopause.
    B. The ozone layer causes thunderstorms to spread laterally.
    C. Ozone is destroyed by chlorine (e.g., from Freon and other CFCs).
    D. Ozone is the most important greenhouse gas.

25. The clouds in thunderstorms rise until they reach
    A. the carbon dioxide layer
    B. the CFC layer
    C. the top of the atmosphere
    D. the ozone layer

26. The original name for the microwave oven was
    A. convection oven
    B. radar range
    C. radio oven
    D. microwave

27. Microwave ovens use radiation with a wavelength
    A. about one micron
    B. same as radar
    C. same as lasers
    D. same as AM or FM radios

28. The microwaves in a microwave oven mostly heat
    A. carbon
    B. air
    C. water
    D. oxygen

29. MRI maps the distribution of which element to produce an image?
    A. iodine
    B. carbon
    C. calcium
    D. hydrogen

30. UV light (choose all that are correct)
    A. is responsible for sunburns and windburns
    B. is used to kill bacteria in the water of remote villages in India
    C. is also called *black light*
    D. is also called *heat radiation*

31. Which of the following is best for seeing the calcium in bones?
    A. x-ray
    B. MRI
    C. PET scan
    D. EEG

32. A black light is designed to emit
    A. UV
    B. IR
    C. radio waves
    D. x-rays

33. A positron is
    A. an antielectron
    B. the same as an electron
    C. a kind of quark
    D. a hypothesized star

34. Sunburn is usually caused by
  A. UV radiation
  B. IR radiation
  C. X radiation
  D. gamma radiation
  E. visible light
  F. all of the above

35. Which of the following uses antimatter?
  A. CAT scan
  B. MRI
  C. PET scan
  D. x-ray

36. To image the hydrogen in the body, use
  A. PET
  B. MRI
  C. CAT
  D. x-rays

37. What is the cause for "windburn"?
  A. cold from the wind
  B. friction from wind
  C. solar UV
  D. solar IR

38. Remote control of a TV is usually done using
  A. UV
  B. IR
  C. microwaves
  D. X-rays

39. Pit vipers and mosquitoes can sense
  A. UV radiation
  B. IR radiation
  C. Gamma rays
  D. Microwaves

40. Infrared radiation can (choose all that apply)
  A. help snakes locate prey
  B. create ozone
  C. cause cancer
  D. cause sunburn

41. Cancer is most likely to be caused by
  A. infrared radiation
  B. microwave radiation
  C. ultraviolet radiation
  D. white light

42. The advertising slogan "whiter than white" means that the material
  A. emits IR
  B. emits visible light

    C. absorbs IR

    D. emits UV

43. To make clothes appear extra clean, manufacturers apply

    A. infrared radiation

    B. electron beam

    C. ultraviolet radiation

    D. fluorescent dyes

44. Which of the following is a typical energy for an x-ray?

    A. 1 eV

    B. 1000 eV

    C. 50,000 eV

    D. 1,000,000 eV

45. In total darkness, you can see people if you have a camera that is sensitive to

    A. ultraviolet radiation

    B. infrared radiation

    C. gamma radiation

    D. far ultraviolet radiation

46. Cell phones emit

    A. microwaves

    B. x-rays

    C. gamma rays

    D. P waves

47. Each of these make use of infrared radiation *except*

    A. capturing illegal immigrants

    B. military "night vision"

    C. measurement of wave velocities in the ocean

    D. using trees to keep morning dew off campers

48. Bats navigate in caves by using

    A. ultrasound

    B. IR

    C. black light

    D. radar

49. The phosphor of a fluorescent lightbulb turns which of these into visible light?

    A. x-rays

    B. infrared light

    C. ultraviolet light

    D. alpha rays

50. "Cool roofs"

    A. must be white in the visible

    B. must reflect infrared

    C. must be brown in the visible

    D. do not absorb UV

51. The best kind of radiation to purify water is
A. visible
B. IR
C. UV
D. microwave

52. Which of the following are electromagnetic waves? Choose *all* that are appropriate (no partial credit).
A. radar
B. visible light
C. UV
D. x-rays

53. To image hydrogen in the brain, the best technique is
A. MRI
B. CAT
C. PET
D. x-ray

54. Which of the following is an advanced method that uses x-rays?
A. MRI
B. PET
C. CAT
D. SAR

Climate Change

Global Warming

Look at the plot in figure 10.1. It shows the average temperature of the Earth from 1850 to 2006. The steep temperature rise is what is called *global warming*.

Newspapers and politicians talk about global warming nearly every day. Sometimes you'll hear a report of a new scientific study, but more often a news report will mention global warming in the context of disaster—a hurricane or a series of tornadoes or drought somewhere or crop failure—that scientists say could be a consequence of global warming. The evidence appears to be so "overwhelming and incontrovertible" that many people pronounce that the "debate is over." Yet the debate continues.

In fact, much of what you hear every day is exaggerated, often on purpose. People feel so passionately about climate change, and they are so frightened about what is coming, that they overstate their case (either pro or anti) in an attempt to enlist proselytes. The heat of the debate can sometimes overwhelm the heat of global warming (which, incidentally, *is* real). In this section, I'll try to give the cool description of global warming. I will try not to exaggerate, either way.

The temperature of the Earth (averaged over the last decade) is now the warmest that it has been in 400 years. Figure 10.1 shows that the change since 1850 was almost 2°F (about 1°C). That doesn't seem like a lot, and in some sense it isn't. The reason so many people worry is that they fear that this is just a portent of what is to come. A substantial part of this rise is very likely a result of human activity, particularly by the burning of fossil fuels. If that is truly the cause, then we expect the temperature to keep rising. Although cheap oil is getting scarce, at high prices ($60 per barrel and above), there seems to be lots available. (I'll show the numbers later in this chapter.) And the countries that need lots of energy appear to have huge amounts of coal. Burn a fossil fuel, and you dump carbon dioxide into the atmosphere, and that's the problem.

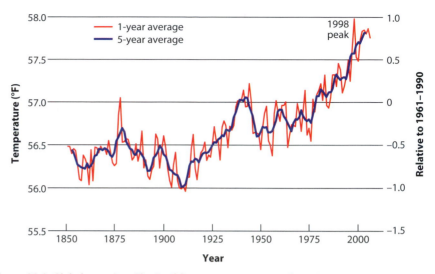

Figure 10.1 Global warming. The Earth's average temperature from 1850 to 2006, as measured with thermometers.

Carbon dioxide is very likely to cause significant warming, and as we burn more fossil fuels, the temperature is very likely to continue to go up. In the next 50 years, the best estimates are that the additional increase will be between 3°F and 10°F. That *is* a lot. Already, warming in Alaska from 1900 to the present has been enough to cause significant portions of the permafrost to melt. A 10°F rise would be enough to make fertile regions in the United States arid and trigger large-scale economic disruption around the world. There is also good reason to believe that the warming will be more intense in the polar regions.

The IPCC—The Scientific "Consensus"

Every few years, a prestigious international committee makes a new assessment of the status of climate change: what we know about it, what the consequences are likely to be, and what we can do. This organization is commissioned by the United Nations and the World Meteorological Organization, and is called the *Intergovernmental Panel on Climate Change*, or just the *IPCC*. The IPCC is important; you can't talk about climate change without knowing it, any more than you can talk about world affairs without knowing the letters *UN*.

The IPCC attempts to do the impossible: reach a consensus among hundreds of scientists, diplomats, and politicians. As a result, its conclusions are often muted and mixed, but its reports contain a wealth of data that help everyone evaluate what is going on. The IPCC shared the 2007 Nobel Peace Prize with former Vice President Al Gore. You need to know the initials. Memorize them: *IPCC*. You don't have to remember what the letters stand for.

It is important to know that there is a scientific consensus—even if not everyone agrees with it. It is also important to know what that consensus actually is, since people on both sides will exaggerate it or distort it and mislead you into thinking that they are describing the IPCC conclusions.

Many people have reported a large number of anomalous weather conditions. In his movie and accompanying book, *An Inconvenient Truth*, Vice President Gore showed increases in the intensity of hurricanes, tornadoes, and wildfires. Much of what he says is exaggerated; I'll discuss the details in this chapter. When such exaggerations are exposed, some people are tempted to dismiss the danger altogether, but that is false logic. Incorrect reasons put forth to substantiate a hypothesis do not prove the hypothesis false. There is plenty of reason for concern. Of course, the actions must be driven by an understanding of what is real and what isn't. Some proposed actions are merely symbolic; others are designed to set an example; others have the purpose of being a first step. Few of the proposals (and virtually none of those presently being put forth by major politicians) will really solve the problem. You need to know the difference between symbolic gestures and effective action.

To make matters worse, the burning of fossil fuels has another effect beyond global warming—one that gets attention in scientific circles but is not widely appreciated by the public. About half of the carbon dioxide emitted from fossil fuels winds up in the oceans, and that makes the oceans more acidic. The problem is not as immediate as acid rain, but still it can affect life in the oceans in potentially disastrous ways. The acidification of the oceans may be a bigger danger to the ecosphere than a few degrees of additional warmth. I won't discuss this problem further in this chapter, but you should not forget about it.

A Brief History of Climate

The most accurate data on the history of climate come from thermometer records covering the period from 1850 to the present; these were shown in figure 10.1. Creating that graph was not trivial. The Northern and Southern Hemispheres don't show exactly the same behavior, probably because two-thirds of the land mass is north of the equator. Care must be taken not to give too much emphasis to cities. Cities are often referred to as *heat islands* because human materials such as asphalt on the streets absorbs more sunlight than the flora they replaced, so cities are hotter than the surrounding countryside. Hot cities are more of a local effect than a sign of global warming. There is still some controversy about that IPCC plot, but I think it is the best one that anyone has produced.[1]

The bold line in figure 10.1 is a running average; that means that each point on it is actually the average of the nearby points on the lighter line. It helps to guide your eye so that you can see the trend. These thermometer data reveal several very interesting things. The average world temperature from 1860 to 1910 (left side of the plot) was about 2°F cooler than it is now. Don't forget that this 2°F number represents an average; some areas did not warm up as much (for example, the contiguous United States), and others warmed more.

---

[1] The data were compiled by the Hadley Center for Climate Prediction and Research, located in the UK. They take exceeding care in compiling the data. A different plot is published by the National Oceanographic and Atmospheric Administration (NOAA), but it is essentially identical.

The coolness of the previous centuries in Europe was enough to repeatedly freeze the Thames in England, and to ice over the canals of Holland during most of the winter. Without such cold, we wouldn't have stories like *Hans Brinker, or, the Silver Skates*. The canals rarely freeze over these days. The chill was the lingering end of the "Little Ice Age," a cold spell that took place all around the world. The length of this period is disputed, but it was likely preceded by a "medieval warm period" that lasted until about 1350.

Some people think that global warming is not caused by human activity, but that the Earth is simply still recovering from whatever natural phenomenon caused the Little Ice Age. The IPCC can't rule out that possibility; in fact, it is very possible that the rise from 1850 to 1950 was natural, perhaps due to changes in the Sun. The subsequent warming, from 1957 until now, is different; the IPCC says that this warming was very likely caused, at least in part, by human activity. They give the rise of the past 50 years only a 10% chance of being natural—that is, not caused by humans. If the rise is natural, then we are lucky; if past records are an indication of the limits to the natural variability, then the rise in temperature is unlikely to continue much further.

Many people misunderstand what the scientific consensus on global warming really is. Here is a summary: in its latest study (2007), the IPCC found that it is 90% likely that humans are responsible for at least some or possibly most of the observed global warming of the last 50 years. Many people are surprised, because they thought that the consensus statement was that there was overwhelming and incontrovertible evidence that humans are responsible for all of the global warming of the past 120 years! That may be true—or maybe not—but that is not the IPCC consensus. The exact wording of the consensus report appears in the footnote.[2]

Even though it will be expensive to act, a 90% chance is something that a president can't ignore. Even in its nonexaggerated form, the IPCC consensus conclusion is strong.

Look at the temperature in figure 10.1 again. Notice that the warmest year on record was 1998. It may seem odd that with all the global warming taking place, the warmest year was actually in the last century, not this one. But that is not a proper concern. The temperature change is not smooth but bumpy, with peaks and dips that depart from the average. We don't know what causes such fluctuations. The source may be natural variability in cloud cover. If you flip a coin 100 times, you don't always get 50 heads and 50 tails. Likewise, if the climate is changing, some years will still be warmer and some cooler than

---

[2] Here is the actual quote from the IPCC 2007 report: "The observed widespread warming of the atmosphere and ocean, together with ice mass loss, support the conclusion that it is extremely unlikely that global climate change of the past fifty years can be explained without external forcing, and very likely that it is not due to known natural causes alone." The IPCC defines *very likely* as 90% confidence. That is equivalent to saying that there is a 10% chance that none of the global warming is caused by humans. In other parts of the report, however, a similar statement is made with the word *most* added, implying that there is a 90% chance that humans are responsible for *most* of the warming. The IPCC report doesn't explain the discrepancy in words. If we include the word *most*, then its conclusion implies that there is a 10% chance that *most* of the observed warming is due to something other than humans, such as solar variability or natural fluctuations in clouds.

the trend. The figure shows that the natural variations fluctuate typically 0.2°F to 0.4°F away from the curve that represents the average. That's why scientists prefer to look at trends.

What if we look back further in time? Alas, good thermometers didn't exist in earlier eras, so we have no good record. However, indicators of climate can be found in ancient ice records, and based on that we can deduce something about ancient temperatures. This is a subject I studied deeply; I wrote many scientific papers and a technical book about it. I show one such plot below, figure 10.2, taken from the ice records of Greenland. In the ice, we measure the presence of different kinds of oxygen (heavier oxygen isotopes) that, based on other experience, appear to reflect temperature differences. Then we add an approximate temperature scale that is based on the known temperature measurements from the recent past.

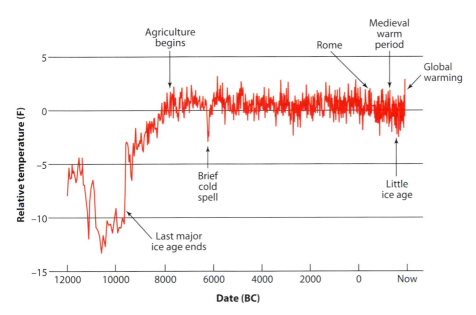

Figure 10.2 Temperatures from 12,000 BC to the present, estimated from Greenland ice measurements of oxygen isotopes.

On this plot, global warming looks small—it is only the slight upturn in the last tiny part of the curve on the rightmost side, nearly invisible in the clutter of data. But remember, it is not the present 2°F warming that concerns us, but the potential of a future 10°F warming. Look at the region marked *Little Ice Age*. It shows as a slight dip, about 1 or 2°F below the level set in the previous thousand years. There is also a *Brief Cold Spell* at about 6000 BC; we don't understand the cause.

The most dramatic thing on this plot was the period of extreme cold that started beyond the left side of the plot and suddenly ended about 9000 BC. That was the last ice age. Although it doesn't show on the plot, it had started about 80,000 years earlier. That's much longer than the time that expired since it ended. The temperature was more than 10°F colder than the present. That cold period makes the Little Ice Age look tiny.

Big ice ages are known to return in a regular way. The pattern is this: about 80,000 to 90,000 years of extreme cold, followed by a short 10,000 to 20,000 year *interglacial* warm period. Agriculture was invented at the beginning of the current warm period, as indicated in figure 10.2. All of civilization was based on agriculture because efficient production of food is what allows a minority to provide for the sustenance of the majority, and that means that there will be food for merchants, artists, and even physics professors.

The fact that the big ice ages recur scared some people in the late 1940s, when dropping temperature made people fear that a big ice age was about to start. Some scientists speculated that the cooling might have been triggered by nuclear bomb tests polluting the atmosphere. (The United States and the Soviet Union ended atmospheric testing in 1963, France continued until 1974, and China ended in 1990. Linus Pauling won the Nobel Peace Prize for his role in bring about this cessation.)

I was in elementary school at the time, and one of our textbooks had a drawing of the consequences to New York City, with 1000-foot glaciers toppling skyscrapers. Figure 10.3 shows a similar image that appeared on the cover of *Amazing Stories* magazine. It shows the Woolworth Building in New York being toppled by a returning glacier.

Figure 10.3 *Amazing Stories* magazine cover, showing the consequences of an ice age returning to Manhattan. In the last ice age, a huge glacier of roughly this size did cross the city; the debris it left behind is now called "Long Island."

To the relief of many people, temperatures began to rise again after 1970. The ice age was not imminent. Even though the cooling ended, no scientist today believes that the nuclear tests were at fault. Correlation does not imply causality. Many experts now attribute the brief cooling spell to an unusual number of volcanic eruptions that took place during those decades and spewed dust high into the atmosphere. Such material tends to reflect sunlight and thereby reduce the insolation—the solar energy reaching the ground. Once the dust settled and the volcanic activity ceased, the Earth began to warm again.

The rise in temperature continued, and now we are worried about warming. Is this a continuation of the prior trend, the finale to the Little Ice Age? Or is it the beginning of something more ominous? Our now deeper understanding of climate leads most scientists today to believe the latter. We'll now discuss how the burning of fossil fuels could be the cause of global warming. It is wise, however, to retain some humility, and to recognize that even a theory that explains what is happening may not be correct.

Carbon Dioxide

Carbon dioxide is created whenever carbon is burned. The chemical symbol for carbon dioxide is CO_2. The C stands for carbon, and the O_2 stands for two molecules of oxygen. (That's why it's called *di*oxide; the *di* means "two.") Burn carbon in oxygen, and you release energy and make CO_2. We can separate the carbon dioxide back into its components, but only by putting back in the energy we took out. If we've used the energy—for example, to make electricity—we are stuck with the CO_2.

Carbon dioxide is a tiny constituent of the atmosphere, only 0.038%. Oxygen, in contrast, is about 21%. But CO_2 is enormously important for life. Virtually all of the carbon in plants, the source of our food, comes from this tiny amount in the air. Plants use energy from sunlight to combine CO_2 with water to manufacture hydrocarbons such as sugar and starch, in a process called photosynthesis. As their name suggests, hydrocarbons are mostly hydrogen and carbon. Hydrocarbons are the building blocks of our food and fuel. Photosynthesis also releases oxygen into the atmosphere, extracted from water (H_2O). When we breathe in oxygen and combine it with food, we get back the energy that the plants absorbed from sunlight and remake the CO_2.

Scientists traditionally write 0.038% as 0.000380 = 380 parts per million = 380 ppm. Figure 10.4 shows how the level of CO_2 in the atmosphere has changed over the past millennium. Notice that the amount was pretty constant from AD 800 until the late 1800s, at a level of 280 ppm. In the last century, it has shot up to 380 ppm—an increase of 36%—due to increased burning of coal, oil, and natural gas; some of the rise came from the extensive burning of rain forests to clear the land for farms. If we continue to burn fossil fuels, we expect the carbon dioxide to keep rising.

If you see this plot elsewhere, it usually has a suppressed zero, so the y axis starts at 260 ppm. I don't do that here because it makes it harder to see that the increase in the last century has been about 36%. It's the recent rise that concerns people. Other measurements (not shown) tell us that the carbon dioxide level now is higher than it has been at any time in the last 20 million years. That fact is not disputed; it is astonishing but not surprising. We know how much carbon we are burning, and that is plenty to account for the increase. (Some of the CO_2 dissolves in the oceans, making them more acid, and some is taken up by increased biomass.)

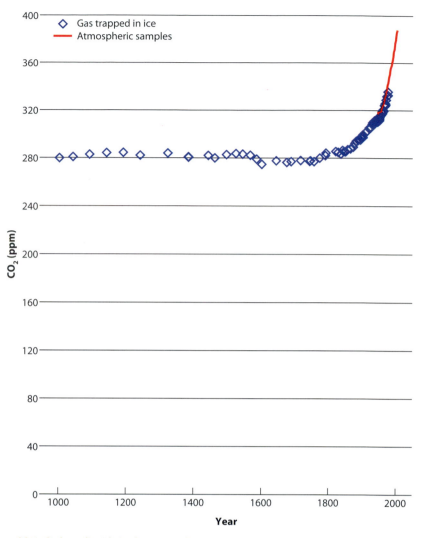

Figure 10.4 Carbon dioxide in the atmosphere from AD 800 to the present in parts per million (PPM). The rapid 36% rise in the past 100 years is due primarily to the burning of fossil fuels and the burning of wood from the clearing of rain forests.

Here is a summary of what the plot says: For about 800 years, the carbon dioxide in the atmosphere was pretty steady at about 280 parts per million (ppm)—i.e., it made up 0.000280 of the atmosphere. Then, sometime after 1800, it began to go up, as a result of the increased use of fossil fuels and the burning of the rain forests of Brazil. We used coal for heating and railroads; then we used "coal gas" to light streets and homes. Then oil was discovered, and that was used for lighting and heating. The development of the automobile occurred simultaneously with the discovery of even greater oil reserves. (The Rockefeller oil fortune was made even before the automobile.)

Look again now at the thermometer record in figure 10.1. There is an increase in the CO_2 right about the time that the Earth began to warm. Is the CO_2 responsible? The IPCC says it is very likely that the CO_2 is responsible for much of the warming of the past 50 years. The physics relating CO_2 to warming is called the *greenhouse effect*.

The Greenhouse Effect

The Earth is warmed by light from the Sun. It would just get warmer and warmer if it didn't have a way to lose that absorbed energy, but it does: infrared emission. If we assume that all the radiation that hits the Earth is absorbed and is equal to the radiation that the Earth emits, we can calculate the following surprising result: the temperature of the Earth is approximately 1/20 the temperature of the Sun. (For the optional calculation, see the footnote.[3]) The Sun has a temperature of 6000 K, and that means that the temperature of the Earth should be about 6000/20 = 300 K = 80F. And that's not too far off.

In the calculation, I assumed that all of the sunlight that hits the Earth is absorbed and turned into heat, but in fact only 60% is absorbed; the rest is reflected. (The reflected amount is called the Earth's *albedo*.) When we do the calculation with more care, we find that the temperature of the Earth should actually be about 26°F, well below freezing. If that were true, the oceans would be frozen, and life as we know it could not exist on Earth. But it isn't that cold; the average temperature is about 57°F (see figure 10.1 again). The extra warmth comes from something called the *greenhouse effect*, illustrated in figure 10.5.

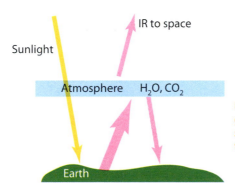

Figure 10.5 The basic atmospheric greenhouse effect. Note that the Earth's surface is warmed by both the Sun and the atmosphere.

Sunlight hits the Earth, the Earth gets warm, and it emits IR (infrared) radiation. But, as you can see in the diagram, most of that radiation does *not* go directly to space but is absorbed by the atmosphere. That's because water vapor (H_2O) and carbon dioxide (CO_2) are effective absorbers of IR. The atmosphere gets warm and it radiates its own IR; half of that goes to space and half comes

---

[3] The following optional calculation is for those who have taken other physics courses. T_S = Sun temperature; T_E = Earth temperature; R_s = Sun radius; R_e = Earth radius; D = Earth–Sun distance; σ is the constant that comes into the T^4 law. Power radiated by the Sun = $P_s = 4\pi R_s^2 \sigma T_S^4$. Power absorbed by the Earth is $P_e = P_s (\pi R_e^2)/(4\pi D^2)$. Power emitted by the Earth is $P_e = 4\pi R_E^2 \sigma T_E^4$. Solving these equations gives $T_E = T_S \sqrt{R_s/(2D_s)}$. We know from looking at the size of the Sun that $R_S/D \approx 1/200$. So $T_E = T_S/20 = 6000/20 = 300$ K.

back to Earth. So the Earth is warmed both by the Sun and by the atmosphere. That's what warms the Earth back up to a livable temperature.

Note that it is mostly *visible* light that warms us from the sun. The Earth then emits *infrared*, and that is what keeps the surface cool. When the IR is absorbed by the atmosphere and reemitted to the Earth, that's the greenhouse effect.[4]

A similar thing happens in a garden greenhouse. Sunlight comes through the glass and warms the soil; the soil emits IR, but the IR cannot escape through the greenhouse glass. So the inside gets warmer. That's why this phenomenon is called the greenhouse effect. These days, more people have experience with this same effect in automobiles in parking lots. Perhaps a more up-to-date name for the phenomenon (at least in the United States) would be "the car-in-the-parking-lot effect."

The greenhouse diagram (figure 10.5) shows all the IR from the Earth being absorbed by the atmosphere, but that was an exaggeration. Some leaks through, so the greenhouse warming is not as much as it would otherwise be. That is shown in the more accurate diagram (although a bit more complex) in figure 10.6.

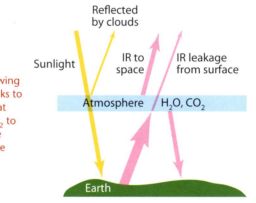

Figure 10.6 The greenhouse effect, showing leakage and clouds. Because some IR leaks to space, the warming is not as great. If that leakage is plugged (by adding more CO_2 to the atmosphere), then the surface of the Earth warms. If cloud cover increases, the surface temperature goes down.

Notice that some sunlight is reflected from clouds and that some of the IR radiation from the Earth leaks directly to space and is not reflected back. The net result is that the Earth is a bit cooler than it would otherwise be—and the average world temperature settles down to about 57°F.

Enhancing the Greenhouse Effect

If we would like to warm up the surface a bit more, all we have to do is make the atmospheric blanket a little more effective by stopping the leakage of IR.

---

[4] To be a little more accurate, we have to talk not just about visible and IR, but about three categories of light: visible, near IR, and far IR. Near IR is invisible light that is *almost* visible; it is just a little longer in wavelength than visible. The sun emits visible and near IR; these are absorbed. The Earth then emits far IR; this is radiation that has a wavelength 20 times longer than that of visible. It is the far IR that gets absorbed by the atmosphere.

We can do this by adding a gas to the atmosphere that absorbs IR. Then less will leak out, the blanket will be more effective, more IR will be radiated back to the surface of the Earth, and the surface of the Earth will get warmer.

That is exactly what we are doing, although not completely on purpose. CO_2 is a good absorber of IR, and it tends to plug the leakage in the atmospheric blanket. The effect of the CO_2 is amplified by the fact that a little bit of warming makes more water evaporate from oceans and damp soil. Water (H_2O) also helps plug the leaky blanket; that's why it is labeled in the diagram along with CO_2. A little CO_2 causes a little warming; that makes H_2O evaporate, and that also causes warming; the total is about twice what you would get without the H_2O. This total warming is believed to account for some of the 1°F warming of the Earth that we experienced in the last 50 years.

Now for some more uncertainty. If CO_2 leads to more water in the atmosphere, then maybe that would increase the number of clouds. Clouds reflect sunlight, so that could cause cooling! It's hard to know, because we have not succeeded in inventing a good way to calculate cloud cover. In fact, it is the possibility that clouds might cancel most of the greenhouse increase that led the IPCC to be cautious in its analysis. It is the uncertainty in cloud increase that led them to conclude that they can be only "90%" confident that the warming of the previous 50 years (1957–2007) was caused, at least in part, by humans.

The Poles Warm the Most

One of the features of greenhouse warming that all the models predict is that the polar regions will have more warming than the equator. Part of this comes about because of the melting of ice and snow, both of which have very high albedo—that is, they reflect sunlight.

The Exaggerations

Once again, I emphasize that there is a consensus among scientists about global warming. It is represented by the IPCC reports. The 2007 report state that it is 90% likely that humans are responsible for at least some of the 1°F observed global warming of the previous 50 years—that is, the warming since 1957. The effect is real, and currently small. As I said previously, the real concern is that it is expected (with a 90% probability) to grow enormously over the next 50 years.

One of former Vice President Al Gore's most famous statements is that the evidence that humans are responsible for global warming is "overwhelming and incontrovertible." It is worthwhile recognizing that he is not representing the consensus of the scientists (i.e., the IPCC) when he says that, unless by overwhelming and incontrovertible he means that there is only 10% chance that the conclusion is wrong.

There is, perhaps, a valid purpose behind the hype. A 90% chance of disaster is very bad. Yet small effects—a one-degree F temperature rise—tend not to excite the public. If the threat isn't imminent and obvious, can't we put off our worry until later? The answer to this question is *no*, because carbon dioxide tends to stay in the atmosphere a very long time. Even though some dissolves in the oceans immediately, the rest is thought to remain in the air. Whatever harm we are doing now will last. And it is cumulative.

Nevertheless, the public did not pay much attention until advocates of action exaggerated the evidence. The exaggerators looked over recent climate records, picked everything that was bad, ignored those things that could be interpreted as good, and attributed all the bad effects to global warming. This approach, called *cherry-picking* (pick only the impressive cherries and tell people that they are representative of the whole crop) can be politically effective in the short term, but it runs the risk of an eventual backlash. The public may eventually decide that scientists exaggerated, or lied, and they lose trust in science.

In this book, I want to give an accurate picture of what is really known. I assume that you are interested, so I don't have to exaggerate. To cover the story accurately means that I have to point out not only the facts (such as the temperature records), but lots of things that you may have been told are true, but aren't. As we go through this list, please remind yourself that the fact that so many of the claims about warming are false does not mean that warming is absent. It just means that the effect so far has been subtle, not as dramatic as some people portray.

The IPCC bases its conclusions about warming on three main effects: the thermometer records, the rise of sea level (most of which comes from the expansion of ocean surface waters as they warm), and the melting of the ice pack in the Arctic. Let's now look at some of the things that are popularly attributed to global warming, examples that got the public excited, but that are really just cases of cherry-picking.

Hurricane Katrina

You will hear it said that the devastating category 5 (the most intense kind) storm that destroyed much of New Orleans in 2005, Hurricane Katrina, is a consequence of global warming, and that as additional warming proceeds, we can expect "many more Katrinas."

In fact, Hurricane Katrina was not a category 5 storm when it hit New Orleans; it was only category 3, a far weaker kind. It is widely called a category 5 storm because it was strong when out at sea, but not when it hit the city. Any medium-size (category 3) hurricane that hit New Orleans any time in the last 40 years was likely to destroy the city. The vulnerability came from the fact that much of New Orleans was built on land that was below sea level, and the levees used to hold back the sea were poorly designed, poorly built, and poorly maintained. But New Orleans was a small target, and fewer than two hurricanes each year hit the United States on average, so it was not surprising that the city was missed—until 2005. The destruction of New Orleans is not an indicator of any change over the past 40 years; it only illustrates that unlikely events (a category 3 storm hitting the small target) do happen if you wait long enough.

You will sometimes hear news reporters (and even scientists) state that both the number and the intensity of hurricanes has increased in recent decades, and that this increase is due to global warming. In fact, the number of hurricanes has *not* been increasing. The IPCC does not claim they have, and that's the consensus report. It is true that more hurricanes are being discovered now than in previous years, but that is likely due to the fact that we now use satellites and automatic reporting systems on sea buoys to report wind velocities far out at sea.[5] More reported hurricanes does not mean that there are actually more hurricanes; it may mean that we are just better at looking.

For an unbiased look at hurricane rates, we can use a standard scientific trick: look at the number of hurricanes that hit a region such as the U.S. coastline. When a hurricane hits there, it is *always* noticed, whether it took place in 1900 (when there were no satellites) or now. A plot of such hurricanes is shown in figure 10.7, based on a report by Marlo Lewis using data from the National Hurricane Center.

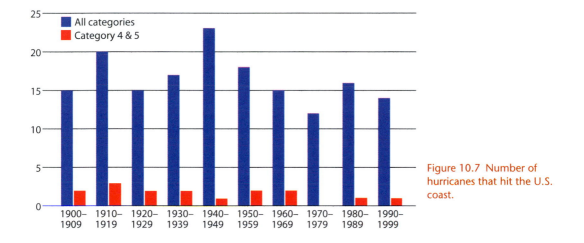

Figure 10.7 Number of hurricanes that hit the U.S. coast.

The tall bars show the number of hurricanes each decade that hit the U.S. coast; note that about 15 hit each 10 years, or about 1.5 each year. The small bars show the number of intense ones that hit (categories 4 and 5); they average one or two per decade. There is a slight trend downward for both kinds of storms, but it is not statistically significant. What can be said is that there is certainly no evidence that hurricanes hitting the United States are increasing, either in total number or in intensity. Yes, there are more hurricanes observed every year, but that is mostly an artifact of our improving ability to detect storms deep at sea, where they previously would have been missed.

Remember: There is good evidence that the climate is warming, with a human contribution (so far) of about 1°F. Would you expect that 1°F warming to cause an increase in violent storms? Maybe, maybe not; there are

---

[5] This fact that reported increases in hurricanes was an illusion caused by more complete hurricane detection was shown by Christopher W. Landsea in his article, "Counting Atlantic Tropical Cyclones Back to 1900," published in the scientific journal *EOS* **88** on 1 May 2007, pp. 197–200.

good arguments on both sides. One argument is that the increased energy (from the higher heat content) provides energy for storms. The other argument is that warming is expected to be greater in the Arctic (sea ice has been decreasing north of Canada), but the effect of that is to reduce the temperature differences between north and south, and by evening out the temperature, storms are less likely to grow; that's because hurricanes feed off temperature differences.

Will storms increase or decrease? We don't know. That's why the IPCC suggests that storms *might* increase, but makes no definitive prediction. They might decrease. Those who claim that increased hurricanes are evidence for global warming are not being careful, nor scientific. But they do get the attention of the public, particularly after the tragedy of Katrina. That may have some benefit.

What about other kinds of storms, such as tornadoes? In his movie *An Inconvenient Truth*, Al Gore claims not only that hurricanes are increasing (a fact we just showed to be in great doubt) but also that tornadoes are increasing.

Tornadoes

Every year the U.S. government publishes a plot of strong-to-violent tornado activity in the United States. The plot for years 1950 through 2006 is shown in figure 10.8.

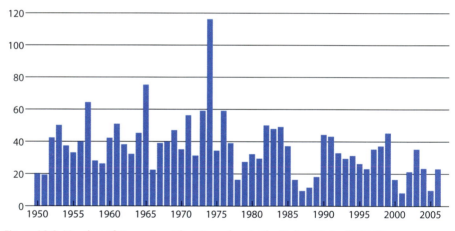

Figure 10.8 Number of strong-to-violent tornadoes in the United States (NOAA).

This plot shows that there is no increase in the number of tornadoes versus time. In fact, there has been a statistically significant decrease—look at the total number of storms on the left side of the plot, compared to those on the right side. So why does Al Gore say the storms are increasing? He doesn't explain the basic for his claim (he presents no data, only his conclusion), but he might have reached that conclusion if he made a plot of total number of tornadoes, including those that hit but never touched the ground. Thanks to radar, we now detect many more storms than we did in the past, so such an increase would really be an observational bias. Figure 10.8 shows that tornadoes that do damage to the United States are *decreasing*; those are the ones shown in the plot. It is even possible that this decrease could be due to global warming, since such warming decreases the temperature between north and south, and might weaken the gradient responsible for violent storms. We don't know. But it does not make good propaganda to suggest that tornadoes are decreasing from global warming; it might make some people mistakenly think that global warming is good.

Global warming is supposed to reduce the gradient of temperatures. One of the bits of evidence that the IPCC uses in support of global warming is the melting of ice in the Arctic Ocean. The disappearance of sea ice is a real effect (although it has *not* led yet to deaths of polar bears). The melting of Alaska is real too.

Alaska

Alaska is melting, literally. Much of Alaska is built on frozen ground called *permafrost*, a soil condition that results when the yearly temperature averages below freezing. But across most of the state, that criterion is just barely met, by a few degrees Fahrenheit. A small bit of warming can make a big difference.

When I drove Alaska's Highway 4 in the summer of 2003, the landscape looked flat, but the ride felt like I was on rolling hills. The road undulated up and down, thanks to partially melted permafrost; costly road repairs had to be done every summer. Along the sides were "drunken trees" (a local term), leaning on each other's shoulders like thin, inebriated giants, their shallow roots loosened by melting soil. There were also drunken homes, leaning and sinking into the ground, and sunken meadows, 10 feet lower than the surrounding forest. Sunken meadows result when trees are cleared and a little bit of extra warmth reaches the ground in the form of direct sunlight.

The ecology itself seems to have a meltdown when temperatures rise above 32°F. Warm weather in Alaska encouraged an infestation of bark beetles that killed 4 million acres of spruce forest. This has been called the greatest epidemic of insect-caused tree mortality ever recorded in North America.

Alaska is frequently cited as the early warning evidence that disastrous global warming is on its way. Now look at figure 10.9, the actual temperature record published by the respected Alaska Climate Research Center, an institute I visited during my 2003 trip.

The first thing to note on the plot is that the temperature averages between 25 and 29°F. That's below freezing—and that's why the ground is frozen into permafrost at Fairbanks. If the average temperature rises above freezing, then the permafrost melts. That hasn't yet happened at Fairbanks (which is north of much of the distressed region), but even at this city there are pockets of land in which the temperature is a bit higher.

Figure 10.9 also shows that the warming of Alaska is real and documented. Look at the left side of the chart, and note that it tends to average about 26°F. Now look at the right side, and you'll see it is a bit warmer, averaging about 28°F, two degrees warmer. A careful mathematical average verifies these results. Alaska has warmed about 2°F over the twentieth century. That's about the same as for the entire United States, shown in figure 10.1.

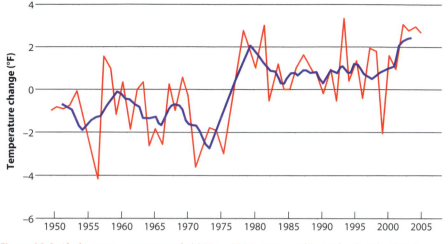

Figure 10.9 Alaska temperature record, 1950 to 2005, measured in Fairbanks. The thin line shows yearly average; the thick line shows the average over several years. (Courtesy of Alaska Climate Research Center.)

Why is a 2°F change so bad? The data above were taken in Fairbanks, and if you go a few hundred miles south, the average temperature is a bit warmer. That is where the greatest damage is being done—in the regions of Alaska where the average temperature used to be just below 32°F, but now is just above. Your house, your highway, is no longer on solid ground, but on mushy marsh. But even in Fairbanks, the slight extra warmth has allowed the bark beetles to spread their damage.

Look at figure 10.9 again, and you may notice something that looks peculiar. If you had only the data from 1950 to 1975, would you have concluded that Alaska was warming? Or does it appear to be cooling? Do it and see what you think.

Now cover the left side, and look only at the record from 1980 to the present. Has the climate of Alaska warmed in those past 28 years? It doesn't seem like it has.

Now look again at the entire plot. Notice that the warming appears to have taken place in a very short time, from 1975 to 1980. Before 1975, it was fairly level, averaging 26°F, and on the right side, it was also level, averaging 28°F.

Compare this plot to the global warming graph in figure 10.1. The patterns of warming in Alaska and on the Earth as a whole seem quite different. Compare the Alaska plot to the increase in CO_2 (figure 10.4). Most of CO_2 increase, and most of the global warming (figure 10.1) has taken place in the past 28 years, from 1980 until now. Yet the temperature of Alaska (figure 10. 9) has been remarkably stable in that same period.

Do the strange pattern and the fact that it doesn't seem to follow the global trend of carbon dioxide show that the melting of Alaska is not due to global warming? No, not at all. The temperature trends of Alaska could well consist of a rise due to global warming, with a downward fluctuation in the last decade caused by something else. (There have been serious scientific papers suggesting that soot from Chinese coal power plants is responsible, and to the possibility of a decadal El Niño kind of sea variation that takes place naturally in the Arctic

Ocean.) However, advocates of action are unlikely to show you this temperature plot, because it raises awkward questions about the cause of the melt. That's another kind of cherry-picking. Show only the data that wows the audience, and avoid anything that seems to contradict the simple picture.

Scientists who are trying to figure out real causes must not let themselves cherry-pick; they have to see all the evidence. That's why I show it, even though I agree with the IPCC that global warming is real, and very likely caused (at least in part) by humans. And, of course, continued warm weather is just as bad for Alaska as increasing warm weather; once the temperature is above freezing, the ground melts. The problem is not so much that Alaska is getting warmer, but the fact that it is staying warm, after a rise that took place before 1980.

It is also interesting that the Alaska plot seems to suggest that the warming of Alaska was about the same as, or perhaps only slightly greater than, the warming of the whole Earth. The Fairbanks data do not yet show evidence for the widely predicted effect that warming in Alaska will be much greater than in the continental United States.

Antarctica

Antarctica is melting too. The numbers are rather dramatic. Measurements of the ice mass have been accurately achieved from a satellite known as GRACE (you can look it up online) that measures the ice's gravitational effects on the satellite orbit. They show that the glaciers of Antarctica are losing 36 cubic miles of ice every year! This appears to be a dramatic and worrisome demonstration of the magnitude of global warming.

Remarkably, the appearance does not reflect reality. In the year 2000, the IPCC knew that the GRACE satellite measurements were imminent, and so they had scientists calculate how much ice change was expected from global warming. The surprising result was that all the scientists predicted that global warming would *increase* the ice of Antarctica, not decrease it. The reason for this is not hard to see. Even with 1 or 2 degrees warming, most of Antarctica remains very cold. Loss of ice mass comes not from melting but from calving, the breaking off of ice as it flows to the sea. With additional warming, the main effect (according to the calculations) was additional evaporation from the sea; warm weather evaporates water. When this water vapor drifts over the continent of Antarctica, it results in added snow, which compresses to ice—and the glaciers were expected to grow. So global warming was predicted to increase the Antarctic ice mass. Had they seen the ice mass increase, scientists might have concluded that their prediction was verified, and that the increase was additional evidence for global warming.

The opposite is what was observed. Does this disagree with the global warming picture? Yes. Does it disprove global warming? No—the temperature evidence is very strong. It does show that our current understanding of the warming is not good enough even to predict huge ice changes in Antarctica. Confusion on these subtleties is so great that as recently as June 2009, a global warming report published by the U.S. government mistakenly cited the melting of Antarctic ice as evidence in favor of the global warming models.

What about the Arctic Ocean? Does it mean that we have to be cautious about interpreting reduced ice in that ocean as evidence for global warming? Yes.

What's the bottom line? Answer: Global warming is observed to be about 2°F since the late 1880s. Since 1957, some (maybe most) of the 1°F rise is very likely due to humans. The evidence is strong, but the rise is not so great (yet) that it can be easily seen in individual locations, such as Alaska or Antarctica. Rather, it is the totality of the evidence, particularly the temperature evidence that gives us cause for concern. Don't attribute a hot day (or reduced Antarctic ice) to global warming; the effect is more subtle. But it is real, and it is very likely that at least some of the warming of the past 50 years has been caused by humans, primarily from our burning of fossil fuels.

Every now and then, you'll hear a news report about a big chunk of ice that breaks free from Antarctica. That will usually accompanied by a scientists stating that it could be evidence for global warming. Indeed it could be. Or maybe not. Increases in ice (and parts of Antarctica are growing glaciers) are not as dramatic, and don't make the news. But given the poor understanding we have of the region, *any* change in conditions will often be accompanied by a statement that the change could be do to global warming. And it could be. Or maybe not.

Fluctuations

There is another kind of exaggeration that allows proponents to attribute *every* bad bit of weather to human-caused global warming, even cold weather, based on the argument that even if the warming is small, the added heat will increase the variability of climate. This effect is suggested in some of the computer-based climate models, but it is certainly not established to be true. In fact, fluctuations might decrease due to the reduced temperature difference between latitudes as the northern regions heat more than the equator.

The IPCC says that some or most of the 1°F temperature rise in the last 50 years is due to humans. Of course, temperature on any given day may vary by 20°F or more. It is extremely unlikely that tomorrow's weather will be within 1°F of today's. So global warming is tough to spot. It is a small effect in the midst of huge fluctuations. The only way you can even detect global warming is by making careful and extensive averages of lots of data.

Climate too varies enormously. Fluctuations (some due to volcanoes) are probably responsible for the cooling in the 1950s that led to the fear of a returning ice age; nobody thinks that was caused by global warming. Weather is famous for being *variable* On any given day, the temperature is unlikely to be at the average for that date. Look again at figure 10.1 at the temperature peak that took place in 1998. That high caused a great deal of consternation in the following years. Look at the variations up and down in that plot. Those variations are what make it hard to detect global warming, and they can easily be played by politicians (and well-meaning scientists) to make the public worry about global warming. Local variations are much larger than those seen in figure 10.1, since that figure represents an average over hundreds of locations. Measurements taken at one location, such as the data for Fairbanks (figure 10.9) show even larger fluctuations. Note that about year 2000,

Fairbanks had the coldest average temperature it had experienced for 35 years! But that is likely just a fluctuation, not an indication that Alaska is cooling; in fact, the running average shown in figure 10.9 illustrates that Alaska has indeed warmed.

In fact, 1998 remains the warmest year, at least through 2008. And the coldest year of the new millennium (from 2001 onward) has been the most recent, 2008. (This book was written in 2009.) Is that evidence *against* global warming? No. It is just a reflection of the fact that when looking at the records, when all of the global warming in 50 years amounted to 1°F, that fluctuations tend to obscure trends. If the cooling of the past decade continues for another 5 to 10 years, then that will throw doubt on the global warming models, and maybe suggest that cloud cover is playing a bigger role than we thought.

Paleoclimate

By digging deep into glaciers, and by looking at sedimentary rock (laid down every year on the ocean floor primarily from coccoliths—the remains of microscopic animals), we can detect evidence of huge climate change in the past. An example of this is shown in figure 10.10, adapted from one shown by Al Gore in his movie/book *An Inconvenient Truth*.

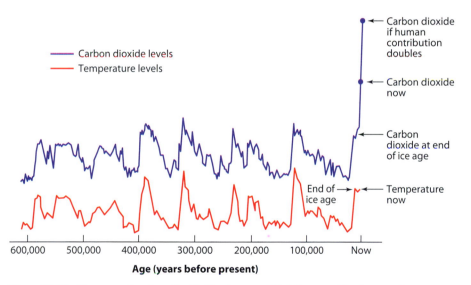

Figure 10.10 Climate and carbon dioxide for the past 600,000 years. The close tracking suggests that they are related, but which is the cause and which is the effect?

First look at the temperature changes (lower curve). There is no scale on the plot, but the swings up and down probably indicate changes of 10 to 15°F. The low regions are the ice ages, and the high points are the warm interglacials. Note that the very end (on the right) is the current warm interglacial, the brief (on this plot) period when farming and civilization developed. It is only 12,000 years in duration (so far), and on a plot that covers 600,000 years, that doesn't take much space. Notice that the warm interglacials take place roughly every 100,000 years. This is the cycle of the big ice ages that I spoke of earlier. They return in a fairly regular way. These cycles are believed to be due to changes in the orbit of the Earth as it is perturbed by Venus, Jupiter, and other planets. The orbital explanation is generally referred to as the *Milankovitch theory*.[6]

Now look at the upper curve, which shows the relative amounts of carbon dioxide in the atmosphere. That varies too, in apparent lock-step with the temperature. In his movie, Al Gore gives the impression that this verifies that CO_2 causes climate change. In fact, even though that is the conclusion that most people watching the movie come away with, he never actually says that. He says that the situation is "complicated." And indeed it is. He summarizes the plot by saying that every time there is a lot of carbon dioxide, it is warm, and whenever the carbon dioxide is low, it is cool.

But most geophysicists believe that it is the temperature that is causing the CO_2 to change, not the other way around. Most of the CO_2 in the biosphere is actually dissolved in ocean water. When the water warms, the CO_2 is driven out; gas doesn't dissolve as well in warmer water. The possibility that warming is causing the CO_2 change is verified by other measurements that indicate that the CO_2 changes *lag* the temperature changes by about 800 years. In other words, the temperature changes first, and then it takes 800 years for the CO_2 to finish coming out of the ocean. That's a reasonable number, because we know that deep ocean water takes about that long before it works its way to the surface, where the CO_2 can escape.

Something else in the plot suggests that the CO_2 is a result of warming, not the other way around. Look at the recent CO_2 rise, at the right side of the plot. The recent increase is about as much as the increases at the ends of the ice ages. If the CO_2 were causing the warming, we would expect to see a 10 to 15°F warming, not the 1 to 2°F warming that we have actually experienced.

Some scientists disagree, and think that CO_2 may have indeed been responsible. The situation is "complicated." Perhaps the 800-year lag has been misinterpreted. There really is a reasonable controversy here. What you really need to know is that the paleoclimate plot (figure 10.10) is not clear and incontrovertible evidence that CO_2 has driven climate in the past. Even so, you should not conclude that therefore the evidence for CO_2 greenhouse warming is weak. It just cannot be based on this plot. The evidence is based on the observed increase in temperature (1°F in the past 50 years), plus our understanding of the likely effects of CO_2 on greenhouse warming.

"Global Warming" versus "Human-Caused Global Warming"

Another common problem in the news is the semantic confusion between the terms *global warming* and *human-caused global warming*. They are often taken to be synonymous. But remember that the IPCC concludes only that some or most of the warming since 1957 was very likely (90% confidence) caused by humans.

[6] My book on this subject, written with coauthor Gordon MacDonald, is titled *Ice Ages and Astronomical Causes*, published in 2000 by Springer-Praxis.

They do not conclude that the warming from 1850 to 1957 was human caused, because they can't rule out the possibility that it was a natural recovery from the Little Ice Age, perhaps caused by changes in the intensity of the Sun. In their reports, they show a correlation between sunspot level and Earth temperature. Up until 1957, they seem to be related; only after 1957 do the two plots dramatically diverge. It is important to recognize that when a politician or scientist claims that the warming prior to 1957 is human caused, they are giving their own conclusion, and not representing the IPCC scientific consensus. They may be right, but maybe not.

Is there global warming? Yes. Is it caused by humans? It is very likely that some, maybe most of the warming of the past 50 years was caused by humans. There is a 10% chance (according to the IPCC) that it wasn't. I repeat this so often because so many people think that the consensus is different than this.

Can We Stop Global Warming?

Humans have very likely contributed to global warming, and that suggests that the worst effects are still ahead of us. What can we do? There are lots of feel-good measures; we can use less gasoline, or perhaps turn down our thermostats to save heating fuel. Such measures are so dramatically short of what is needed, that there is a danger that people who do such things think they are leading the way to a real solution. In this section, I'll explain why the problem is so hard to solve.

Figure 10.11 shows the energy use per person (y axis), plotted against the income per person (x axis) for various countries. Several points are plotted for

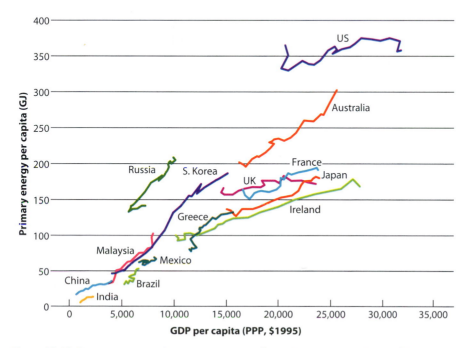

Figure 10.11 Energy use versus income, per person, for various countries. (UN and DOE EIA.)

each country to show how the values changed from 1982 to 2004. Note that the United States uses more energy per person than any of the other major countries. This is in part because the United States is spread out geographically, and in part because energy in the United States has been very cheap and so we have not had to conserve. Australia is high for similar reasons. Note also that the United States has not increased its energy use per person very much in recent years. The curve for Russia actually went down, as its economy collapsed in the late 1980s; the hook near the bottom of the Russia curve shows some recovery.

Look at the general trend in this plot. Every country is near the diagonal, meaning that every country seems to be within a factor of two of using 25 megawatt-hours of energy use for every $10,000 of income. We don't know why this is true. Some people speculate that energy is necessary for income, although the relatively unchanged energy use in the United States for the last 20 years, while the economy grew, shows that isn't a strict law of economics. Others suggest that wealthy people can afford to use more energy, since they value light, heat, high-tech goods, and travel, so the energy use might be a consequence rather than a cause.

What makes the data worrisome is the fact that most of the population of the world is at the lower left corner: poor in both energy and dollars. As poor as they are, these economies are growing rapidly. In recent years, the Chinese gross domestic product (or GDP, the total sum of goods and services) has been growing at an astonishing rate of 10% per year. Look at the curve for China, and although the numbers are small, you can still see that the $ per person has more than tripled over the past 20 years! India is also showing amazing growth; for it, the $ per person has doubled. That's why these countries are called *developing* nations! Most caring people look forward to the end of poverty in these countries. But if they follow the general trend (and so far they seem to be doing that), then the energy use of the world will grow enormously in the coming years.

Will there be enough energy available to allow these countries to stay on the trend line? Many people think we are running out of fossil fuels, but that isn't really true; what we are running out of is cheap oil. We will not run out of expensive oil for a long time. Let's begin by looking at the price at which oil is sold on the world market. That is shown in figure 10.12.

This plot shows the price of oil from 1970 until 2006, plotted in "constant dollars" (that is, it is adjusted for inflation; the fact that a dollar in 1970 was worth about five times as much as a dollar in 2006). The plot shows the cost to *buy* oil. The price to drill it is quite different. In Saudi Arabia, oil can be drilled for only $3 per barrel; when sold at $100 per barrel, the profit is enormous. But in some other places, oil costs $20 or more to drill. The high price of oil right now is determined more by the limited supply. The recent growth in the economy of China has made a demand on oil that the world's producers can't match, and so the price has risen.

Now take a look at the complicated chart in figure 10.13. It is worth studying, because it gives important insights for the future of oil for the next few decades. Here's the way to read the plot. The horizontal axis represents the amount of oil available; the vertical shows the cost to obtain that oil. (This is not the consumer price to buy the oil, but the oil company cost to get it from the ground.) The rectangle in the lower-left corner labeled *Already produced* shows that we

Figure 10.12 Price of oil (per barrel) adjusted for inflation.

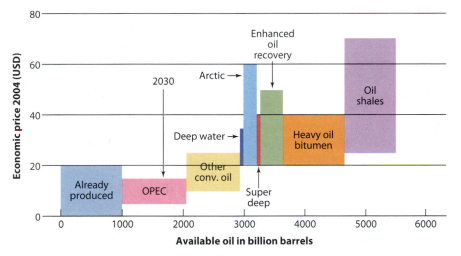

Figure 10.13 Availability of oil as a function of price. Notice that when the price of oil is in the range of $20 to $70 per barrel, the amount that is available is much larger than when the price is below $20 per barrel. (IEA, 2005.)

have already produced, in the history of the world, about 1000 billion barrels of oil (horizontal axis) at a cost that ranged from $0 to $20 per barrel (vertical axis). The arrow points to the amount of oil we expect we will need by 2030. That oil can still be provided by OPEC. *EOR* stands for "enhanced oil recovery". That means the oil that you couldn't afford to recover when the selling price was $20 per barrel, but that you can afford to get out when the price is between $20 and $50 per barrel.

In the olden days (1990s), people assumed that nobody would pay more than $20 per barrel. At that price, there is only a tiny bit more than 2000 barrels total—only about twice what we have already pumped. That's why people thought that we would run out by 2050. But oil prices have shot up to $145 per barrel and then dropped again. At high prices, all of the oil shown on the plot is available. That's why I say we are not running out of oil, but only out of cheap oil. That's actually bad news for the environment; there is lots of carbon in all that oil.

It is worth looking at some of the blocks in the plot. The block labeled *OPEC* refers to the oil available from the monopoly (technically a cartel) known as the Organization of Petroleum Exporting Countries. These countries get together to decide on the price to sell oil; they do this to avoid competition that might otherwise lower the price. The next block is *Other conventional oil*, including standard drilling in Texas and around the world. Then we get to *Deep water* (with oil rigs that float above) and the Arctic—both more expensive to recover. *Heavy oil* refers to places such as Alberta in Canada where the oil is thick, almost tarlike, and can be extracted only by sending hot steam down into the ground to heat the oil and make it less viscous. The price of oil is now sufficiently high that this is currently being done. The final rectangle on the plot is for oil shale—there is much of this in the United States, but to extract it requires heating the rock above 600°F, a temperature at which the heavy bitumen molecules break down into lighter and more fluid ones.

Fossil Fuel Resources

When we consider fossil fuels other than oil, the situation is even more dramatic. Coal and natural gas appear to be abundant, particularly in the countries that will need them. Table 10.1 shows the fossil fuel reserves for several key countries. The amounts are described in billions of barrels of "oil equivalent." So coal and natural gas are given in terms of how many barrels of oil it would take to give a similar amount of energy. Oil from shale is the most expensive of all, but as you can see in figure 10.13, it can be recovered at prices well below $100 per barrel.

Carefully examine the table. Mark the page so that you can find it easily, because you will probably want to refer back when you discuss the world energy situation. Note that the United States has low reserves of oil, but enormous reserves of coal. In fact, the fossil reserves of the United States are greater than that of any other country; if you include oil shale, the United States is enormously far ahead. Other large countries with big populations (and these are the ones that will be using lots of energy in the future) include Russia, India, and China, and they are all near the top of the list.

From figure 10.13, you see that the total amount of OPEC oil available for the future is about 1000 billion barrels. From table 10.1, you see that the United States has 3737 billion barrels of oil equivalent—3.7 times as much as OPEC, but in the form of coal and shale, not liquid oil.

So we are not about to run out of fossil fuels, only out of cheap oil.

Note that Canada has the second largest reserves of oil in the world. Actually, that is true only if the price is high enough. Its oil sands cost about $60 per barrel to recover. So if the selling price of oil is well above $60 per barrel, then Canada can recover this oil at a profit. But whenever the price drops below $60 per barrel, this oil suddenly becomes "unrecoverable." The bouncing of oil

Table 10.1 Fossil Fuel Reserves (Billions of Barrels of Oil Equivalent)

	Oil	Coal	Natural gas	TOTAL	Oil shale	TOTAL with shale
United States	21	1284	200	1505	2500	4005
Russia	60	831	280	1171	250	1626
China	48	442	13	503	16	519
Australia	130	418	5	553	—	553
India	5	489	7	502	—	502
Iran	136	—	157	293	—	293
Saudi Arabia	260	—	—	260	—	260
Canada	179	32	9	220	—	220
Qatar	15	—	152	167	—	167
Brazil	8	49	2	59	80	139
Iraq	115	—	18	133	—	133
UAE	97	—	35	132	—	132
Kuwait	99	—	9	108	—	108
Venezuela	80	2	25	107	—	107
Mexico	12	17	5	34	100	234

Note: The numbers in this table oversimplify the complexity of the situation. Recoverable reserves are always estimates; they are uncertain and the amounts are a strong function of the price that you are willing to pay to recover them.

prices in recent years has made the amount of recoverable oil from that country bounce up and down in an equally dramatic way.

Part of the reason that we import so much oil is that we can't burn coal in our autos. Or can we? Think about the history of the locomotive. In the early United States, locomotives ran on biofuels: the trees that grew along the tracks. As these were used up, and coal discoveries were made in Pennsylvania, they converted the engines to burn coal. Nobody worried about the fact that coal creates about twice as much CO_2 as oil, because nobody was worrying about global warming. Eventually, these coal burners were replaced by diesel engines, using *diesel fuel*, a kind of gasoline. Diesel was much easier to use, because it left no residue, unlike coal, which left ash behind.

In fact, coal can be used to manufacture oil, including diesel fuel and gasoline, through chemistry.

Fisher-Tropsch: Coal and Water Make Oil

Coal can be turned into diesel fuel by a chemical factory known as a Fisher-Tropsch plant. The carbon in coal (C) is combined with water (H_2O) to make oil (long chains of CH_2) and CO gas. The CO (carbon monoxide) can be burned as fuel to run an electric generator. Fisher-Tropsch plants were developed by countries that had lots of coal but no access to oil; these include Germany during World War II, and South Africa when they were denied oil imports during the apartheid era. It was not used in the rest of the world because oil has been cheap. Oil from a Fischer-Tropsch plant costs between $40 and $60 per barrel. When the cost of oil gets above this, then oil from coal becomes economical.

We are running out of oil in the United States and paying exorbitant prices for oil from OPEC, so why don't we start making our own oil from our abundant (and cheap) coal?

The primary reason seems to be the cost of building a plant. Estimates vary, but numbers I have heard range from $100 million to $1 billion per chemical factory. That could be a good investment, but it could also be a bad one—if the price of oil drops. And you can see from figure 10.13 that there are other sources of oil available at lower prices. If the price of oil drops below $40 per barrel, it could make a billion-dollar Fisher-Tropsch investment worthless. Some investors had decided to go ahead and build Fisher-Tropsch plants, but only if the U.S. government will promise to buy oil at a minimum price of about $60 per barrel, even if OPEC drops the price below this.

To many people, the greater concern is the CO_2 production that would come from exploiting these coal resources.

Energy Security

Aside from price, there is another enormous pressure driving us to greater use of coal. That is "energy security." Right now, over half of the oil used in the United States is imported. The demand for such oil drives up the price, and that funds countries such as Iran that support terrorists. Some people say that our "love affair" with oil means that we are funding *both sides* of the war on terror.

The desire for energy security—reducing our imports and operating our economy only our U.S. reserves, is powerful. But if we move in that direction, we will use more and more coal (it is cheaper than oil shale), and coal produces about twice as much CO_2 as does oil, for the same energy. That means that there is a conflict between the desire to reduce CO_2 and our wish for energy security.

Kyoto—and Copenhagen

Even if human-caused global warming is not certain, the consensus of the IPCC indicates that the risk is substantial. Many experts conclude that we must exercise *risk management*, even before the danger is proven to be 100% real.

What can we do? Many people suggest that we need to cut back strongly on our CO_2 emissions. In 1998, then Vice President Al Gore signed a proposed amendment to a treaty called the United Nations Framework Convention on Climate Change, in Kyoto Japan. Ever since, this document has been referred to variously as the *Kyoto Protocol*, *Kyoto treaty*, *Kyoto accord*, or simply *Kyoto*. Senate ratification of the treaty would commit the United States to a reduction in carbon dioxide emissions to 7% below our 1990 level. Because emissions have grown since 1990, the actual cut required works out to about 29% of the levels expected in 2010. The Kyoto protocol expires in 2012, and countries met at the end of 2009 in Copenhagen to forge its replacement treaty.

The Kyoto treaty has been ratified by 164 countries—almost the entire world—but not by the United States, to the embarrassment of many U.S. citizens. In fact, it was not even submitted for ratification by Presidents Bill Clinton or George W. Bush, presumably because they knew it would fail to pass in the Senate. But recently, thanks in large part to Al Gore, U.S. interest in the Kyoto/Copenhagen protocols has increased.

The main objection to the Kyoto accord is that it places no limits whatsoever on developing countries such as China and India. All of the reductions are supposed to come from the developed world, not the poor. Take a look at figure 10.14. It shows various groups of countries, including Western Europe and Eastern Europe (including Russia), but the United States is not grouped—it stands by itself. The United States dumps more CO_2 into the atmosphere than all of Western Europe. That's why many people demand that the United States cut back.

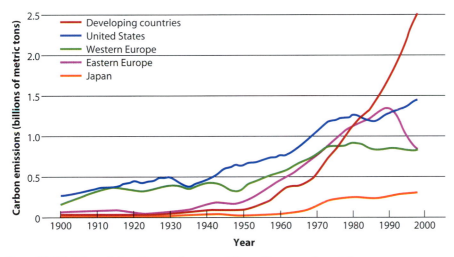

Figure 10.14 CO_2 emissions from various countries and groups of countries.

Based on numbers such as these, we can indeed conclude that the United States was responsible for about 1/4 of the 1°F temperature rise of the past 50 years. We will very likely *not* be responsible for the predicted 3°F to 12°F temperature rise of the next century, however. The main causes for that increase will be India, China, Russia, and the rest of the developing world. Even now, the major increases are coming from these developing countries. In 2006, the CO_2 production of China surpassed that of the United States. Look at the plot, and imagine that the United States reduces its use down to the 1990 level. That would reduce 0.2 billion tons of carbon per year. Now assume that we stay there, but that the developing countries continue their growth. They would add in an additional 0.2 billion tons, enough to undo the U.S. cutback, in less than 3 years. If the United States keeps its CO_2 emissions low, but the developing world continues to develop (in CO_2 emissions), then we have delayed the potential global warming by only 3 years, even if the U.S. emissions stay low.

Why, then, are so many people so enthusiastic about the Kyoto treaty? Many Kyoto proponents say that the United States must set an example, in the hopes that one day China and India will follow. Opponents say that China and India can't follow such an example unless the means of doing it are very inexpensive. They will likely develop their economies as rapidly as possible, just as we did, and then, when their people are as wealthy as ours, they will have the economic means of controlling their emissions, just as we are now doing. Although the carbon dioxide that they now produce exceeds that of the United States, their production per capita is less than one-fourth that of ours. If you were president of China instead of the United States, would you cut back now? With a population that still suffers from poverty, malnutrition, poor health, lack of opportunity, widespread illiteracy, and periodic famines, would you slow economic growth in order to keep the temperature from going up a few degrees? Add to this the facts that China has plenty of coal, certainly enough to meet the worst scenarios of the global-warming models, and that it is accelerating its exploitation of that resource.

Many people fear that cutting back on U.S. emissions, if done with expensive technologies, would not persuade the developing countries to follow suit. In fact, one vote has been taken by the U.S. Senate on the Kyoto accord, although not to ratify it. The vote was for the Byrd–Hagel Resolution, and it passed in a very bipartisan 95-to-0 vote. The resolution states that the United States should not ratify Kyoto until the treaty is rewritten to include binding targets and timetables for developing nations. The Senate simply did not trust that our CO_2 example would be sufficiently persuasive.

The fact that China surpassed the United States in CO_2 emissions in 2006 has made it easier for opponents to argue that the United States is no longer the main problem. Most people believe that there is no easy solution, but that to manage the risk, many different options will have to be used, all at the same time.

Solutions

The discussion seems to suggest that even if the United States were to cut back to the levels described by the Kyoto protocols, the growth of the developing world will still result in enormous CO_2 increases. Is the situation hopeless? I don't think so, but unless our methods of reducing CO_2 can be afforded by developing countries, they are unlikely to do anything other than delay the predicted warming by only a few years. The emphasis must be on CO_2 reductions that can be used by all. By far, the easiest is energy efficiency and conservation.

Energy Efficiency and Conservation

Ponder the following physics question: How much energy *should* it take to drive from San Francisco to New York City? The answer, from physics, is surprising. In principle, it could be done with *no* energy. After all, you can slide across ice effortlessly; the only energy it takes is to overcome friction. What if you eliminated friction? Is any other energy necessary? With an efficient hybrid engine, you can recover the energy used in accelerating a car and put it back into a battery when you slow down. That's called *regenerative braking*; it uses the motion of the car to turn a generator that charges the battery. Likewise it takes energy to go up a hill, but the same principle can be used to recover the energy when you coast down. The basic conclusion: with better auto design, we can *enormously* reduce the energy consumption of autos. My Prius already gets 50 miles per gallon on long round-trips; there is no reason why an auto could not achieve 100 mpg or more. Of course, an expensive hybrid like the Prius is not really an option to developing countries. But they could use the principle in their trucks and buses.

Similarly for heating our homes. The only reason we have to heat them is that energy is lost to the outside world primarily through convection (open or leaky windows, chimneys), and conduction (through glass windows and uninsulated walls and roofs). With good insulation (including double-paned windows), the amount of heat we need can be made tiny. In fact, recent analysis (by the respected firm of McKinsey) has shown that you save money by doing that; the cost of putting in better insulation is recovered in just a few years, and after that, it is pure profit. That makes it one of the best investments you can make; invest your money into insulation, and take as your "interest" in the form of money saved keeping your home warm. And you don't have to pay taxes on this kind of interest.

Use less energy for the same function; that's called *energy efficiency* and *conservation*. Conservation is the easiest way to reduce greenhouse emissions, and because it is the least expensive way to do so, it will be particularly valuable in the developing countries such as China and India since it pays back whatever investment is needed in a short time.

A lot can be said about conservation. It has a bad reputation among consumers because they associate it with discomfort; in the 1970s, President Jimmy Carter encouraged people in winter to live in a cold house (65°F) and to put on a sweater in order to save energy. But comfortable conservation should be more attractive. Put some insulation in your walls (rather than on your bodies), and turn up the thermostat to whatever temperature you want. Save energy by not letting it leak out. And it is a great investment; your return on the money you put into conservation will pay higher interest than a savings account.

Similar principles work to save the energy used to air condition houses in summer. About half of the solar radiation hitting our roofs is in the form of infrared. If you use roofing material that reflects this, then the heating of the home is drastically reduced, and that lowers the energy needed to air condition. And to the human eye, which does not see IR, the "cool roofs" can still be a pleasant brown color, or whatever the homeowner wants.

A whole chapter or even a book can be written on comfortable conservation. It is the cheapest way to avoid carbon dioxide in the atmosphere. But we need to discuss other possibilities.

Clean Coal

Over half of the electric power in the United States is produced by burning coal and creating CO_2. If this CO_2 is produced at a central power plant, then in principle it is possible to capture it and store it. This is called *carbon capture and storage*, or *CCS*—an acronym that you *do* need to know, since it is already becoming an important national issue. Another related term is *sequestration*; that refers to the process of pumping the CO_2 underground, where it is either stored in cavities or dissolved in deep brines (salt water).

According to the IPCC, this process seems to be feasible. Sequestration is already being tried at several locations around the world, although with a different purpose: pumping CO_2 into oil wells can help bring up additional oil. There is debate over how safe CCS will be. Will the CO_2 really stay there for thousands of years, or will it eventually leak out? The IPCC has written an extensive report on this subject. Most experts believe that if it stays down for a few years, it is very likely to stay down for thousands, and so we will quickly learn how reliable sequestration can be.

In the olden days, when people referred to *clean coal*, they meant coal that did not pollute the nearby regions with soot, nitrous oxides (which produce smog), sulfur dioxide (which produces acid rain), and mercury. But today, when the term is used, many people include CO_2, and when they refer to clean coal, CCS is included. But when you read about clean coal, make sure you know which definition is being used.

The biggest issue may not be whether CCS can be done, but whether it can be done economically; remember, China has huge coal supplies and is actively building new plants. In 2007, it built more than one new gigawatt coal-burning plant every week, on average! CCS will make the power cost more, perhaps 50% to 100% more. China may argue that it should not have to pay the premium until its citizens live at Western standards of living. If they do, then perhaps the wealthier countries will have to pay the difference. That could be accomplished through carbon *cap and trade*, a topic we'll discuss shortly.

The United States had a major clean coal demonstration power plant under construction called *FutureGen*, and its goal was to show that an efficient and inexpensive clean coal production of electricity with CCS was feasible. The program was cancelled in 2008. Some people think that it was a terrible mistake to cancel a demonstration of this essential technology. Others think that it was a good idea; the demonstration plant was rushing ahead so fast that it was proving to be expensive, and proponents of CCS worried that it would give a misimpression that clean coal was far more expensive than it need be.

Biofuels

Biofuels are fuels that are made by plants; they include wood, pulp, and liquid fuels such as ethanol that are made from plant products. Early railroads used the wood that grew along the track; that was an example of a biofuel. When biofuels burn, they emit CO_2, but no more than they took out of the atmosphere when they grew. Therefore, they are described as *carbon neutral*. That's not always true, however, since many plants grown on farms require fertilizer and machinery to grow them, and oil to run the tractors; these are often made using fossil fuel. Transporting the fuel also creates CO_2. Ethanol made from corn is particularly bad; you use almost as much fossil fuel to make the ethanol as you get in biofuel. In contrast, ethanol made from sugar cane is a good biofuel, provided that you don't have to cut down a forest in order to clear a growing region. But a new generation of biofuels is under development, using materials such as tall grasses (switch grass and miscanthus) that require little water, grow fast, and truly are carbon neutral. To use these grasses as liquid fuels, we need to develop methods to convert the cellulose in the plant stalks to ethanol. Ways of doing that are under active development around the world; that is the goal of a large new Berkeley Energy Biosciences Institute funded in part by BP (British Petroleum).

Some states have already passed legislation requiring that autos use a mixture of gasoline and bioethanol, a combination often called *gasohol*. Much of this legislation was passed prior to the analysis that showed that ethanol-based alcohol is not really energy neutral. (Or it was done to subsidize farmers in the critical presidential caucus state of Iowa.) But with the new crops and new technology, biofuels could make a significant contribution to both CO_2 reduction and energy independence.

Nuclear Power

There are 104 nuclear power plants in the United States, producing on average about 1 gigawatt of electric power each, and providing about 20% of the U.S. electric power. Construction of new nuclear power plants came to a virtual halt in the United States after the 1986 Three Mile Island accident, although plants that had been under construction were eventually commissioned (such as the Watts Bar nuclear power plant, in 1996).

The reasons for the halt were fear of accidents, concerns about waste storage, and the high cost of operation. The fears are now being reevaluated, thanks to the recognition that fossil fuels also have risks, and because there are nuclear reactor designs that have little or no accident probability. (Look up "pebble bed reactors" on the Web.) Many people think the waste storage issue was exaggerated; I discussed this in chapter 5. Moreover, the cost of operation of nuclear plants has come down, largely through better management. The *capacity factor* (the fraction of time that a nuclear reactor is actually working and delivering power) was barely above 50% in 1980 and now it is nearly 90%. This has made nuclear power much cheaper than it had been.

Some environmentalists now argue that coal is so bad in its CO_2 production that a larger part of our electric power should come from nuclear. China has been building about two new nuclear power plants each year (small {replace 'small' with: a small number} compared to the 70 coal burning ones); France gets about 80% of its power from nuclear reactors; and in the United States several companies are applying for licenses to begin construction of new plants.

Wind

The use of wind to generate electric power has recently grown enormously in the United States. In the last four years, the installed capabilities in the United States have doubled, from a half percent to over one percent of U.S. electric production. It is still small, but that is a huge change, and it is expected to grow even more. The technology is old but innovative; new wind turbines[7] are quiet and efficient. The biggest fields are being installed in Texas, which has the advantage that the wind is close to the population centers. Right now, the growth of this technology appears to be limited by the limited capability of the United States to manufacture the wind turbines.

One problem with wind is that it is irregular, and it may not blow when you most need power. To address this issue, people are studying energy storage methods, such as batteries. One of the most practical may turn out to be one of the most surprising: use excess power to compress air, stored underground; when the power is needed, use the compressed air to run another turbine. Flywheels are also under serious consideration.

Solar

There is a gigawatt of solar power in a square kilometer, and that's as much power as you get from a large nuclear or fossil fuel power plant. So solar power sounds reasonable. There are several difficulties. It is often cloudy; solar isn't as reliable as methods that burn fuel. Not all of the power can be converted to electricity; the efficiency to do this is between 10 and 40%, depending on the price of the cell. And most importantly, it is still more expensive than other methods. Proponents say that isn't true; it is not more expensive when you include the environmental costs of other methods, and if we figure out how to charge coal plants for their CO_2 emissions, then solar becomes competitive.

To install solar cells now costs about $3 to $10 for each watt of capability. Some economists say that the cost must be brought down to about $1 per watt (equivalent to $1 billion for a gigawatt plant). That may happen in the near future, thanks to advances in technology, and the possibility of paying for the installation using *carbon credits*, a topic we will discuss in the next section.

Solar power has many traditional uses, from drying clothes to warming rooms (through windows) to heating water for baths. These uses are important for conservation—for example, for reducing our dependence on fossil fuels. In this section, however, I will limit the discussion to *big* solar—the kind that could be used to replace large fossil fuel–burning electric power plants. Here is a quick rundown of the possibilities for solar.

SOLAR THERMAL

Mirrors are used to concentrate the solar power onto a small area to heat a fluid such as water. (Heat can also liquefy salt, and that is sometimes used.) Steam from heated water can power a turbine that runs an electric generator.

Several of these solar thermal plants are already in operation. One famous one is a "power tower" near Seville, Spain, which consists of a boiler on a high tower with mirrors aimed at it from the surrounding countryside. The mirrors must be redirected as the Sun moves. This plant currently delivers electric power at the relatively high price of 28¢ per kilowatt-hour; that is expensive compared to the

---

[7] The term *windmill* is usually reserved for wind-operated mills—that is, structures that grind flour. For electricity generation, we use the term *wind turbine*.

price in the United States from fossil fuels, which averages 10¢ per kWh). The Spanish government subsidizes the price of electricity from this plant in order to encourage solar construction and to study the costs, and because it is trying to meet its goals under the Kyoto treaty.

There is also a solar energy generating system (called SEGS) in California that works with smaller reflectors placed in long parabolic troughs. Other solar thermal plants are operating in California and Nevada, in desert regions where the sunlight is abundant, but not too far from the factories and cities where the power is used. Transmission line losses in the United States average about 7%, and it would be worse if the lines were longer than a few hundred miles. A disadvantage of the solar concentrator technology is that it works only on sunny days; if the light is diffused by clouds, then the mirrors can't concentrate the light enough to heat the water. But there is optimism that the solar thermal plants could deliver power at a cost cheaper than that of natural gas by about 2020.

The efficiency of these plants is low if you consider that only a fraction of the area is covered by mirrors—so much of the sunlight falls on land. But in many places in the world, the key issue is cost, not land area covered, so that measure of efficiency is not relevant. Of the light that hits the mirrors, typically 20% to 40% is converted to electric power.

SOLAR CELLS (PVS)

Solar cells are also called *photovoltaic cells*, or simple *PVs*; they get this name from the fact that photons (light) interact in the cell to caused electrons to flow to metal plates, so they produce a voltage. We'll discuss how they do this in the next chapter, on quantum physics. Traditional solar cells were based on crystals of silicon, and they typically converted about 10% of the sunlight to electricity. For a 1-gigawatt plant, when the Sun is directly overhead, that would require 10 square kilometers of these cells. Traditional solar cells cost about $10 per installed watt, and so they are not really considered competitive with fossil fuels. These are the kinds being installed by homeowners, sometimes because they think they are saving money, but more often because they want to reduce the CO_2 that they are personally responsible for. But the field is advancing rapidly.

One of the truly hopeful developments in recent years has been the development of highly efficient solar cells. These are complex devices because extracting as much energy as possible out of sunlight requires having separate layers to convert different colors. These sophisticated solar cells are now being built, and one major producer, Boeing (yes, the airplane company; it started producing solar cells when they were needed for space) is selling solar cells that convert 41% of the incident Sun power to electric power. They say that the efficiency should rise to 45% in the near future. Wow!

There is a catch, of course. Even when purchased in large quantities, these special cells cost about $10 per square centimeter; that's equal to $70 per square inch, or about $10,000 per square foot. A foot-sized cell would yield 41 watts—not much for the $10,000 investment. Why do I call this hopeful? The reason is that sunlight (if there are no clouds) can be focused using a lens or mirror. You can make a plastic lens that is 1 foot square for less than $1 and use it to focus the sunlight onto a cell 0.4 inch on a side. A cell of that size costs $10. Your total cost for the 41 watts is now reduced to $10, plus $1 for the lens, plus whatever you spend to build the module. The cell is only 25¢ per installed watt! That sounds *very* attractive. The tricky part is that you have to keep the cell pointed at the Sun, and that requires a mechanical system. If our goal is to spend no more than $1 per installed watt, then the total cost for the square-foot device must be less than $41. Can that be done? It is not obviously impossible, and several companies in California are already building such systems to see if they can be cost-effective. Even if it costs three times that, this system still becomes the cheapest form of solar power. This approach is called *solar concentrator technology*. Its greatest drawback is that it works only on sunny days, when the Sun is visible and its rays can be focused. Imagine now an array of foot-sized concentrator solar cells covering a square mile of sunny Nevada. Since there are 5280 feet in a mile, there would be $5280 \times 5280 = 27,878,400$ modules. Each module would be only a foot high, making the system quite robust against wind. Driven by tiny electric motors, the modules would all point in the same direction: toward the Sun. With 41 watts from each, the total electric power output at midday would be over a gigawatt. Of course, there may be other expenses, such as keeping the reflectors clean. In a recent trip to Nevada, I found that much of the region I visited had over a foot of "bug dust" that whirled around every time the wind blew.

Another hopeful development is cheap solar cells made without growing crystals; these are called *amorphous* (noncrystalline) cells. There is much excitement over a particular kind called *CIGS*. (The letters stand for the elements that go into the material: copper, indium, gallium, and selenium.) CIGS are manufactured using a technique similar to that of an ink-jet printer: they are basically sprayed on a piece of plastic. CIGS have already achieved an efficiency of 19%, and enough people are convinced of their future that factories costing hundreds of million of dollars are under construction to build CIGS cells. For business reasons, many of the details have not yet been released. The price may be determined by competition, since sales of the cells will have to pay back the huge investment being put into these plants. And people worry that with huge numbers of solar cells being built, that the world will not be able to supply enough indium for the cells! But optimism in the solar energy business is rampant. Many investors are jumping in. They believe that the future of solar is sunny.

Carbon Credits: "Cap and Trade"

Clean technologies would be more competitive with traditional ones if the environmental costs of CO_2 pollution were taken into account. One way to do this is by a global treaty, in which power plants that pollute the atmosphere are required to buy carbon credits for the harm they do. Nonpolluting plants, such as solar or wind, would get credits; they could then make additional profit by selling these credits to organizations that emit CO_2. The whole thing would be managed in such a way that there would be a net reduction of CO_2, and the hope is that it could be done in such a way that market forces would make the process economically efficient.

The Kyoto protocol set up such a system, and the countries that endorsed that treaty are now using these credits. The approach is called *cap and trade*. A country is given a limit for the amount of CO_2 that it emits; that's the cap. If they go under that limit, they get credits that they can trade; if they go over the limit, they are required to buy credits to cover the pollution. Other names for this procedure are *carbon dioxide trading* and *carbon trading*. Despite these names, the credits are also used for other gases that contribute to greenhouse warming, such as methane.

Opponents of this system say that it allows too much cheating. A collapsing economy in Russia, for example, enabled that country to sell a large number of credits for carbon dioxide that it never would have produced. Trading of credits, in this case, led to an increase in the carbon dioxide dumped into the atmosphere over that which would have otherwise been emitted.

Developing countries that signed the treaty do not have caps assigned to them; such caps could stunt their growth, and doing that was considered unfair. Opponents of the treaty argue that the major pollution in the coming century will come from those countries, so unless they too have caps, the problem will not be solved. Some people argue that giving generous carbon credits to the developing world may be a politically viable way to subsidize the construction of clean energy in these countries. Suppose that a developing country builds a plant that emits half as much CO_2 as current plants? Should they be given credits for doing this, even though they are adding to the CO_2 problem? Under Kyoto, the answer was *yes*. The process for doing that was called the *Clean Development Mechanism*.

The Kyoto treaty did not allow credits for new nuclear plants, even though they emit essentially no carbon dioxide. The reason was the fear that nuclear plants created a different kind of pollution, radioactivity. However the radioactivity issue needs to be reevaluated. Many people now think that the dangers of CO_2 far exceed those of buried radioactive waste.

Keep in mind all of these complexities. The solution to CO_2 pollution, if one exists, probably involves a very wide range of approaches, including conservation, carbon credits, solar, wind, biofuels, nuclear, and CO_2 sequestration. In the jargon of the global warming community, these many approaches are called *wedges*, and many wedges are needed. No one by itself is enough.

Chapter Review

Over the past 120 years, the world average temperature has risen about 2°F, with about a 1°F change since 1957. The IPCC, a U.N.–sponsored committee that reviews climate change, concluded with 90% confidence that humans are responsible for some or most of the recent 1°F change. This is not a big climate change yet, but the fear is that the rise is caused by greenhouse gases such as CO_2 from the burning of fossil fuels. CO_2 is now at the highest level it has been at for 20 million years, and the recent rise of 36% occurred primarily in the 20th century. Greenhouse warming takes place when infrared heat from the Earth is trapped in the atmosphere; the increased CO_2 helps plug some frequency holes in the atmospheric blanket. The uncertainty in the IPCC estimates about warming comes largely from unknown behavior of clouds.

Although the warming is real, politicians and some scientists exaggerate its effects when they claim that increased hurricanes, tornadoes, and even the melting of Alaska are proven to be a consequence. The melting of Antarctica appears to contradict global warming, but uncertainties are so large, that it is best to base conclusions solely on the temperature record and sea-level rise.

Because the risk of continued warming is large, many people want to cut the production of greenhouse gases. That is made difficult by the fact that fossil fuels are the cheapest source of energy, and perhaps the only kind that can be afforded by developing countries. There is also economic pressure since clean energy may be more expensive. The world is not running out of oil, only out of cheap oil. At $100 per barrel, reserves will last for over a hundred years. Coal is abundant in the big countries that will need energy: United States, China, Russia, India. Coal can be converted to liquid fuels through the Fisher-Tropsch process, but that does not prevent CO_2 pollution when the cars burn the fuel.

The Kyoto protocol was ratified by 164 nations, but not by the United States. The protocol calls for reduction in greenhouse gas emissions. It also sets up a system of carbon limits called *cap and trade* to regulate emissions. Emissions in the past were dominated by the United States and western Europe; in 2006, emissions from China surpassed those of the United States. In the future, the emissions from developing countries are expected to dominate. The Kyoto protocol expires in 2012; discussions continue at Copenhagen to define a replacement.

Solutions are possible. They include extensive conservation (which could be "comfortable"), biofuels, clean coal (with carbon capture and storage, CCS, also called sequestration), nuclear power, wind power (which has doubled in the United States in the past four years), and solar. Solar includes solar thermal and photovoltaics (PVs), also known as solar cells. Efficiencies range from 10% to over 40%, and the main issue is cost: to deliver solar electricity at an installation cost less than $1 per watt. A combination of all these approaches may be necessary to address the CO_2 problem.

Discussion Topics

1. There is uncertainty about the many possible effects of enhanced greenhouse warming. What actions do you think are appropriate? Is there any downside if the warming turns out to be natural? Try to evaluate the risks and benefits of public policy.

2. It is good (right? ethical? necessary?) to exaggerate a technical/scientific issue in order to get the public concerned? What are the dangers of doing that? What are the dangers of not doing that? Describe the various ways that you could exaggerate the global warming issue if you felt it was necessary to do so.

3. What is your preferred solution to global warming? Why do you prefer it? Is it the same solution that you preferred before reading this chapter?

4. What kinds of conservation do you think would prove effective in the developing world? Hybrid autos? Better home insulation? Fluorescent lamps?

5. According to figure 10.11, the United States has shown less of an percentage increase in energy use per person than any other country except Russia. Should we be proud of this? Or ashamed that we produce more per person than in any other country?

6. According to figure 10.11, the average person uses 25 megawatt-hours of energy per year for every $10,000 of income. Does that seem about right? Try doing the calculation for your own home (or that of your parents). Here's a start: if you use 1 kW for a year (which has $24 \times 365 = 8760$ hours), then the home has used 8760 kWh = 8.76 megawatt-hours of energy in one year. What was the income *per person* in that home?

Internet Research Topics

1. Look up the history of FutureGen, the clean coal system canceled in 2008. Can you find arguments on both sides of the issue? Look at the technology. Any surprises? Notice that they planned to separate the oxygen from air prior to burning the coal.

2. Read about "pebble bed reactors." Why are they called that? What is their history? Why were early pebble bed reactors canceled? Why are they considered safe? How would waste be handled?

3. Wind turbines. Where are they being built? How close together are they placed? (You can estimate this from photos.) What are the reasons that some environmentalists oppose them?

4. What is the current status of solar cells? Can you find prices (per watt) for the various kinds? Perhaps find a local installer and see what they charge. How big is the market? What fraction of the world electricity is being delivered by solar power plants?

5. On the Internet or in printed articles, find two or more examples of cherry-picking with regard to climate (or, at least, examples that could be cherry-picking). Try hard to find examples on *both* sides—in which people are picking examples either to exaggerate or to minimize the reality of global warming.

6. Brazilian autos run primarily on biofuels. What can you find out about the Brazilian program? What biofuels do they use? What other countries might match their approach? Did they start their program because of the CO_2 problem, or for energy independence?

7. Much of the CO_2 that is emitted from fossil fuels dissolves in the oceans and makes the oceans more acidic. See what you can find out about this issue. How much variation is there currently in ocean acidity? What are the projections for the future?

8. Some people propose *climate engineering* or *geoengineering* in case we fail to cut back on global CO_2. One approach is to increase reflectivity of the Earth in order to reduce solar heating and thereby reduce warming. Other methods include dumping nutrients into the sea to increase plant growth. What methods have been proposed? How many trees would you have to plant? (The total CO_2 in the atmosphere is about 3 trillion tons.)

Essay Questions

1. Many people use the terms *greenhouse effect* and *ozone layer* interchangeably as though they address the same issue. Explain the differences. Why are the greenhouse effect and ozone layer "controversial" in public debate? Describe the current day concerns for each, and what action is appropriate.

2. Describe the greenhouse effect. What causes it? How big is it? What role does it play in global warming? How can the greenhouse effect be strengthened?

3. What is the strongest evidence for global warming? What part of this is attributed to human activity?

4. Discuss the warming of Alaska. What is known about it? Is it real? Does it match expectations? How can people be misled by the observed fluctuations, either to affirm their prior belief that Alaska illustrates warming, or to deny it?

5. Many people worry that we are about to run out of oil. Is that true? Describe the relationship between available oil and its price. Could coal be used to replace oil?

6. Describe what is meant by *comfortable conservation*. What other type of conservation could there be? Give several specific examples of each kind.

7. What are the various possible solutions to the CO_2 problem? List them, and give the pros and cons for each. Which are most readily adoptable by developing countries?

8. What "evidence" for global warming is not really scientific evidence but only an example of either cherry-picking or exaggeration?

Multiple-Choice Questions

1. Higher carbon dioxide levels than we have now were last seen about
   A. 600 years ago
   B. 2000 years ago
   C. 13,000 years ago
   D. 2 million years ago

2. The country that produces the most carbon dioxide each year is
   A. United States
   B. Russia
   C. China
   D. India

3. The country that produces the most carbon dioxide per person every year is
   A. United States
   B. Russia
   C. China
   D. Saudi Arabia

4. The greenhouse effect currently warms the Earth by approximately (careful: this may be a trick question)

    A. 1°F

    B. 2°F

    C. 4°F

    D. 35°F

5. Which fossil fuel produces the most CO_2, per pound, when burned?

    A. oil

    B. natural gas

    C. coal

    D. gasoline

6. Converting coal to oil is achieved with

    A. pebble bed reactors

    B. Fisher-Tropsch

    C. enhanced oil recovery

    D. CIGS

7. Carbon dioxide in the atmosphere has increased over the past 100 years by about

    A. 0.6%

    B. 6%

    C. 36%

    D. 96%

8. Ice in Antarctica is

    A. melting, as expected from global warming predictions

    B. increasing, as expected from global warming predictions

    C. melting, contradicting global warming predictions

    D. increasing, contradicting global warming predictions

9. A treaty that was signed by the United States but not ratified by the Senate is

    A. IPCC

    B. CIGS

    C. SEGS

    D. Kyoto

10. According to the text, wind power is

    A. a possibility for the distant future, but not practical until the price drops

    B. rapidly growing, with wind turbines delivering twice the power they did four years ago

    C. widely used in Europe, but not suitable for the United States

    D. declining in the United States because the wind turbines have been discovered to kill birds

11. *Clean coal*

    A. is an oxymoron (self-contradicting word), since coal makes more CO_2 than any other fossil fuel

    B. refers to coal that has been converted to gasoline

    C. refers to coal that has had its carbon removed

    D. refers to coal used at a power plant with carbon capture and storage (CCS).

12. In France, nuclear power produces what percentage of the electricity?
   A. 2%
   B. 20%
   C. 37%
   D. 80%

13. In the United States, nuclear power produces what percentage of the electricity?
   A. less than 1%
   B. 3%
   C. 20%
   D. 80%

14. The organization that studies climate change is
   A. IPCC
   B. NAACP
   C. AFL-CIO
   D. IAEA

15. Which of these years had the warmest worldwide climate?
   A. 2008
   B. 2005
   C. 2001
   D. 1998

16. If the United States reduced CO_2 emissions to those suggested by the Kyoto treaty, and the developing nations continued to increase theirs at the current (allowed) rate, then global warming would be delayed by about
   A. 3 years
   B. 10 years
   C. 30 years
   D. 70 years

17. The greenhouse effect involves atmospheric emission of
   A. IR
   B. visible light
   C. UV
   D. microwaves

18. To be affordable without subsidy or carbon credits, the cost of a billion-watt solar plant must be not much more than
   A. $100 million
   B. $1 billion
   C. $30 billion
   D. $10 billion

19. The average temperature of the Earth (including the poles and the equator) is now about
   A. 42°F
   B. 57°F
   C. 65°F
   D. 80°F

20. Which country produces the most greenhouse gas emission per GDP (gross domestic product)?
A. China
B. Russia
C. USA
D. Japan

21. A key feature of the Kyoto treaty is that it
A. required CO_2 reductions for China
B. did not require reductions for Japan
C. reduced CO_2 but ignored other greenhouse gases such as methane
D. set up a method to trade carbon credits

22. Cooling in the last big ice age, which ended about 12,000 BC, was approximately
A. 2°F
B. 4°F
C. 11°F
D. 32°F

23. Cooling during the Little Ice Age (AD 1350 to 1850) was approximately
A. 2°F
B. 4°F
C. 11°F
D. 32°F

24. Which of the following is *not* a major greenhouse gas?
A. water vapor
B. carbon dioxide
C. methane
D. CFCs (such as Freon)

25. The least expensive form of carbon abatement (putting less CO_2 into the atmosphere) is
A. nuclear power
B. solar power
C. wind power
D. conservation

26. Thanks to humans, the carbon dioxide in the atmosphere has
A. decreased by about 35%
B. increased by about 36%
C. doubled
D. reduced to 50% of its previous level

11
CHAPTER

Quantum Physics

What do all the following have in common?

- Lasers
- Solar cells
- Xerox machines
- Transistors
- Computer circuitry (integrated circuits)
- Digital camera CCDs
- Superconductors

The answer is that they all make use of the quantum physics discovered in the twentieth century. In fact, it could be said that most of what we call *high-tech* is founded on quantum physics. But what is quantum physics?

Quantum physics is basically the recognition that there is less difference between waves and particles than we thought. The key insights are these:

1. Everything we thought was a particle also behaves like a wave (or wave packet).
2. Waves gain or lose energy only in *quantized amounts* (the quantum leap)—thus, they have particle-like properties.
3. When a particle is detected or measured, its wave will usually change suddenly into a new wave.

Those simple statements are, of course, tricky to understand, but they lead to all the mysterious behavior of quantum physics. That includes not only the list given here, but also the famous Heisenberg *uncertainty principle*, a result that has had major impact on philosophy and even the way that people unfamiliar with physics think about life. The uncertainty principle showed that physics was theoretically incapable of predicting every detail of what will happen in the future.

But before we get into a discussion of this *wave-particle duality*, let's examine the features of quantum physics that result from it. We begin by looking closely at the consequences of the fact that an electron has wave properties.

Electron Waves

In an atom, the electron orbits the nucleus in much the same way that the Earth orbits the Sun—except, the electron is held in orbit by electric force rather than by gravity. But in quantum physics, we must also think of the electron as having some of the properties of a wave. The frequency of the wave is given by the first key equation of quantum physics, sometimes called the *Einstein equation*. It relates the energy of the wave to its frequency.

$$\text{Einstein equation: } E = hf$$

The constant h is called Planck's constant. Max Planck found this constant when he studied the behavior of light. If f is measured in Hertz (Hz, equal to the number of cycles per second), and E is measured in Joules, then $h = 6 \times 10^{-34}$. You don't need to memorize this number.

For a typical atomic energy of about 1 eV (1 electron-volt = 1.6×10^{-19} joules), this frequency is very high. Put in those numbers and solve for f and you'll get $f = = 2.7 \times 10^{14}$ Hz = = 270,000,000,000,000 Hz.

So electron waves oscillate with an extremely high frequency. That's why you never notice them. They oscillate so fast that you can't perceive the oscillation. Remember that A above middle C on the music scale is 440 cycles per second,[1] and yet even at that low frequency you don't perceive it as a vibration, just as a musical tone. An electron oscillation is similar. With the right instruments, we can now measure these frequencies (by seeing beats with other oscillations), but we don't sense them directly.

Electrons in Atoms: Quantized Energy

Now here is the way that the wave behavior really makes a difference: in an atom, the wave packet is usually longer than the circumference. That means that when orbiting the atoms, the electron wave runs into its own tail. Since it is a wave, it can actually cancel itself out.

If it cancels itself out, then there is no electron. So if the electron is orbiting the nucleus, the only possible orbits are those where, after an orbit, the wave doesn't cancel. That means that only certain frequencies are allowed.

According to the Einstein equation, frequency is related to energy. So if only certain frequencies are allowed, then only certain *energies* are allowed. This is sometimes stated by saying that the electron energy is *quantized*.

The allowed kinetic energies for the hydrogen atom (the electron energies of the waves that circle the atom) have been calculated, and they match the calculated energies for which the orbiting wave doesn't cancel itself. These *allowed* energies include the following: 13.6 eV, 3.4 eV, 1.5 eV, and 0.85 eV. These are the energies that don't lead to self-cancellation. There are other allowed energies

---

[1] Instruments can be tuned differently, since the human ear is more sensitive to relative frequency than absolute. A fairly universal standard now is to tune the A above middle C to 440 Hz. With that convention, then middle C is 262 Hz.

too. These kinetic energies can all be described by a simple formula discovered by Niels Bohr:

$$KE = \frac{13.6}{n^2}$$

where n is an integer ($n = 1, 2, 3, \ldots$), and the energy is in units of eV (electron-volts). Different values of n give the energies in the list. For $n = 1$, we get 13.6 eV. For $n = 2$, we get 3.4 eV, and so on for all the energies possible. Notice that there are an infinite number of allowed energies, but there are many energies (those between the allowed one) that are forbidden. No electron can orbit the nucleus of a hydrogen atom unless its kinetic energy matches one from the Bohr formula.

Although energies of orbiting electrons are always quantized, an electron that is moving through empty space in a straight line can have any kinetic energy. The quantization of energy exists for the hydrogen atom because the electron is moving in an orbit and must not cancel itself. But if it is not in a *closed* orbit, one that comes back upon itself, then there is no such condition. The energy of such a *free* electron (not bound in an orbit) is not quantized.

Quantum Leaps

When one hydrogen molecule collides with another (as it does in a gas), it can knock the electron from one allowed orbit to another one that has a greater energy. Put another way, a collision can alter the energy stored in an atom. If it increases the energy of the electron, we say that the atom is *excited*, and that the electron is in an excited orbit. The electron can lose this energy and fall back into the original orbit by radiating an electromagnetic wave that carries off the energy difference.

In our new language of quantum physics, we say that the electron changes its orbit (a quantum leap) and emits a light wave. If the energy *difference* between the two orbits is E, then E is the energy carried by the light wave.

The frequency of this emitted light wave is given by the Einstein equation

$$E = hf$$

Note that we have now used this equation twice, once for the electron (to put it in an allowed orbit) and once for the light that is emitted (when the electron changes orbit).

SPECTRAL FINGERPRINTS

Quantum leaps between quantized energy levels implies that for a given atom, the frequency of the emitted light has only certain values—those that correspond to the differences in the quantized energy levels. But frequency is color, so that means that the color of the emitted light is also quantized and only certain colors are emitted.

The quantization of color can be seen by putting the light from hot hydrogen gas through a prism to separate out the colors. Instead of a continuous spectrum (as we got from the Sun), we see that only a few colors are present, as in figure 11.1. The scale shows to the wavelength of the light in nanometers (billionths of a meter): 500 nanometers is 0.5 micron.

Look at that pattern of light. Notice that the wavelength scale is backward. Each color (e.g., the red between 700 and 600, the three blue lines between 500 and 400) comes from a quantum leap, from an electron jumping from one allowed orbit to another. There are no colors between, because there are no leaps that produce photons that have those colors (i.e., those energies).

In the spectral image, the colors are spread out in the vertical direction. That makes the colors appear as lines. This is an artifact of the way the images were made. But this is the way they were originally studied, and it is for this reason that the quantized colors were historically, and are often still, referred to as *spectral lines*. Spectral lines really simply refer to the different wavelengths (or frequencies) that the gas emits. A hot hydrogen gas will have lots of collisions, so it will always emit the pattern of spectral lines shown in figure 11.1. If you see this pattern, then you know the gas is hydrogen. That's a good way to identify hydrogen—from the light it emits.

If the gas gets very hot, as on the surface of the Sun, then the electrons get knocked off the atoms. We then have separate protons and electrons, and the gas is called a *plasma*. When that happens, the electrons are no longer circling the protons, and their orbits are no longer quantized. They can have any energy. When they lose or gain energy in collisions, the amounts of energy are not quantized. No quantum leaps are forced. All energies are OK. The result is that sunlight has all the colors of the rainbow—the colors that you get when you shine sunlight on a prism. Sometimes we say that the spectrum is a *continuum*. Such a set of colors appears in figure 11.2. I've moved the image to line up the colors with the hydrogen spectrum shown in figure 11.1

Figure 11.3 shows the spectra from hydrogen and helium compared. They are lined up with the rainbow spectrum in figure 11.2 so that the same colors appear at the same locations.

Notice that the spectrum of helium is different from that of hydrogen. The reason is that helium has two protons in its nucleus, so the electron has to orbit at a higher speed to balance that attraction. That means that the energies of the quantized orbits will be different in helium from those in hydrogen.

Because the colors are unique to the element, we think of these emissions as playing a role similar to that of fingerprints. They are so easy to tell apart that even if the gases hydrogen and helium were mixed, we could tell how much hydrogen and how much helium was present.

The origin of the spectral emissions was once a great mystery. They were finally understood through the theory of quantum physics that I just explained. Einstein came up with his equation in 1905, and Bohr came up with his formula for the colors of the hydrogen atom in 1913. Now the existence of these lines is no longer considered to be a mystery. The fingerprints provide an incredibly powerful way to identify elements and molecules. Measurement of spectra such as these has allowed us to determine the gases on the surface of distant stars, and even to measure the gaseous emissions from smokestacks that may be emitting illegal pollution.

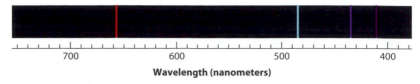

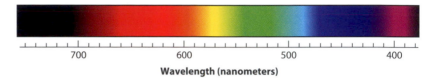

Figure 11.1 Spectrum of hydrogen. The short vertical lines show the colors emitted when hydrogen gas is heated.

Figure 11.2 Colors of the rainbow. These are the colors emitted by the hot gases on the surface of the Sun.

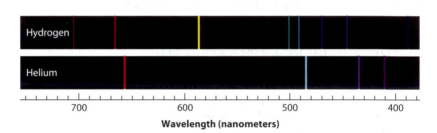

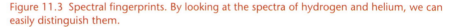

Figure 11.3 Spectral fingerprints. By looking at the spectra of hydrogen and helium, we can easily distinguish them.

The fact that electrons can only change energy by quantum leaps makes possible many of the high-tech quantum physics devices that we will discuss in this chapter, from lasers to transistors to computers.

Photons

When Einstein first came up with his equation, $E = hf$, he wasn't initially applying it to electrons (although it does apply to them). Rather, he was thinking about light. He had concluded (based on the observations of other people) that when a light wave is emitted or absorbed, the change in the energy of the wave is always quantized by an amount given by the equation. But we now know that his equation applies to everything that has energy. Every such object is oscillating like a wave, with its frequency (usually extremely high) given by that equation.

This equation says that light that has a frequency f can only have energy E given by that equation. But what if we turn up the brightness of light? Can't we give a beam of light any amount of energy we want? The answer is yes, but the energy must be a multiple of hf. The lowest energy it can have is just hf. When that is true, we say there is one *photon* present. The beam of light can also have 2 photons, and then its energy is $2hf$, and so on. Think of a photon as a quantum of light. If you have a light wave, it can be absorbed, but only in an amount equal to an integral number of photons.

If you know the frequency of a light wave, then you know the energy of the photons from the Einstein equation. A very bright light has many photons present. A dim light has only a few.

In the language of quantum physics, we say that light is a *quantized wave* (or sometimes a *wave packet*). Every time it gains or loses energy, it does so one photon at a time. It behaves like a wave, but it also behaves like a collection of photons.

Laser—a Quantum Chain Reaction

Lasers are used to burn holes in metal, to send information at enormously high rates over fibers, to read supermarket labels, to measure the exact shape of irreplaceable sculptures, to give spectacular light shows, as convenient pointers, to make holograms, and to find the distance to a remote object (including the Moon). Future uses may include the triggering of controlled nuclear fusion and shooting down military airplanes and ballistic missiles.

Lasers work on the principle of stimulated emission, another effect predicted by Einstein. Recall that an electron emits light when it makes a quantum leap from one allowed energy orbit to a lower one. If there already is light present, then that light could reinforce the emitted light. When that happens, the probability of emitting the new energy is increased.

Let me say the same thing, but in the new language of quantum physics. When one photon is present, that makes the probability higher that another photon will be emitted. This enhanced probability is called *stimulated emission*. A key and important feature of stimulated emission is that the second photon has the same frequency and direction as the photon that stimulated it to be emitted.

The word laser is an acronym:

Light

  Amplification by

    Stimulated

      Emission of

        Radiation

In a typical laser, there are a huge number of atoms present (e.g., 10^{20}). The first step is to make sure the electrons in many of these atoms are in an *excited state*—which means that they are in an orbit with extra energy. This is sometimes done by hitting these atoms with another source of light, or by running an electric current through the material. That gives the electrons more energy.

They are in an excited state. Eventually they will make a quantum leap to a lower energy state, emitting a photon. But in the laser, they don't stay long. They are *stimulated* to make the leap and emit a photon.

Stimulated emission was predicted theoretically by Einstein before it was observed. He calculated that if one photon passed by an excited atom (one with an electron in an excited state), it would stimulate that electron to leap. So one photon led to two. Two then led to four, four to eight, and so on. It was a predicted chain reaction of photon emission. The light would get stronger and stronger, as more and more energy was released. That is the *amplification* part of the acronym. An avalanche develops. The *a* in laser could also stand for "avalanche," but it doesn't.

Here's an amusing story: According to Charles Townes, the original inventor of the laser, he briefly considered an alternative name: electromagnetic radiation by stimulated emission of radiation. If they had kept those words, the acronym today would be eraser.

The first experiments with this effect used microwaves instead of visible light, and the device was called a *maser* (for microwave amplification by stimulated emission of radiation). Lasers now work in the infrared and the ultraviolet, and research has been done to try to get them to work with x-rays. The principle for all these is the same: a photon avalanche, or equivalently, a photon chain reaction.

As with the nuclear bomb, the chain reaction can happen very quickly. When this occurs, the energy in the entire laser is dumped out very quickly (60 to 80 generations), so the pulse of light can be extremely powerful, although very short in duration. When the laser operates in this way, it is called a *pulsed laser*, and these are the most powerful ones. Pulsed lasers are the kinds that are being used at our national laboratories in plans to ignite nuclear fusion without using a fission primary. They are also the kind that the military is developing for laser weapons.

However, it is also possible to operate the laser in a continuous manner in which the light output remains constant. (That makes it analogous to the sustained chain reaction in a nuclear reactor.) To do this, we must constantly excite new atoms at the same rate that they are emitting. This is done in a gas laser by sending a continuous electric current through the gas. *Continuous lasers* are used for laser communications (sometimes through the air, but mostly through fiber optics), and for measurement (range finding and leveling). Continuous lasers are also used for laser pointers and supermarket label readers.

Special Properties of Laser Light

The laser has two important properties that make it unlike the other chain reactions we studied, and contribute enormously to its value. I mentioned these earlier. They are:

- The emitted photons all have virtually identical frequency.
- The emitted photons all have virtually identical direction.

Since the photons have identical frequencies, the light is only one color—i.e., it is *monochromatic*. This is the feature that makes lasers really valuable for communications. Information is carried by a laser beam by modulating its brightness (called *amplitude modulation*, or AM) or its frequency (called *frequency modulation*, or FM). A 1000-cycle-per-second variation in the intensity of the very high frequency light beam could represent, for example, a 1000-cycle-per-second audio tone from a violin. But if there are several frequencies present in light (as there are in light from a tungsten bulb), then these difference frequencies can make beats with each other, and that will give a false brightness modulation. So it is really important that only a single frequency is present.

You'll sometimes hear that the light from a laser is coherent. *Coherent* is a fancy word meaning that only one frequency is present—or at least that the range of frequencies present is very small.[2]

The fact that the emitted photons have identical direction is more important than you might guess. It means that the beam comes out of the laser with all the light parallel, or well collimated. That's why a laser beam doesn't seem to spread very much, unlike a flashlight beam or headlight of a car. Even sunlight has light coming from different directions; since the Sun covers about a half angular degree in the sky, light from different parts comes at slightly different directions. But laser beams are different. Of course, they will spread a little, since they are waves. But that spreading angle can be tiny, since the wavelength is so short.[3] Sometimes it is necessary to spread a laser beam, —for example, if you want it to illuminate a hologram. You can do this by passing it through a lens or bouncing it off a curved surface. But the light originally produced is very well collimated.

Laser Measurements

The collimation of the narrow laser beam makes it useful for measurement that otherwise might be difficult. A pulsed laser beam can be directed at a distance object, and the bright spot on the object can be observed in a telescope. The time it takes the beam to return gives the distance to the object. That is the basis for laser range finding. If you measure distance for many different directions, you get a record of the entire shape of an object. Lasers have been used in this way to measure the changing shape of volcanoes, the interiors of buildings and caves, and historic structures such as the Roman Coliseum. Laser scanners, based on similar principles, are now being used for very detailed measurements of the shapes of objects, including valuable and irreplaceable sculptures.

At construction sites, a laser beam can be made level and placed across the entire construction site to make the structures level with each other. Lasers

---

[2] Optional: The *coherence time* is equal to 1 divided by the bandwidth—i.e., the difference in the maximum frequency present minus the minimum. If the bandwidth is small, then the coherence time is very long.

[3] The equation for spreading is the same one that we discussed in chapter 8: $B = (L/D) \, R$. For a beam of diameter $D = 5$ mm $= 5 \times 10^{-3}$ m, and $L = 0.5$ micron $= 5 \times 10^{-7}$ m, the $(L/D) = 10^{-4}$. So in $R = 100$ meters, it will spread only 1 cm. In 1 km, it will spread to a size of 10 cm. If you see a laser light show, by the time the end of the beam is 1 km away, it still looks pretty small.

are sometimes lined up with boundaries of property, to see what objects are in the property and which are outside. They are particularly useful for hilly land where the surveyor could not lay out a string. They can be used to show construction workers exactly where to lay foundations or columns.

An intriguing use of lasers for measurement was done for the movie *The Lord of the Rings: The Two Towers* (2002). An actor (Andy Serkis) climbed down cliffs, walked on all fours, moved in complex ways—and the positions of his shoulders, head, hands, and other parts of his body were measured using lasers to detect corner reflectors that had been attached to them. A computer then generated a new image, completely computer-generated, of an imaginary creature named Gollum.

Supermarket Lasers for Bar Code Reading

Supermarket lasers emit a very narrow (less than a millimeter across) single-color laser beam. These lasers move the direction of the laser beam rapidly in a complex pattern. The supermarket checker puts the bar code (a series of lines that indicate the price of the item to a computer) under this scanning laser beam.

In addition to the laser, there is a detector that looks at reflected light. The detector is designed to measure only light that matches the frequency of the laser. (A filter is used to eliminate all other light.)

When the beam scans across the bar code on the product, the reflected light changes rapidly, matching the dark and light spots on the code. The detector notices this rapid blinking, and records the pattern. From the pattern, it can look up the price in a catalogue, or just record the fact that the item was purchased. The narrowness of the laser beam is important for being able to record the narrow pattern.

It turns out that the easiest way to point the laser beam is not by moving the laser, but by bouncing it off a spinning hologram. Different parts of the hologram point the beam in different directions. So supermarket scanners make use of two very high-tech devices: lasers and holograms.

CDs and DVDs

CDs (compact disks) and DVDs (digital video disks) make use of the fact that a laser beam can be focused to a small spot. The compact disk has music recorded on a thin layer of aluminum buried inside the plastic. The music has been recorded with small bumps and hollows between the bumps, about 0.5 micron wide and about 1 micron long. Each spot represents a 0 or a 1, and the reflected intensity is measured to read the pattern. The light shines on only one bit at a time. The CD is spun, and about 1.4 million such bits pass the focused laser beam every second. The CD player can distinguish between 0s and 1s from the amount of scattered light that comes back at this megahertz rate.

Because the laser beam can be focused, it is possible to record even more information on a disk by having several layers on one disk. This method, along with smaller bump size, is used for advanced DVD recording to enable them to record long movies. The outermost layers have to be partially transparent so that some of the light passes through it to the deeper layer. To read the bumps on one of the two layers, the light is focused on it. Any light reaching the wrong layer will be out of focus. Because the spot is broad, it takes a longer time for the bumps and lands to pass under it, so the reflected pulses are longer in duration. These longer pulses (from the unwanted layer) can be eliminated by the electronics. Most DVD players today have one layer, but in the near future that may go up to four (they look at two from the top and two from the underside) to store all the information for a movie. Combine that with the smaller bump size, and an advanced DVD can hold seven times as much information as can a CD. The first DVD player was marketed in 1997.

Recordable CDs and DVDs don't use aluminum as the reflector, but instead they use a chemical with reflection properties that can be altered by heat. When you record on one of these disks, you heat up a tiny region enough to alter it. Then, when you "play" the disk, the laser is low intensity, too little to alter the spot, but the amount of reflected light shows whether the spot is shiny or dull. A detector that measures the reflected light then feeds that signal into the computer. The computer turns the 0s and 1s into an audio (or, for DVDs, visual) output.

Laser Cleaning

Lasers are being used to clean old and valuable statues, without damaging the surface. The advantage is in their ability to deliver high power for very short pulses. A laser pulse delivered to the surface can cause very intense heating, enough to vaporize soot and oil, but if the pulse is very short (typically, a laser is used with a pulse that lasts only a few nanoseconds), then only a very thin layer of the statue is heated.[4]

Needless to say, when something is developed for scientific or artistic reasons, someone will figure out how to make money from the process. Laser cleaning and whitening of teeth is already being practiced in the United States, and dentists are looking seriously into the use of lasers for removal of dental caries and for other medical procedures.

Laser Weapons

Ever since the laser was invented, the military has looked for potential weapons applications. Lasers can deliver a lot of energy at the speed of light. This application was limited, at first, by the huge size and enormous weight of the very energetic lasers that were available. But recent development of portable lasers that use carbon dioxide has revived interest. Such lasers could, in principle, be carried on airplanes.

Note that the problem with weight is not from having to point the laser. The laser does not have to be turned. The laser can be pointed by moving a mirror, and that can be light in weight.

---

[4] For a dramatic image showing how lasers can clean old statues, see www.buildingconservation .com/articles/laser/laser.htm.

Lasers have already been used to shoot down small, unmanned aircraft called *drones*. Lasers are frequently mentioned for their possible application for shooting down missiles. This is a situation in which speed is needed, since the missiles travel fast, and you need to destroy them before they hit their target.

The laser beam does its damage by depositing heat on the surface. If the surface is moving, then the laser beam must follow the same spot. A potential problem with such systems is that laser beams can be reflected. If the target has a mirror-like surface, then little light is absorbed, and the laser "weapon" does no damage. It is also difficult to heat the surface of a missile if the missile is rotating, since the spot exposed to the beam is constantly changing. For this reason, laser weapons may have a limited future.

A more serious application for the military is as an anti-satellite weapon (ASAT). A laser can deliver a substantial amount of energy to a satellite over a period of a few minutes, and the satellite has no way to lose that energy except by heating up and radiating it. Most satellites are severely damaged if they are heated only a few tens of degrees Celsius.

Lasers and Controlled Thermonuclear Fusion

In chapter 5, we discussed the National Ignition Facility, or NIF. That is a huge building full of 192 large lasers that will (in 2010, we hope) ignite controlled thermonuclear fusion in small pellets of hydrogen, specifically in the isotopes deuterium and tritium.

The NIF uses two of the key properties of lasers. The first is the fact that laser light can be focused on a very small object. The second is that the laser light can come out of the lasers in a very short pulse. That is important so that the target gets heated very quickly, before it has time to explode from its energy. You'll recall that the same principle was used in the hydrogen bomb. For that, the fission bomb provided the intense and fast delivery of energy to ignite the hydrogen. So in the NIF, the lasers are really taking the place of the fission primary.

Here are the numbers again: the lasers deliver a power of 500 trillion (5×10^{14}) watts; that is 1000 times the electric generating power of the United States, but only for 4 nanoseconds. The energy released in this time is 1.8 megajoules. Recall that 4 kilojoules is the energy in one gram of TNT. So 1.8 megajoules is the energy in 500 grams of TNT, about one pound. That's not much. But if it can be delivered quickly enough to a very small object, it will trigger the fusion of the hydrogen.

Laser Eye Safety

Lasers can be dangerous to eyes for several reasons. The simplest, of course, is that lasers have high power that can be concentrated on a small spot, and the eye is delicate. But there are other reasons. The light from lasers is usually highly collimated, and parallel light is focused by the lens of the eye on the retina. Even relatively weak laser light can become intense when focused in this way. If the laser light is in the infrared or some other invisible wavelength, then the victim may not even know his eye is being damaged. For these reasons, people who work in or visit laser laboratories are usually required to wear special goggles that block out all light of the laser frequency, while allowing other light through.

Laser Eye Surgery

Lasers have found an important use in surgery, particularly for the eye. A broad laser beam can enter the eye and be focused on a tiny spot. Because the power of the beam is not concentrated until it comes to the small focus, there is not much heating except at the target spot. Perhaps the most exciting application of this technique is to "weld" a detached retina to the back surface of the eye. This procedure is now common, and it has prevented thousands of people from going blind. The irony here is that the very aspect of a laser that makes it dangerous is the one that makes it useful for medicine. (Of course, the same is also true of a knife.)

Lasers are also used to cut away parts of the cornea, to reshape it so that it focuses better on the retina. This has given "normal" eyesight to people who otherwise would have to wear glasses or contact lenses. Such surgery can be done in a few minutes in a doctor's office. The most popular kind of this surgery is called Lasik, for "laser-assisted in situ keratomileusis." In this procedure, a knife is used to open a flap on the cornea, and then a laser is used to vaporize and remove portions of the underlying cornea. The flap is put back in place, and the patient walks out of the doctor's office. The patient can see immediately. The eye takes several days for its initial healing, and is not completely normal for several months. Of course, this kind of surgery is not able to cure the loss of accommodation that occurs with age, since that is just a loss of flexibility of the lens.

Lasers are also used for other kinds of eye surgery. One of these is to stop the bleeding of blood vessels in the retina that leads to an illness called *macular degeneration*. The heat from the laser, delivered precisely to the blood vessel, can cause the blood to clot and seal the leak. The effect on the blood is called *laser photocoagulation*. Untreated, macular degeneration leads to loss of most vision.

Lasers are also being tried for other types of surgery. The highly focused beam can cut a very small region, and the heat automatically cauterizes the cut flesh (i.e., stops it from bleeding). Moreover, there is no need to sterilize a laser beam, in the way that people have to sterilize knives.

A Story from the Inventor

The list of laser applications seems to go on and on. Look up lasers on the Web for even more. Charles Townes, the original inventor of the maser/laser[5] was

---

[5] Charles Townes successfully built a maser—that is, a laser that operates at microwave frequencies. The first person to build a high-frequency (light) maser was his student Gordon Gould. Since the light device has a different acronym, credit is often given to Gould for the invention of the laser. Had Townes used the acronym *eraser* (mentioned in the text), then the two inventions would have been considered just variations of each other.

asked whether he ever imagined how important the laser would become. He replied by telling a story. He said he felt like a beaver who was standing next to the great Grand Coulee Dam. A deer asks him, "Did you build that?" The beaver replies, "No. But it was based on my idea."

The Photoelectric Effect

When light hits a surface, each photon can have enough energy to knock an electron out of an atom. Because it involves a photon and an electron, it is called the photoelectric effect. When it does this, the liberated electron can be used to create an electric current. This is the basis of the operation of many devices. To remember the word *photoelectric*, just think of what is happening: a photon hits a material to make an electron move. It's a very fancy name for a very simple process. But much of high-tech uses the photoelectric effect, including solar cells, digital cameras, night-vision glasses, laser printers, and photocopy machines.

Solar Cells

A solar cell consists of a crystal with two metal plates. When a photon hits the crystal, it gives an electron enough energy to jump to one of the electrodes. That electron can return to the other electrode by flowing through a wire. Thus, a solar cell uses sunlight to create electric current. The energy of sunlight is converted directly into useful electric energy, which can be used to run a motor or any other electric device. Such solar cells are often called PVCs, for photovoltaic converters, or just PV cells. They convert solar energy to electricity.

Solar cells are used in many applications, from satellites to battery charges, but they are not yet a major source of the power that runs the world. The main problem is their expense. The sunlight is free, of course, and the main expense comes from building, installing, and maintaining the cell. Some people do this at home and save money, because they don't include the cost of their own time. But for it to become an industrial process, you have to pay the workers, and that increases the cost. In some areas (California, Germany), solar power is highly subsidized by the government; that makes using the cells economically feasible.

An economist might say that the cost of solar cells is not much higher than that of fossil fuels if you include environmental costs (such as pollusiton and global warming). But right now, most people who use "dirty" power do not have to pay the environmental costs.

The future of solar cells can become brighter if their cost can be brought down, or if the cost of fossil fuels increases significantly. A promising technology for cheap solar cells uses amorphous silicon; do an Internet search for recent news on these and other cheap photocells. Another method is to use solar concentrators. These take sunlight from a large area and focus it onto a small solar cell. And one of the most promising of all is the CIGS technology, which uses solar cells created in a machine that looks much like an ink-jet printer.

Digital Cameras

Digital cameras work in exactly the same way as solar cells: they use photons to make electrons move. In a digital camera, the light is focused on an array of photocells; there is one photocell for each pixel—i.e., for each picture element. An 8-megapixel camera has 8 million of these photocells. When a photon hits any one of these cells, the emitted electron results in an electric current, and that can be read by the small computer that is part of every digital camera. There are two kinds of photocell arrays in common use. The first is called the *CCD array* (for charge-coupled device), and the other is the *MOSFET* (for metal-oxide semiconductor field effect transistor—you don't have to know this). The differences between these are not important for this course; the key thing to know is that they are arrays of photodetectors that use the photoelectric effect.

Some of the first digital cameras ever used were aboard U.S. spy satellites. They could take an image and use radio signals to send the image back down to the ground. At that time, even the fact that this could be done was highly classified.

A problem with digital cameras comes from the fact that the pixels are very small. In a photograph in low light, you may get only a few electrons per pixel. The number you get depends on chance; for a certain level of light, you may get 100 electrons, or maybe 110, or maybe 90. (They typical variation is the square-root of the number; so if the expected number for a give light level is 100 electrons, it can easily be 10 electrons higher or lower.) But this kind of variation causes some pixels to have stronger signals than others, so in the final photograph, two pixels that should look the same will have different brightness. The camera review magazines will tell you that the camera takes "noisy" photos in low light. In fact, that noise comes from the random fluctuations in the number of electrons that are excited by the photoelectric effect. More sensitive cameras can be made, but they usually require larger pixels so that more light is likely to hit each one. If you want a low-light-level digital camera, you have to buy one that has a physically large CCD. That usually means that the lenses and every thing else is larger too. If you buy a small camera, you will not get good low-light sensitivity.

Similar digital noise also affects laser communication using fiber optics.

Fiber Optics Noise

In chapter 9, I explained that light is an extremely good way to send signals because of its high frequency. (Recall that Shannon's information theorem says that the bits-per-second rate is approximately equal to the frequency.) But now we can get an interesting result from quantum physics: high frequency isn't enough, we also need high power.

Here is why: a 1-milliwatt laser beam (typical for a laser pointer) has 10^{-3} J/s = 6×10^{15} eV/s. Since each photon is 2.4 eV, this means that the light has a little over 2×10^{15} photons every second. The frequency for green light is 6×10^{14} Hz. So there are only about 3 photons per cycle, on average.

That is pretty low. You cannot send signals faster than the photon rate, and even 3 per cycle is low. The value of 3 is only an average number, and statistical calculations show that if the average is 3, then there will often be 5, 4, 2, or only 1 photon. In fact, about 5% of the time there will be no photons at all in a given cycle, even when the cycle is supposed to have 3. That means that if you send a bit for a 1, there is a 5% chance that it will show no photons, and be interpreted as a 0 bit. That is an error. The conclusion is that enough power must be used to avoid this *photon limit*. To avoid high error rates, you need many more than 1 photon for each cycle.

So fiber optics communication requires high frequency (so there can be a large number of bits per second) and high power (so there can be many photons per bit).

Image Intensifiers and Night Vision

The human eye and brain aren't sensitive enough to sense the light from a single photon. Neither is a typical digital camera. But there is an electronic device that can, called an *image intensifier*. It works on the photoelectric effect, just like photocells, by using photons to create an electric current. But in an image intensifier, the light is used to create an immediate image that is bright enough for the eye to see.

A modern image intensifier consists of a large number of narrow tubes, tightly packed together in a configuration called a *multichannel plate*. When a photon enters the end of one of the thin tubes, it typically hits the wall of the tube before going very far. Visible photons, even in very dim light, have an energy of 2.4 volts, and that is enough to knock an electron off the surface. (This is another example of the photoelectric effect.) An electric field then accelerates the electron down the tube, and it soon crashes into the side and knocks out additional electrons. This process continues as an avalanche, and by the time all the electrons reach the end of the tube, the electron signal can be very large, a billion electrons or more. These electrons hit a phosphor, and if there are enough of them, they make a bright spot.

The entire stack of tubes can be placed at the back of a camera. The photons hit the multichannel plate instead of film, and they trigger the electron avalanche that eventually emerges from the end of the tube and hits a piece of glass coated with a phosphor. If a photon entered the front end of the stack, then there will be a bright spot at the other end, bright enough for a human to see. Multichannel plates are used in most of the inexpensive image intensifiers that can be purchased on the Internet.

The original image intensifiers were used to make devices called *starlight scopes*. These were camera-like systems that could be worn on a person's head (like attached binoculars) and used to see things in dark places. But they do require some light—that's why the word *starlight* appears in the name. They are one of the technologies used in night vision. Contrast them with IR night-vision scopes.

IR imagers can work in total darkness, using only the IR light emitted by the warmth of the person or object. Starlight scopes require a little bit of visible light to be present, and then they only amplify that.

Xerox Machines and Laser Printers

The original Xerox machine (the generic term is *photocopier*) took advantage of the unusual properties of the element selenium. If you put charge on a selenium surface, the charge stays there; selenium is not a good conductor of electricity. However, if you shine light on a region of the selenium surface, the energy of the absorbed photons is sufficient to eject the charge. Again, this is the photoelectric effect.

The Xerox machine works by first putting an electric charge on a plate or drum of selenium. Because selenium doesn't conduct, the charge stays stuck on the surface. The selenium is then exposed to an image of the thing you are copying. Where the light is bright, electrons are ejected, leaving no charge behind. Where the image is dark, the charge remains.

If the selenium is then exposed to a cloud of carbon soot, the soot will be attracted by the electric field to the charged regions. The result is that the surface stays clean wherever there is no charge—that is, where bright light hit it—and is darkened at all the places where there was no light to eject the electrons.

Once the sooty selenium is ready, a piece of paper is brought into contact with it, and it picks up the carbon. That dirty paper becomes the Xerox copy, or photocopy. The soot contains a binding material, and when the paper is heated (on the way out of the machine), the soot is permanently bound to the paper.

If the paper gets stuck before it is heated, and you have to open the machine to extract it, you'll find that the soot didn't stick to the paper, and your hands, and anything else that touches the paper, become dirty.

A laser printer is a Xerox machine in which a laser is used to expose the selenium instead of an optical image. The laser scans across the surface with a fine beam with varying brightness in just the way needed to produce the image.

Manufacturers have found materials other than selenium that work just as well and sometimes better. But the Xerox patent on using selenium in this way has expired, so any company can now use that material for this kind of photocopying.

Quantum Physics of Gamma Rays and X-Rays

Let's use the Einstein equation to compare the energy of a gamma ray photon to that of a visible photon.

In chapter 9, I said that the frequency of a typical gamma ray is about 3×10^{21} Hz. Visible light has a frequency of about 6×10^{14} Hz. So the gamma frequency is higher by a factor of $3 \times 10^{21}/6 \times 10^{14} = 5$ million.

That means that the gamma energy is also 5 million times greater than the visible photon energy. A typical visible photon has an energy of 2.5 eV. A typical gamma ray energy will be 5 million times greater, about 10 MeV. That's enough energy to break a deuterium nucleus into its constituent proton and neutron. Radioactive decays, which release energies typically in the MeV range, often emit gamma rays.

Let's do a similar thing for x-rays. According to table 9.1 in chapter 9, typical x-rays have frequencies of about 10^{19} Hz. That is about 20,000 times greater than the visible light frequency. According the Einstein equation, the energy for x-ray photons will be 20,000 times greater. Since visible light is typically 2.5 eV per photon, that means that an x-ray photon will have energy 20,000 × 2.5 = 50,000 eV = 50 keV.

X-rays are often made by taking electrons with kinetic energy of 50 to 100 keV and slamming them into tungsten or other metals. This calculation shows that most of the energy of the electron comes out as the energy of the x-ray photon.

Suppose that we have a beam of visible photons with a total energy of 10 MeV. Since each photon has only 2.5 eV, that means that there are 4 million photons present. Each photon will lose 2.5 eV when it hits a molecule. But a gamma ray beam with the same energy will have only one gamma ray photon. When it is absorbed, it will depose all its energy at one place. Gamma rays will never be absorbed bit by bit.[6] In that sense, they appear to be more like a particle than does visible light, which normally has so many photons that the quantum absorption is not noticed.

It is this large energy per photon that gives gamma rays and x-rays many of their distinctive properties. A single gamma ray can deposit enough energy in a cell to destroy it. In contrast, a single UV photon can, at most, mutate the DNA. Five million is a big factor. Because they dump so much energy in a single burst, gamma rays seem to be more particle-like than any other electromagnetic wave.

Semiconductor Transistors

Essentially all of modern electronics is based on the fact that electrons are waves. Their wave nature is very important when they flow through crystals known as semiconductors. (The word *semiconductor* comes from the fact that the material is not as conductive as a metal, yet it still conducts.) The most important semiconductors are silicon and germanium, often with small amounts of aluminum or phosphorus mixed into their crystals. Important applications of semiconductors include the microprocessor that runs your computer, laser diodes in your compact disk players, and virtually all other modern electronics in your TV, your car, and even in screw-in fluorescent lightbulbs.

The key feature of semiconductors, the one that makes it so important, is the fact that electrons cannot have all energies. There is an energy gap just as there is in the hydrogen and helium atom.

This energy gap is a result of the fact that electrons are waves. When electron waves move through crystals, they can bounce off the atoms. Two bounces, and they are traveling in the original direction. When they do that they can interfere with (cancel or reinforce) the original wave. That turns out to mean that certain

---

[6] For the experts only: There are exceptions, such as Compton scattering. But even in Compton scattering, the gamma gives a substantial fraction of its energy on each scatter.

energy electrons cancel themselves—so these energies are impossible. That makes an energy gap between the levels that are allowed. As with a hydrogen atom, the typical energy gap is a few electron volts.

Semiconductors become useful for electronics when you take two semiconductors with different properties (they may, for example, have different energy levels) and you put them in contact with each other. Such a combination is called a *transistor*. Two such materials in contact make a *diode*, and three make a *triode*. There are lots of variations on this basic scheme. Semiconductor physics is a vast subject, and it has spawned a huge industry. We will touch only on the most basic aspect of this, the use made of the energy gap.[7]

Diode Transistors

Diode transistors are very simple devices that can convert AC (alternating current) to DC (direct current). When they do this, they are called *rectifiers*. The reason rectifiers are so important is that almost all electronics require DC, in which the electrons only flow one way. Yet, as we discussed in chapter 6, the electricity that comes to our homes is AC. A diode can turn AC to DC by letting through only the half of the current that is flowing in the right way. That's why it is called a rectifier—it allows current to flow only the "right" way.

To make a diode, you put two semiconductors that have mismatched energy levels into contact with each other.[8] As soon as that is done, electrons begin to flow from the one that has higher energy to the one that has lower energy. The flow finally stops when enough electric charge builds up to repel additional electrons. The same thing would happen with balls rolling down a hill. If they were charged, eventually the repulsion would keep other balls from rolling down.

The electrons that have accumulated create a strong electric field near the junction. This field prevents additional charges from flowing. If you weaken this field by applying an opposite voltage (e.g., from a battery), then additional electrons will flow to rebuild it. But if you strengthen the field by applying an additional voltage in the same direction, then no current can flow. This is the basic idea that makes the semiconductor diode work. It lets current flow one way, but not the other. Put AC into it, and current will flow only half of the time—the half when it is going in the right direction. That's the way it turns AC into DC. (The magnitude of the DC will still vary with time, going up and down, but it will never go the other way. To smooth it out takes other electronics.)

Diodes have a long history. Like superconductors, they were used before they were understood. One of the earliest diodes ever used was the "cat's whisker"

---

[7] Note for the experts: Most introductions to semiconductors emphasize the importance both of electrons and of objects called *holes*"—which can be thought of as bubbles in the electron sea, i.e., the absence of electrons. Holes behave much like positively charged electrons. Another important issue is *doping*, a method of adding extra electrons or holes to regions that normally have very few. Diffusion of electrons and holes are often responsible for mismatched energy gaps.

[8] Optional, for the experts only: The two materials are often both made out of silicon, but they have different impurities purposely mixed in—for example, aluminum in one and phosphorus in the other. This creates *donor levels*, and it is the energy difference between these donor levels that plays the key role. Diffusion of charge carriers brings these donor levels to the same energy, and that diffusion creates the electric field at the junction.

of the old crystal radio sets that amateurs (and even the author of this book) used to make as a hobby. A thin wire was delicately placed on a crystal and moved around until a spot was found that conducted electricity in just one direction. The wire was supposed to be as thin as a cat's whisker, so these were called cat's whisker diodes. But real cat's whiskers were never used. It was a metaphor.

Light-Emitting Diodes (LEDs)

A *light emitting diode* (LED) is a semiconductor that emits light when a voltage is applied across it. Those little red lights that let you know your computer (or anything else) is on are usually LEDs. The large TV displays used at stadiums and for some street displays are large arrays of red, blue, and green LEDs. LEDs light up your watch when you push the button. Many traffic lights are being replaced by LEDs because the LEDs are more efficient; they don't produce waste heat and they don't burn out like tungsten filaments. In the near future, most flashlights will use LEDs instead of small tungsten bulbs. (Expensive flashlights already use them.) Infrared LEDs on your TV remote control send a burst of invisible light to the TV to tell it to turn on, off, or to change the channel.

An LED works in a simple way: an applied electric voltage gives an electron extra energy.[9] Because of the energy gap, it can't lose little bits of this energy, but only the entire amount, all at once. It does this by emitting a photon. The color of the photon is related to the energy gap by the Einstein formula $E = hf$. An LED with a small energy gap gives red light; an LED with a large energy gap gives blue light.

Look around and see how many LEDs you come in contact with every day. You may have to look at a traffic light up close to notice. In 2009 (when this is being written), many red traffic lights have LEDs, but the green ones still use incandescent lamps. Look on computers and stereo sets for the little light that comes on indicating that power is present. LEDs are also used to create the light on large TV screens seen in public places, such as Times Square in New York and football stadiums.

LEDs are finding more and more important uses for lighting. They are beginning to replace fluorescent bulbs as the source of light for computer screens and flat-panel TVs They may soon provide the most convenient light for people in developing countries, since they are nearly as efficient as fluorescent lights and much more robust.

Diode Lasers

A *diode laser* is the kind that is used in supermarket scanners, in laser pointers, and in CD and DVD players to create the light that is reflected off the disk. It is very similar to an LED; it is a small semiconductor in which the electron is

---

[9] Most books will describe in detail how energy is delivered to the electron. This usually involves a junction between two semiconductors that have different doping, which leads to different energy levels. But the key reason that light is emitted is because of the energy gap.

excited from its low energy to a higher one. The main difference between LEDs and diode lasers is that the diode laser takes advantage of stimulated emission—i.e., the fact that one emitted photon can stimulate the emission of another photon. To achieve this, scientists had to find a semiconductor in which the photon would not be emitted spontaneously before it could be stimulated to emit.

Because the diode laser is a laser, the photons that come out are all going in approximately the same direction. This collimation is not quite as good as for a large size laser, but it is much better than in the LED, in which the light comes out in all directions. The collimation allows the light to form a very narrow beam that does not spread as much, and which can be focused to a very small spot. This is important for most of the applications mentioned above.

It is the diode laser that has really transformed the laser into an everyday device. Prior lasers looked like fluorescent lightbulbs, but with the light coming out the end. They required special power supplies, and had very limited lifetimes.

Transistors

The heart of most electronics (radios, TVs, computers, stereos, iPods) is a tiny device called a *transistor*. In much modern usage, what used to be called a transistor radio is now often called a transistor; but that's not what I am referring to. By a transistor, I mean the little electronic devices that make up the components of the transistor radio.

A transistor is a modern version of an amplifier, something that takes in a weak signal (maybe from the reflected light from a CD, or from a weak electrical voltage from an antenna) and puts out a very strong signal, enough to drive a loudspeaker (if it is a sound signal) or a TV picture.

The word *amplifier* can be misleading, because an amplifier doesn't really take a weak signal and make it stronger. (Doing that might violate the conservation of energy.) What it really does is take a weak signal and create a strong one that oscillates in an identical way. You could say that it makes a clone of the original weak signal. The energy for that clone comes from a battery or other source of power, such as electricity from a wall plug.

An amplifier is very similar, in some aspects, to a water valve. Think of the way that a water valve operates. You turn a small handle, and with very little energy, you control a huge flow of water.

The original electric amplifier used a device called a *vacuum tube*. Some old TV sets, and some stereos, still use vacuum tubes. (Some audiophiles claim tube sets sound better.) Vacuum tubes were based on the fact that a large electric current flowing through a vacuum could be controlled by a very small voltage.[10] (The vacuum was located inside a glass tube, and that's the origin of the

---

[10] In the vacuum of the tube, a tungsten filament heated a piece of metal called a *cathode*, which emitted electrons when it was hot. The electrons would flow through the vacuum to another piece of metal called an *anode*. Between was a grid of wires. Small voltages applied to this grid could make large changes in the current flowing through the grid. This design is still used in CRT (cathode ray tube) displays. The heated filament made the vacuum tube hot and used a lot of power. Tubes needed a lot of space for the electrons to flow through the vacuum, so they were large.

term *vacuum tube*.) It was just like turning a valve to control water flow. In fact, what we call vacuum tubes in the United States, are called electric valves in England.

In the 1960s, devices called transistor amplifiers began to replace the tubes. The new radios were called transistor radios. This was the beginning of a great revolution in high-tech. Transistors don't use vacuum or metal electron emitters. Instead, they use semiconductors.

The original transistor amplifier consisted of two transistor diodes placed back to back, with a thin layer wedged in between that was made very thin. Very small voltages placed on the central layer would control very large currents that flowed from one side to the other. Again, as with diodes, the reason it works is that the three regions have energy gaps, and this makes charges flow that oppose further current. Small changes in the voltages can affect the relative energies of the gaps. It is analogous to having a large dam, with a huge reservoir. Small changes in the height of the dam can cause huge changes in the amount of water that flows over.

Transistors are far more reliable, much faster, take less power, and generate less waste heat than vacuum tubes. They had been invented back in 1947.[11] In the 1960s, a good (and expensive) portable radio contained, typically, eight transistors. But the size and cost of transistors has continued to decrease.

Computer Circuits

An important breakthrough came when engineers figured out how to put many transistors on a single chip of silicon, creating a device called the *integrated circuit*, or IC. The Nobel Prize in 2000 was awarded to Jack Kilby for this invention.

The integrated circuit really made Moore's law start to operate (see chapter 5). As the transistors were made smaller, the complexity of circuits could grow. The first full *microprocessor* (computer that has all of its complex circuitry on one chip of silicon) was created in 1971. Now we have over a billion of the quantum devices called transistors on a single chip of silicon. That is a common number for the chip in a home computer. Such numbers were inconceivable just a few years ago.

All "computation" done in a computer is done with special transistor amplifiers called switches. They are like water valves, but with only two positions: fully on, and completely off. Linked switches can be used to add numbers, multiply them, divide them, or make logical conclusions.

These switch have traditionally been called *gates*. A gate is like a valve: open it, and things can pass through. A set of switches connected to make a logical decision is called a *logic gate*. (Note: The name *gate* was not named after Bill Gates, any more than the logical procedure used to operate them, the algorithm, was named after Al Gore.)

Computers have become very complex. The transistors are made very small (a few microns in size), so they require very little power and can be put together

[11] In 1956, the Nobel Prize in physics was awarded to W. Shockley, J. Bardeen, and W. Brattain of Bell Telephone Laboratories for their invention of the transistor.

into very small chips. The small size is important for speed, since the velocity of signals that travel down wires is always less than the speed of light. In one typical computer cycle (1 nanosecond, or a billionth of a second), light can travel only 30 centimeters, or 1 foot. So to be able to exchange information with other parts of the computer, the whole thing must be small.

OPTIONAL: A LITTLE MORE ABOUT LOGIC GATES

It is amazing that a collection of switches can be used to compute logical results. Here's a simple example that illustrates how they can do it. Consider two possible statements: (A) Leslie is Mary's father; and (B) Leslie is male. Note that if A is true, it implies that B is true. But if B is true, it doesn't necessarily imply that A is true. In computers, this logic is performed with logic gate. If connector A gets a positive voltage (because A is true), then that puts a voltage on B (showing that B is also true). Some gates (switches) require two inputs to be turned on; others will be turned on if either of two other switches is turned on.

All computation done in computers is done by connection of gates such as this. One of the most fascinating theorems of computer science is that every known computation (those doable using logic or math) can be accomplished by using such switches, in conjunction with other devices that work as the memory.

Superconductors

When we talked about superconductors, I never explained why the electrons can move through the superconducting metal with no loss of energy. In fact, superconductivity was discovered by H. K. Omnes in 1911 (he was awarded the Nobel Prize in 1913), and yet the phenomenon was not understood by him or anyone else for many decades. For much of the twentieth century, it was the outstanding failure of the quantum theory that nobody could figure out why superconductors were superconductors!

The reason did turn out to be quantum physics, and just as with spectral fingerprints and semiconductors, the answer was the existence of an energy gap. Just as in the other materials, the behavior of superconductors comes about because electrons are waves, and in certain crystals this can lead to an energy gap—certain energies that an electron cannot have when it travels through these crystals. The energy gap for superconductors is only 0.001 eV, but that is enough to give zero resistance.

Now here is the key reason that superconductors have their remarkable properties: the flowing electrons all have a low amount of energy. In ordinary metals, such as copper wires, they would collide with impurities in the metal and gain or lose energy. Imagine that you are traveling with the electrons. (In a superconductor, all the electrons are moving together at the same speed.) In this frame of reference, you have no energy, but the nuclei of the atoms are moving. In this frame of reference, there is an energy gap—and that means that you can't gain just a little bit of energy at a time. Even if you collide with a nucleus (or impurity or a crack "dislocation" in the material), you must gain at least one energy gap worth of energy, or you will gain nothing at all. In a superconductor, the amount of energy that you would gain from such collisions is less than the gap.

In the almost perverse logic of quantum mechanics, that means that the electrons can't collide with impurities because losing that amount of energy is impossible! So the electrons will continue to flow without energy loss. That's what superconductivity is.

The strange concept in the last paragraph is not completely new. Physics often states that behavior which contradicts the conservation of energy cannot happen. What seems most weird in superconductors is that the impurities are invisible to the electrons, because if the electrons bounce off the impurities, they would have illegal (impossible) energies.

Not all metals become superconductors at low temperatures. We now understand that for a metal to become superconductor, an interesting thing must happen: the electrons have to move in pairs. This happens when slow-moving electrons pull the positive charges of the metal close to them, and that distortion of the metal tends to attract another electron. The net result is that electrons attract other electrons. The electrons never get very close, and it doesn't take very much energy to break them apart. That's why this happens only at low temperatures. The two electrons are called *Cooper pairs* after the name of the physicist who first predicted their existence.

The full quantum theory of superconductivity was worked out in 1957 by John Bardeen, Leon Cooper, and Robert Schrieffer. Their theory is called the *BCS theory* after the letters of their names. They were awarded the Nobel Prize for this work in 1972. From their theory, they could predict which metals can become superconductors and which can't, and at what temperatures it happens. One of the strangest results of their theory is that hydrogen can become a superconductor at high pressure. That means that the core of the planet Jupiter may be a superconductor, as well as the stars known as *pulsars*. These predictions have not yet been verified.

Despite the fact that we understand superconductivity at low temperatures, once again, superconductivity has a mystery. This time, it is high-temperature superconductors, the ones that become superconducting at temperatures up to 150 K. That's nearly 150 C below freezing, so it doesn't really seem like high temperature, but it is far away from absolute zero. The BCS theory does not predict their existence, and nobody has been able to figure out why these compounds become superconductors at such high temperatures. (There are actually several theories, but none of these has been accepted as the clear winner.) We do know that the flow involves Cooper pairs. But nobody has a good theory to predict which materials will become high-temperature superconductors, or how high the superconducting temperature can go. If a material is found that is superconducting at room temperature, then we will see a new technology revolution, bigger than the one involving transistors. Energy transport will become very easy, energy loss in computers will become very small, and we may switch from AC in our homes to DC.

Electron Microscope

Objects that are less than a micron in size cannot be resolved with ordinary light because you cannot focus a beam on a spot smaller than the beam's wavelength. If you want to look at something smaller than that, you need a wave with a shorter wavelength. X-rays are sometimes used, but x-rays tend to go right through objects, especially objects that are only a few microns thick. A more widely used option is electrons. Electron beams with an energy of 50 keV have a wavelength smaller than the size of atoms.[12]

There are several kinds of electron microscopes, but the most interesting one is the *scanning electron microscope*, or SEM. In an SEM, a beam of electrons is scanned across the object, much as a beam of electrons scans on the surface of a material, and the number of electrons that bounce off in a particular direction is measured. This number is called the *brightness*, and based on that, an image is created in a computer. These images look remarkably like ordinary photos. That's probably because shadows make it look realistic. (The shadows are there because electron beams hitting the back side of an object reflect away from the detector, and fewer of them are collected.) The image shown in figure 11.4 is an SEM image of an insect.

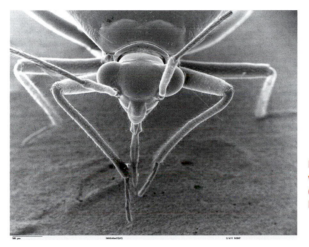

Figure 11.4 Image of an insect taken with a scanning electron microscope. (Courtesy of Dartmouth College Electron Microscope Facility.)

Look on the Internet for more SEM images. You'll find things that are truly amazing. Search for images that have the words *SEM* or *electron microscope* associated with them.[13]

Deeper Aspects of Quantum Physics

In this section, we will address the issues that most physics majors study, the abstract issues of quantum physics. These include the nature of the electron wave, the particle-wave duality, what happens when you make a measurement,

---

[12] Optional: Since electrons have mass, according to the theory of quantum physics (not all of which we have discussed), the wavelength L has to be calculated from the deBroglie equation, $L = h/(mv)$. The momentum mv can be calculated if you know the energy $E = 1/2\ mv^2 = (mv)^2/(2m)$.

[13] One good source is remf.dartmouth.edu/imagesindex.html.

and the uncertainty principle. These are fascinating topics, but we relegate them to this section because they are not really important for future presidents to master. They are important, however, in one possible future application: the quantum computer.

Incidentally, most physics majors do not study much of the preceding material (e.g., Xerox machines, LEDs, SEMs, laser printers and semiconductor computers). That's because they are trying hard to learn how to do calculations.

Do Photons Really Exist?

We've been talking about photons as if they are particles. Yet we know that they are electromagnetic waves. So how can we do that? Do photons really exist? Are they particles, waves, or both? Now we are discussing the heart of quantum physics. Don't expect simple answers to these simple questions. The answers are bound to be confusing. But let's give it a try.

Light behaves as a wave—except when it is emitted or absorbed. All the quantum features showed themselves only during these times. Of course, that is when we interact with them, so that is important. But in between, after emission and before absorption, the "photon" nature of light doesn't seem to exist.

If that strikes you as weird, then I am glad. It is weird, and it still bothers many physicists. Let me illustrate what it means with a simple example: the soap bubble.

Recall that the colors of the soap bubble come about because some of the light wave bounces off the inner surface of the bubble, and some bounces off the front surface, and when these two waves come together, the waves interfere. Some colors (the ones that come out in phase with each other) are made stronger, and some (those that cancel) are made weaker or nonexistent.

How does this interference fit in with the picture of photons? Let's imagine that we turn down the intensity of the light until only one photon every minute is detected reflecting off an oil slick, or a soap bubble. You might think that the photon is reflected off the outer surface of the bubble, or off the inner surface, but obviously it couldn't be reflected off both. So at very low levels of light, you would think that all the colors that arise from wave cancellation would disappear. You can't possibly have beats when only one photon is present! Right? Wrong.

The experiment has been done, not with soap bubbles, but with mirrors. In fact, it is not hard to do, and can be done by undergraduate physics majors in a junior- or senior-level laboratory. The results are unambiguous. It is as if the photon split in half and bounced off both surfaces. So the photon behaves like a wave, right up to the point where you detect it. Only then is the particle behavior evident.

In fact, the best way to think of light is as a wave that can be emitted or absorbed only in quanta—but that in between, it is a wave. It moves like a wave, diffracts like a wave, bends like a wave, and interferes like a wave. But it is not emitted and absorbed like a wave, but like a particle. This is, as I mentioned previously, the famous *wave-particle duality* of quantum physics that mystifies many people. But it mystifies them only because they think particles and waves are different things. I like to use the term *particle-wave* or *wave-particle* because real things have some of the properties of both. Another name might be *quantum wave*. We'll get back to these issues when we discuss the Heisenberg uncertainty principle.

Are Ordinary Waves Quantized?

Light waves are quantized. Electron waves are quantized. Do you think that water waves are also quantized? Are there only certain allowed energies that a water wave can have? What do you think?

The surprising answer is yes. How can that be? Why don't we notice?

Let's look at a typical water wave. It might have a frequency of $f = 1$ cycle per second, (Such a wave has a wavelength of about 1.6 meters.) The energy will be quantized by the Einstein equation $E = hf = 6.6 \times 10^{-34}$ J. That is tiny. If a wave hits you, and delivers an energy of 1 J, then that consists of $1/(6.6 \times 10^{-34}) = 1.5 \times 10^{33}$ quanta. With that many wave quanta, the fact that the energy is quantized is impossible to notice. This is a case in which a *quantum leap* is very tiny indeed.

In fact, the quanta are so small that they have never been observed for water waves. They don't even have a name. (Waterons? Hydrolons?) But according to quantum theory, they are there.

A similar quantization happens for all those waves that appear to us as real waves: sound waves, low-frequency radio and TV waves, rope and Slinky waves. They are quantized, but in the limit of low frequencies, the quanta are so small that we would never notice.

Is the Earth's Orbit Quantized?

Yes. At least we think so. Why did we never notice? The reason is that the gaps between the energy levels are tiny. A calculation using quantum mechanics shows that the energy gap between different orbits of the Earth is about 0.001 eV—that is, it is a thousand times smaller than the typical energy gap in a single atom. Compared to the energy in the motion of the Earth, this "gap" is only 1 part in 10^{55}. That's why it hasn't been observed, and probably can't be observed. We think it is quantized because every orbit where the prediction is observable verifies the prediction of quantum physics.

Is the entire Earth really a wave? Again, the answer is yes, but it is a complicated wave, with many parts. Speech is also complicated, and waves in the ocean during a storm are complicated; light waves from the page that you are reading are complicated. They consist of waves of many wavelengths all mixed together. Because of this complexity, it doesn't turn out to be very useful to think of the Earth as a wave, even though it is. In the atom, where the energy gap is comparable to the electron energy, then the wave concept is not only useful, but essential.

Particle-Wave Duality

Particles are really waves? Waves are really particles? Why can't quantum physicists make up their minds? Which is it? This is one of the most confusing issues for people beginning to learn quantum physics.

The real answer is that particles are not waves, and waves are not particles. (Now are you even more confused? Be patient.) In fact, there are no such things as particles or waves. There is only one kind of "thing," and it has no name. (Maybe that's why you are confused.) We could make up a name. How's this: a *warticle*. Alternatively, we could call it a *pwave*. Let's use that. No other book uses that term, so this won't help you read other books on quantum physics, but it might help you understand.

Everything is a pwave. An electron is a pwave, and so is a photon. Pwaves have some of the properties of waves and some of the properties of particles. Pwaves can interfere with each other, just as waves do. Because of this, their energy levels are quantized when they are in circular orbits (that is, when they interfere with themselves by circling back).

Uncertainty

The fact that a particle has wave properties implies that its position is uncertain. Waves don't exist at a single location; they are spread out. This is the heart of the famous *uncertainty principle*. The particle does not have a well-defined position. If the position is actually measured, the particle will be found somewhere in the wave region.

Waves can be extended, or they could consist of only a few cycles. A short burst of oscillations is called a *wave packet*, and a typical example is shown in figure 11.5.

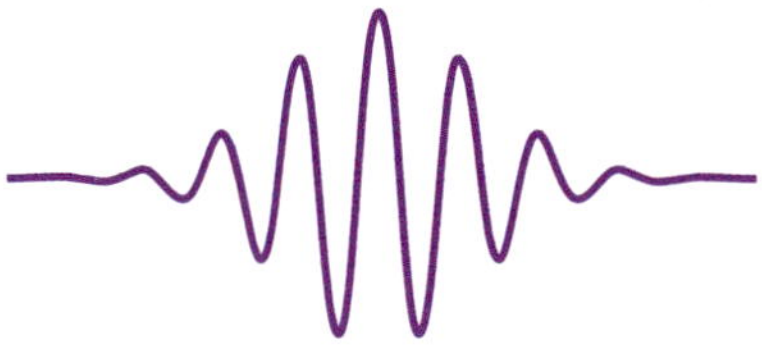

Figure 11.5 A typical wave packet.

In this wave packet, the particle will not be found at the two extreme ends, region, but somewhere in the middle where the oscillations are strong.[14] But quantum physics doesn't require the wave packet to have the nice shape shown in figure 11.5. It could easily have the form shown in figure 11.6.

---

[14] Optional: In the math of quantum physics, we express the wave packet as a complex function usually called $\psi(x)$. Then the probability that the particle will be found at any particular location will be $|\psi(x)|^2$.

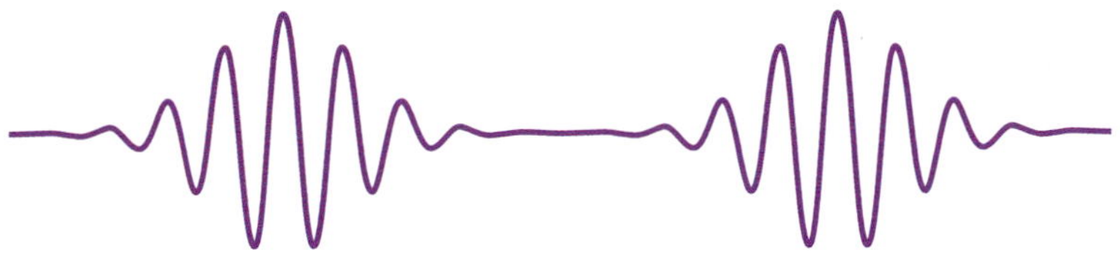

Figure 11.6 Another typical wave packet. This wave represents only one particle, despite the fact that the wave is broken into two discrete segments.

That looks like two particles, right? But it can also be one particle, but with a very uncertain position. In fact, the particle is not likely to be found in the middle (where the amplitude is near zero), nor at the two extreme ends, but either in the oscillating region on the left or in the oscillating region on the right. But this is a valid wave function for a single particle, even though it is split into two parts. In fact, any shape function is valid, at least for a particle moving in empty space, where we don't have to worry about circular orbits canceling themselves out. We are now far away from classical thinking. It is weird. Let's look at things that are even weirder.

Uncertainty in Energy

I showed in the last example that the wave of an electron could be in two disconnected pieces. If you are willing to accept that, then how about this: put an electron into two separate circular orbits around two different protons.

That sounds crazy. But it is no crazier than the two separated waves shown in Fig 11.6. And nothing in quantum physics prohibits it. In fact, it happens frequently in quantum experiments. It is a consequence of the fact that electrons are waves—or at least have wavelike properties.

Could the two orbits have different energies? Sure. But then what is the energy of the electron? Answer: It is uncertain, just as the position is uncertain.

Could we put the electron into two different orbits around the same proton, at the same time? Answer: Sure. Again, its energy would then be uncertain.

There are lots of allowed orbits. Could we put the electron into 17 of them at the same time? Answer: Guess, and then look in the footnote.[15]

Measurement of the Quantum Wave

Suppose that a very broad wave packet of a photon hits the CCD (the light-sensitive detector) in a digital camera. You can even assume that the width of the wave packet is larger than the entire camera. Because the packet is so large, does it give a signal in every pixel of the CCD? No. It knocks an electron out of only one pixel because it has only enough energy to affect one.

Which one? Quantum physics supplies a curious answer. The electron knocked out will be random, but only in the region where the wave packet hits.

[15] Sure. It could even be in an infinite number of orbits.

The places where the wave packet is stronger (greater amplitude) have a higher probability. In fact, according to the math of quantum physics, the probability is proportional to the square of the amplitude.

So quantum mechanics is inherently a random theory. There is no way to predict where the electron will be knocked out.

Some people like to say that the position of the knocked-out electron represents the "true" position of where the photon "really was." But that's wrong. The wave is spread out over space, and the photon is not hidden within this range. We know that because of soap bubbles—the photon can bounce off both the front and the back of the soap bubble because it is a wave, not a localized particle.

We came across this before, when we discussed radioactive decay. Two identical nuclei of tritium will emit their beta particles at different times, despite the fact that the nuclei are identical. Quantum physics can give only the probability that the beta will be emitted.

Is Energy Still Conserved?

If energy is uncertain, does that mean we can lose some? Good question. Remarkably, the answer is no. Energy, even though uncertain, is still conserved. (Note how often people use the term "remarkable" or "weird" or "strange" in the discussion of quantum physics. That represents the fact that most physicists who think deeply about these issues still find them peculiar.)

Suppose that we have a definite amount of energy and give it to two electrons. Assume that each electron has uncertain energy. The electrons move in orbits about different protons, perhaps a mile, or a million miles, in apart from each other. Finally, we detect each electron. The energy we get could turn out to be anything within the uncertainty. But when we add together the two energies we measure, the sum will be the same as we started with. Energy is conserved.

Does that mean that the electron really had a certain energy, but we just didn't know what it was? No, that possibility is called *hidden variable theory* (look it up on the Internet). It makes other predictions that have been tested and found to be wrong. There have been many versions of hidden variable theories, but they have been tested and shown to be wrong.

Einstein didn't like quantum physics, because he sensed that the uncertainty principle was in fundamental contradiction to his theory of relativity. Specifically, he thought that the sudden changes in waves that take place whenever a particle was detected could not be as sudden as quantum physics required, because the changes seemed to take place at speeds faster than the speed of light. He always favored hidden variable theories until they were proven wrong by experimental tests. Alas, as of today, every tested hidden variable theory has been proven wrong. Maybe someday a new hidden variable theory will be invented that works, and then Einstein's spirit will smile.

OPTIONAL: THE PRECISE STATEMENT OF THE UNCERTAINTY PRINCIPLE

You can make the position of a particle certain by making a very small wave packet. But in quantum mechanics, such a wave is made up of many different waves traveling at different velocities. Yet only one particle is present. So once you detect the particle, all the other waves have to suddenly disappear. This sudden disappearance has a special name in quantum physics; it is called the *collapse of the wave function.*

So although the position of the electron may be well known, the velocity is uncertain. That means that its energy is uncertain too. These relations are at the heart of the famous *Heisenberg uncertainty principle.* The principle states that if you create an electron with a very well-defined position, so it is known to an accuracy of Δx (which can be smaller than an atom, or larger than the sun), then the velocity is uncertain by an amount at least equal to

$$\Delta v = \frac{h}{2\pi m \Delta x}$$

In this equation, read Δv as one term; it means "the uncertainty in the velocity v." Also in this equation, m is the mass of the electron, and h is Planck's constant, and Δx is the uncertainty in x. So if you improve your knowledge of the position (i.e., make the wave packet, and Δx, small by passing the wave through a small hole), that makes the uncertainty in velocity greater.

A similar thing happens with light. Let's talk about knowing the position of the photon in the x direction for a beam of light that is traveling in the y direction. To determine the position of x, you let the light wave pass through an opening of width D. But in doing so, you make the wave spread out from diffraction. That means that the velocity in the x direction is no longer certain; part of the wave is moving to the left, and part to the right. When the photon is detected, it could be moving (at least in part) sideways. It turns out that the blurring equation that we gave in chapter 8 is also the Heisenberg uncertainty relation, but in the special form that is appropriate for light.

Tunneling

Tunneling is one of the more famous phenomena in quantum physics. It says that particles can travel to places where they appear to violate the conservation of energy. Tunneling is a consequence of the uncertainty principle—in particular, the fact that for a wave packet the energy of a wave is uncertain. The name *tunneling* comes about because, in effect, a particle can go from one side of a hill to the other, even though it doesn't have sufficient energy to get over the hill.

Tunneling is relatively easily to calculate when you know the height and width of the hill. We teach junior physics majors how to calculate it. Like other things in quantum mechanics, calculations give probabilistic results. You can't say for sure that something will tunnel, but you can calculate the probability that it will tunnel in a given time. We won't do the calculation here, but instead we will discuss the consequences of tunneling, and the practical application in the tunnel diode.

Alpha Radioactivity

Remember alpha particle radioactivity, from chapter 4? It turns out that this kind of radioactivity occurs because of tunneling. The alpha particle is inside the nucleus prior to the decay, but there are forces that prevent it from coming out. According to ordinary physical laws, it doesn't have enough energy to overcome the attraction of the nuclear force. But, thanks to the uncertainty principle, there is some chance that it will tunnel out anyway. Its energy is uncertain, and therefore there is some small chance at any moment that it will have enough energy to escape. Because nobody can calculate when it will come out, but just the likelihood that it will come out in any time period, the decays occur randomly.

It is worth pointing out that not all radioactivity is due to tunneling. In beta decay, the electron and neutrino are both created at the time of their emission. They are like the sound waves that you create when you speak; they didn't exist until they were created at the moment of decay, but once they are created, they can carry away energy. Likewise, x-ray and gamma ray radioactivity are not examples of tunneling.

So every time you see an alpha decay, you know that tunneling has taken place. Energy conservation was violated, but only for a short period of time. Once the alpha particle is out, the energy it has is identical to the energy it had inside the nucleus. We never actually see it violate energy conservation. We just calculate that it must have done so, but only for a very short period of time. So, in the end, energy is conserved. There is no more energy than there was before the decay. Somehow the alpha particle snuck through. We say that it tunneled.

Scanning Tunneling Microscope (STM)

One of the newest and most powerful microscopes, one that allows us to detect the positions of individual atoms, is the *scanning tunneling microscope*, or STM. Recall the image of *IBM* shown at the beginning of chapter 4 (figure 4.1). That image was taken with an STM.

An STM consists of a small sharp needle point with a electric charge on it. The needle point is brought very close to the surface that will be examined, but it doesn't quite touch. Then the point is scanned (that's the *S* part of STM) just above the surface. It moves back and forth, eventually over the entire surface. It picks up small electric current through tunneling (we'll come back to that). The amount of current is then recorded. Last, the computer puts together a map of the current. More current might be made whiter, and less current darker. The resulting map is the image.

The process is similar to that of a blind person feeling a statue. His fingers move over it, and after they have covered the whole surface, he knows that statue as well as (maybe better than) someone who looks at it from every angle. (The sighted person may know more about the color, if there is any, but the blind person knows more about the texture.)

The key to the STM is the tunneling process. Normally, electrons cannot leave the surface of the needle tip because they don't have enough energy. However, if the tip is brought very close to another atom, then the distance is so small that tunneling occurs. The smaller the distance is, the larger the tunneling. So measuring the current flowing gives the shape (ups and down) on the surface.

Care must be taken so that the needle never actually touches the surface. So an STM can be operated in a different mode. The needle tip is attached to a piezoelectric crystal, a crystal whose thickness can be adjusted to very high precision by applying a voltage across it (the crystal voltage). As this point is brought closer to the surface (by adjusting the crystal voltage on the piezoelectric), electrons begin to tunnel across from the tip to the surface. When a certain amount current flows, the tip is brought no closer. Then the tip is moved across the surface.

Here is the tricky part: the position of the tip is adjusted as much as necessary, in order to keep the tip current constant. That means that the tip is being moved up and down (by changes in the crystal thickness) to keep it a constant distance above the surface. The crystal thickness needed to do this is recorded, and it becomes a record of the height of the surface at every location. That record can then be made into a map of the surface height.

The STM was also used to place the xenon atoms in the IBM photo. By moving the tip extra close and adjusting the voltage, the atoms could be picked up and put down. So for that manipulation, the needle tip actually does touch some atoms.

STMs are the best way to get images of atoms and their positions. Right now, their main use is to study the properties of surfaces and the way atoms are bound to those surfaces. In the near future, STMs may be used to scan across DNA molecules to read the genetic code. Some people think that they may be used to store information by adjusting the positions of individual atoms, but I am guessing that that is unlikely, at least for the next 10 years.

Tunneling in the Sun

As I described in chapter 5, the Sun is powered by nuclear fusion. At high enough temperatures, protons, deuterons, and other positively charged nuclei have enough kinetic energy to overcome their electric repulsion, so they get close enough for the nuclear force to bring them together and they fuse.

However, calculations show that the Sun is not hot enough to bring the nuclei that close. Their thermal energy brings them near each other, but not enough to fuse. Yet they fuse anyway. The reason is tunneling. Once the nuclei get close, there is a high probability that they can tunnel right through the barrier of repulsion (it is completely analogous to pushing a weight up a hill) and get close enough for the fusion to take place. In that sense, essentially all of the energy we have on the Earth is produced using tunneling. The same process takes place in all stars.

Tunneling is also important in nuclear fission. Calculations show that the forces holding the two fission fragments together are quite strong, too strong to ever let them break apart. But because the fission fragments behave as quantum-mechanical waves, they can overcome this energy deficiency if they do it fast enough. Without tunneling, we would not have fission and its applications (reactors and bombs).

Quantum Computers

Unlike most of the other technologies described in this book, real quantum computers don't yet exist. The only kind that have been made perform extremely simple operations, such as factoring the number 15 (into 3×5). Nobody knows whether they will ever prove practical. Yet there is a great deal of interest in them, and so they are worth mentioning.

All computers use quantum mechanics, in that the energies of the electron flow in semiconductors are quantized. The random memory of a computer is based on the storage of electrons on small pieces of metal on the chip surface. Electric charge is quantized—that is, it is always present in some multiple of the electron charge. But despite all these ways that ordinary computers are "quantized," none of them can be described by what we mean by quantum computers.

In ordinary computers, charge is stored, and it flows though switches. Every computation consists of changing the stored charge by regulating the flow of electric current. But in a quantum computer, the idea is fundamentally different. All manipulations are done with the electron wave rather than with the current. This can be done by changing the wave with an electric field or some other external force. No particle is measured or stored until the computation is all finished.

Quantum computers take advantage of the fact that the particles (such as electrons) can be in different orbits simultaneously. As a result, a very large number of computations can be carried out simultaneously using very simple circuits, and much less energy can be used in computation. In a sense, the quantum computer will take advantage of the uncertainty principle. By being careful not to detect the electron, the spread-out wave can carry more information than the simple presence or absence of the electron. Each electron can, in principle, carry the equivalent of many bits of information. They are called *quantum bits*, or *qubits* for short.

At least that is the theory. Nobody knows whether quantum computing will ever prove practical for really large and difficult computation. Part of the problem is, of course, that we already have pretty good computers for most pretty hard problems, so the quantum computer has to make a lot of progress before it would be used for any practical purpose. For the latest developments, do an Internet search on "quantum computing."

It is important to recognize that most new technologies never become practical in the way that people who try to look far ahead, sometimes called "futurists," speculate. For example, in the 1920s, and every decade since, futurists have predicted that ordinary people would soon be driving their own airplanes instead of cars. They predicted that this was such a certainty that it surely would happen by the 1940s. Yet, it hasn't happened yet. In the 1940s, futurists predicted that we would have robots helping us in our homes, certainly by the 1960s. Yet that hasn't happened. Other things that were not predicted (like laptop computers) have happened. The future is hard to predict. Quantum computing has lots of obstacles before it can become practical, and some of them are fundamental (such as keeping noise out of the computation). They may never become useful. But, then again, they may.

Chapter Review

Electrons, protons, and all other particles are quantum waves, in the same sense that a light wave is a wave. Their particle properties refer to the way they behave when they are detected or measured. The wave nature is most evident in the way these objects move from one place to another. One important consequence of their wave nature is the quantization of energy levels, both in atoms and in crystals.

Lasers depend on the fact that the presence of a photon will trigger "stimulated emission" of another photon, as the electron that had that energy changes energy, and this results in a chain reaction of photons. The emitted photons have the same frequency as the one that stimulated them, as well as the same direction. That means that a laser beam spreads very little, and it can be focused to a small spot, and that means that the energy it delivers can be strongly concentrated. That feature allows lasers to be used for laser cleaning and for surgery. Laser applications that make use of one or more of these properties include CD and DVD sensors, supermarket scanners, weapons, and laser printers.

The relation between the frequency of the light and the energy of the photon is given by the Einstein relation, $E = hf$.

X-rays and gamma rays have very large values of photon energy; that's why they seem to be more like particles than do other electromagnetic waves. Visible light photons hitting a surface can give this much energy to an electron, a process called the *photoelectric effect*. For some materials (e.g., selenium, used in Xerox machines), that energy is enough to eject the electron from the surface. For other materials (e.g., solar cell silicon), the same amount of energy isn't enough to eject the electron, but it does give it enough energy that it can be conducted away. The photoelectric effect is used in solar cells, digital cameras, laser printers, Xerox machines, and image intensifiers.

Semiconductors such as silicon also have special properties because of quantum mechanics. Put two different semiconductors together, and if their energy levels are different, then some electrons will flow from one to the other. This can be used to make a semiconductor diode (which lets current flow in only one direction, thereby converting AC to DC) or a transistor amplifier. Virtually all modern electronics is based on diodes and transistors. A switch, the basic computation element of computers, is a version of the transistor. Integrated circuits consist of thousands to millions of transistors on one piece of semiconductor.

Superconductors work because at low temperature the electrons form Cooper pairs, and the motion of these pairs has an energy gap that prevents the pairs from losing small amounts of energy from collisions. As a result, they lose none.

Electron microscopes work by focusing a very fine beam on the object that is being observed. They can see things much smaller than can visible-light microscopes. That's because the wavelength of an electron is smaller than the wavelength of light.

All waves are quantized, but for low-frequency waves (water waves, radio waves, sound waves), the quantum of energy is so small that it is impossible to detect the tiny quantum leaps.

The Heisenberg uncertainty principle is a consequence of the fact that electrons and other "particles" behave as waves. Not only is location uncertain, but so is velocity and energy. Uncertainty in energy allows electrons to "tunnel" through regions that appear to violate the conservation of energy. Such tunneling is responsible for alpha radioactivity. Tunneling finds practical applications in the tunnel diode and in the scanning tunneling microscope.

Quantum computers make use of the wave nature of particles to store information in qubits. Nobody knows if they will prove practical.

Discussion Topics

1. Why does quantum physics seem so much more alien than other physics? Do you find it hard to believe that a particle can be in two different energy levels at the same time? What else in quantum physics seems so impossible to our everyday minds?
2. What would be the effect on our ordinary lives if superconductors could operate at room temperatures? Would we change the way we deliver power from AC to DC? Would long transmission lines become more practical? What else can you think of?
3. Can you think of applications of lasers other than those mentioned in this text? What about for entertainment? What else? Do lasers make good flashlights? Why?
4. Discuss the uncertainty principle. Does it make sense? Can you think of ways it might affect your life?

Internet Research Topics

1. Find images taken with electron microscopes. For each one, either find out what the magnification is, or estimate it by finding the size of the object imaged. Find some objects that have images both from ordinary visible light and from electron microscopy. What can be learned best from each kind of image?
2. Find commercial applications of lasers other than the ones I described in this chapter.
3. What is the current status of lasers as weapons? Are any being tested? Are any being deployed? Can you find discussion of their potential use for anti-ballistic-missile (ABM) systems?
4. Find applications of spectral lines. Which of the systems you find can be used remotely (e.g., in the open atmosphere) and which require laboratory measurements? Find uses for environmental measurement, industrial measurement, and purely scientific (e.g., determining the composition of stars).
5. Find out what you can about high-temperature superconductors. Do they involve Cooper pairs? Have scientists estimated the size of the energy gap? What is the highest temperature that people have obtained? Are scientists optimistic about reaching higher temperatures?
6. Look up "crystal radio" and "cat's whisker diode" on the Web, and see what you can learn about the early days of radio reception, including the use of vacuum tubes as diodes.
7. What is the current status of quantum computing? What is the most complex calculation that anyone has done? What are the problems

that are too difficult for ordinary computers that might be solvable if quantum computers become viable?

8. Look up "Schrodinger's cat." Read about it, and ponder the paradoxes it raises. Discuss it with other students. Are you bothered by the paradoxes it raises? If not, why not? (It bothers the author of this book.)

Essay Questions

1. Lasers are one of the most widely used technological inventions of the twentieth and twenty-first centuries. What does the word *laser* stand for? What properties of laser light make it useful? What are the main parts of a laser? What principles of quantum physics make a laser work? Give examples of devices that use lasers in your answer.

2. Light is a wave. But according to quantum mechanics, it is also *quantized*. What does that mean? Describe some practical applications of light that make use of the fact that both it and energy levels are sometimes quantized.

3. Many people think that *quantum physics* is used only in the laboratory. In fact, many devices that depend on quantum physics for their operation are used in business, industry, and our everyday lives. Give examples of four such uses, and for each one, describe how quantum physics is essential for its operation. Try to pick four devices that appear, to the nonphysics student, to be as different from each other as possible.

4. Describe the key features of quantum physics that make it different from "ordinary" physics. Give examples showing how these features are used in important applications.

5. Superconductivity sounds super, yet people do not use it in everyday life. Explain the nature of superconductivity, the physics behind it, and why it is not more widely used. Discuss its present-day applications, and the possibilities for the future.

6. Lasers have several properties that make them special. Describe what these properties are. For each special property, describe practical applications that take advantage of it.

7. Quantum physics has properties that seem very strange to someone who doesn't recognize that particles have wavelike behavior. What behavior of electrons would be impossible to understand based on classical (nonquantum) physics? What behavior of light would be impossible to understand based on the classical theory of light (i.e., light as a wave, not a quantum wave)?

8. Describe phenomena that depend on the presence of an energy gap.

9. What methods can be used to identify gases—for example, to tell whether a gas is hydrogen or helium or a mixture? Describe the principles on which the method is based.

10. What phenomena can be understood as a consequence of tunneling? What practical applications are there to this behavior?

Multiple-Choice Questions

1. The energy of a photon depends on
   A. color
   B. direction
   C. velocity
   D. width

2. Stimulated emission is important for
   A. integrated circuits
   B. superconductors
   C. LEDs
   D. lasers

3. Transistors use semiconductors with different
   A. frequencies
   B. energy gaps
   C. densities
   D. wavelengths

4. Compact disk readers use
   A. x-rays
   B. lasers
   C. LEDs
   D. spectra

5. To tell hydrogen from helium, look at
   A. their x-ray emission
   B. their visible spectrum
   C. the photoelectric effect
   D. their amplification

6. Quantum computers are
   A. used mostly by the military
   B. used by Google for searches
   C. used in everyday laptops
   D. not yet very good at anything

7. The energy gap is important for (choose all that are correct)
   A. spectral fingerprinting
   B. superconductivity
   C. transistors
   D. lasers

8. Quanta are not observed from water waves because
   A. such waves are not quantized
   B. the energy of the quanta are too small
   C. water atoms are too small
   D. water atoms are too large

9. Xerox machines make use of
   A. the photoelectric effect
   B. Cooper pairs

C. stimulated emission

D. a chain reaction

10. *High-temperature superconductors* operate at approximately
    A. room temperature
    B. 4 K (liquid He)
    C. 150 K (liquid nitrogen)
    D. 1200 C (tungsten filament)

11. Lasers are useful because (choose all that are correct)
    A. their light is single frequency
    B. they can be well collimated
    C. they can be made intense
    D. they require no power

12. A laser operates on the principle of
    A. nuclear magnetic resonance
    B. fluorescence
    C. quantization of charge
    D. photon avalanche

13. A transistor is
    A. a form of superconductor
    B. a device that emits gamma radiation
    C. a strong absorber of radio waves
    D. a device that can amplify or turn AC to DC

14. The quantum chain reaction refers to
    A. a semiconductor
    B. a superconductor
    C. a laser
    D. the hydrogen spectra

15. An electron is in orbit around a proton. What is the maximum number of orbits
    it can be in at the same time?
    A. 1
    B. 2
    C. 0
    D. any number

16. Which of the following depend on an energy gap? (Choose
    all that apply.)
    A. semiconductors
    B. superconductors
    C. lasers
    D. gas spectrum analysis

17. The photoelectric effect is important for all of the following *except*
    A. digital cameras
    B. Xerox machines
    C. solar cells
    D. polarized sunglasses

18. Superconducting materials allow electrons to move without loss of energy because
A. there are no impurities in the metal
B. there is an energy gap
C. cold electrons move slowly
D. high pressure prevents collisions that cause resistance

19. Computer circuits are based on which material?
A. samarium cobalt
B. sodium chloride
C. silicon
D. strontium-90

Relativity

A Dialogue

The student's words are in plain text. My responses are in *italics*.

What is time?
I don't know.

Someone told me it was the fourth dimension.
That's just a physicist's way of confusing you. It's true, but it is a much less deep statement than you would guess.

Time moves. I know that. But I don't understand it. What does it mean that time moves?
I don't know. Neither does anybody else.

Does time ever slow down?
Yes.

How can you say that when you don't know what the motion of time is?
Because we can measure relative rates. We can make time slow down in the laboratory. We see it in the stars.

Can we travel in time?
Sure. We're doing it right now. We are both going forward in time.

I meant, can we go back in time?
Nothing in physics prevents that. But I don't believe we can.

Believe? I thought that we were discussing physics.
Nothing in physics prevents going backward in time. But backward time travel violates my belief in my own free will.

What determines the direction of time?
Some people will tell you it is determined by entropy. But that is controversial, and not proven. There is no way to test the idea, so it too is more a belief than it is solid physics.

Nothing about time is obvious. Yet, given the mysterious nature of time, you may be surprised at some of the things we do know about it. For example, we know that if two twins are exactly the same age and one travels while the other stays at home, then when the traveler returns, the traveler will have experienced less time than the stay-at-home twin!

There is nothing odder about time than that. Yet Albert Einstein gave us a formula that tells us precisely how much less time the moving twin experienced. And that fact has been experimentally verified with very sensitive clocks flown on airplanes. Even radioactive atoms, when they move, experience less time than those that are stationary. That fact is verified every day at accelerator laboratories where such atoms are sent near the speed of light, and physicists note that their radioactive decays slow down.

The nature of time (and space) is at the heart of the theory of relativity. That's what this chapter is about. Einstein created the theory of relativity in the early 1900s. The theory of relativity consists of two parts. The first is called the *special theory of relativity*, and it has to do with the nature of time, space, energy, and momentum. It was in this work, published in 1905, that Einstein presented his famous equation $E = mc^2$. The second part was published in 1916 and is called the *general theory of relativity*. It is really a theory of gravity. It "explains" all of gravity as due to a bending of space and time. This theory is needed to understand some of the recent discoveries in cosmology about the nature of the Universe.

This chapter departs a bit from my previous philosophy. Future presidents don't really need to know about the theory of relativity. It is important, however, to physicists, to philosophers, to those who plan trips to other planets, and to anyone who wants to have their mind stretched beyond what this course has already done.

Events—and the "Fourth Dimension"

Time is often called the "fourth dimension." That turns out to be a useful definition, not an observable fact. And it is not something super-mysterious or deeply abstract. In fact, when used in that way, the word *dimension* is being used in a very technical and narrow way: the dimension of a quantity is the number of different numbers you need to describe it.

Suppose that you want to specify a location on the Earth. You can do that with three coordinates, such as latitude, longitude, and altitude. Or you could use a system with x, y, and z. The key thing is that you need only three numbers. Any two objects that have the same set of three numbers must be at the same location. In math, we say that a location is a three-dimensional number. That's all that the fancy word *dimension* actually means. Space is three dimensional.

If you want to specify an *event*, rather than a location, then it is sufficient to give the location and the time of the event. Suppose that I were to tell you that there is an event at my house at 8 PM tonight. Then there is no confusion; you might not know what is going to happen, but you have located it in both time and space. The event can have a name, such as "Elizabeth's birthday party" or "Melinda goes to bed." But to be unique (Elizabeth has a birthday each year, and Melinda goes to bed almost every night), you also specify the time. Events are specified by four numbers. So we say that events are four dimensional. That's not deep. It is trivial. That's is the entire meaning of saying that time is the fourth dimension.

That is not what is interesting about time. What is interesting is that the amount of time can change depending on the velocity that an object is moving in the three dimensions of space. That idea is deep, and requires some explanation.

Time Dilation

I described in the opening of this chapter how two twins can experience different amounts of time. That seems to violate common sense. How can it be true? The answer is that the effect is very small unless the velocity is very fast—that's why you never notice it. Common sense is based on experience, and that kind of *time dilation* is not part of our normal lives, so it violates our common sense. It also violated the common sense of ancient scholars to think that the Sun has a million times the volume of the Earth. Does that violate your common sense? Sometimes, all it takes to incorporate something into your common sense is to hear it many times, or to gain familiarity with it. Maybe after you have read this chapter, time dilation will start to become part of your common sense.

Time dilation is so small that it's difficult even to measure unless the velocity is near the speed of light. For airplanes moving at 675 miles per hour, the effect is about 5×10^{-13}. That means that if you traveled at this speed for one day, you would lose 43 nanoseconds.[1] That is the time it would take light to travel about 43 feet. If you fly for a *year*, you will experience 16 microseconds less time than your twin who doesn't travel.

This small effect becomes large if the velocity approaches the speed of light. At 60% of the speed of light, the time dilation factor is 0.8. Let me show you how to do the calculation yourself. Suppose that one object is moving at a velocity v. Let the speed of light be called c. In science fiction, the ratio of v to c is called the *light speed*. If you are moving at 60% the speed of light, your light speed is 0.6. In physics, we usually call the light speed "beta" and use the Greek letter β (which looks like a B with a tail).

$$\beta = v/c = \text{light speed}$$

Einstein gave an exact formula for calculating this. Although the term is not usually given a name, I like to call it the *Einstein factor*. Time will slow down by

The Einstein factor:
$$\sqrt{1 - \beta^2}$$

You don't have to memorize this, but you might want to anyway, because then you can do real relativistic calculations.

---

[1] We get this number by multiplying 5×10^{-13} by the number of seconds in a day. The number of seconds in a day is $24 \times 60 \times 60 = 86{,}400$.

Let's get back to our example. If the light speed $\beta = 0.6$, then the equation gives the Einstein factor to be

$$\sqrt{1 - \beta^2} = \sqrt{1 - 0.6^2} = \sqrt{1 - 0.36} = \sqrt{0.64} = 0.8$$

If a man named John stays at home, and his fraternal twin Mary travels at 0.6 light speed, then her time will go slower at a rate that is only 0.8 that of John's time. If John ages 1 year, then Mary will age only 0.8 year. When Mary returns, and they compare ages, John will be 0.2 years older than Mary (i.e., a little more than 2 months older). Yet they are twins, born at the same time.

The effect gets much more dramatic as Mary's velocity increases. Suppose that she travels at light speed 0.99999—i.e., at 99.999% the speed of light. If you plug that into the time equation, you'll find Mary's time progresses at a rate only 0.0045 the rate of John's time. If John ages 1 year, Mary will age 0.0045 year. To convert that to days, multiply by the number of days in a year: 0.0045 yr × 365 days/yr = 1.6 days.

Not only that, but she will *experience* only 1.6 days, while John experiences a full year. If they began as freshmen, Mary will still be a freshman, but John will be a sophomore.

The fastest that any astronaut has ever traveled is approximately Earth escape velocity, about 11 km/s. This is equivalent to light speed $\beta = 0.0037$. Plug this into the time equation (use a calculator), and you'll find that the astronaut time goes at a rate that is 0.99999933 slower than Earth time. That isn't a big change (since the number is so close to 1). If he travels for 1 year (that is, 365 days × 24 hours × 3600 seconds per hour = 3.16×10^7 s), then he will experience 0.02 second less than if he stayed at home. That's not enough for him to notice unless he is carrying a very accurate clock.

We have sent radioactive atoms at velocities close to the speed of light, and their radioactivity does slow down, by exactly the predicted factor.

Suppose that we go faster than the speed of light—for example, we try light speed $\beta = 2$. Try plugging that into the equation and see what happens. We'll discuss faster-than-light particles later in this chapter.

I've worked out the value of the time factor for different values of the light speed. These are listed in table 12.1. I've added a third row for a term called the *Lorentz factor*, γ. The symbol γ is the Greek letter *gamma*. The Lorentz factor is equal to the 1 divided by the Einstein factor.

Table 12.1   Light Speed Table

$\beta = v/c$	0	0.25	0.5	3/5	4/5	0.9	12/13	0.99	0.999	0.99999	1
$\sqrt{1 - \beta^2}$	1	0.97	0.87	4/5	3/5	0.44	5/13	0.14	0.045	0.0045	0
γ	1	1.03	1.15	5/4	5/3	2.3	13/5	7.1	22.4	224	infinity

OPTIONAL: EXERCISE FOR THOSE WHO LIKE TO PLAY WITH MATH

Maybe you noticed that some of the values in table 12.1 are given as exact fractions, rather than in terms of their decimal approximations. You can verify that if you have an exact right triangle—e.g., a 3:4:5 triangle or a 5:12:13

triangle—then values of velocity given by fractions made of these numbers will give exact fractions for the Einstein factor E and the Lorentz factor γ.

But . . . But . . . How Can Time Depend on Velocity?

It sounds absurd. It goes against intuition. It goes against everything we experience. Or does it? Why do you believe that time is independent of your path? Did you always believe it? I'll bet that you didn't believe it when you were a child. An hour at the dentist's office didn't seem to go as fast as an hour in a swimming pool. But you were trained to watch clocks and to "be on time," and you finally learned that there is a "universal" time that you can follow in order to get to appointments on time. But it was never intuitively obvious.

Nor is it true. It is *almost* true, however, since for everyday velocities, the Einstein function is very close to having a value of 1, the value for which there is no time dilation. Even in the airplane example, the factor was very close to the value of 1. At 675 mph, v/c is 10^{-6}. The factor $(v/c)^2$ is 10^{-12}. The time factor f, on most calculators, will come out to be exactly 1, since they don't handle enough decimal places! If calculated on a sufficiently accurate computer, the Einstein factor $f = 0.9999999999995$. (There should be 12 nines in that number, if I entered it correctly.) It can be written as $1 - 5 \times 10^{-13}$. That is pretty close to 1, so it is hard to notice the difference.[2]

Nevertheless, in 1972, two scientists recognized that clocks had become accurate enough that the twin effect could be measured even in an airplane. Their results were published in *Science Magazine*.[3] Their results confirmed that even at the "slow" velocity of airplanes, the equation works. The "moving" clock experienced less time than the stationary clock on the ground.[4]

Not All Motion Is Relative

Isn't all motion relative? Who is to say which clock is moving? I raise this issue only because it is a favorite complaint made by people who have studied a little bit of relativity theory. In fact, in relativity theory, it is *not* true that all motion is relative. The clock that is moving is the one that had to be accelerated to make it return home.[5]

---

[2] Here is a way to do the calculation without a computer. This is only for those who love math: when v is very small, we can write sqrt$[1 - (v/c)^2] \approx 1 - (v/c)^2/2$. [To prove that, multiply $1 - (v/c)^2/2$ by itself, and ignore all terms that have $(v/c)^4$ in them.] In our example, $(v/c)^2 = 8 \times 10^{-13}$. So $f = 1 - 4 \times 10^{-13}$. That is the number I gave in the text.

[3] I don't expect you to read this, but here is the reference: J. C. Hafele and R. E. Keating, "Around-the-world atomic clocks: predicted relativistic time gains," *Science* **177**: 166 (1972).

[4] In their work, Hafele and Keating had to calculate the gravity effect, something we haven't discussed yet, as well as the twin effect. Their result matched the prediction for the combined effects.

[5] I've written a more detailed article on this. It was published in the *American Journal of Physics* **40**: 966–969 (1972).

OPTIONAL: INERTIAL FRAMES

Here are some details, for those interested. The twin equation is good only if applied in a single "inertial" frame of reference—i.e., one in which there is no acceleration without applied forces. The frame of the stay-at-home twin can be used. The frame of the traveling twin cannot be used, because that twin must change direction halfway through the trip. Which twin changed velocity? That is unambiguous, since it takes force to accelerate. So the inertial frame is the one in which the astronaut felt no force. Einstein's time dilation formula works only if applied within an inertial fram. It is not correct when used in frames that are accelerating. Objects can be accelerating—but not the frame that you are using as reference.

If you are willing now to believe that time depends on velocity (or at least accept the concept for a while), then you are ready to move on to a few more of the astonishing facts of relativity theory. You need no more math. All the results use the same functions, the Einstein factor and γ.

Lorentz Contraction

When an object moves to a velocity v, it gets shorter. Its new length is the old length multiplied by the Einstein factor. This effect is called the *Lorentz contraction*. It is named after the person who first proposed it, even before Einstein. (Einstein published his theory of relativity in 1905. But it was based on two decades of work preceding it, including that of H. A. Lorentz.) Just to make it explicit, if a stationary object has rest length L_S, then when the object moves, its new length is L_M given by

$$L_M = L_S \sqrt{1 - \beta^2}$$

The moving object is shorter, since its length is the rest length multiplied by the Einstein factor.

This contraction turns out to be tricky to measure. If your ruler is moving with you, then it shortens too, so you think your object hasn't gotten any shorter! To see the effect, you have to have a stationary ruler measuring a moving object.

Remember, the factor f is so close to 1 at everyday speeds, that this effect is difficult to notice. It becomes really important only when v approaches c, which occurs for particles emitted from radioactive nuclei, in accelerators (popularly called *atom smashers*) and in cosmology where distant galaxies are moving away from us at speeds that approach the speed of light.

Relative Velocities

Suppose that you are moving at half the speed of light, $c/2$, and that you crash into someone coming from the opposite direction, who is also moving at half the speed of light. What is your relative velocity? You would probably expect it to be c, the sum of the two velocities. Suppose that you were moving at 0.75 c, and so was the other person. Was your relative velocity 1.5 c?

Here comes another surprise: the answer is no. The correct answer was calculated by Einstein, taking into account both the Lorentz contraction and time dilation. If you are moving at velocity v_1 and the other person is moving a velocity v_2, then he showed that the relative velocity V is given by:

$$V = \frac{v_1 + v_2}{1 + \frac{v_1 v_2}{c^2}}$$

You are not required to know this formula, but the consequences are important. Watch what happens when you put $v_1 = 0.5$ c and $v_2 = 0.5$ c. You get

$$V = \frac{\frac{c}{2} + \frac{c}{2}}{1 + \frac{\left(\frac{c}{2}\right)\left(\frac{c}{2}\right)}{c^2}}$$
$$= (c)/(1 + c^2/4c^2)$$
$$= c/(1 + 1/4)$$
$$= 0.8c$$

So the relative velocity is 0.8 c, eight-tenths of light speed, less than the speed of light.

Suppose that we try $v_1 = 0.9$ c and $v_2 = 0.9$ c. Then the relative velocity is still less than c. Try it yourself. The answer is $V = 0.994$ c. In fact, no matter how close the individual velocities are to c, their relative velocity is still less than c. (If you like math, you might enjoy trying to prove that.)

One consequence of this is that relative velocities are always less than c. Here's another example. Suppose that you have a multistage rocket, with the first stage getting to $0.9\,c$, and the second stage getting to $0.9\,c$ relative to the first, the total velocity that you will get for the second stage is only 0.994 c. That's one of the reasons you can't ever reach light speed. We'll discuss another one shortly, in the section "Energy and Mass."

Invariance of the Speed of Light

Suppose that a photon is coming toward you, traveling at the speed of light c. You move with a velocity v toward that photon. With what velocity will the photon hit you when you are moving? It is natural to assume that it will be greater than c, but that is not true. In fact, the velocity of the photon, measured from a moving frame, is still c! I'll show that in the following optional section.

OPTIONAL: CALCULATION

I'll use the velocity equation, with my velocity $v_1 = v$, and the photon velocity $v_2 = c$. This gives the new photon velocity as follows:

$$V = \frac{v_1 + v_2}{1 + \frac{v_1 v_2}{c^2}}$$

$$= \frac{v + c}{1 + \frac{vc}{c^2}}$$

$$= \frac{v + c}{1 + \frac{v}{c}}$$

Now I multiply the numerator and denominator by c:

$$= \frac{c(v + c)}{c\left(1 + \dfrac{v}{c}\right)} = \frac{c(v + c)}{c(c + v)} = c$$

So the speed of the photon is still c, even in the moving frame.

The photon hits me with velocity c, even when I am approaching it. This surprising property is sometimes called the *invariance of the speed of light*. In fact, in many books it is taken as a fundamental assumption, and then all the rest of relativity theory can be derived from it.

Energy and Mass

Einstein noticed another consequence of the velocity equation. The old concept of momentum conservation no longer worked. If two objects have equal and opposite momentum (mass times velocity; see chapter 3), then on a collision the result will be at rest. Momentum is conserved. But when Einstein calculated the momentum for objects in which both were moving, that was no longer true. Einstein guessed (correctly, it turns out) that momentum conservation was still valid and that the velocity equation was right. The mistake was that the mass of an object is not really constant, but that it depends on velocity. If we let m_0 be the mass of the object when it is at rest, then its mass when it is moving, its *kinetic mass*, also called its *relativistic mass*, is

$$m = \gamma m_0 = \frac{m_0}{\sqrt{1 - \beta^2}}$$

Note that the Lorentz factor γ is always greater than 1, just as the Einstein factor is always less than 1. That means that the kinetic mass is *bigger* than rest mass by the same factor that tells us how much length and time are *smaller*.

When something gets a bigger mass, that means that it gets harder and harder to accelerate, and it also means that the pull of gravity will increase. But it also has an important consequence for energy.

Here are three key facts that you should learn. When an object is moving:

- Its time slows down.
- Its length contracts.
- Its mass increases.

The amount in each case can be calculated by multiplying or dividing by the Einstein factor γ.

Now we'll get to the most famous equation of the twentieth century.

$E = mc^2$

Einstein took these calculations one step further and calculated the energy of a moving object. He deduced that it is given by:

$$E = mc^2$$

At first look, it appears that the energy of an object does not depend on velocity. But that isn't correct, because the mass depends on velocity. (The mass $m = \gamma m_0$.)

Nevertheless, the equation looks very different from the old kinetic energy equation:

$$E = 1/2 \; m_0 v^2$$

Was the old equation completely wrong? How can the two equations be reconciled? At first look, it seems impossible. At zero velocity, the kinetic energy equation gives $E = 0$, whereas the Einstein's equation gives $E = m_0 c^2$. Those are very different numbers. Because the value of c is huge, the zero-velocity energy $m_0 c^2$ is huge.

Yet, even though they disagree, the two equations are more similar than you might think. For small velocities, it is possible to show that if the velocity is low, then the Einstein equation can be approximated as follows:

$$E \approx m_0 \; c^2 + 1/2 \; m_0 v^2$$

(If you are mathematically inclined, you might try to prove this. I give some hints in a footnote.[6])

REST ENERGY

The equation is an approximate version of Einstein's energy equation, valid only at low velocities. It says that the energy is equal to the old term $1/2 \; mv^2$

---

[6] If you want to try deriving this, here are the key equations you will need. Assume that $(v/c) = \beta$ is a very small number. There are two key approximations you must make. The first is that $\sqrt{1 - \beta^2} \approx 1 + \beta^2/2$. You can check this equation on a calculator by putting in some small numbers (e.g., 0.01) for β. The second approximate equation is $1/(1 - \beta^2/2) \approx 1 + \beta^2/2$. Check this one too. Both of these approximations can be derived using algebra; calculus is not required. You can check the first equation by squaring it; you can check the second by cross-multiplication; in both cases, throw away the tiny β^4 terms.

plus a new term that doesn't depend on velocity: $m_0 c^2$. This constant term has a famous name. It is called the *rest energy*. The smaller part, the $1/2\ mv^2$ term, is still called the *kinetic energy*. We now say that the total energy is the sum of the rest energy plus the kinetic energy.

Note that we get the rest energy term directly from the exact Einstein equation if we put $v = 0$. From table 12.1, at zero velocity we have $\gamma = 1$, so the Einstein equation gives $E = m_0 c^2$. That's the rest energy.

This result suggests that even a particle that is at rest stores enormous energy, roughly the same energy it would have (by the previous equation) if it were moving near the speed of light.

But how do you extract this energy? Part of it is extracted when we have a radioactive decay. The mass of the debris is less than the mass of the original particle, since some of the mass has been turned into energy. But in typical chemical changes, the mass doesn't change. In chemistry, this is called *conservation of mass*. Since the mass is constant, you can ignore it when looking at energy conservation. That's why it was never noticed.

But it can be noticed when you release huge energies in the radioactive decays. When that happens, the sum of the masses of the pieces is less than the mass of the original atom. The energy of the radioactive explosion came from converting mass to kinetic energy.

For a proton, the rest energy is 938 MeV. That is huge compared to the typical 1 MeV released in radioactive decay. The rest energy of the electron is smaller, only 0.511 MeV. That's because the rest mass of the electron is 2000 times smaller than the rest mass of the proton.

Some particles have all of their energy in kinetic energy. An example is the photon. All of the energy of a photon can be absorbed when that photon hits an object.

ANTIMATTER ENGINES

Could we release the rest energy of the electron and turn it into kinetic energy? Yes, there is one way: use antimatter. An antielectron, also called a *positron*, has the same mass as the electron but opposite charge. Bring it together with an electron, and the charges will cancel, and all of their mass energy will be released as photons—wave packets of light. Two photons will emerge in equal and opposite directions, and all the energy of photons is kinetic (since they have zero rest mass, as we will see in the next section). Therefore all that energy can be turned into heat. This process was used in a practical way for the medical PET scanner described in chapter 9.

You can do much better if you bring a proton together with an antiproton. The process of releasing all this energy is called *annihilation*. When a proton is annihilated, virtually all 938 MeV of its rest energy is released. That's why antimatter drives are so popular in science fiction stories. Matter and antimatter may constitute the ultimate energy fuel.

Kinetic energy can also be turned into mass. When a gamma ray passes close to a nucleus, we often observe a phenomenon called *pair production*. The energy of the gamma ray is suddenly converted to the mass of a particle and an antiparticle, such as an electron and a positron. This is a fairly common occurrence for high-energy gamma rays, and it is seen, for example, in cosmic radiation. The first positron ever detected was one that had been produced by a cosmic gamma ray. Other collisions (such as between electrons) can also create such pairs. The first antiproton ever detected was created as part of a proton–antiproton pair in the Berkeley atom smasher known as the Bevatron located at the Lawrence Berkeley Laboratory.

Zero Rest Mass

Physicists usually say that the *rest mass* of a photon is zero. That is an odd statement, since you can't bring a photon to rest. But if you take energy away from a photon (perhaps by scattering it off an electron), then the energy can be made to get smaller and smaller. Eventually, the photon has energy approaching zero, and that could be true only if it had no rest energy—i.e., had zero rest mass.

Another way to deduce that a photon has zero rest mass begins with the equation $E = mc^2 = \gamma m_0 c^2$, which is true for all particles. When the velocity $v = c$, then γ is infinite. This seems to say that any particle that travels at the speed of light, such as a photon, must have infinite energy! But, no, that result isn't right. The equation contains the rest mass m_0. The rest mass of the photon must be zero. That is the only number which, when you multiply it by infinity, gives a noninfinite answer. In fact, zero times infinity is "indeterminate." That means that it could be any number.[7] You can't tell what it is unless you have another equation. For the photon, we do have such an equation. From chapter 11, we have $E = hf$, where h is Planck's constant and f is the photon frequency.

Likewise, we conclude that if a particle has mass = 0 but has energy, then it must move at the velocity c. It can't move faster, and it can't move slower. Its energy is not related to its velocity, but only to its wave frequency.

But what about light in glass? Doesn't that move at c/n, where n is the index of refraction. Isn't that slower than c?

The answer is that the equation for the energy did not include the glass; if you put that term in, then you no longer conclude that light must travel at c. In fact, you can no longer conclude that a photon has zero rest mass in glass! Why is that? Remember that a photon is an electromagnetic wave. In a vacuum, all that is "waving" are the electric and magnetic fields. But in glass, there are electrons and atoms moving too. When the electric field changes, it makes these particles move. In a sense, the wave is no longer massless because it must include the mass of everything that is oscillating, and that includes atoms. But in a vacuum, a zero mass particle must move at c, all the time.

You might notice that the kinetic mass of a photon,

$$m = \frac{E}{c^2} = \frac{hf}{c^2}$$

is never zero. Yet the rest mass always is.

The fact that the photon has nonzero kinetic mass suggests that it feels the force of gravity. In fact, Einstein's theory of gravity, called *general relativity*, predicts that photons will be affected by gravity.

---

[7] One way to see that is the fact than any number x divided by zero is infinite: $\infty = x/0$. If you cross multiply, you get $0 \, \infty = x$, true for any x.

MASSLESS PARTICLES DON'T AGE

Look at table 12.1 for a particle, such as a photon, moving at the speed of light. The Einstein factor for such a massless particle is zero. That means that while an hour passes for you and me, the time that passes for a massless particle is zero. If this bothers you, then let's just assume that the particle has a very tiny mass. It is moving at *almost* the speed of light, and f is very close to zero. So the particle experiences very little time.

If a massless particle does not experience time, can it undergo radioactive decay? Think about this for a moment. What would you guess? The answer is no, it can't decay.[8] Or to put it another way, if a particle has a rest mass very very close to zero, then its time dilation is so great that to us it will take a very very long time to decay.

DO NEUTRINOS HAVE MASS?

The *neutrino* is a particle that we always thought had mass zero. Neutrinos are emitted in many nuclear decays, and they travel through most matter without being noticed. They are emitted from the sun, but sensitive experiments to detect them have shown that we detect only 1/3 of the number we expect. What is going on?

New experiments have been done that suggest an answer: the neutrinos are changing from ordinary neutrinos into exotic neutrinos called *muon neutrinos* and *tau neutrinos*. (Ordinary neutrinos are often called *electron neutrinos*, since they are never created by themselves but always in combination with electrons.)

But if the neutrinos are changing, they must be experiencing time. That means that they aren't traveling at the speed of light. That means that they aren't truly massless. So by observing a low number of solar neutrinos, physicists were led to conclude that neutrinos must have some mass. That was a conclusion that surprised nearly everyone.

WHY YOU CAN'T GET TO LIGHT SPEED

Suppose that you take a massive particle (such as an electron, or a person) that has a nonzero rest mass m_0, and you accelerate it to the speed of light. Then its energy is

$$E = \gamma \, m_0 \, c^2 = \infty$$

Here ∞ is the traditional symbol for "infinity." For any particle that has nonzero rest mass (e.g., you), if you accelerate it to the speed of light c, then your energy will be infinite. That is the fundamental reason why you can't travel at light speed.

---

[8] Optional: Even without invoking the time dilation argument, you can prove they don't decay from the more formal approach—that is, by showing such a decay cannot simultaneously conserve energy and momentum. That is a good exercise for a physics major to try.

$E = MC^2$ AND THE ATOM BOMB

Many people think that Einstein's discovery of the equation $E = mc^2$ led to the invention of the atomic bomb. In fact, the discovery of this equation made no difference to the invention of the bomb. In the late 1800s, it was discovered that radioactive decay released a million times more energy than chemical explosions. That discovery was the key to nuclear energy. You did not have to know that the source of the energy was the rest mass of the nucleus. To make a bomb, all you needed was a way to make a large number of nuclei explode all within a millionth of a second. That was discovered in the late 1930s, when scientists found that two neutron from one fission could induce two more fissions.

Einstein's equation (published in 1905) showed that the enormous release of energy would be accompanied by a slight disappearance of mass. But nobody needed to know that in order to build an atomic bomb.

BEYOND LIGHT SPEED: TACHYONS

I showed that you can't send particles at the speed of light because that would take infinite energy. But can you send them faster? The surprising answer is *maybe*. If such particles exist, we call them *tachyons*. They have a surprising property: they must travel faster than light. They have the lowest energy when they move at infinite speed, and infinite energy when they approach light speed c.

As I show in the following optional section, the equation for the energy of a tachyon is

$$E = \frac{m_0 c^2}{\sqrt{(v/c)^2 - 1}}$$

It looks just like the relativity energy equation, except you'll notice that the terms under the square-root symbol are reversed. As a result, this has positive energy only if v is greater than c!

How can you get it going that fast? Doesn't it have to start at rest, and so it would take infinite energy to get just up to c and, therefore, it could never go beyond? No, there is a loophole. Just as photons are moving at c immediately when they are created, we have to assume that tachyons are *born* moving faster than c. If you study the tachyon equation, you'll see that the faster the tachyons go, the lower is their energy. They can lose energy by speeding up, or gain energy by slowing down. (That is backward from the way ordinary particles, such as electrons, behave.) Tachyons get infinite energy when they travel at c, just like ordinary particles.

Tachyons may exist and, from time to time, physicists set up experiments to search for them. However, I do not think that they will find the particles. The reason comes from another property of relativity having to do with the simultaneity of events.

OPTIONAL: THE MATH OF TACHYONS

Look at the Einstein energy equation: $E = mc^2 = \gamma m_0 c^2$. The Lorentz factor γ is

$$\gamma = \frac{1}{\sqrt{1 - \left(\dfrac{v}{c}\right)^2}}$$

Note that γ goes to infinity when $v = c$. That means that the particle would have infinite energy. Now suppose that v is greater than c. Then f is the square root of a negative number. That makes the energy factor γ imaginary, right? Right. Doesn't that make the energy E imaginary too? No, not if the tachyon had imaginary mass. Does a particle with imaginary mass mean that the particle doesn't really exist? No, such a particle can really exist. (If you've never felt comfortable with "imaginary" numbers, don't worry. Stick with me, and learn the conclusions, if not the logic that we use to get there. This section will be followed by an optional section on imaginary numbers.)

The two imaginary numbers, the mass and γ, when multiplied together, create a real number. That means that everything works out okay. We can rewrite the energy equation in terms of real numbers. Let the mass of the tachyon be $(i\, m_0)$, where i is the square root of -1, and m_0 is a real number. Then, for a tachyon, the two values of i will cancel, and we'll get the following:

$$\gamma = \frac{m_0 c^2}{\sqrt{(v/c)^2 - 1}}$$

This is the tachyon energy equation. When v becomes infinite, the denominator becomes infinite, and the energy E goes to zero. When $v = c$, the energy E is infinite.

If you think this is weird, be assured that everyone else thinks so too. But that doesn't prevent it from being true. Of course, as of the writing of this book, no tachyons have been discovered, so they may not exist.

OPTIONAL: IMAGINARY NUMBERS

If you were confused by imaginary mass, the likely reason is that imaginary numbers were named badly. They really do exist. If they were called "complex numbers," you wouldn't be bothered so much. Then the mass of the tachyon would have been a "complex mass," and that doesn't sound so bad.

The square root of -1 does exist. It has a name: i. If you are confused, it's because a teacher once told you that i doesn't exist, but suggested that you "pretend" that it did. That teacher was mistaken. Just because it can't be written in terms of fractions doesn't mean it doesn't exist. After all, π can't be written as a fraction either—does that mean that it doesn't exist?

In elementary school, I had a teacher who told the class that negative numbers don't exist. After all, she said, you can't have a negative number of apples. I sat there listening and decided she was wrong. If you owe somebody apples, you have a negative number of them. Much of the rest of the class believed her, and their math education stopped.

Simultaneity

Different people can experience different amounts of time. That means that there can be no such thing as a universal clock that tells everyone what time it is. Your time depends on your motion.

One consequence of this fact is that it is fundamentally impossible to determine whether two events are simultaneous.[9] And it gets even worse. The order in which two events happened can depend on the motion of the observer.[10] This happens with tachyons. Suppose that you emit a tachyon at point A, and it travels to point B, where it is absorbed. To a different observer, one moving along the line between the two events with high speed, the order will be reversed. In that moving reference frame, event B will occur before event A.[11]

This fact may bother you more than anything else in this chapter. It means that, to one observer, event A precedes B, but to another observer, event B comes first. Suppose that the tachyon was shot from a gun, and used in a murder. The shooter is arrested. In his defense, he points out that, in a moving reference frame, the victim was killed before the gun was fired.

There is nothing in that story that violates the laws of physics. But it does violate our sense of free will. If the victim dies, can we still choose not to fire the gun? Physics doesn't answer this question. But this violation of causality is enough to cause many physicists to believe that tachyons probably do not exist.

The same can be said for time travel. Can you go "backward" in time? If you could travel faster than the speed of light, then according to the equations of relativity, the order of events can reverse. Does that mean you can travel back in time? Well, the first problem is getting yourself to go faster than the speed of light. It takes infinite energy just to get up to c. But, could you tunnel across, go right to hyper light speed without ever being at light speed, by doing some sort of tunneling? Physics can't rule that out. But the problem, as with the tachyon gun, may be the havoc such events would play on our own concept of causality and free will. How can I chose to do something if, later on, someone came back from the future to stop me? That paradox is enough to get some physicists (including me) to guess that traveling backward in time will turn out to be impossible. But that is not a proof. And even physicists that believe it

---

[9] There is an exception. If the two events occur at the same location, then the order of events is unambiguous.

[10] To be precise, it depends on the velocity of the frame of reference, not on the velocity of the observer.

[11] The following math is only for those who have studied the Lorentz transformation in a different course. Let the two events be (x_1, t_1) and (x_2, t_2). In the moving frame, they occur spaced by a time interval $\Delta t' = \gamma(\Delta t - u\Delta x/c^2)$, where u is the relative velocities of the two frames. $\Delta t'$ will have a different sign from Δt only if $\Delta x/\Delta t > c^2/u > c$. That is possible if the two events have a tachyon moving from one to the other that has $\Delta x/\Delta t > c$.

is impossible still can enjoy science fiction movies (such as *Back to the Future* and *Terminator*) in which it happens.

General Relativity—a Theory of Gravity

Einstein's theory of general relativity is really a theory about gravity. It gets its name from Einstein's approach. Einstein was able to explain gravity in a most remarkable way. In his theory, mass and energy don't create a *gravitational field*—rather, they change space and time in a way similar to the way velocity changes space and time. There is both time dilation and space contraction. Einstein wrote a simple equation relating mass–energy to space–time, and it looks something like this:

$$G_{ab} = k\,E_{ab}$$

That's it. Of course, I haven't explained to you want G and E are, and they turn out to be mathematically complicated. G is a mathematical structure called a *tensor*. It actually consists of 16 functions, and the combination describes the contraction of space and time. E is likewise a tensor, and it describes the energy and momentum density. The math is sufficiently complex that it is normally taught only at the graduate level.

There is one simple and important result that can be deduced from this equation; in fact, it is simpler to derive that the full equation itself, and you can find a plausibility argument in some elementary physics books. The result is the following: if you go to a higher altitude, then your time will progress *faster*! The equation is very simple. The rate of time at an altitude h (in meters) is faster by a factor

$$f = 1 + gh/c^2$$

where g is the gravitational acceleration (from chapter 3), equal to about 10 m/s per second, and c is the speed of light ($= 3 \times 10^8$). So if you go up 10 meters (33 feet), then your time goes faster by a factor of

$$f = 1 + (10)(10)/(3 \times 10^8)^2$$
$$= 1 + 10^{-15}$$

That is a small change, but it has been measured, using a tower. Remarkably, the time factor becomes very important for keeping the clocks correct in a GPS satellite.

Practical Uses of Relativity: GPS

The time dilations of relativity, at everyday speeds and altitudes, seem to be tiny, and so you might think that they have no practical use. But let's consider a clock that is in a satellite orbiting the Earth. The velocity of a GPS satellite in MEO is 3.9 km/sec. (You can look that up on the Web.) Convert this to 3900 meters per second, and calculate the Einstein factor, to get

$$\gamma = 1 + 8 \times 10^{-11}$$

That is a small change (a value of 1 would mean no change at all), but a day contains 86,400 seconds (60 seconds per minute $\times$ 60 minutes per hour $\times$ 24 hours per day). So after a day, the clock on the satellite will be in error by $86400 \times 8 \times 10^{-11} = 7.3 \times 10^{-6}$ seconds $= 7.3$ microseconds. That, it turns out, is a lot! In that time, light travels 2.2 kilometers, about 1.4 miles. So if the clock sends you the wrong time, and your GPS receiver uses that time to calculate the distance to the satellite, it could be off by 1.4 miles. And your calculated position will be in error by this much.

It gets worse. The satellite has an altitude of 26,560 km = 26,560,000 meters. This, according to general relativity, speeds the clock up. From the altitude, we would calculate that the factor for speed-up is

$$f = 1 + gh/c^2$$
$$= 1 + (10)(26{,}560{,}000)/(3 \times 10^8)^2$$
$$= 1 + 3 \times 10^{-9}$$

That calculation is not quite correct, because the satellite is so far away that the acceleration of gravity is weaker; that means that g is not constant. If we do the calculation taking that into account, the number we get is a little closer to 1:

$$f = 1 + 5 \times 10^{-10}$$

But this is still a bigger effect than time dilation from the velocity. After 1 day, the clock will be incorrect by $86{,}400 \times 5 \times 10^{-10} = 4.5 \times 10^{-5}$ second $= 45$ microseconds.

If you combine the two effects, the net time error (after one day) is $45 - 7 = 38$ microseconds. Light goes 1000 feet per microsecond, so this is an error of 38,000 feet $= 7.2$ miles. And after two days, the error would be 14.4 miles.

Of course, the designers of GPS knew all this, so they put in a computer that calculates the time effect and adjusts the satellite clock so that you will get the right distances. So no need to worry. The effect is so large that it can't be missed. If Einstein had never predicted the velocity and gravity effects, they certainly would have been discovered as soon as we tried doing careful timing with clocks on satellites.

Questions about Time

Having gotten this far in the theory of relativity, you have earned the right to speculate along with the best physicists. Just to get you started, here are some questions to ponder:

> Is it possible that before about 14 billion years ago, when time began, there was no time? In fact, the statement "before the Big Bang" may be meaningless, since there was no "before" if there was no time.
>
> Could time stop? What if the Universe collapses into zero space (from the mutual gravitational attraction of all the galaxies)? Would space vanish? And would time vanish along with it?

What is the meaning of "now"? Can you explain it to someone? Can you write a paragraph explaining what it means? Does it mean the same thing to different people? Is my "now" the same as your "now"? What are you doing right now?

Why do we remember the past? Could the universe be such that we would remember the future instead? Or would we then call it the past? Could we remember some of the past, and some of the future—perhaps a 50/50 mix?

What is the "pace" of time? Is the rate of time set by our heartbeats and our rate of thinking? Is there any meaning to the "passage of time"? Does time "move"? If time sped up and slowed down, would we notice?

Space exists in three dimensions (at least). Could there be two dimensions of time? Could time run simultaneously in two separate ways? Or three? What would life be like with two dimensions of time?

Answers to the Questions

Actually, I don't know the answers to any of them. I have some speculation about remembering the past, rather than the future, but that is about it. Most of the questions that I have given in the preceding list are not considered part of physics, but I think that is only because we have made so little progress in understanding them.

Chapter Review

An event can be specified by four dimensions: three positions, and the time of the event. But the time interval between two events depends on the frame of reference. The amount of time experienced by a moving traveler is less than that by someone who is stationary, by the Einstein factor

$$\sqrt{1 - \beta^2}$$

In this equation, β is the light speed, equal to the velocity divided by the velocity of light. Another useful quantity is the Lorentz factor γ, equal to 1 divided by the Einstein factor:

$$\gamma = \frac{1}{\sqrt{1 - \beta^2}}$$

The difference in experienced time has been verified by experiments with clocks on airplanes and with accelerated radioactive particles. For low velocities, the Einstein factor ≈ 1, and that is why we don't usually notice the effect. But as v gets close to c, the value can be much less than 1. The same factor also describes length contraction, also known as the *Lorentz contraction*. According to this result, objects get shorter along the direction of motion.

Velocities don't add in the usual way. No matter who observes light, it will appear to be moving at c. That is the *invariance of the speed of light*. If the object is moving at a speed less than c, then it will still be moving at less than c for all observers, no matter how fast the observer is moving.

The energy of a moving object is given by $E = mc^2$. This includes both rest energy and kinetic energy. The m in this equation is the kinetic mass, which grows at high velocity according to the equation $m = \gamma m_0$ (where m_0 is a number called the *rest mass* that doesn't change). At low velocities, the total energy is the rest energy plus $1/2\ mv^2$. Rest mass can be turned to kinetic energy. This is done in nuclear decay, and in annihilation. A photon has rest mass zero. Such particles cannot undergo decay. We used to think that neutrinos also have rest mass zero, but since they can change into other neutrinos, they must have some mass. Einstein's equation $E = mc^2$ was not important in the invention of the nuclear reactor or the atomic bomb.

Tachyons are hypothesized particles that only travel at speeds faster than light. They reach zero energy at infinite velocity, and infinite energy when they travel at c.

Because of the variability of time in relativity theory, it is impossible to define *simultaneous* in an absolute way, unless two events occur at the same location.

Discussion Topics

1. How is our concept of time affected by our culture? Do different cultures treat it differently? Has the existence of accurate watches and clocks affected our thinking?

2. What can you say about the "flow" of time? Why does time appear to change? What is the meaning of "now"? Can these ideas be formulated in a clear way, clear enough that you could explain them to an extraterrestrial creature via a radio communication? Try doing that with a friend, but assume that the time it takes the signal to travel is a day.

3. Space has three dimensions. What would the world be like if time had more than one dimension? Could it have two dimensions of time?

4. Discuss some of the other questions posed near the end of the chapter: Could time have begun at the Big Bang? Could time stop? What is the possible meaning of "remembering" the future?

Internet Research Topics

1. Can you find movies (YouTube?) or Web pages that illustrate effects of relativity visually? You might find depictions of the Lorentz contraction, time dilation, or maybe the visual effects of looking at something that is moving near light speed.

2. Can you find Web sites that dispute the truth of relativity theory? What are their arguments? Are they right? If they aren't, can you figure out what mistakes they are making? (Hint: They often ignore the fact that events that are simultaneous in one frame of reference are not always simultaneous in other frames.)

Essay Questions

1. What usual concepts of time are upset by the theory of relativity? What do most people accept as obvious that turns out not to be true?
2. Energy can be converted to mass, and mass to energy. Describe how. Give specific examples.
3. Physicists often say that people will never be able to travel at the speed of light. Explain why they believe that.
4. Describe the peculiar phenomenon of the *twin effect* in relativity.
5. If relativity theory is correct, doesn't that mean that all the old equations we learned are wrong? What about the equation $E = 1/2\ mv^2$? If that is wrong, why did we learn it in an earlier chapter?

Multiple-Choice Questions

1. Which of the following quantities do not depend on velocity?
   A. m
   B. m_0
   C. kinetic energy
   D. total energy

2. A particle travels at the speed of light. We conclude that (be careful, only one answer is correct)
   A. its energy is infinite
   B. it violates special relativity
   C. its energy is zero
   D. its rest mass is zero

3. In everyday life, we don't see the effects of relativity because
   A. they only occur when near c
   B. they are too small to detect easily
   C. we have become accustomed to them, so we don't notice

4. In the twin effect, the traveling twin is
   A. younger
   B. lighter
   C. longer
   D. older

5. Neutrinos are believed to have mass
   A. zero
   B. small, but not zero
   C. infinite
   D. imaginary

6. Tachyons are particles that
   A. travel at c
   B. have zero mass

C. travel faster than c

D. have infinite energy

7. The photon has a rest mass approximately equal to

A. the mass of an electron

B. 0.000003 gram

C. 0

D. 3.14 grams

8. A moving object is

A. shorter and younger

B. longer and older

C. shorter and older

D. longer and younger

9. If Muller discovers a new, energetic particle (the "Mulleron") that has zero mass, then we can conclude that the Mulleron

A. can pass directly through matter

B. is a black hole

C. travels at the speed of light

D. is radioactive

10. Which of the following is not a zero-mass particle?

A. neutrino

B. graviton

C. gamma ray

D. beta ray

11. Tachyons

A. have never been observed

B. were discovered in the last 10 years

C. are used for medical imaging

D. are the components of protons

12. The speed of a massless particle (careful: this may be a trick question)

A. is the same to all observers

B. is faster to people moving in the same direction

C. is slower to people moving in the same direction

13. When a particle approaches the speed of light, its mass approaches

A. zero

B. infinity

C. m_0

14. Identify which are the particles that pass easily through the Earth. (Choose all that apply.)

A. cathode rays

B. x-rays

C. neutrons

D. neutrinos

15. The Lorentz contraction says that a football thrown past your face

A. is shorter by a factor of gamma

B. is longer by a factor of gamma

C. is the same length as it is at rest

D. appears different, but is not truly different.

16. The observation of neutrino decay proves that
    A. neutrinos do have mass
    B. neutrinos violate relativity theory
    C. neutrinos have charge
    D. neutrinos travel at the speed of light

17. A neutrino can have which of the following values for its energy?
    (Careful, this is a tricky question.)
    A. 0 joules
    B. 200 joules
    C. either of the preceding two values
    D. −200 joules (negative)
    E. any of the preceding three values

18. An astronaut Alice makes a round-trip at 3/5 the speed of light. Her twin Bob stays on the Earth. When Alice returns, Bob is
    A. younger than Alice
    B. older than Alice
    C. the same age as Alice

19. A radioactive particle has a half-life of 1 second. If it moves at 3/5 the speed of light, its new half-life will be
    A. 3/5 second
    B. 5/3 seconds
    C. 4/5 second
    D. 5/4 seconds

20. The Lorentz contraction refers to the fact that
    A. metals contract when they are cooled
    B. ice contracts when it melts
    C. plastics contract when they are cooled
    D. moving objects get shorter

21. Jim and Mary are both from Berkeley and are the exact same age. Jim travels at a velocity of 65 miles per hour to Los Angeles and waits there. Mary drives the next day at 70 mph. When she gets to Los Angeles, who is older?
    A. Jim
    B. Mary
    C. They are the same age.
    D. It depends on the distance.

22. If one person travels at 99% the speed of light for 7 years, how much younger will he be than his identical twin who didn't travel?
    A. 1 year
    B. 8 years
    C. He will be the same age.
    D. He will be older.

23. The most accurate equation for energy is
    A. $E = 1/2\ mv^2$
    B. $E = m_0 c^2 + 1/2\ mv^2$

C. $E = mc^2$
D. $E = m_0c^2$

24. Jane and Tom synchronize their watches on Earth. Jane remains on the Earth while Tom takes a round-trip to Pluto at near light speed. When they get back together, whose clock is correct?
    A. Jane's clock (on the Earth)
    B. Tom's clock (on the spaceship)
    C. Both clocks are correct.
    D. Neither clock is correct (since the Earth is always moving too).

25. We measure that a particle travels at the speed of light. We conclude that (be careful, only one answer is correct)
    A. its energy is infinite
    B. it violates special relativity
    C. its energy is zero
    D. its rest mass is zero

26. When the velocity of an object with rest mass m_0 approaches the speed of light, its energy approaches:
    A. m_0c^2
    B. $1/2\ m_0c^2$
    C. $m_0c^2 + 1/2\ m_0c^2$
    D. infinity

27. When a person moves at a velocity near the speed of light, which of the following are true—compared to a person at rest? (Choose all that apply.)
    A. He doesn't age as rapidly.
    B. His length gets longer.
    C. His mass increases.
    D. His energy increases.

28. If you travel at 50% of the speed of light, your mass is increased by a factor of
    A. 1.15
    B. 1.50
    C. 2
    D. 1

29. According to relativity theory, the energy of an object of mass m_0 that is moving at velocity v is
    A. $1/2\ m_0v^2$
    B. m_0c^2 (where c is the velocity of light)
    C. $\gamma\ m_0c^2$
    D. m_0c^2

30. As the velocity of a moving object approaches the speed of light, what happens to its energy?
    A. It goes to zero.
    B. It goes to infinity.
    C. It becomes mc^2.
    D. It becomes $1/2\ mc^2$.

31. Electrons cannot go as fast as light because
   A. of the uncertainty principle
   B. it would violate free will
   C. they would expand to infinite size
   D. it would take infinite energy

32. At very high speed, all of these change *except*
   A. length
   B. mass
   C. time
   D. electric charge

33. The effects of special relativity must be taken into account for
   A. semiconductor chips
   B. GPS
   C. lasers
   D. MRI

34. If the velocity is $(12/13)c$, then the gamma function is
   A. sqrt(5/13)
   B. 13/5
   C. 5/3
   D. none of the above

35. When the velocity of a proton is half of the speed of light, then the gamma function is approximately equal to
   A. 0
   B. 0.5
   C. 0.7
   D. 1
   E. 1.4
   F. 2

36. When a radioactive rock travels near the speed of light, its mass
   A. increases and there are fewer decays
   B. increases and there are more decays
   C. decreases and there are fewer decays
   D. decreases and there are more decays

The Universe

Puzzles

By the *Universe*, physicists mean "everything." It is expanding. But if it is everything, then what is it expanding into?

All galaxies are getting further away from our own. Doesn't that mean that we are at the center of the Universe? Isn't that weird?

The Universe was created about 14 billion years ago. How could we know that? And if it is true, what came before that?

How can the Universe be infinite? And if it isn't, how can it be finite?

To answer these questions, you first need to know some facts about the Universe, and many of these were discovered only recently. By the word *Universe*, we mean everything physical, all the places that we can see, probe, or somehow learn about. It is space, as far as we can see, as far as we will ever see, and everything in it. *Cosmology* refers to the study of these very large scale things. The Universe is made of atoms, stars, and bunches of stars known as galaxies, and clusters of galaxies. To understand the answers to the preceding questions, we begin with understanding the Solar System.

The Solar System

By the *Solar System*, we mean all those objects that are attached to the Sun through gravity. That includes the planets, asteroids, comets, and other things that we haven't discovered yet (such as the hypothetical companion star Nemesis, dimly lurking at a distance of one light-year).

We believe that the Solar System was formed about 4.6 billion years ago from interstellar debris, the ashes of a blown-up star called a *supernova*. That makes the Sun a "secondary" star. The reason that we think it is secondary is that the Earth contains elements such as carbon and oxygen that are created inside stars. Somehow those elements had to escape if they were to form planets. Moreover, primary stars don't contain heavy elements like lead and uranium, but the Sun does. These heavy elements are created only when stars explode as supernovas.

So here is what we think: 4.6 billion years ago, a region of space was filled with the ashes of an exploded star. These ashes started to attract each other through gravity, and they started to form a clump. Apparently, the ashes had a little bit of spin to them, because when the material gathered together, it started spinning faster (just as an ice skater does when she pulls in her arms; see the section "Angular Momentum and Torque" in chapter 3). Because the matter was spinning, it splayed itself out into a disk. The big mass at the center became the Sun. The smaller masses out in the disk did not fall inward because of their circular motion. They clumped into planets and asteroids.

The Sun was so large that the center became compressed and hot, and that ignited thermonuclear fusion. The planets were too small, so their cores never reached that temperature. But the early planets were molten. They emitted infrared radiation, and eventually a crust formed. Some planets completely solidified, but not the Earth—its core is still molten.

Even though the material of the Solar System is mostly hydrogen and helium, those gases were lost from the early Earth because of their high velocity (see chapter 2). The only hydrogen which remained was that which combined with other elements to make water, hydrocarbons, etc.

Light from the Sun takes about 8 minutes to travel the 93 million miles (150 million kilometers) to the Earth. We say that the distance to the Sun is 8 light-minutes. The nearest known star to the Sun is the Alpha Centauri–Proxima Centauri double star system. These stars are about 4.3 light-years away. The *light-year* is not a unit of time, but of distance. It is the distance that light travels in a year. Stars are, typically, separated by light-years.

Comets

Farther out than the planets, the Solar System is inhabited by comets. These are small, icy bodies that orbit the Sun at great distances. They are mostly too small to see unless their orbit brings them close. A few of them have orbits that bring them close to the Sun. When that happens, they heat up, and much of their frozen gases evaporate, creating the "tail" that gives comets their name: *comet* is Greek for "tail." (The word *comma* also derives from that word.)

Astronomer Jan Oort realized that we saw only a tiny fraction of the comets, and he calculated that there were over 10^{10} of them. Even so, their total mass is less than that of Jupiter. We now call this bunch of comets the *Oort comet cloud*. Only a few of these get close enough to the Sun to be visible to the unaided eye—typically, one every 10 years or so. That is rare enough that ancient people considered their arrivals to be very significant, usually a bad omen. It is interesting to look up the history of cometary fears on the Internet. Of course, every now and then, such an object might hit and cause devastation. It was either a comet or an asteroid that was responsible for the death of the dinosaurs.

Nemesis

Most stars, when they form, form in pairs or triplets, like in the Alpha Centauri–Proxima Centauri system. Most people believe that the Sun is an exception, a lone star, with only planets to keep it company.

In 1984, two colleagues (Marc Davis and Piet Hut) and I published a theory that postulates another star orbiting the Sun. We playfully gave it the name Nemesis. Strictly speaking, the Sun and Nemesis are orbiting each other. The reason it hasn't been noticed is that Nemesis is about a light-year away.

The Nemesis theory was devised to account for paleontological evidence that great extinctions occurred every 26 million years. The orbit of Nemesis is elliptical, and when it comes close to the Oort comet cloud every 26 million years, it triggers a comet storm. The comet storm brings a few billion comets into the inner Solar System. The Earth is small, so the chances that it will be hit by any one comet are one in a billion. But with several billion comets coming in, there are likely to be several hits.[1]

There is no direct evidence that Nemesis exists, so most astronomers assume that it doesn't. However, upcoming surveys of stars (e.g., the Pan-Starrs project; see the Wikipedia article) are likely to discover Nemesis in the next few years; if the surveys do not find it, that indicates that it does not exist.

Planets around Other Stars

We once speculated that the Solar System is unique, but now we know that most stars have planets orbiting around them. Most of the known extra-solar planets were discovered by Geoffrey Marcy, a Berkeley professor.[2]

The planets around other stars are dim, but we could probably see them in telescopes if they weren't so close to bright stars. They were discovered by scientists seeing their effects on the stars they orbit. They make these stars wobble, and we can detect that by observing small changes in the frequencies of their spectral lines as the velocity of the star oscillates, from the Doppler shift. (The Dopper shift is explained in chapter 7.)

Galaxies

On a clear winter night, look straight up. If you are in a sufficiently dark place, you may be able to see a small fuzzy spot, no larger in angular size than the Moon, that looks like a tiny scrap of the Milky Way that was torn off. But it is much farther away than the Milky Way. In fact, the nebulous spot is the most distant object that you can see with your unaided eye. It is three million light-years away. A lot more detail can be seen in a long exposure taken with a telescope. The galaxy is so large in the sky that the best photos are composites of the different parts, such as shown in figure 13.1.

---

[1] For more details, see my Nemesis Web page at www.muller.lbl.gov/pages/lbl-nem.htm.
[2] His Web site is http://astro.berkeley.edu/~gmarcy/.

Figure 13.1 Andromeda Galaxy. The most distant object you can see without a telescope. It contains over 10 billion stars. (Photo courtesy of NASA.)

It was Edwin Hubble (after whom the Hubble space telescope was named) who discovered that this fuzzy spot actually consists of more than a billion individual stars, forming a flat circular plate, all spiraling around each other in a gravitational whirlpool that we now call a *galaxy*. The one shown in figure 13.1 is called the Andromeda Galaxy.

The Milky Way is also a galaxy. It doesn't look to us like the picture in figure 13.1 only because we are inside it, looking out. The milky path that can be seen across the sky (best seen in summer) is actually the light from millions of stars that you see when you look out toward the edge of the plate. When you look straight up or down, you see out through only a thin layer of stars. But everything you see, every star, is part of the Milky Way—except for that fuzzy patch Andromeda.

The Andromeda and Milky Way galaxies are like solar systems. They are held together by their own gravity, but they don't collapse, because the stars within them are moving in circles. The velocity of the Earth in the Milky Way galaxy is about a million miles per hour.[3] We don't particularly notice, since nearby stars are orbiting our galaxy along with us. It takes the Sun and the Earth about 250 million years to make one loop around the center. (You might call that a "galactic year".) We now know that at the center of our galaxy is a giant black hole. No light emerges from it, but we can see its enormous gravity from the rapid orbits that nearly stars make around it.

---

[3] This value was first measured by the author of this book, along with colleagues George Smoot and Marc Gorenstein. For details, see www.muller.lbl.gov.

Other Galaxies

Look at the photo of the Andromeda Galaxy in figure 13.1 again. There is a small fuzzy patch just below and to the right of the center—that is another, but smaller, companion galaxy. You can also see a lot of bright points—those are stars. But those stars are not part of the Andromeda Galaxy. Stars in Andromeda are too far away to be seen in this image. The stars that you see are nearby, in our own Milky Way. Although they look like background stars, that's an optical illusion based on your experience that stars are always in the background. They are really in the foreground.

There are many more galaxies than Andromeda, its companion, and the Milky Way. Examine the photograph in figure 13.2, made of a combination of images taken by the Hubble space telescope. It looks like it is full of stars, and in some sense it is. But almost all of the bright spots that you see are not individual stars. They are galaxies, spinning clumps of a billion stars each. The Hubble team has counted over 1500 galaxies in this image. (Not all are visible in this print.) Some of them are 4 billion times dimmer than can be seen with the unaided human eye. This composite image took 10 days to expose, since the galaxies were so far away. The picture was taken in the direction of the Big Dipper. That region was chosen since there were not very many foreground stars to obscure the distant galaxies. One star is evident—the one with the bright spikes. Those are artifacts of the telescope optics.

Figure 13.2 The Hubble Deep Space photo. Virtually all the objects in the image are galaxies. (Photo courtesy of NASA.)

The image covers only a tiny part of the sky, about as much as would be covered by a dime that is 75 feet away. Although it is a small part of the sky, we believe that it is typical. Based on this image, we can estimate that the total number of such galaxies that are visible to such an instrument is about 40 billion.[4] That means that there are more galaxies in the Universe than there

---

[4] A dime has an area of about 2.6 square centimeters. At a distance of 75 feet = 23 meters, a sphere would have a surface area of 65 million square centimeters, so it would take about 25 mil-

are stars within our own Milky Way. These galaxies fill up the observable Universe, although there is a lot of empty space between.

Dark Matter

Our star, the Sun, is moving around the Milky Way, held in by the gravitational attraction of the other stars. But here is a serious problem: if we estimate the number of stars, and the mass of each one, and add them all together, there is *not* enough mass to hold the Sun. Yet we can determine that we are in a circular orbit. What is holding us in?

The guess is that there is some kind of object that has all that mass, but doesn't glow like stars, and so we don't see it. The material is called *dark matter*. Look up this term on the Internet, and you'll find a million hits.

Moreover, if we look at clusters of galaxies, those that are swirling around each other, we find again that there is not enough mass unless we postulate a huge amount of dark matter—more than the amount of matter that exists in all the stars of the galaxy.

Think about that for a moment. The startling conclusion is that most of the mass of the Universe is in dark matter. And we don't know what dark matter is. Put even more dramatically, we have not figured out what the Universe is made out of! There are two serious candidates for dark matter: WIMPs and MACHOs.

WIMPS

WIMPs stands for Weakly Interacting Massive Particles. These are ghostlike particles, like neutrinos, that pass through the Earth and stars without hitting anything. The reason is that they have no electric and no nuclear "charge," and they feel only the weak interaction (that's where the *W* comes from) and gravity. They are "massive" (not zero rest mass) so that their gravitational pull can be important. If WIMPs exist, they are everywhere, and sensitive detectors to find them are being built at laboratories around the world. Physics professor Bernard Sadoulet of Berkeley is one of the world leaders in this search. WIMPs could exist, but maybe they don't.

MACHOs

MACHOs are MAssive Compact Halo Objects. These may be large planets, black holes, or other massive but compact objects. The *H* is in there because these objects must fill up the galactic "halo," a region that extends above and below the galactic disk. Some experiments have detected MACHOs by looking at their effects in blocking starlight. But so far nobody has found enough of them to account for the dark matter. MACHOs and WIMPs form the cutest acronym pair in all of physics.

---

lion dimes to obscure our view. If each had 1500 galaxies covered by each dime, the total number would be $25 \times 10^6 \times 1500 = 38 \times 10^9 \approx 40 \times 10^9$.

Extraterrestrial Life and Drake's Equation

In 1971, Frank Drake tried to estimate the number of planets in our galaxy that would have intelligent life trying to communicate with us. To do this, he wrote down an equation that has subsequently become famous, and is now called *Drake's equation*. Look it up on the Internet. The equation is:

$$N = G\, f_p\, f_e\, f_l\, f_i\, f_c\, f_L$$

This equation may look impressive, but it is nothing more than a bunch of numbers multiplied together. It simply says that the number of nearby stars with intelligent life, N, depends on seven factors. G is the number of stars in our galaxy, about 10^{10}. Then all of the f numbers are simply fractions that meet various criteria needed for intelligent life. f_p is the fraction of those starts that have planets. You'll see that it won't matter that we don't know this number very well. The variable f_e is the fraction of those planets that can sustain life, f_l is the fraction of these in which life evolves, and f_i is the fraction in which intelligent life evolves. Out of this intelligent life, f_c is the fraction that chooses to communicate, and f_L is the fraction that survives at the right time to be communicating to us right now. Multiply all these together, and N will be the number of civilizations in planets around other stars whose signals we should be able to pick up.

Put in some reasonable numbers, and do the calculation yourself. One Web page suggests the values $f_e = 1$, $f_p = 0.5$, $f_l = 0.5$, $f_i = 0.2$, and $f_L = 10^{-6}$. That gives $N = 1000$. So we expect 1000 stars out there to have planets sending us signals!

Belief in the Drake equation has been the inspiration for *SETI*, the Search for Extraterrestrial Intelligence. Look up SETI on the Internet. You'll learn that you can participate, by doing some analysis on your home computer through the SETI-at-Home program.[5]

The large number of planets with intelligent life is a surprise to many people. If you look at the equation, you'll see that it comes about because G is so large. It hardly matters what numbers you put in for the other factors, as long as they aren't too small. But that's the catch. One or more of those numbers could, in fact, be tiny.

I remain skeptical of the use of the Drake equation. Suppose that the probability of life evolving on a suitable planet, instead of being 0.5, were $f_l = 10^{-9}$. Then there would be no extraterrestrial signals. Some SETI advocates argue that f_l must be very high, since it happened so quickly here on the Earth. But that is a weak argument for this important number. After all, if life hadn't developed quickly on the Earth, we wouldn't be here thinking about such problems. So we may be the exception, the rare planet on which life happened to develop early. The error is similar in nature to the error made by a person living in the middle of New York City extrapolating the human population density he observes locally to the rest of the world, including the deserts and oceans. What we see locally is not necessarily typical.

[5] Their Web site is http://setiathome.ssl.berkeley.edu/.

There is great uncertainty because we really don't understand how life began on the Earth. Simple organic molecules such as amino acids (the building blocks of proteins) can form accidentally, or be created in lightning strikes. But we have no idea how complex molecules such as RNA and DNA formed. It is conceivable that the probability of them forming is 10^{-15}. If that's the case, then we would be the only intelligent life in the entire Universe.

A fascinating book skeptical of extraterrestrial intelligence is *Rare Earth* by Peter Ward and Don Brownlee (Springer, 2000), both of whom are famous and highly esteemed scientists. If you find this subject interesting, you would enjoy this book. It presents the strong possibility that we are truly alone.

I can't say there isn't intelligent life out there. I just don't believe that the Drake equation demonstrates that it is likely within our own galaxy.

Looking Back in Time

The galaxies in the Hubble Deep Space photo (figure 13.2) are so far away that the light we are observing was emitted about 10 billion years ago. That means that we are not seeing the galaxies the way they are now, but the way they were back then. In fact, careful measurements show that the galaxies are somewhat different from ours, and this can be attributed to the fact that we are seeing them when they were young. They contain fewer of the heavy elements (such as iron) that take time for stars to manufacture. We believe that the galaxies, by now, should have generated enough of these elements, that by now they are similar to the Milky Way and to Andromeda. We think that this image shows galaxies the way they used to be, soon after they were created.

This sounds strange, but you look back in time every time you see the Sun. It took the light about 8 minutes to reach you. You see the Sun the way it looked 8 minutes ago, not the way it looks now. We often measure distances in terms of light-travel time, as in the table 13.1. Note that the unit *light-year* measures distance.

Table 13.1   Distances to Astronomical Objects

Object	Distance
Moon	1.3 light-seconds
Sun	8 light-minutes
Sirius	8.6 light-years
Andromeda Galaxy	3 million light-years
Hubble Deep Space galaxies	10 billion light-years

Expansion of the Universe

There is another difference seen in figure 13.2. The galaxies back 10 billion years ago were closer to each other, on average, than they are now. The space between galaxies has been increasing. This fact is called the *expansion of the Universe*.

The first indication that the Universe was expanding came when Edwin Hubble measured the velocity of galaxies. He did this by looking at the Doppler shift in the frequencies of the spectral lines emitted from the stars. Remember how the Doppler shift works? If something is moving away from you, then it takes each cycle a little longer to reach you, so each period is longer. That means that the frequency is lower.

Hubble found that all distant galaxies had shifted spectral lines. Moreover, the lines were all shifted to lower frequencies, i.e., toward the color red. That's why some people say that Hubble discovered the *red shift* of distant galaxies.

Where is the needed space coming from? Since space is infinite, there's no real need to find more space; infinity is pretty large. Imagine that the Universe is like a rising (expanding) loaf of raisin bread. As the bread bakes, the whole thing grows, and all the raisins (galaxies) get farther apart from each other. Note that if you are sitting on any raisin, all the other raisins will be getting farther away from you. It doesn't matter which raisin you are on.

For the real Universe, there is no obvious crust, and no obvious location that you can say is at rest. Infinite space expands into infinite space. The galaxies keep on getting farther apart, but there is plenty of room.

I'm not surprised if that still bothers you. So here is yet another example: imagine the numerals from one to infinity spread out on a line:

1 . . . . . 2 . . . . . 3 . . . . . 4 . . . . . . 5 . . . . . 6 . . . . . 7 . . . . and so on forever.

They can go on forever because the Universe is infinite (at least let's assume that it is). But now lets stretch this infinite string, so the letters get spread out even more:

1 . . . . 2 . . . . 3 . . . . 4 . . . . 5 . . . . 7 . . . . and so on forever.

They are all farther apart. We didn't have to find new space to do this, because even when they are separated by greater distances, they still just go out to infinity. There is lots of room at infinity.

Another thing to notice is that no matter which number you are, all the other numbers are getting farther and farther away from the one you picked. Suppose that, for example, you are the number 4. You don't think of yourself as moving You notice that 5 moves farther away, and so does 3. So just the fact that everything is moving away from you doesn't mean that you are in the center.

Hubble's Law

Hubble's discovery is now summarized by a formula known as Hubble's law:

$$v = HR$$

It says that the velocity v at which distant objects are receding from us is proportional to R, their distance from us. H is a number called *Hubble's constant*. If we measure the velocity v in km/s, and the distance to the galaxy R in km, then Hubble's constant H is approximately 2.3×10^{-18}. This number is so small only because the Universe is so large.

You will find a lot of people who are bothered by Hubble's law. They will think that it means that we are in the center of the Universe. You can have a lot of fun explaining to them why it does not imply that at all. Tell them that no matter which galaxy they are sitting on, all other galaxies will appear to be receding from them. It's just like expanding raisin bread—or the expanding list of numbers.

Dark Energy

In the Hubble expansion, galaxies are flying apart from each other. No force is needed for this; they are just sailing from their initial velocity. Where did this initial velocity come from? Some people attribute it to a quantum fluctuation, but I find that explanation implausible. My answer is simpler: I don't know.

But as they fly apart, the galaxies do experience a force of mutual gravitational attraction, and that should slow them down. The dark matter will contribute too.

I started a project at Berkeley to measure this slow down; it was eventually taken over by my former student Saul Perlmutter. The idea was to make a very careful measurement of galaxies at different distances, and look for a change in the Hubble law as galaxies are decelerated. We measured distances to galaxies by looking at the brightness of exploding stars (supernovas) in those galaxies, and we measured their velocities by the Doppler shift of their spectral lines.

Under the direction of Perlmutter, the project finally succeeded; it was also successfully done at about the same time by another international group including Alex Filippenko of the UC Berkeley Astronomy Department. The results were completely shocking. The expansion of the Universe was not slowing down. It was speeding up!

Why is the universe expansion accelerating? Nobody knows, so they did what people do when they don't understand something—they gave it a name, *dark energy*. The word *energy* sounds plausible, because the galaxies seem to be picking up kinetic energy. But it was unexpected.

Some physicists think that the acceleration might be due to quantum mechanics effects. But when they do the calculation, they get an answer that is wrong by a factor of 10^{120}. That's 1 followed by 120 zeros. This disagreement is thought to be the biggest ever in the history of science.

It is also possible that the dark energy is something simpler, such as the self-repulsion of the vacuum. This is not currently part of quantum mechanics (which purports to describe the vacuum), but its absence in the equations might be because measurements of the properties of the vacuum over distances as large as billions of light years never previously existed.

The Beginning

In the past, galaxies were closer together. How close? Hubble's law tells us that the distance grows with time. Was there a time when the distance was zero?

According to Hubble's law, the answer is yes. A galaxy at distance R, moving at velocity v, in a direction that is away from us, must have been here in the past. To get the time, divide the distance by the velocity:

$$T = \frac{R}{v}$$

Now substitute v from Hubble's law to get

$$T = \frac{R}{v} = \frac{R}{HR} = \frac{1}{H} = \frac{1}{2.3 \times 10^{-18}}$$
$$= 4.4 \times 10^{17} \text{ seconds}$$
$$= 14 \times 10^{9} \text{ years} = 14 \text{ billion years}$$

This says that 14 billion years ago the distance between all galaxies was zero. Could that be true? We think that the answer is *yes*. That was at the moment of the Big Bang.

The Big Bang

What was the Universe like 14 billion years ago? It didn't consist of galaxies right on top of each other, because we don't think that galaxies (or even stars) had yet formed. But the mass (hydrogen and helium gases) would have all been right on top of itself, at an extremely high density.

This idea sounds ridiculous. The first person who took it seriously was George Gamow (who had been the first person to understand alpha-particle emission in radioactivity). He worked in collaboration with Ralph Alpher and Robert Hermann to analyze what would have happened if the compressed early Universe really had happened.

You can image what it would be like by running the Hubble expansion backward. Stop the expansion, and let the galaxies all fall into each other. As they fall, they will collide, and their kinetic energy will turn into heat. So the early Universe must have been very hot and dense.

Gamow realized that these were the conditions that could induce nuclear fusion and that might answer a mystery: Why do most stars consist of 90% hydrogen and 10% helium? (Other elements, such as carbon and oxygen, make up less than 1%.)

Gamow's answer: The very early universe consisted of nothing other than protons, electrons, neutrons, and a few other elementary particles. But the conditions were so hot and so dense that the hydrogen underwent fusion, and created helium. Modern calculations show that this would have happened in the first four minutes of the explosion.

This theory was wild, and other astronomers mocked it. In particular, the famous British astronomer Fred Hoyle (who made some wonderful discoveries himself about how stars worked) made fun of Gamow's theory by calling it "The Big Bang." The name stuck.

One of the more amusing anecdotes on this history came when Gamow and Alpher wrote a paper together, Gamow had a little fun. He added the name of his friend, the renowned physicist Hans Bethe, to the paper without asking Bethe. His reason was that he wanted to make a pun. The authors of the paper became Alpher, Bethe, and Gamow—which reminded him of the first three letters of the Greek alphabet, alpha, beta, and gamma. Even today, this famous work is referred to as the αβγ paper.

The 3-K Cosmic Microwave Radiation

Gamow realized that a hot early universe would have emitted visible light. Even after a half-million years, the entire Universe would be at the temperature of the Sun (6000 K) and full of matter radiating intensely bright light. When the Universe cooled, electrons and protons would form neutral hydrogen atoms, and the universe would suddenly become transparent. (Except for a few stars and planets, but they don't fill up much of the sky.) But the light that already filled the universe would continue to move through space.

Radiation emitted from a region 14 billion light-years away would just be reaching us now. Because the matter that emitted it was moving away from us at very high speed, it would have its frequency shifted. (That's the Doppler shift again.) The effect would be so large that the light, in our frame of reference, would be microwave radiation. So the Big Bang theory predicted that the Universe would be full of microwave radiation. It would have the same black-body spectrum (see chapter 9) that an object would emit if it's temperature were 3 K.

That radiation was finally discovered by Arno Penzias and Robert Wilson in 1965. They found the radiation coming from all directions in space, just as the Big Bang theory predicted. It had a *black body spectrum* corresponding to a temperature of 3 K, again, just as the theory predicted. They were given a Nobel Prize for their discovery. (Gamow had died before the prize was awarded.)

Starting with this discovery, people took the Big Bang theory seriously. Now we can calculate what happened in the first 3 microseconds of the Big Bang. We postulate that in the very early Universe, even protons didn't exist, only a quark–gluon plasma. Of course, we have no idea if these calculations are true or not, except for the success of the Big Bang theory.

Creation of the Elements—and Life

According to the Big Bang theory, the early Universe (say after it was one second old) contained protons, electrons, and light, but no helium or other elements. But because the Universe was so hot and dense, fusion took place in the first few minutes and created most of the helium found in the Universe. Indeed, about 25% of the mass of stars is helium, and we believe that was created by fusion in the Big Bang. But calculations show that no oxygen or nitrogen or other elements would be created. It was too hot, and they would have been burned apart. Where did they come from? Gamow didn't know, and the answer came many years later. We now understand that these elements were created in the fusion inside stars. Three helium nuclei fused to form carbon. Carbon fused with hydrogen to form nitrogen. And in this manner, the elements necessary for life were created.

According to the calculations, this "slow" fusion in the stars was not enough to create all the heavier elements. Many of the elements with nuclei heavier than iron would not form in this way. We now think that the heavy elements were created in yet another fusion process that took place when the star exploded in what is called a *supernova event*. In the few minutes of such an explosion, most of the elements beyond iron in the periodic table were created.

The supernova explosion served another purpose, as far as we are concerned. Elements buried deep inside a star are not very useful for creation of life; it's just too hot. (Some science fiction, such as Arthur C. Clarke's *2001: A Space Odyssey*, denies this and postulates that advanced life takes place inside stars.) When the star exploded, the elements were ejected into space, where they cooled (through IR emission) into dust. Eventually, the dust particles attracted each other (through gravity) and formed a new star. The debris left behind orbited that star, and formed planets. In the plants, conditions were cool enough for water to form, and water serves as a catalyst (a chemical that enhances chemical reactions), and that led to life. So for us to be here, we had to have helium created in the Big Bang, the elements formed inside stars, the stars explode to release these elements, a second star to form to give sustained energy, and relatively cool planets to form around that star with a temperature that allowed chemical reactions—and life.

Black Holes

Black holes occur when the gravity of a star is so strong that enormous energy must be used to throw the object off its orbit around the star. But when you give it that much energy, it increases the gravitational attraction, so the energy is not enough. If the object is a black hole, then you can never win: more energy always increases the attraction so much that the object will not leave.

The reason for this has to do with the relativistic mass increase. If you are sitting on the surface of a black hole and want to shoot an object to infinity, you might try to give it a very high velocity. But when it gets going fast, its mass increases, and that in turn increases the force of gravity on it. For a black hole, the increase in gravity always outweighs the increase in velocity. So no matter how high a velocity you give the object, it will not escape. Another way to put this is that, for a black hole, the escape velocity is greater than c. That's why nothing can escape.

Note that this explanation for a black hole seems to be different from the one in chapter 3. Back there, I said that a black hole was simply an object with so much mass in such a small volume that the escape velocity exceeded the speed of light. But it turns out that the two explanations are mathematically equivalent. So they are both right—they are just two ways of thinking about the consequences of relativity theory.

We could turn the Sun into a black hole if we squeezed all of its mass into a sphere with a radius of 3 km. Such squeezing does take place in a supernova explosion, and we think that the celestial x-ray emitter called Cygnus X-1 is such a black hole. X-rays are emitted by particles falling into the black hole.

We could turn the Earth into a black hole if we squeezed all of its mass into a sphere with a radius of 1 cm.

We also believe that the center of the Milky Way contains a black hole, as do the centers of many other galaxies. This is a relatively recent discovery, and we do not yet know how they were formed.

Finite Universe?

In the theory of relativity, space is "flexible" in the sense that the distance between two objects depends on the frame of reference. In the general theory of relativity, gravity is included by allowing for accelerated reference frames. This winds up making space even more variable.

The most fascinating part of general relativity is the possibility of curved space. In this section, I am not going to try to explain what this means, but I'll give some examples.

Suppose that there were no dark energy, and the mass density of the Universe was fairly high—high enough so that the expansion of the Universe would eventually stop, and turn around, and ultimately result in a big crunch. In that case, the equations of general relativity predict that the Universe would be finite. It would be finite in the same way that the surface of the Earth is finite. Now here is the amazing analogy: the geometry of the Universe, under those circumstances, would be analogous to the geometry of the surface of a four-dimensional sphere. (The surface of that sphere is the three-dimensional space that we are accustomed to.)

That means that an object moving in a straight line would eventually come back to the same point, just as an object moving on the surface of the Earth would eventually return, after going all the way around the Earth. The Universe curves around, but only in the hidden fourth dimension that we don't see or experience. (If there are four spatial dimensions, then time will be considered the fifth, rather than the fourth dimension.)

This is what we mean when we say that the Universe is finite, but has no boundaries. If you traveled enough, you could visit everywhere. There would be no new places to go. You could even write down the number of cubic meters in the Universe.

Why don't I tell you how many cubic meters? The reason is that the recent evidence shows that the Universe is not finite, but appears to be infinite. The general theory of relativity, when you include the dark energy, says that the current Universe goes on forever.

That doesn't mean that the Universe really is infinite, since the general theory of relativity may not be valid for really large distances. So maybe someday we'll conclude that the Universe is finite. But, for now, we believe it is infinite.

Here's a philosophical issue: some people are bothered by the thought that the Universe is infinite and some are bothered by the possibility that it may be finite. Clearly, no matter what the Universe turns out to be, it will bother people.

Before the Big Bang

Many people wonder what caused the Big Bang. What came before it? There is a view of the Big Bang that is not shared by all physicists, but is held by many.

The Big Bang, in this view, was not an explosion of matter within space, but it was an explosion of space itself. In the Big Bang, space was created. The galaxies are not really moving, they are staying stationary, but the space between them continues to expand.

Remember, another name for *space* is *the vacuum*. An ancient name for it is the *Aether*. Space is not made of particles; rather, particles are waves in the medium of space. Don't worry if this makes no sense to you. This is the last chapter, and this kind of knowledge is hardly needed by presidents.

In relativity theory, space and time are often treated as different dimensions in the *space–time continuum*. So if space was created in the Big Bang, maybe time was created too. If so, then the question "What happened before the Big Bang?" makes no sense, because time did not exist. It's like asking, "What is shorter than a line of zero length?" or "What happens when molecules go below absolute zero?" (i.e., "What happens when they move slower than no movement at all?"). These questions cannot be answered because they make no sense. If time didn't exist, then there would be no time before the Big Bang.

If that is true, then we may never be able to answer the question of what caused the Big Bang. The answer may lie outside of our sense of reality.

Theory of Everything

In the newspapers, you will sometimes read that somebody has devised a new "theory of everything." It is worth knowing what physicists mean when they use this term.

Back in the 1600s, when Newton was discovering the basic laws of physics, gravity was seen as a tendency of objects to accelerate toward the Earth. At that time, the motion of the Moon around the Earth was known, but nobody realized it was gravity that held it in circular motion. Newton's success in creating a theory of gravity to combine these two apparently different phenomena (see chapter 3) was the first successful combination of two forces into one.

The next real success came in the late 1800s, when James Maxwell succeeded in combining electricity with magnetism. Magnetism was seen to be just the force of electric charges that happens when those charges are in motion.

In the early 1900s, Einstein worked hard to combine the theory of electromagnetism with the theory of gravity. He called his goal the creation of a *unified field theory*. But he never succeeded in his goal.

The next success came in the 1970s, when the weak force (responsible for some kinds of radioactivity) was discovered to be a version of electromagnetism. This combination was called the *electroweak theory*, and it could be understood only in the context of quantum physics, an approach that Einstein had ignored. Moreover, he had been trying to unify the wrong forces: electromagnetism and gravity.

We have had no further true success. There has been a lot of speculation about combining the electroweak force with the strong force that hold the protons in the nucleus together, but there is no compelling evidence that any of the proposed theories are really true. More recently, the goal is to include not only the strong force, but also gravity. If all of these forces can be seen as different aspects of the same fundamental force, then people will claim that we truly do have one theory that covers everything, rather than separate theories that have to all be accepted independently.

One of the exciting developments has been something called *string theory*. The reason people are so encouraged is that string theory combines all the forces (electroweak, strong, and gravity), and it takes fully into account both quantum physics and relativity. But the theory is so mathematically complex that nobody has been able to use it to make predictions that can be tested. In string theory, all the elementary particles, including electrons, quarks (the constituents of protons and neutrons), neutrinos, and everything else, are all versions of an elementary object called a *string*. The differences between these particles are just a consequence of the way the string vibrates.

But whatever the ultimate theory of everything turns out to be, it will be quite different from the kind of unified theories that people once imagined. String theory makes no pretense of being able to predict the future perfectly, provided we had perfect knowledge of the present. That's because it is a quantum theory, so it only predicts probabilities. And it cannot predict things very far into the future because the probabilities compound and grow. The quantum theory of everything still will not be able to predict when any radioactive atom will explode. An atom of potassium-40, the theory says, will probably explode sometime in the next billion years. Beyond that, it cannot say. If the decay of that atom was used to trigger a nuclear explosion, then the theory of everything would only be able to say that such an explosion would occur sometimes in the next billion years. By its very probabilistic formulation, the theory of everything can do no more. Therefore, unlike the original goal of Einstein, the new theory of everything, based on quantum physics, is not in contradiction with the idea of free will.

Many people love super-string theory, and many people are skeptical that it will succeed. It actually does make lots of qualitative predictions (e.g., the existence of twice as many particles as we know of) that turn out to be wrong, but are excused away by changing the theory slightly. It is possible that in the future, the main interest in string theory will come from mathematicians rather than from physicists, unless it ultimately proves to be able to predict some things that turn out to be right. And string theorists may be making the same mistake that Einstein made when he tried to combine the wrong forces. Perhaps there are other forces, such as the one that gives rise to dark energy, forces that we haven't discovered yet, and until we have them all, any attempt to combine the forces will be doomed to failure.

Chapter Review

Our solar system was formed about 4.6 billion years ago, from the remains of an exploded star (supernova). It consists mostly of the Sun; the planets are tiny. Light takes 8 minutes to go from the Sun to the Earth. Stars are similar to the sun, and are dim because of their great distance (the closest is over 3 light-years away). Most stars come as part of binary star systems. Many stars have planets. Galaxies are disks with 10 to 100 billion stars. Our galaxy is called the Milky Way. It takes us 250 million years to go around. At the center is a large black hole. There are more galaxies in the observable universe than there are stars within the Milky Way.

Dark matter is material that has gravity but is not seen with light. We don't know what it is, but there is more of that than there is of ordinary matter. It could be made of WIMPs or of MACHOs. Drake's equation is used to predict the existence of intelligent life outside the Earth, but some of the factors in the equation are very uncertain. When we look in space, we are looking backward in time. The Universe is expanding; since space appears to be infinite, it does not require anything more to expand into. Hubble's law is the observation that the Universe is expanding uniformly. There is no center. The beginning was about 14 billion years ago, the *Big Bang*.

Dark energy is the name given for the force that is making the expansion accelerate. We don't know what it is. Helium was created in the Big Bang. So was radiation that is observed today as microwaves. Gravity affects time, so there is a gravitational twin paradox. In general relativity, space can be curved, so that the distance to objects is not what you would expect from simple geometry. Nobody knows what happened "before" the Big Bang; maybe time did not exist. Theories of everything (if they are ever formulated) would try to explain all forces (gravity, electricity, nuclear, weak) as different aspects of a single force. String theory is one attempt to do this, but many people think it is just math and not really representative of reality.

Discussion Topics

1. Which is harder to believe: an infinite universe (in which space, not necessarily matter, goes on forever) or a finite universe (in which space has a big but finite volume)?
2. Do you think that extraterrestrial intelligence exists? Do you think it is smart to try to communicate with such beings? Why? It is out of curiosity—or do you think we night learn something important?
3. Suppose that a physicist succeeds in creating a "theory of everything" along the lines described in the text. Would you consider it to be a theory of everything? What other things would you like to see in a theory that has such an exalted name. (For example, should a real theory of everything be able to predict when a radioactive atom will decay?)

Internet Research Topics

1. Look at the SETI-at-Home program. How many people are currently participating? Does it make sense to you? Consider joining.
2. Find the site for the Astronomy Picture of the Day. What instruments are used to take these images? How many of them do you find intriguing? What questions do they bring to mind when you look at them? Save some for your computer.
3. What can you find out about currently known black holes? How big are they? Where are they located in the sky?

4. What can you find out about current searches for dark matter and dark energy? Have they ruled out any theories yet? What could the dark matter and the dark energy be?

5. Do some Web sites dispute the reality of the Big Bang? Why? Could they be right?

6. What is the difference between astronomy, astrophysics, space science, and cosmology? Are they synonymous, or are the words used differently? Do they usually describe different aspects of science? Look up each one, and see what kind of topics they cover.

Essay Questions

1. Discuss Drake's equation. How is it used to argue that there must be intelligent life outside of the Earth? What are the weaknesses of this argument?

2. How can the Universe expand, if it is already infinite? Give a simple example that might help the president understand this, if he or she were to ask you.

3. What does it mean when we say that astronomical observations "look back in time"? Can we really see the past? Can we use this technique to see our own past?

4. What is Hubble's law? Why do some people think that it means that we must be at the center of the Universe? Are they right? If not, explain the fault in their reasoning. If they are right, explain why we are at the center.

Multiple-Choice Questions

1. Most of the mass of the Universe is (be careful)
   A. hydrogen
   B. helium
   C. starlight
   D. dark matter

2. Dark matter is made of
   A. ordinary stars
   B. MACHOs
   C. WIMPs
   D. We don't know.

3. The Doppler shift for galaxies
   A. is usually toward red
   B. is usually toward blue
   C. is blue and red about equally
   D. has not yet been observed

4. The discovery of dark energy means that
   A. the Universe is heading toward a "Big Crunch"
   B. most stars have burned out

C. the expansion of the universe is accelerating

D. the expansion of the universe is slowing down

5. In the Big Bang theory
   A. the Earth is at the center of the Universe
   B. the Earth is near the center of the Universe
   C. the center of the Universe is not known
   D. there is no center to the Universe

6. What was made in the first 4 minutes after the Big Bang? (Choose all that are correct.)
   A. hydrogen
   B. helium
   C. carbon
   D. iron

7. To become a black hole, an object must
   A. have a mass much greater than that of the Earth
   B. have a size much smaller than that of the Earth
   C. have enough mass that the escape velocity exceeds c
   D. have been formed in a supernova explosion

8. Dark energy makes the Universe
   A. slow down
   B. speed up
   C. expand
   D. maintain constant size

9. When you look at a galaxy 5 billion light-years away, you
   A. see this galaxy as it was 5 billion years ago
   B. see this galaxy as it was about a million years ago
   C. see the galaxy the way it is now
   D. You can never see a galaxy 5 billion light-years away.

10. Who is famous for his estimate of extraterrestrial intelligence?
    A. Moore
    B. Hubble
    C. Drake
    D. Heisenberg

11. The expansion of the Universe is accelerating. Physicists attribute this to
    A. the Hubble expansion
    B. dark matter
    C. dark energy
    D. antimatter

12. To make a black hole, we need
    A. high mass in a small radius
    B. high mass in a large radius
    C. low mass in a small radius

13. Most of mass in the universe is made of
   A. planets
   B. stars
   C. dark matter
   D. comets

14. Oxygen and nitrogen were created
   A. in the first few million years of the Earth's existence
   B. within a star
   C. in the Big Bang
   D. in a supernova explosion

15. What is Nemesis?
   A. a recently discovered planet that orbits the Sun
   B. a theoretical star that orbits the Sun, making comets and asteroids veer toward Earth
   C. an asteroid found near Pluto, not considered a planet
   D. a nearby galaxy that we can view with the unaided eye

16. Light from the Andromeda Galaxy reaches us in
   A. several minutes
   B. several years
   C. several millions of years
   D. several billions of years

17. The age of the Universe is closest to
   A. 100 million years
   B. 1 billion years
   C. 10 billion years
   D. 1 trillion years

18. For the Earth to become a black hole, it would have to be reduced to the size of
   A. Berkeley
   B. an automobile
   C. a coin
   D. a pin head

19. If the Sun were made into a black hole, its radius would be approximately
   A. one centimeter
   B. one meter
   C. a few kilometers
   D. the size of the Earth

20. If we turn the Sun into a black hole, then the force on the Earth will
   A. increase by a factor less than a million
   B. increase by a factor much greater than a million
   C. be unchanged

21. Choose all the following that are *not* part of our Solar System:
   A. Jupiter
   B. asteroids

C. the Andromeda Galaxy

D. Halley's Comet

22. A galaxy is

A. a large collection of stars

B. a region of gas

C. an exploding star

D. a planet-sized object

23. A light-year is

A. dilated time

B. a distance

C. longer than a year

D. the same as an ordinary year

Epilogue (a poem)

The Creation

At first there is nothing
no Earth, no Sun
no space, no time
nothing

Time begins
and the vacuum explodes, erupts
from nothing, filled with fire
everywhere
furiously hot and bright

Fast as light, space grows,
and the firestorm grows
weaker. Crystals appear
droplets
of the very first matter. Strange matter
fragile bits
a billionth of the Universe
overwhelmed in turbulence
of no importance
they seem
as they wait
for the violence to subside

The Universe cools and the crystals shatter
and shatter again,
and again and again
until they can shatter no more. Fragments

electrons, gluons, quarks,
grasp at each other, but are burned back apart
by the blue-white heat, still far too hot
for atoms to endure

Space grows, and the fire diminishes
to white to red to infrared
to darkness.
A million-year holocaust has passed.
Particles huddle in the cold and bind themselves
into atoms—hydrogen, helium, simple atoms
from which all else is made.

Drawn by gravity, the atoms gather
and divide
and form clouds of all sizes
stars and galaxies
of stars, clusters of galaxies. In the voids
there is empty space
for the first time.

In a small star cloud, a clump of cool matter
compresses and heats
and ignites
and once again there is light.

Deep within a star, nuclei
are fuel and food, burning and cooking
for billions of years, fusing
to carbon and oxygen and iron, matter of life
and intelligence, born slowly, buried
trapped
deep within a star

Burned and burdened, a giant star's heart
collapses. Convulses. A flash. In seconds
energy from gravity, thrown out
overheats, explodes, ejects
the shell of the star. Supernova! Growing brighter
than a thousand stars. Still brighter, brighter
than a million stars, a billion stars, brighter
than a galaxy of stars. Cinders of carbon, oxygen, iron
expelled into space
escape
free! They cool and harden
to dust, the ashes of a star
the substance of life

In the Milky Way Galaxy at the edge of the Virgo Cluster
(named five billion years later, for a mother),

the dust divides and gathers and begins to form
a new star. Nearby a smudge of dust begins to form
a planet. The young Sun
compresses, and heats
and ignites
and warms the infant Earth

—Richard A. Muller

Index